Fractal Control and Its Applications

Shu Tang Liu · Yong Ping Zhang · Chang An Liu

# Fractal Control and Its Applications

Springer

Shu Tang Liu
College of Control Science and Engineering
Shandong University
Jinan, Shandong, China

Yong Ping Zhang
School of Mathematics and Statistics
Shandong University (Weihai)
Weihai, Shandong, China

Chang An Liu
Chinese University of Hong Kong
Hong Kong, China

ISBN 978-981-15-5461-2 ISBN 978-981-15-5459-9 (eBook)
https://doi.org/10.1007/978-981-15-5459-9



This Springer imprint is published by the registered company Springer Nature Singapore Pte Ltd.
The registered company address is: 152 Beach Road, #21-01/04 Gateway East, Singapore 189721, Singapore

# Preface

Fractal theory is one of the important contents in nonlinear sciences and there are abundant results for fractal theory. The authors bring together the team's latest research and make up chapters of this book. The important contents of this book include the fractal behaviors of classical fractal system, time delay system, physical system, economic system, ecosystems, thermal fractal diffusion systems, SIRS systems and global warming, designing the controllers and coupling items, achieving the control and synchronization of fractals, computing the fractal dimensions to depict the complexity of systems.

Applications in various fields about fractal are related to many contents of fractal theory. In this book, classical fractal Julia set and Mandelbrot set are taken to control including the planar J-M sets, spatial J-M sets, stochastic J-M sets, fractional J-M sets and so on.

The book is divided into 11 chapters. Chapter 2 discusses the new properties of the fractal control theory. Fractal control and synchronization of classical systems, time delay systems, spatial fractal are introduced, respectively, in Chaps. 2, 4, 5. Fractal phenomena and control in economic systems are studied in Chap. 6. Control and synchronization of Julia sets of physical systems are researched in Chap. 7 and the fractal dimensions of controlled Julia sets are computed. Fractal applications in biological systems are analyzed in Chap. 8. And control and fractal dimensions of the Julia sets of integer order and fractional order biological models are given. Control of thermal fractal diffusion system is discussed in Chap. 9. The applications of fractal control in SIRS models are given in Chap. 10. Applications in other fields of fractal control are given in Chap. 11, such as global warming.

In addition, this book can be used as a textbook for college students, postgraduates and doctoral students in the polytechnic university, also be used as the reference book for engineers and professional researchers. This book will promote the development of theoretical research and applied research in this academic field.

We would like to express our appreciation for the great efforts of Drs. Li Zhang, Yupin Wang, Weihua Sun, Da Wang, Wen Wang and Wei Qiao, and graduate students Changquan Hu, Feng Liang and Kexin Liu.

Finally, we are grateful for the sequential support of the National Natural Science Foundation of China -Shandong joint fund (No. U1806203) and the Key Program of the National Natural Science Foundation of China (No. 61533011), the Foundation for the Author of National Excellent Doctoral Dissertation of China (No. 200444), the National Natural Science Foundation of China (No. 60472112, No. 60372028, No. 600874009, No. 10971120, No. 61273088), the Shandong Province Natural Science Foundation (No. ZR2010FM010, No. Y98A02005), the China Postdoctoral Science Foundation (No. 2004035036), Project supported by the Scientific Research Staring Foundation for the Returned Overseas Chinese Scholars, Ministry of Education of China, and the K. C. Wong Education Foundation, Hong Kong, China, which ensure the smooth developments of this research project.

Due to the influence of scientific research level and the limitations, there are inevitably omissions and deficiencies in the book, and readers are welcome to put forward valuable suggestions.

Jinan, China Shu Tang Liu
Weihai, China Yong Ping Zhang
Hong Kong, China Chang An Liu

# Contents

# Abbreviations

| | |
|---|---|
| inf | infimum |
| lim | limit |
| max | maximum |
| min | minimum |
| sup | supremum |
| CML | coupled map lattice |
| DLA | diffusion limited aggregation |

# Symbols

| | |
|---|---|
| $\mathbb{R}, \mathbb{R}^n, \mathbb{R}^{n\times m}$ | set of real numbers, set of $n$-dimensional real vectors, and set of $n \times m$ real matrices |
| $\mathbb{C}, \mathbb{C}^n, \mathbb{C}^{n\times m}$ | set of complex numbers, set of $n$-dimensional complex vectors, and set of $n \times m$ complex matrice |
| $\mathbb{R}^+$ | set of non-negative real numbers |
| $\mathbb{Z}^+$ | set of non-negative integers |
| $Re(s)$ | real part of $s \in \mathbb{C}$ |
| $Im(s)$ | imaginary part of $s \in \mathbb{C}$ |
| $A^T$ | transpose of matrix $A$ |
| $A^{-1}$ | inverse of matrix $A$ |
| $A^{-T}$ | shorthand for $(A^{-1})^T$ |
| $I_n$ | $n \times n$ identify matrix (the subscript is omitted if no confusion will occur) |
| $diag\{A_1, \ldots, A_n\}$ | diagonal matrix with $A_i$ as its $i$th diagonal element |
| $\begin{bmatrix} X & Y \\ * & Z \end{bmatrix}$ | symmetric matrix $\begin{bmatrix} X & Y \\ Y^T & Z \end{bmatrix}$ |
| $A > 0(<0)$ | symmetric positive (negative) definite matrix |
| $A \geq 0(\leq 0)$ | symmetric positive (negative) semi-definite matrix |
| $det(A)$ | determinant of matrix $A$ |
| $Tr\{A\}$ | trace of matrix $A$ |
| $\lambda(A)$ | eigenvalue of matrix $A$ |
| $\lambda_{max}(A)$ | largest eigenvalue of matrix $A$ |
| $\lambda_{min}(A)$ | smallest eigenvalue of matrix $A$ |
| $\sigma_{max}(A)$ | largest singular value of matrix $A$ |
| $\lvert \cdot \rvert$ | absolute value (or modulus) |
| $\lVert \cdot \rVert$ | Euclidean norm of a vector or spectral norm of a matrix |
| $\lVert \cdot \rVert_\infty$ | included $l_\infty-$norm |
| $\mathscr{C}([a,b], \mathbb{R}^n)$ | family of continuous functions $\varphi$ from $[a,b]$ to $\mathbb{R}^n$ |
| $\lVert \varphi \rVert_c$ | continuous norm $sup_{a\leq t\leq b}\lVert\varphi(t)\rVert$ for $\varphi \in \mathscr{C}([a,b], \mathbb{R}^n)$ |

| | |
|---|---|
| $\varepsilon$ | mathematical expectation |
| $\left[\begin{array}{c\|c} A & B \\ \hline C & D \end{array}\right]$ | shorthand for state space realization $C(sI - A)^{-1}B + D$ for a continuous-time system or $C(zI - A)^{-1}B + D$ for a discrete-time system |
| $\forall$ | for all |
| $\in$ | belongs to |
| $\exists$ | there exists |
| $\subseteq$ | is a subset of |
| $\cup$ | union |
| $\cap$ | intersection |
| $\rightarrow$ | tends toward or is mapped into (case sensitive) |
| $\Rightarrow$ | implies |
| $\frac{\partial V}{\partial t}$ | partial derivative of function $V$ |
| $\mathrm{grad} V$ | gradient of function $V$ |
| $\prod_{i=1}^{n} \lambda_i$ | product of $\lambda_1, \lambda_2, \ldots, \lambda_n$ |
| $\Delta A$ | amount of change in matrix $A$ |
| $\square$ | end of proof |

# Chapter 1
# Introduction

With the continuous development of natural science and the need of engineering technology, as the classical contents of nonlinear theory, fractal theory, fractal control theory and fractal application in practice continue to deepen and advance [1–156].

Prior to this work, we have published a special work of "Fractal Control Theory" in Springer Publishing Company, which systematically summarizes and discusses our previous work about fractal theory and control. On this basis, the authors' team has again synthesized our latest research and inform the content of this book, which involves:

I the theoretical achievements of fractal;
II the newest control achievements of fractal;
III the application achievements in practices of fractal theory and fractal control.

The content of part I includes the connectivity theory and properties of real and complex dynamical systems, the sparse and dense of fractal, the formatted analysis and characteristics of Julia (J) set and Mandelbrot (M) set, the latest theories of plane J-M sets, spatial J-M sets, random J-M sets and the fractional order J-M sets.

The content of part II includes the latest work about control and synchronization theory of fractal behaviors of complex polynomial system, exponential function system, trigonometric function system, rational function system, rational function, switching system and so on, as well as the Discriminative Local Alignment (DLA) fractal behavior and control results.

The content of part III is the applications of fractal theory in the engineering technology, which illustrates that some works are difficult to analyze with the traditional scientific, but the fractal theory often shows its infinite charm and efficacy. The contents of this part are related to the growth and diffusion of marine life, the spatial and temporal structure of the condensed fractal of marine life, the DLA model and phase transformation, the fractal analysis in economic field, the fractal analysis of some physical characteristics of the atmosphere warming and the prediction.



S. T. Liu et al., *Fractal Control and Its Applications*,
https://doi.org/10.1007/978-981-15-5459-9_1

For the applications and analysis of fractal control, we discuss the restriction and applications of fractal control behaviors in some economic system, physical system, biological system, infectious disease system and so on.

We believe that the publication of this book will certainly play the positive role and important application values in fractal theory, fractal control, applications of fractal control and other fields. And also it will have a positive role in promoting social benefits and practical significance and play a positive leading role for science and technology workers and researchers.

# Chapter 2
# New Characteristics About the Fractal Control Theory

The natural process can be depicted as discrete dynamic systems which reveal how a process runs from one state to the next via certain rules. Due to the randomness and complexity of nature, it is more accurate to model a natural process by applying the alternated iteration of different systems than a unique system. Thus it is meaningful to compare the properties of the alternated systems with the properties of the original separated systems.

## 2.1 Characteristic Analysis on "disconnected plus disconnected equal connected" in Fractal Geometry

In [157, 158], Parrondo et al. proposed that two games with losing gains? if combined in an alternate way, can paradoxically become a winning game. This is the well known Parrondo's paradox, "losing + losing = winning" phenomenon, which aroused a research fever on it in the physics and mathematics areas [159, 160].

In the chaotic research area, Almeida et al. studied an interesting phenomenon "chaos + chaos = order" which shows that the chaotic behaviors of two systems completely change into ordered via iteration alteration [161]. In some sense this phenomenon can be regarded as a new chaos control or anticontrol strategy which has been investigated in complex chaotic systems [162] and fractional chaotic systems [163].

It is noted that chaos is closely related with fractal geometry which is regarded as one of the best tools to describe the nonlinear problems [154, 165]. As one of the branches of fractal geometry, the M-J sets generated from discrete complex map have attracted the attentions from scholars in recent years.

The planar M-J sets have been developed in the properties analysis [166–170], effect of noise [171–173] and even applications [165, 174–179]. For instance, some


S. T. Liu et al., *Fractal Control and Its Applications*,
https://doi.org/10.1007/978-981-15-5459-9_2

new boundedness results were obtained for the Julia sets with Jungck Mann and Jungck Ishikawa iterations [168, 170]. The graphical properties of Julia sets and the parameter sets generated from the generalized complex map $z_{n+1} = z_n^{\alpha} + c$, $\alpha < 0$ were given in [166, 167]. The effect of noise of planar M-J sets have been investigated in the classical map [172] and some modified maps [171, 173, 176, 178]. In addition, these theoretical researches, the applications of planar M-J sets have also been widely studied, especially in the physical field. Bech [174] firstly provided the physical meaning of the Julia set by considering a kicked charged particle moving in a double-well potential. What's more, this research was carried on in [175, 177] by pointing out that the fractal structure of the generalized M-J sets may visually reflect the change law of particles velocity. In the biological field, [165] pointed out that the filled Julia set is similar in appearance and several important characteristics to those of real organisms.

By extending the definition domain into multicomplex space, some interesting results about the spatial M-J sets [180–183] are obtained. In [180], the author investigated a class of bicomplex dynamics and proposed the spatial Mandelbrot set the set of points where the associated generalized "filled-Julia" set is connected.

In these researches of property analysis on M-J sets, the connectivity research is one of the hottest topics for it is an useful measurement in the description of the morphologies of the natural processes or phenomena such as cellular morphology [165], particle moving [174, 175, 177] and network structure [184]. In [185, 186], M-F. Danca combined "Parrondo's paradox" with "Fractal" by graphically investigating the properties of alternated Julia sets which are obtained by alternating the iteration of two complex maps. Subsequently, in [187], the author gave the strictly mathematical proof of the some visually observations in [186] about the properties of alternated Julia sets. However, these researches [185–187] were both limited in analysing the properties of the combined system. There is a lack of comparison between the properties of the combined systems with the properties of the original separated systems as [157, 158, 161] did.

Inspired by the above researches, we further extend the Parrondo's philosophy into fractal research by giving a preliminary study on the phenomenon: "disconnected plus disconnected equal connected", that is, the combination of two complex maps with disconnected Julia sets can generate a connected Julia set.

### 2.1.1 Definitions and Lemmas

In this section, some necessary definitions and lemmas are given.

**Definition 2.1** ([164]) Consider the following system

$$P_c: \quad z_{n+1} = z_n^2 - c. \tag{2.1.1}$$

The filled Julia set of system (2.1.1) is defined as $K(P_c)$ which satisfies that

$$K(P_c) = \{z_0 | P_c^n(z_0) \nrightarrow \infty, n \rightarrow \infty\},$$

where $P_c^n$ denote the $n$-th iteration of $z_0$. The Julia set of system (2.1.1) is the boundary of $K(P_c)$, *i.e.* $J(P_c) = \partial K(P_c)$.

**Definition 2.2** ([164]) The Mandelbrot set of system (2.1.1) is defined as

$$M(P_c) = \{c| \ The \ Julia \ sets \ \ J(P_c) \ \ is \ connected\}.$$

These are the classical definitions of M-J sets of simple quadratic map. Here in Fig. 2.1 we give an example that the Julia sets exhibit different connectivity properties via choosing different parameters.

**Definition 2.3** ([185]) Consider the following alternated complex map

$$P_{c_1,c_2}: \quad z_{n+1} = \begin{cases} z_n^2 + c_1, & \text{if } n \text{ is even}, \\ z_n^2 + c_2, & \text{if } n \text{ is odd}, \end{cases} \tag{2.1.2}$$

The filled Julia set of the system (2.2.2), denoted as $K(P_{c_1,c_2})$, is the set of all the values of initial conditions $z_0$ such that

$$\lim_{n \to \infty} |P^n(z_0, c_1, c_2)| \nrightarrow \infty, \tag{2.1.3}$$

where $P^n(\cdot)$ represents the $n$-$th$ iteration of the initial point $z_C$. The Julia set of $P_{c_1,c_2}$ which is the boundary of the filled Julia set is denoted as $J(P_{c_1,c_2}) = \partial(K(P_{c_1,c_2}))$.

**Lemma 2.1** ([185, 186]) *Consider the following system*

$$Q_{c_1,c_2}: \ w_{n+1} = (w_n^2 + c_1)^2 + c_2. \tag{2.1.4}$$

*Thus* $J(P_{c_1,c_2})$ *and* $J(Q_{c_1,c_2})$ *are the same for given* $c_1$ *and* $c_2$ *parameter values.*

**Lemma 2.2** ([185, 187]) *The connectivity properties of the Julia set for a complex polynomial of degree* $d > 2$ *can be identified based on the following cases:*

*(1) The Julia set is connected if and only if all the critical orbits are bounded.*
*(2) The Julia set is totally disconnected, a cantor set, if (but not only if) all the critical orbits are unbounded.*
*(3) For a polynomial with at least one critical orbit unbounded, the Julia set is totally disconnected if and only if all the bounded critical orbits are aperiodic.*

Then from Lemmas 2.1 and 2.2, the connectivity of the $J(P_{c_1,c_2})$ or $K(P_{c_1,c_2})$ ($J$ and $K$ have the same connectivity properties [164]) relies on the boundedness of the three critical points of $Q_{c_1,c_2}$: $0, \pm\sqrt{-c_1}$. The following definition is given.

**Definition 2.4** ([185, 186]) The Connectivity Loci, denoted as $\mathscr{M}(\mathscr{P})$, which represent the three states of $J(P_{c_1,c_2})$: connectedness, disconnectedness and totally disconnectedness are respectively denoted as:

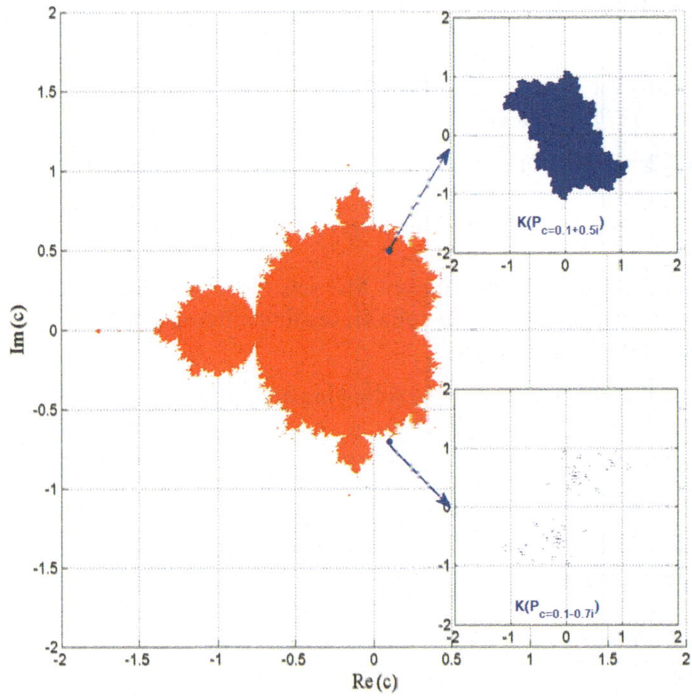


**Fig. 2.1** The classical Mandelbrot set $M(P_c)$ and the filled Julia sets with $c = 0.1 - 0.5i$ and $c = 0.1 - 0.7i$

1. $CL(P_{c_1,c_2})$ (The orbits of 0 and $\sqrt{-c_1}$ are bounded).
2. $DL(P_{c_1,c_2})$ (The orbit of 0 is bounded but the orbit of $\sqrt{-c_1}$ are unbounded and vice versa).
3. $TDL(P_{c_1,c_2})$ (The orbits of 0 and $\sqrt{-c_1}$ are unbounded).

**Lemma 2.3** ([187]) $J(P_{c_1,c_2}) \subset \{z_0 \mid |z_0| \leq \max\{|c_1|, |c_2|, R\}\}$, *where* $R$ *is the real root of the complex equation* $\lambda^3 - 2\lambda^2 - \lambda - 2 = 0$. *i.e.* $T \approx 2.659$.

**Lemma 2.4** ([187]) $CL(P_{c_1,c_2}) \subset \{(c_1, c_2) \mid |c_1| \leq R, |c_2| \leq R\}$, *where* $R$ *is the same as Lemma 2.3.*

### 2.1.2 Graphical Exploration of the Phenomenon "disconnected plus disconnected equal connected"

This section aim to give a graphical exploration of phenomenon "disconnected plus disconnected equal connected" (in the sentence "disconnected plus disconnected equal connected", disconnected means totally disconnected), which we hope could represent a guide for the future theoretical research.

Consider the following two individual systems:

$$P_{c_1}: \quad z_{n+1} = z_n^2 + c_1 \quad z,\ c_1 \in \mathbb{C}, \tag{2.1.5}$$

$$P_{c_2}: \quad z_{n+1} = z_n^2 + c_2 \quad z,\ c_2 \in \mathbb{C}, \tag{2.1.6}$$

The alternated system of the two system is system (2.2.2). It is worth noting that 0 is only critical point for system (2.1.5) and system (2.1.6), so there are only two possibilities, connected and totally disconnected, for their corresponding Julia sets. These two possibilities rely on whether the parameters $c_1$, $c_2$ belong to the classical Mandelbrot set. Our purpose is to find two proper parameters $c_1$, $c_2$ which give rise to disconnected $J(P_{c_1})$, $J(P_{c_2})$ and connected $J(P_{c_1,c_2})$. From Fig. 2.1, the parameter $c_1$ can be taken as $c_1 = 0.1 - 0.7i$ to make sure that $J(P_{c_1})$ is totally disconnected. From Definition 2.4, $\mathcal{M}(\mathcal{P})$ belongs to a 4-D space. Thus when $c_1 = 0.1 - 0.7i$ is fixed, one can get a 2-D slice of the $\mathcal{M}(\mathcal{P})$.

In [185, 187], the region where $\mathcal{M}(\mathcal{P})$ is located in was "empirically" chosen. Then in [188], the author gave a proper outer boundary of $CL(P_{c_1,c_2})$ and $DL(P_{c_1,c_2})$. With the help of the graphical method proposed in [186] and Lemma 2.4 [185, 187], $\mathcal{M}(\mathcal{P})$ with $c_1 = 0.1 - 0.7i$ and $c_1 = 0.1 + 0.5i$ are illustrated in Fig. 2.2. The partial enlarged details of $\mathcal{M}(\mathcal{P})$ with $c_1 = 0.1 - 0.7i$ is shown in Fig. 2.3.

As shown in Fig. 2.3, the blue region is the location of $c_2$ which makes the Julia set $J(P_{c_1=0.1-0.7i,c_2})$ connected. The gray region is the location of $c_2$ which makes the Julia set $J(P_{c_1=0.1-0.7i,c_2})$ disconnected. The white region is the location of $c_2$ which makes the Julia set $J(P_{c_1=0.1-0.7i,c_2})$ totally disconnected.

Generally speaking, the key point is finding a proper $c_2$, which is also located outside the classical Mandelbrot set, to guarantee that $J(P_{c_1,c_2})$ is connected. Our solution is adding the boundary of the classical Mandelbrot set to the Fig. 2.3. Based on the above analysis, the equivalent condition of the phenomenon "disconnected + disconnected = connected" is that $c_2$ should be outside of the red boundary and inside of the blue region.

From Fig. 2.3, a proper $c_2$ can be chosen as $G: c_2 = 0.48 - 0.32i$. The three Julia sets $K(P_{c=0.1-0.7i})$(totally disconnected), $K(P_{c=0.48+0.32i})$ (totally disconnected), and $K(P_{c_1=0.1-0.7i,c_2=0.48+0.32i})$ (connected) are shown in Fig. 2.4.

Similarity, the phenomenon "connected + connected = disconnected" is analyzed as follows. From Fig. 2.2, one can choose $c_1 = 0.1 + 0.5i$ to get a connected $J(P_{c_1})$. The 2-D slice of $\mathcal{M}(\mathcal{P})$ with $c_1 = 0.1 + 0.5i$ and its partial enlarged detail are shown in Fig. 2.6 in which we also add the boundary of the classical Mandelbrot set.

Based on the above analysis, the $c_2$ which satisfies the phenomenon "connected + connected = disconnected" should be outside of the blue and gray region and inside of the red boundary. From Fig. 2.5, one can choose a proper point $H: c_2 = 0.14 - 0.44i$. The three Julia sets $K(P_{0.1+0.5i})$(connected), $K(P_{0.14-0.44})$ (connected), and $K(P_{c_1=0.1+0.5i,c_2=0.14-0.44i})$ (totally disconnected) are shown in Fig. 2.5.

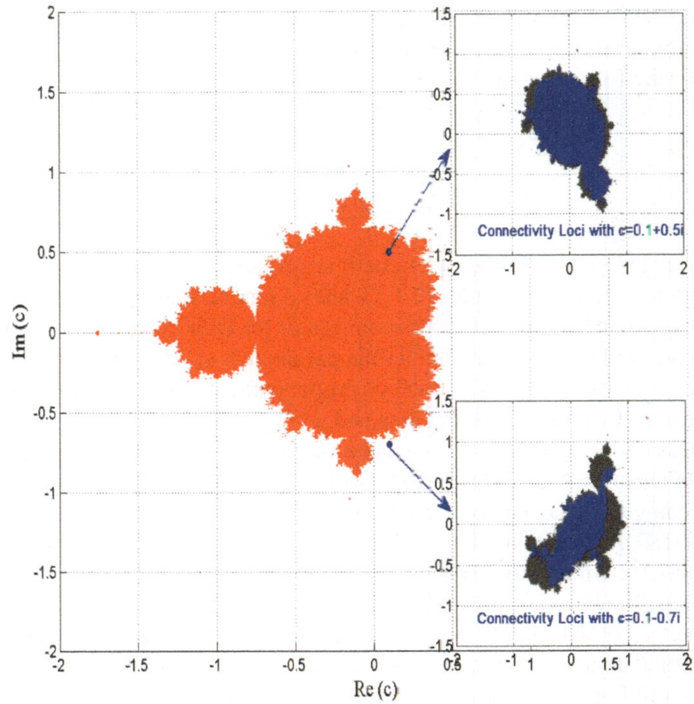


**Fig. 2.2** The classical Mandelbrot set $M(P_c)$ and the 2-D slices of $\mathscr{M}(\mathscr{P})$ with $c = 0.1 + 0.5i$ and $c = 0.1 - 0.7i$

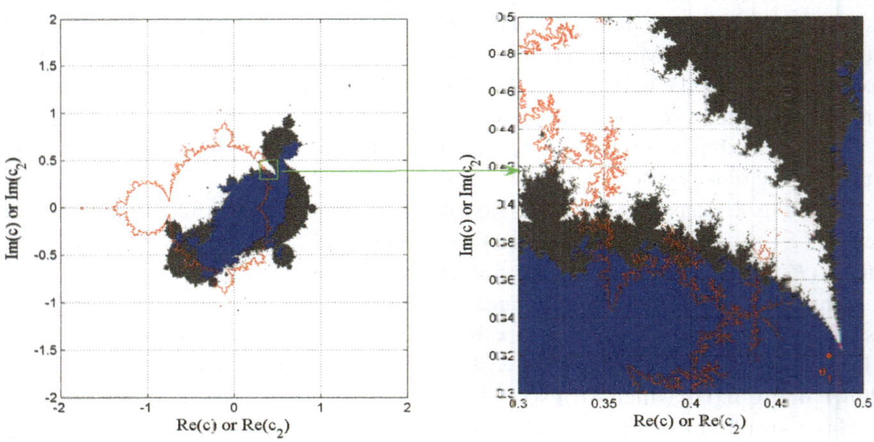


**Fig. 2.3** The partial enlarged details of $\mathscr{M}(\mathscr{P})$ with $c_1 = 0.1 - 0.7i$

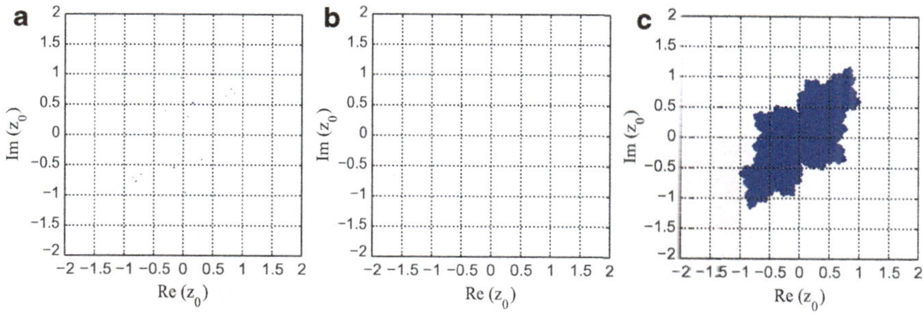


**Fig. 2.4** **a** $K(P_{c=0.1-0.7i})$. **b** $K(P_{c=0.48+0.32i})$. **c** $K(P_{c_1=0.1-0.7i,c_2=0.48-0.32i})$

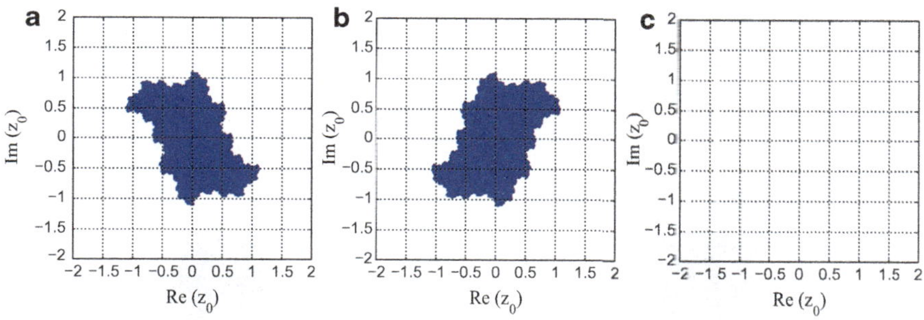


**Fig. 2.5** **a** $K(P_{c=0.1+0.5i})$. **b** $K(P_{c=0.14-0.44i})$. **c** $K(P_{c_1=0.1+0.5i,c_2=0.14-0.44i})$

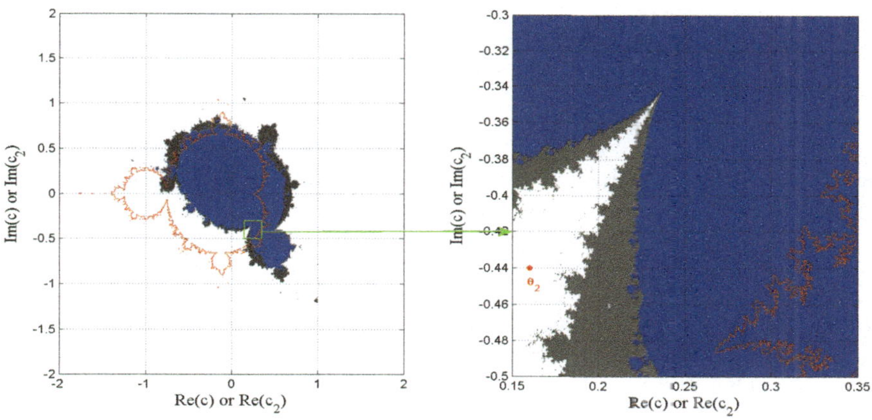


**Fig. 2.6** The partial enlarged details of $\mathscr{M}(\mathscr{P})$ with $c_1 = 0.1 + 0.5i$

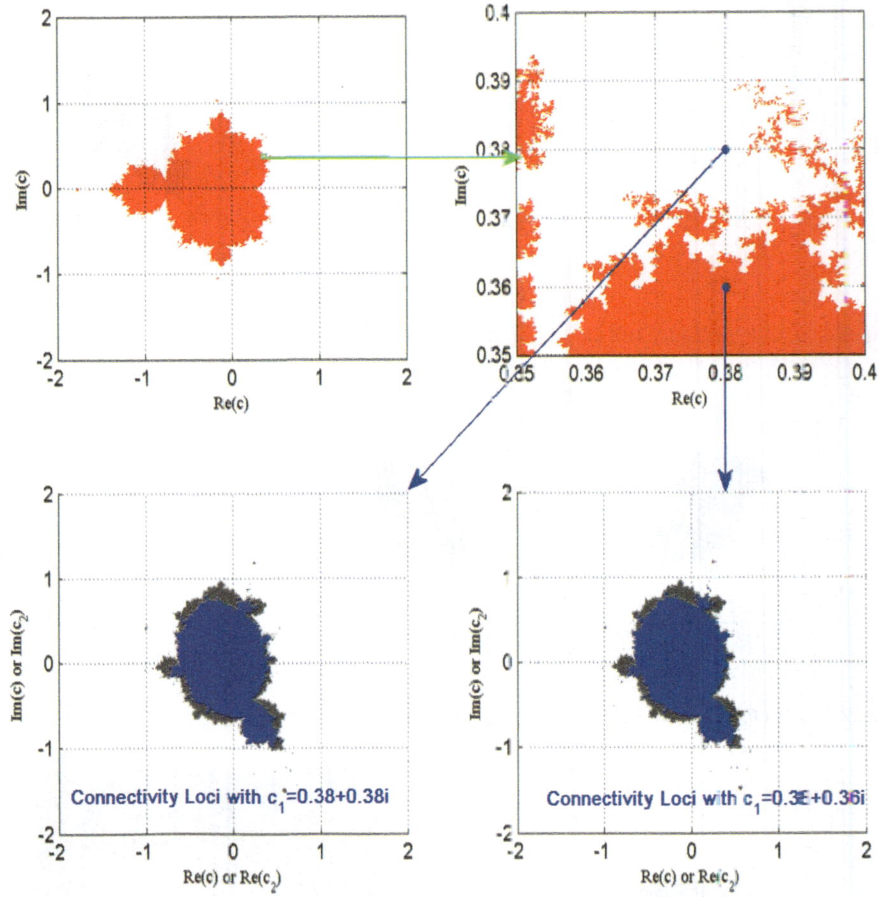


**Fig. 2.7** The classical Mandelbrot set $M(P_c)$ and the 2-D slices of the $\mathscr{M}(\mathscr{P})$ with $c = 0.38 + 0.38i$ and $c = 0.38 + 0.36i$

To further verify the above analysis, we enlarge the local details of the classical Mandelbrot set and take two more close points $0.38 + 0.38i$, $0.38 + 0.36i$ which give rise to Julia sets with totally different connectivity. That is, $J(P_{c=0.38-0.38})$ is totally disconnected but $J(P_{c=0.38+0.36})$ is connected. Then Fig. 2.7 illustrates $\mathscr{M}(\mathscr{P})$ with $c_1 = 0.38 + 0.38i$ and $c_1 = 0.38 + 0.36i$.

To find a proper $c_2$ (denote as point $P$) in $\mathscr{M}(\mathscr{P})$ with $c_1 = 0.38 + 0.38i$, so we can ensure that $J(P_{c_2})$ is disconnected and meanwhile $J(P_{c_1=0.38+0.38i,c_2})$ is connected. On the other hand, we will find a proper $c_2$ (denote as point $Q$) in $\mathscr{M}(\mathscr{P})$ with $c_1 = 0.38 + 0.36i$ to ensure that $J(P_{c_2})$ is connected and meanwhile $J(P_{c_1=0.38+0.36i,c_2})$ is totally connected. Furthermore we hope that points $P$ and $Q$ could be also close to each other.

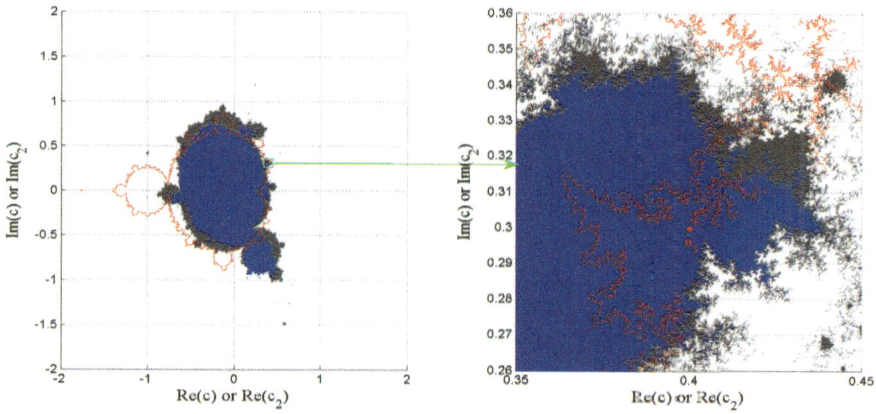


Fig. 2.8 The partial enlarged details of $\mathscr{M}(\mathscr{P})$ with $c_1 = 0.38 + 0.38i$

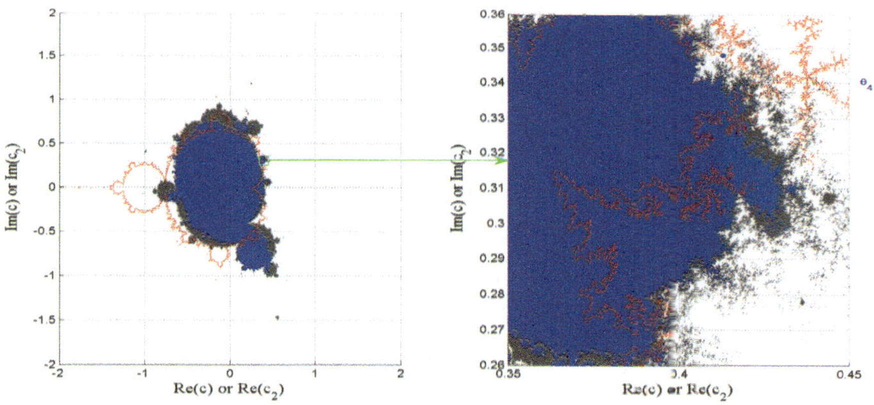


Fig. 2.9 The partial enlarged details of $\mathscr{M}(\mathscr{P})$ with $c_1 = 0.38 + 0.36i$

To achieve these goals, the partial enlarged details of $\mathscr{M}(\mathscr{P})$ with $c_1 = 0.38 + 0.38i$ and $c_1 = 0.38 + 0.36i$ are respectively shown in Figs. 2.8 and 2.9. Based on the previous analysis, the proper points $P$ and $\mathcal{Q}$ can be taken as $c_2 = 0.4 + 0.3i$ and $c_2 = 0.413 + 0.348i$. Then $J(P_{c=0.38+0.38})$, $J(P_{c=0.4+0.3})$ and $J(P_{c_1=0.38+0.38i, c_2=0.4+0.3i})$ are illustrated in Fig. 2.10. $J(P_{c=0.38+0.36})$, $J(P_{c=0.413+0.348})$ and $J(P_{c_1=0.38+0.36i, c_2=0.413+0.348i})$ are illustrated in Fig. 2.11.

The above "disconnected + disconnected = connected" and "connected + connected = disconnected" results can be summarized in Fig. 2.12 which shows that a tiny difference of the parameters can result in totally different connectivity changes.

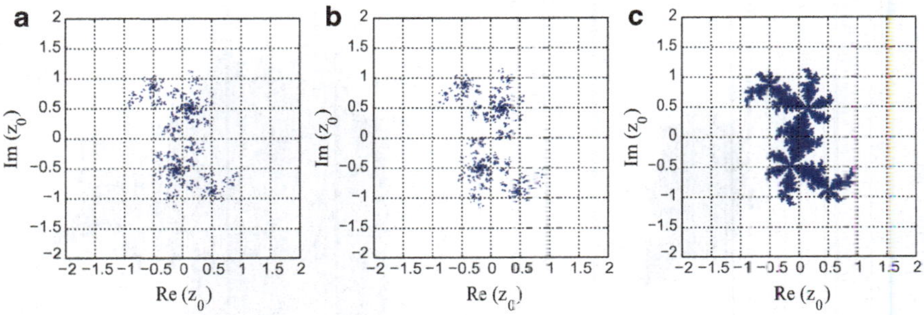


**Fig. 2.10** **a** $K(P_{c=0.38+0.38i})$. **b** $K(P_{c=0.4+0.3i})$. **c** $K(P_{c_1=0.38+0.38i,c_2=0.4+0.3i})$

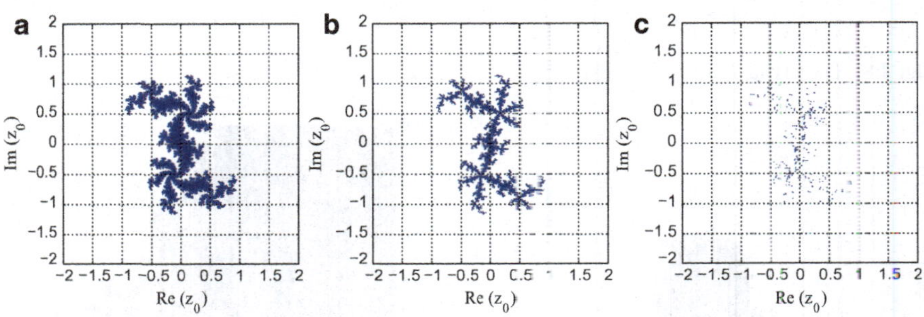


**Fig. 2.11** **a** $K(P_{c=0.38+0.36i})$. **b** $K(P_{c=0.413+0.348i})$. **c** $K(P_{c_1=0.38+0.36i,c_2=0.413+0.348i})$

| Parameters | $c_1 = 0.38 + 0.38i.$ | $c_2 = 0.4 + 0.3i.$ | $c_1 = 0.38 + 0.38i, c_2 = 0.4 - 0.3i.$ |
|---|---|---|---|
| Julia sets | $J(P_{c_1})$ | $J(P_{c_2})$ | $J(P_{c_1,c_2})$ |
| Connectivity | Totally disconnected | Totally disconnected | Connected |

| Parameters | $c_1 = 0.38 + 0.36i.$ | $c_2 = 0.413 + 0.348i.$ | $c_1 = 0.38 + 0.36i, c_2 = 0.413 + 0.348i.$ |
|---|---|---|---|
| Julia sets | $J(P_{c_1})$ | $J(P_{c_2})$ | $J(P_{c_1,c_2})$ |
| Connectivity | Connected | Connected | Totally disconnected |

**Fig. 2.12** The summarization of connectivity of Julia sets with different parameters

## 2.2 On the Boundedness and Symmetry Properties of the Fractal Sets Generated from Alternated Complex Map

Julia set [189], one of the most attractive fractal sets, is proposed by Gaston Julia when he studied the following complex quadratic polynomial

$$P_c: \quad z_{n+1} = z_n^\alpha + c \quad c \in \mathbb{C},\ \alpha = 2. \tag{2.2.1}$$

Since it was proposed, Julia set has attracted significant interests in the topics of property analysis [167, 190–192], superior iteration [162, 193, 194], time-delay [195], control [196, 197] and so forth. Due to its good explanation for lots of complicated phenomena, Julia set has also been widely applied in the fields of physics [174], biology [198], image cryptography [199] and so on.

As mentioned in the research of [161, 200], it may not be quite accurate to depict the evolution of the natural process mentioned above just by employing a unique system. In some cases, the alternated iterations can be used to have a better understanding of the natural process. In particular, for system (2.2.1), Danca et al. [185, 186] proposed its alternated version as follows:

$$P_{c_1,c_2}: \quad z_{n+1} = \begin{cases} z_n^2 + c_1, & \text{if } n \text{ is even}, \\ z_n^2 + c_2, & \text{if } n \text{ is odd}, \end{cases} \tag{2.2.2}$$

where $z_n, c_1 = p + iq, c_2 = r + is \in \mathbb{C}$, $C = (c_1, c_2)^T$ ($T$ means transposition).

Based on the Fatou-Julia theorem [188] (see Theorem 2.2), the Julia set generated from $P_{c_1,c_2}$, which is denoted as $J(P_{c_1,c_2})$, can be connected, disconnected and totally disconnected via choosing different parameters $c_1, c_2$ [185]. To test numerically which values of the parameters $c_1, c_2$ give rise to connected, disconnected, or totally disconnected $J(P_{c_1,c_2})$, the auxiliary complex quadratic polynomial $Q_{c_1,c_2}$ was proposed in [185]:

$$Q_{c_1,c_2}: \quad w_{n+1} = (w_n^2 + c_1)^2 + c_2, \tag{2.2.3}$$

where $w_n = x_n + iy_n$. Furthermore, they proved that the Julia sets generated from $P_{c_1,c_2}$ and $Q_{c_1,c_2}$ are the same with given $c_1, c_2$, i.e.s $J(P_{c_1,c_2}) = J(Q_{c_1,c_2})$. Thus, the connectivity of $J(P_{c_1,c_2})$ relies on the critical orbits of system (2.2.3). In [186], the graphical results of the connectivity loci which represent the three connectivity states of the corresponding $J(P_{c_1,c_2})$ were unveiled by analyzing the critical orbits of (2.2.3). In concrete terms, the fractal sets which represent the connected, disconnected, totally disconnected Julia sets were respectively denoted as the Connectedness-Locus, Disconnectedness-Locus and Totally connectedness-Locus [185, 186]. Based on these graphical results, some observations about the boundedness and symmetry properties of these connectivity loci were summarized in Sect. 4 of [186]. These unproved properties are just the original purpose and the starting point of this article.

Actually, there have been some researches on the boundedness and symmetry properties for the fractal sets of complex maps which have only one critical point (see the classical M-J sets [190, 201, 202] and the generalized M-J sets [167, 201, 203]. For the systems like (2.2.3) which have more than one critical point, the results mainly focused on the properties of their Julia sets (see the Julia sets of cubic polynomials [159] and the superior Julia set [162]). To our knowledge, few researches have addressed the problems about the boundedness and symmetry properties of their connectivity loci (or Mandelbrot set) since all the critical orbits should be considered.

Thus, as a supplementary research of [185, 186], the purpose and achievement of this section are no other than giving the mathematical proof about the two properties of the Connectedness-Locus and the Julia sets generated from (2.2.2) via analyzing all the critical orbits of system (2.2.3).

### 2.2.1 The Fractal Sets Generated from Alternated Complex Map

In this section, some necessary definitions and theorems are recalled and given.

**Lemma 2.5** ([185]) *$J(P_{c_1,c_2})$ and $J(Q_{c_1,c_2})$ are the same for given $c_1$ and $c_2$ parameter values.*

**Lemma 2.6** ([188]) *The connectivity properties of the Julia set for a complex polynomial of degree $d > 2$ can be identified based on the following cases:*
*(1) The Julia set is connected if and only if all the critical orbits are bounded.*
*(2) The Julia set is totally disconnected, a cantor set, if (but not only if) all the critical orbits are unbounded.*
*(3) For a polynomial with at least one critical orbit unbounded, the Julia set is totally disconnected if and only if all the bounded critical orbits are aperiodic.*

**Definition 2.5** ([186]) The Connectedness-Locus of the alternated complex map (2.2.2), denoted as $M_e(P_{c_1,c_2})$, is composed by all the initial constant $c_1$, $c_2$ such that the orbits of 0 and $\sqrt{-c_1}$ are bounded (since $\pm\sqrt{-c_1}$ behave the same boundedness properties).

In [186], $M_e(P_{c_1,c_2})$ is named as "Connectedness-Locus" or "Connected zone". The reason we denote it as $M_e(P_{c_1,c_2})$ in this part is that it has the same efficacy with the classical Mandelbrot set [204] which gives rise to the connectedness property. Thus, in this part, we could also call $M_e(P_{c_1,c_2})$ the Mandelbrot-efficacy set.

As shown in Definition 2.5 $M_e(P_{c_1,c_2})$ is formed by the initial constants $c_1, c_2$, which illustrates that it belongs to a four-dimensional space $(p, q, r, s)$. By fixing $q_0 = 0$, the 3-D slice of $M_e(P_{c_1,c_2})$ is shown in Fig. 2.13a (also see the red part of Fig. 2.1 in [186]). The internal slice of $M_e(P_{c_1,c_2})_{q_0=0}$ with $p_0 = 0$ is revealed in Fig. 2.13b. The partial enlarged views of Fig. 2.13b are successively shown in

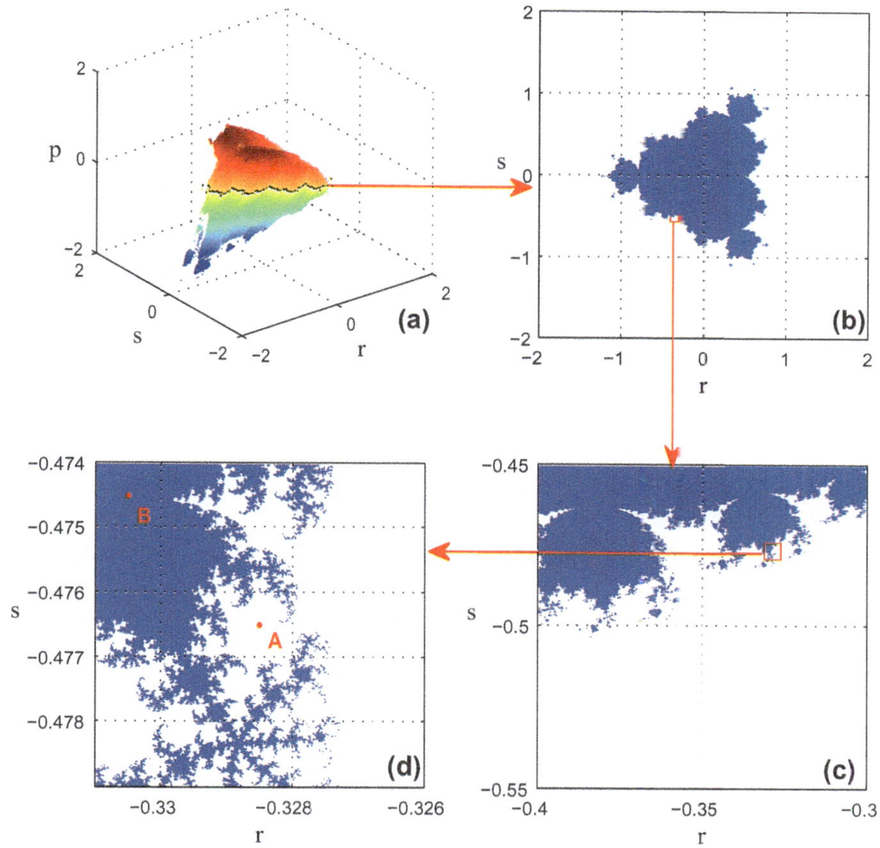


**Fig. 2.13** **a** The 3-D slice of Mandelbrot-efficacy set: $M_e(P_{c_1,c_2})_{q_0=0}$. **b** The 2-D slice of $M_e(P_{c_1,c_2})_{q_0=0}$ with $p_0 = 0$. **c, d** The partial enlarged views of $M_e(P_{c_1,c_2})_{p_0=0,q_0=0}$

Fig. 2.13c, d which fully display the beautiful and complex fractal structure of $M_e(P_{c_1,c_2})$.

The Julia sets $J(P_{c_1,c_2})$ with different parameters $A$ and $B$ which are located in Fig. 2.13d are shown in Fig. 2.14. It can be seen that $J(P_{c_1,c_2})$ is connected with $c_2 = -0.3305 - 0.4745i$ but disconnected with $c_2 = -0.3285 - 0.4765i$.

### 2.2.2 Boundedness of the Fractal Sets Generated from Alternated Complex Map

In this section, by analysing the orbits of the three critical points of system (2.2.3), the boundedness properties of $J(P_{c_1,c_2})$ and $M_e(P_{c_1,c_2})$ are given in the next two theorems.

**Theorem 2.1** *$J(P_{c_1,c_2}) \subset \{z_0 \mid |z_0| \leq \max\{|c_1|, |c_2|, R\}\}$, where $R$ is the real root of the complex equation $\lambda^3 - 2\lambda^2 - \lambda - 2 = 0$. i.e.s $T \approx 2.659$ (Solution of $R$ is based on the Cardan's formula [205]).*

***Proof*** If $\exists w_0$ such that $|w_0|>\max\{|c_1|, |c_2|, R\}$, one can get

$$\begin{aligned}|w_1| &= |w_0^4 + 2w_0^2c_1 + c_1^2 - c_2| \\ &>|w_0|^4 - 2|w_0|^2|c_1| - |c_1|^2 - |c_2| \\ &>|w_0|^4 - 2|w_0|^3 - |w_0|^2 - |w_0| \\ &= |w_0|(|w_0|^3 - 2|w_0|^2 - |w_0| - 1).\end{aligned}$$

Since $|w_0|^3 - 2|w_0|^2 - |w_0| > 2$ when $|w_0| > R$ there exists a $\epsilon > 0$ such that

$$|w_1| > (1+\epsilon)|w_0|.$$

Then, $|w_1| > |w_0|>\max\{|c_1|, |c_2|, R\}$, one gets

$$\begin{aligned}|w_2| &= |w_1^4 + 2w_1^2c_1 + c_1^2 - c_2| \\ &> |w_1|(|w_1|^3 - 2|w_1|^2 - |w_1| - 1). \\ &> |w_1|(|w_0|^3 - 2|w_0|^2 - |w_0| - 1). \\ &> (1+\epsilon)|w_1|.\end{aligned}$$

Finally, one can get $|w_n| > (1+\epsilon)^n|w_0| \to \infty$ when $n \to \infty$. Then from Definition of Julia set, one knows that $J(Q_{c_1,c_2}) \subset \{w_0 \mid |w_0| \leq \max\{|c_1|, |c_2|, R\}\}$. As $J(Q_{c_1,c_2}) = J(P_{c_1,c_2})$, the theorem is obtained.

**Theorem 2.2** *$M_e(P_{c_1,c_2}) \subset \{(c_1, c_2) \mid |c_1| \leq R, |c_2| \leq R\}$, where $R$ is the same as Theorem 2.1.*

***Proof*** Derive the $max\{|c_1|, |c_2|, R\}$ in Theorem 2.1 into three cases as follows:

$$\begin{cases} \max\{|c_1|, |c_2|\} > R : \begin{cases} ①: & |c_1|>\max\{|c_2|, R\} \\ ②: & |c_2|>\max\{|c_1|, R\}. \end{cases} \\ ③ : \max\{|c_1|, |c_2|\} \leq R. \end{cases}$$

(①): For case ①, one knows that $|c_1|>|c_2|$ and $|c_1|>T$, the orbit of the critical value 0 satisfies that

$$\begin{aligned}|Q(0)| &= |c_1^2 + c_2|\geq|c_1|^2 - |c_2| \\ &>|c_1|^2 - |c_1| > |c_1|(R-1) \\ &> |c_1|.\end{aligned}$$

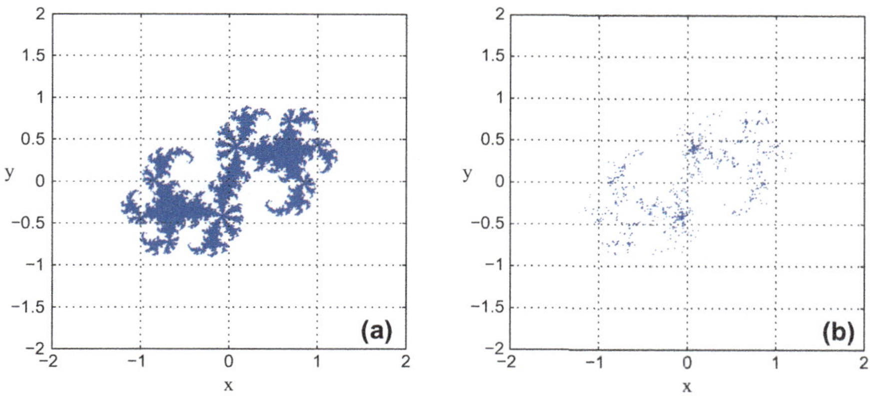


**Fig. 2.14** **a** $K(P_{c_1,c_2})$ with $c_1 = 0,\ c_2 = -0.3305 - 0.4745i$ (A in Fig. 2.13d). **b** $K(P_{c_1,c_2})$ with $c_1 = 0,\ c_2 = -0.3285 - 0.4765i$ (B in Fig. 2.13d)

Then, $|Q(0)|$ satisfies the condition of Theorem 2.1, it is clear that the orbit of 0 is diverging.

(②): For case ②, one knows that $|c_2|>|c_1|$ and $|c_2|>R$, the orbit of the critical value $\sqrt{-c_1}$ satisfies that:

$$\begin{aligned}|Q^2(\sqrt{-c_1})| &= |c_2^4 + 2c_2^2c_1 + c_1^2 + c_2| \\ &\geq |c_2|^4 - |2c_2|^2|c_1| - |c_1|^2 - |c_2| \\ &> |c_2|^4 - |2c_2|^3 - |c_2|^2 - |c_2| \\ &> |c_2|(R^3 - 2R^2 - R - 1) \\ &> |c_2|.\end{aligned}$$

Thus, $|Q^2(\sqrt{-c_1})|$ satisfies the condition of Theorem 2.1, one can know that the orbit of $\sqrt{-c_1}$ is diverging.

Taken together, for all the three critical values, their orbits are bounded only when $max\{|c_1|, |c_2|\} \leq R$. (case ③). From Definition 2.5, one gets: $M_e(P_{c_1,c_2}) \subset \{(c_1, c_2) \mid |c_1| \leq R, |c_2| \leq R\}$.

**Remark 2.1** When $c_1 = c_2 = c$, system (2.2.2) becomes the classical map (2.2.1). The complex equation $\lambda^3 - 2\lambda^2 - \lambda - 2 = 0$ in Theorem 2.1 becomes $\lambda - 2 = 0$.

- Then Theorem 2.1 becomes: $J(P_c) \subset \{z \mid |z| \leq max\{|c|, 2\}\}$.
- Theorem 2.2 becomes: $M(P_c) \subset \{c \mid |c| \leq 2\}$.

$J(P_c)$ and $M(P_c)$ mean the Julia set and Mandelbrot set of $P_c$. The conclusions in this remark are consistent with the classical results [204].

### 2.2.3 Symmetry of the Fractal Sets Generated from Alternated Complex Map

Symmetry [167, 190, 203] is an important property of fractal sets. This section focuses on the symmetry properties of the Julia set $J(P_{c_1,c_2})$ and the 3-D slice of the Mandelbrot-efficacy set $M_e(P_{c_1,c_2})_{q_0=0}$. The following two theorems are given and proved.

**Theorem 2.3** *The Mandelbrot-efficacy set $M_e(P_{c_1,c_2})_{q_0=0}$ is symmetric around the $pr$-plane.*

***Proof*** For the critical points $0, \pm\sqrt{-c_1}$, we give them a unitary expression $w_0'$. Then, by choosing any two parameters values

$$(c_1, c_2)^T = (p_0, 0, r_0, s_0)^T, \quad (c_1, \overline{c_2})^T = (p_0, 0, r_0, -s_0)^T,$$

the orbits of all the three critical points $w_0'$ preserve complex conjugacy when $n > 1$. That is:

$$(w_n'^2 + c_1)^2 + c_2 = w_{n+1}',$$

$$(w_n'^2 + c_1)^2 + \overline{c_2} = \overline{w_{n+1}'}.$$

Thus, the critical orbits $w_n'$ behave the same boundedness property by choosing any two parameters which are symmetric around the $pr$-plane. Then, based on Definition 2.5, the two parts of $M_e(P_{c_1,c_2})_{q_0=0}$ divided by the $pr$-plane are symmetric. □

It is illustrated in Fig. 2.15 that the 2-D slices $M_e(P_{c_1,c_2})_{q_0=0,s_0}$ have the same morphological structure with $M_e(P_{c_1,c_2})_{q_0=0,-s_0}$. The simulations verify the validity of the Theorem 2.3.

**Theorem 2.4** *The Julia set $J(P_{c_1,c_2})$ is centrally symmetric around the origin.*

***Proof*** When $c_1, c_2$ are given, one gets a corresponding Julia set. For any two initial points $w_0$ and $-w_0$, they map to the same $w_1 = (w_0^2 + c_1)^2 + c_2$.

Thus, for any two points which are centrally symmetric around the origin, their orbits behave the same boundedness properties. From Definition 2.6, one knows that $J(P_{c_1,c_2})$ is centrally symmetric around the origin. □

With the help of the graphical method proposed in [186], we supplement the Disconnectedness-Locus (gray zone) and the Totally disconnectedness-Locus (white zone) to $M_e(P_{c_1,c_2})_{q_0=0,s_0=-0.75}$. Then three different points are picked out and their corresponding filled Julia sets are shown in Fig. 2.16.

The simulations of both Figs. 2.14 and 2.16 verify the correctness of Theorem 2.3.

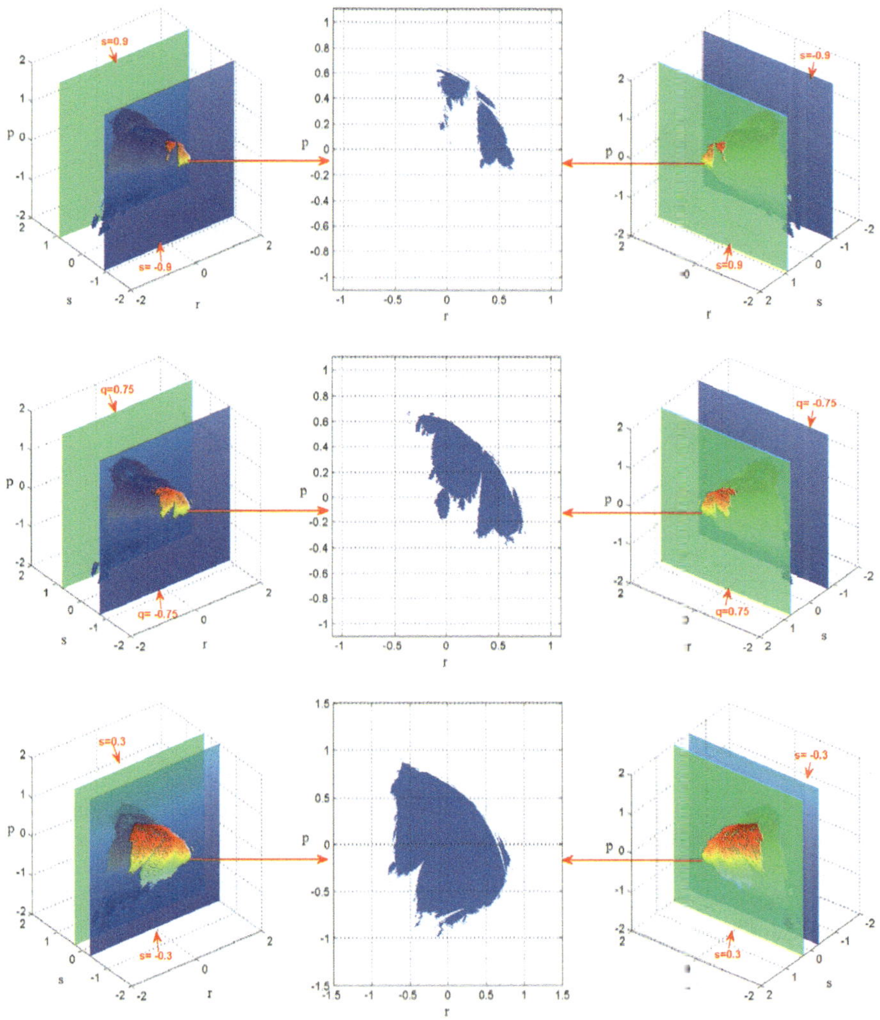


**Fig. 2.15** **a** $M_e(P_{c_1,c_2})_{q_0=0,s_0=-0.9}$ and $M_e(P_{c_1,c_2})_{q_0=0,s_0=0.9}$. **b** $M_e(P_{c\ ,c_2})_{q_0=0,s_0=-0.75}$ and $M_e(P_{c_1,c_2})_{q_0=0,s_0=0.75}$. **c** $M_e(P_{c_1,c_2})_{q_0=0,s_0=-0.3}$ and $M_e(P_{c_1,c_2})_{q_0=0,s_0=0.3}$.

## 2.3 The Formats of Julia Sets for Complex Dynamic Systems

In this section, the formats of Julia sets for a class of nonlinear complex dynamic systems with variable coefficients were studied under certain conditions. For the complex dynamic systems in piecewise cases, we proposed some methods to control the forms of their Julia sets and stable domains analytically. What's more, we

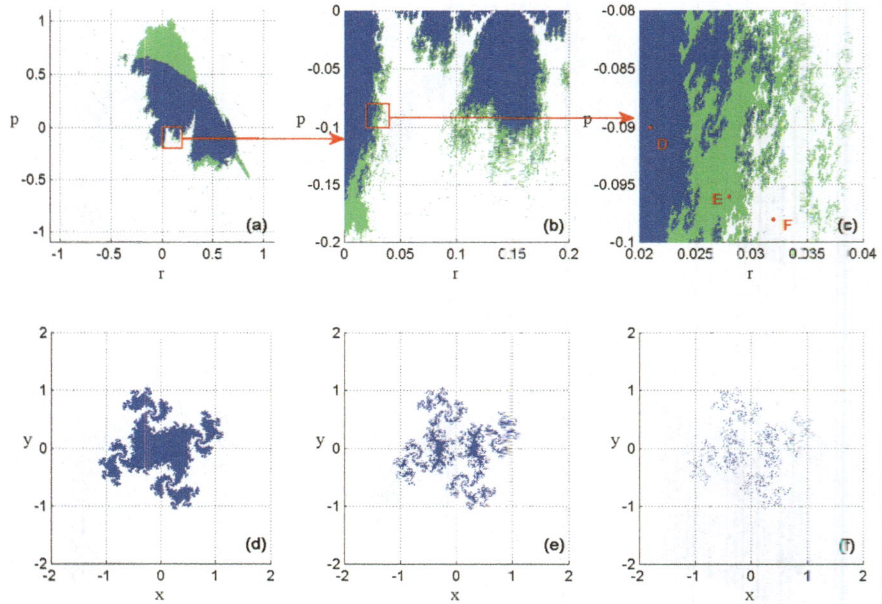


**Fig. 2.16** **a–c** The connectivity loci of system (2.2.2) and its partial enlarged views. **d** Connected $K(P_{c_1,c_2})$ with $c_1 = -0.09,\ c_2 = 0.021 - 0.75i$ (D point in **c**). **e** Disconnected $K(P_{c_1,c_2})$ with $c_1 = -0.096,\ c_2 = 0.028 - 0.75i$ (E point). **f** Totally disconnected $K(P_{c_1,c_2})$ with $c_1 = -0.098,\ c_2 = 0.032 - 0.75i$ (F point)

illustrated that our methods worked well by computational simulations. Our work provides a better understanding about how to control the Julia sets of certain complex dynamic systems.

Here, we give the definition of the formats of complex dynamic systems and introduce Julia set as follows.

**Definition 2.6** For the nonlinear time delay complex dynamic system:

$$z_{n+1} = f(a_n, c_n, z_n, z_{n-k})\ , \tag{2.3.1}$$

where $a_n$ and $c_n$ are known sequences of complex numbers, $k$ is a quantity of time delay and a nonnegative integer. The formats of complex dynamic systems are the process which makes the stable domain of system (2.3.1) to achieve some enacted aims by adjusting $a_n$, $c_n$ and $k$.

Especially, when analyzing system (2.3.1) through the definition of Julia sets, the stable domains are the formats of Julia set of system (2.3.1).

For (2.3.1), let $C$ is a set of all complex numbers, and we call $\partial(C)$ the boundary of $C$. The definition of Julia sets is as follows [206].

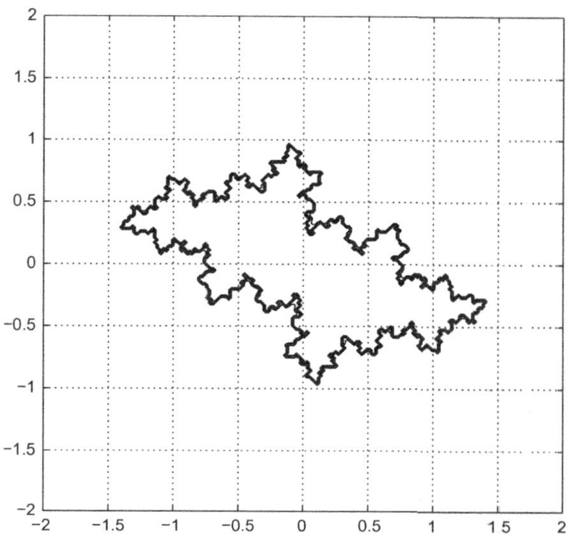


**Fig. 2.17** The Julia set of system (2.3.3) when $c = -0.5(1 - i)$

**Definition 2.7** For complex polynomial $f$, let $\mathscr{K}(f) = \{z \in C : f^k(z) \nrightarrow \infty, k \to \infty\}$, $\mathscr{J}(f) \triangleq \partial(\mathscr{K}(f))$. We call $\mathscr{J}(f)$ the Julia set of system (2.3.1).

In this part, firstly, we will study the formats of Julia sets for above system in general conditions. Then we analyze the formats of Julia sets of special conditions of the system above, which are generated through modifying the complex sequences $a_n$ and $c_n$. At last, we study the formats of Julia sets for the system above in the piecewise cases. We will illustrate our results through simulations.

For (2.3.1), we consider that $f(\cdot) = a(z+b)^2 - c$ . For the complex dynamic system:

$$z_{n+1} = a_n(z_n + c_n)^2 - c_{n+1} , \tag{2.3.2}$$

it becomes to the well-known classical system

$$z_{n+1} = z_n^2 + c \tag{2.3.3}$$

when $a_n \equiv 1$, $c_n \equiv 0$ and $c_{n+1} \equiv -c$, which was used to study its Julia set originally.

Setting $c = -0.5(1 + i)$, the Julia set of system (2.3.3) has the form showed in Fig. 2.17.

We can see that this form of the Julia set of system (2.3.3) is irregular, which implies that its stable domain is also irregular. What's more, we should note that system (2.3.3) is a special form of system (2.3.2).

For achieving certain requirements of stability for the systems, we need to obtain the regular domains of their Julia sets. As a result, we will focus on studying the regular formats of Julia sets for system (2.3.2).

### 2.3.1 The Formats of Julia Sets for System (2.3.2)

We will focus on the formats of Julia sets for system (2.3.2). Here we use the symbol $\mathscr{J}(2)$ to denote the Julia sets of system (2.3.2).

**Theorem 2.5** *For the complex dynamic system (2.3.2), assume that* $||a_n|| \le a$ *and* $||c_n|| \le c$ *for* $\forall n$, *then* $z_n$ *will be bounded when* $||z_0 + c_0|| \le \frac{1}{a}$ *in the case of* $n \to \infty$, *and*

$$\mathscr{J}(2) = \partial\left\{ z_0 : ||z_0 + c_0|| \le r,\ r \ge \frac{1}{a} \right\}.$$

***Proof*** For system (2.3.2), we note that

$$\begin{aligned} z_n &= a_{n-1}(z_{n-1} + c_{n-1})^2 - c_n\,, \\ z_{n-1} &= a_{n-2}(z_{n-2} + c_{n-2})^2 - c_{n-1}\,, \\ &\vdots \\ z_1 &= a_0(z_0 + c_0)^2 - c_1\,. \end{aligned}$$

Through iterating in sequence, we obtain

$$\begin{aligned} z_{n+1} &= a_n[a_{n-1}(z_{n-1} + c_{n-1})^2 - c_n + c_n]^2 - c_{n+1} \\ &= a_n a_{n-1}^2 (z_{n-1} + c_{n-1})^4 - c_{n+1} \\ &= a_n a_{n-1}^2 [a_{n-2}(z_{n-2} + c_{n-2})^2 - c_{n-1} + c_{n-1}]^4 - c_{n+1} \\ &= a_n a_{n-1}^2 a_{n-2}^4 (z_{n-2} + c_{n-2})^8 - c_{n+1} \\ &\vdots \\ &= a_n a_{n-1}^2 \cdots a_1^{2^{n-1}} a_0^{2^n} (z_0 + c_0)^{2^{n+1}} - c_{n+1}\,. \end{aligned}$$

Then we have

$$\begin{aligned} ||z_{n+1}|| &= ||a_n a_{n-1}^2 \cdots a_1^{2^{n-1}} a_0^{2^n} (z_0 + c_0)^{2^{n+1}} - c_{n+1}|| \\ &\le ||a_n|| \cdot ||a_{n-1}||^2 \cdots ||a_1||^{2^{n-1}} \cdot ||a_0||^{2^n} \cdot ||z_0 + c_0||^{2^{n+1}} + ||c_{n+1}||\,. \end{aligned}$$

For $||a_n|| \le a$ and $||c_n|| \le c$ for $\forall n$, we have

$$\begin{aligned} ||z_{n+1}|| &\le a \cdot a^2 \cdot a^4 \cdots a^{2^{n-1}} \cdot a^{2^n} \cdot ||z_0 + c_0||^{2^{n+1}} + c \\ &= a^{(2^0 + 2^1 + \cdots + 2^{n-1} + 2^n)} \cdot ||z_0 + c_0||^{2^{n+1}} + c \\ &= a^{\frac{1-2^{n+1}}{1-2}} \cdot ||z_0 + c_0||^{2^{n+1}} + c \\ &= a^{2^{n+1}-1} \cdot ||z_0 + c_0||^{2^{n-1}} + c \\ &= ||z_0 + c_0|| \cdot (a||z_0 + c_0||)^{2^{n+1}-1} + c\,. \end{aligned}$$

According to the proof above, we obtain that $||z_{n+1}|| \leq \frac{1}{a} + c$ when $n \to \infty$ and $||z_0 + c_0|| \leq \frac{1}{a}$, therefore $z_n$ is bounded. So $||z_n||$ is bounded for $||z_0 + c_0|| \leq r$ when $n \to \infty$, and $r \geq \frac{1}{a}$.

From Definition 2.7, the Julia set of system (2.3.2) is the boundary of all the $z_0$ which satisfies $||z_0 + c_0|| \leq r$. As a result,

$$\mathscr{J}(2) = \partial \left\{ z_0 : ||z_0 + c_0|| \leq r,\ r \geq \frac{1}{a} \right\} .$$

This implies that it is a circle with the center of $-c_0$ and the radius of some value equal to or greater than $\frac{1}{a}$. □

**Remark 2.2** According to the appearance of $a$ in the proof above, we can make sure that $z_n$ is bounded in the case of $n \to \infty$ if we choose $||z_0 + c_0|| \leq \frac{1}{a}$.

**Example 2.1** In system (2.3.2), let $c_n \equiv 0$ for $\forall n$, $a_0 = 4$ and $a_n \equiv 1$ when $n \geq 1$. Then it becomes to

$$z_{n+1} = \begin{cases} 4z_n^2\ , & n = 0\ ; \\ z_n^2\ , & n \geq 1\ . \end{cases} \tag{2.3.4}$$

Note that $||a_n|| \leq ||a_0|| = 4$ and $c_n \equiv 0$ for $\forall n$, from Theorem 2.5, we have

$$\mathscr{J}(4) = \partial \left\{ z_0 : ||z_0|| \leq r,\ r \geq \frac{1}{4} \right\} .$$

This implies that its Julia set is a circle with the center of $-c_0 = 0$ and the radius of some value equal to or greater than $\frac{1}{4}$.

We can proof it from the following analysis. Note that

$$z_{n+1} = a_n a_{n-1}^2 \cdots a_1^{2^{n-1}} a_0^{2^n} z_0^{2^{n+1}} = 4^{2n} \cdot z_0^{2^{n+1}} = (4z_0^2)^{2n},$$

then

$$||z_{n+1}|| = (4||z_0||^2)^{2n} .$$

So in the case of $n \to \infty$, $z_{n+1}$ is bounded if and only if $4||z_0||^2 \leq 1$. It implies $||z_n||$ is bounded if and only if $||z_0|| \leq 0.5$. According to Definition 2.7, its Julia set is the boundary of $||z_0|| \leq 0.5$. This means that its Julia set is a circle with the center of origin and the radius of 0.5, which satisfies the Theorem 2.5. Its Julia set is demonstrated in Fig. 2.18.

We can see that the result of the simulation matches the Theorem 2.5 well from above figure.

**Example 2.2** In system (2.3.2), let $a_n = \frac{1}{2(n+1)} + \frac{\sqrt{3}}{2(n+1)}i$ and $c_n = -\frac{2}{(n+1)^2} + \frac{3}{n+2}i$ for $\forall n$, then it leads to

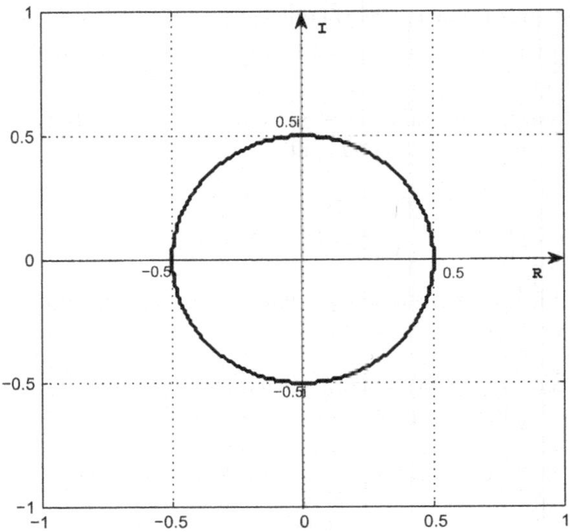


**Fig. 2.18** The Julia set of system (2.3.4)

$$z_{n+1} = \left[\frac{1}{2(n+1)} + \frac{\sqrt{3}}{2(n+1)}i\right]\left[z_n + \left(-\frac{2}{(n+1)^2} + \frac{3}{n+2}i\right)\right]^2 - \left(-\frac{2}{(n+2)^2} + \frac{3}{n+3}i\right). \quad (2.3.5)$$

Because $||a_n|| \le ||a_0|| = 1$ and $||c_n|| \le ||c_0|| = 2.5$ for $\forall n$, according to Theorem 2.5, we have

$$\mathscr{J}(5) = \partial\{\, z_0 : ||z_0 - 2 + 1.5i|| \le r,\ r \ge 1 \,\}\,.$$

This implies that its Julia set is a circle with the center of $-c_0 = 2 - 1.5i$ and the radius of 1. Its Julia set is demonstrated in Fig. 2.19.

We can see that the result of the simulation matches Theorem 2.5 well from Fig. 2.19.

### 2.3.2 *The Formats of Julia Sets for System* (2.3.2) *When* $a_n$ *Is a Constant*

For system (2.3.2), let $a_n \equiv a_0 \ne 0$, then it becomes to the following system

$$z_{n+1} = a_0(z_n + c_n)^2 - c_{n+1}. \quad (2.3.6)$$

We will analyze the formats of Julia sets for system (2.3.6).

**Theorem 2.6** *For the complex dynamic system* (2.3.6)*, assume that* $||c_n|| \le c$ *for* $\forall n$*, then* $z_n$ *is bounded if and only if* $||z_0 + c_0|| \le \frac{1}{||a_0||}$ *in the case of* $n \to \infty$*. Moreover,*

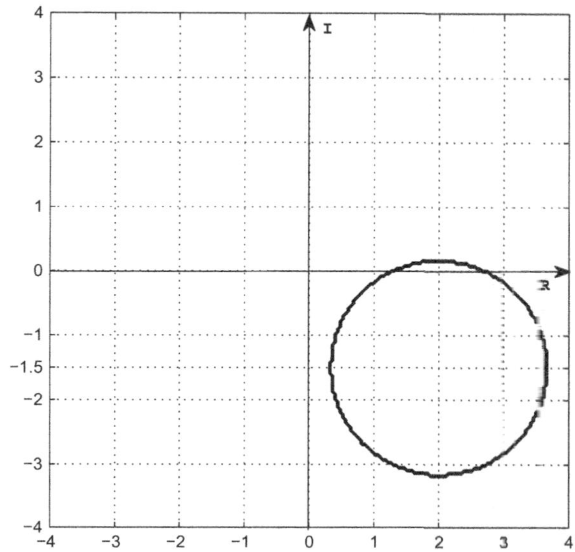


**Fig. 2.19** The Julia set of system (2.3.5)

$$\mathscr{J}(6) = \partial \left\{ z_0 : ||z_0 + c_0|| \leq \frac{1}{||a_0||} \right\} .$$

***Proof*** According to the proof of Theorem 2.5, we have

$$z_{n+1} = (z_0 + c_0)[a_0(z_0 + c_0)]^{2^{n+1}-1} - c_{n+1},$$

$$||z_{n+1}|| \leq ||z_0 + c_0||(||a_0|| \cdot ||z_0 + c_0||)^{2^{n+1}-1} + ||c_{n+1}||.$$

Because $||c_n|| \leq c$ for $\forall n$, we have

$$||z_{n+1}|| \leq ||z_0 + c_0||(||a_0|| \cdot ||z_0 + c_0||)^{2^{n+1}-1} + c .$$

Then $||z_{n+1}|| \leq \frac{1}{||a_0||} + c$ if and only if $||a_0|| \cdot ||z_0 + c_0|| \leq 1$ in the case of $n \to \infty$. This implies $z_n$ is bounded if and only if $||z_0 + c_0|| \leq \frac{1}{||a_0||}$ in the case of $n \to \infty$.

From Definition 2.7, we have

$$\mathscr{J}(6) = \partial \left\{ z_0 : ||z_0 + c_0|| \leq \frac{1}{||a_0||} \right\} ,$$

which means the Julia set of system (2.3.6) is a circle with the center of $-c_0$ and the radius of $\frac{1}{||a_0||}$. □

**Example 2.3** In system (2.3.6), let $a_n \equiv 2i$ and $c_n = -\frac{0.5+i}{n+1}$ for $\forall n$, then it becomes to

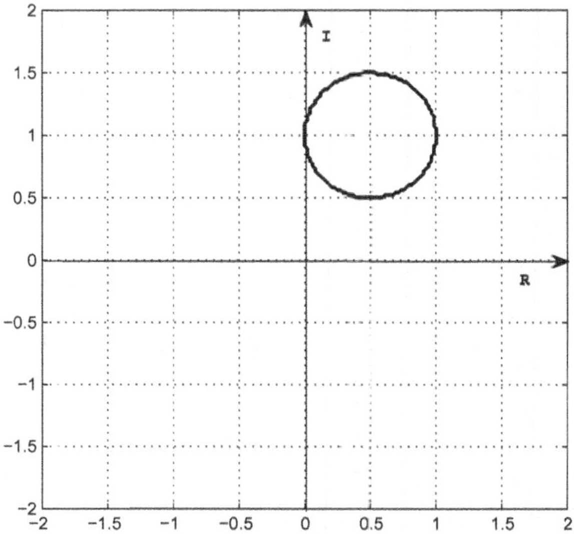


**Fig. 2.20** The Julia set of system (2.3.6)

$$z_{n+1} = 2i\left(z_n - \frac{0.5+i}{n+1}\right)^2 + \frac{0.5+i}{n+2}\,. \tag{2.3.7}$$

Because $||c_n|| \le ||c_0|| = \sqrt{1.25}$ for $\forall n$, from Theorem 2.6, we have

$$\mathscr{J}(7) = \partial\{\, z_0 : ||z_0 - 0.5 - i|| \le 0.5 \,\}\,.$$

This means that its Julia set is a circle with the center of $-c_0 = 0.5 + i$ and the radius of $\frac{1}{||a_0||} = 0.5$. Its Julia set is illustrated in Fig. 2.20.

We can see that the result of the simulation matches Theorem 2.6 well from the Fig. 2.20.

**Remark 2.3** The reason of studying the particular case of system (2.3.2) is that, from (2.3.6) and Theorem 2.6, we can construct a specific system that its Julia set is a circle with the center and the radius of any given value on the complex plane.

We will use the following example to illustrate it.

**Example 2.4** Finding out a system that its Julia set is a circle with the center of $5 + 6\sqrt{2}i$ and the radius of 10 on the complex plane.

From Theorem 2.6, the system we want is one of specific cases of system (2.3.6). It requires $c_n$ is bounded, $c_0 = -5 - 6\sqrt{2}i$ and $\frac{1}{||a_0||} = 10$ in system (2.3.6). So a specific case of system (2.3.6) which satisfies these three conditions is just the system we need. Let $c_n \equiv c_0 = -5 - 6\sqrt{2}i$ and $a_0 = 0.1$, system (2.3.6) becomes to the specific case $z_{n+1} = 0.1(z_n - 5 - 6\sqrt{2}i)^2 + 5 + 6\sqrt{2}i$. It is one case of the system we want.

For system (2.3.6), let $c_n \equiv c_0$, it becomes to a special system

$$z_{n+1} = a_0(z_n + c_0)^2 - c_0 \,. \tag{2.3.8}$$

From Theorem 2.6, we have the following corollary.

**Corollary 2.1** *For the complex dynamic system* (2.3.8), $z_n$ *is bounded in the case of* $n \to \infty$ *if and only if* $||z_0 + c_0|| \le \frac{1}{||a_0||}$. *What's more,* $\mathscr{J}(2.3.8) = \partial\{\, z_0 : ||z_0 + c_0|| \le \frac{1}{||a_0||} \,\}$.

## 2.3.3 *The Formats of Julia Sets of Piecewise Cases for System (2.3.2)*

In this section, we will study the formats of Julia sets of piecewise cases for system (2.3.2):

$$z_{n+1} = \begin{cases} a_{1n}(z_n + c_{1n})^2 - c_{1(n+1)} : & ||z_0 + c|| < r_1 \,; \\ a_{2n}(z_n + c_{2n})^2 - c_{2(n+1)} : & r_1 \le ||z_0 + c|| < r_2 \,; \\ \cdots, \cdots & \cdots \\ a_{sn}(z_n + c_{sn})^2 - c_{s(n+1)} : & r_s \le ||z_0 + c|| \,. \end{cases}$$

Firstly, we give the following theorem.

**Theorem 2.7** *For the complex dynamic system* (2.3.2), *assume that* $0 < a' \le ||a_n|| \le a$ *and* $||c_n|| \le c$ *for* $\forall n$, *then* $z_n$ *is bounded when* $||z_0 + c_0|| \le \frac{1}{a}$ *in the case of* $n \to \infty$. *Moreover,*

$$\mathscr{J}(2.3.2) = \partial \left\{ z_0 : ||z_0 + c_0|| \le r, \; \frac{1}{a} \le r \le \frac{1}{a'} \right\} .$$

***Proof*** We have proved that $z_n$ was bounded when $||z_0 + c_0|| \le \frac{1}{a}$ in the case of $n \to \infty$ and $r \ge \frac{1}{a}$. Here, we deal with $r \le \frac{1}{a'}$.

According to the proof of Theorem 2.5,

$$||z_{n+1}|| = ||a_n a_{n-1}^2 \cdots a_1^{2^{n-1}} a_0^{2^n} (z_0 + c_0)^{2^{n+1}} - c_{n-1}|| \,,$$

we have

$$||z_{n+1}|| \ge ||a_n|| \cdot ||a_{n-1}||^2 \cdots ||a_1||^{2^{n-1}} \cdot ||a_0||^{2^n} \cdot ||z_0 + c_0||^{2^{n+1}} - ||c_{n+1}|| \,.$$

Because $||a_n|| \ge a' > 0$ and $||c_n|| \le c$ for $\forall n$, we can get

$$\begin{aligned}||z_{n+1}|| &\geq a' \cdot (a')^2 \cdot (a')^4 \cdots (a')^{2^{n-1}} \cdot (a')^{2^n} \cdot ||z_0 + c_0||^{2^{n+1}} - c \\ &= (a')^{(2^0+2^1+\cdots+2^{n-1}+2^n)} \cdot ||z_0 + c_0||^{2^{n+1}} - c \\ &= (a')^{\frac{1-2^{n+1}}{1-2}} \cdot ||z_0 + c_0||^{2^{n+1}} - c \\ &= (a')^{2^{n+1}-1} \cdot ||z_0 + c_0||^{2^{n+1}} - c \\ &= ||z_0 + c_0|| \cdot (a'||z_0 + c_0||)^{2^{n+1}-1} - c\,.\end{aligned}$$

Then we have that $||z_{n+1}|| \geq \frac{1}{a'} - c$ when $a'||z_0 + c_0|| \leq 1$ in the case of $n \to \infty$. This implies that $z_n$ may be bounded only when $||z_0 + c_0|| \leq \frac{1}{a'}$ in the case of $n \to \infty$. As a result, $||z_n||$ is bounded when $n \to \infty$ if $||z_0 + c_0|| \leq r$ and $r \leq \frac{1}{a'}$.

From Definition 2.7, the Julia set of system (2.3.2) is the boundary of all the $z_0$ which satisfies $||z_0 + c_0|| \leq r$. According to Theorem 2.5 and the proof above, we have

$$\mathscr{J}(2.3.2) = \partial\left\{ z_0 : ||z_0 + c_0|| \leq r,\ \frac{1}{a} \leq r \leq \frac{1}{a'} \right\}.$$

This means that it is a circle with the center of $-c_0$ and the radius of some value between $\frac{1}{a}$ and $\frac{1}{a'}$. □

**Remark 2.4** According to the appearance of $a'$ in the proof above, we can make sure that $z_n$ is bounded in the case of $n \to \infty$ if we choose $||z_0 + c_0|| \leq \frac{1}{a'}$.

**The Cases that Julia Sets Are Concentric Circles**

Let system (2.3.2) be the following system

$$z_{n+1} = \begin{cases} a_{1n}(z_n + c_{1n})^2 - c_{1(n+1)} : & ||z_0 + c|| < a\,; \\ a_{2n}(z_n + c_{2n})^2 - c_{2(n+1)} : & ||z_0 + c|| \geq a\,. \end{cases} \tag{2.3.9}$$

In system (2.3.9), $a_{1n},\ a_{2n},\ c_{1n}, c_{2n},\ c \in C \neq \varnothing$, and $a > 0,\ n = 0, 1, 2, \cdots$.

**Theorem 2.8** *If piecewise system* (2.3.9) *satisfies the following conditions (i)* $0 < a'_1 \leq ||a_{1n}|| \leq a_1$ *and* $0 < a'_2 \leq ||a_{2n}|| \leq a_2$ *for* $\forall n$;

*(ii)* $||c_{1n}|| \leq c_1$ *and* $||c_{2n}|| \leq c_2$ *for* $\forall n$;

*(iii)* $c_{10} = c_{20} = c$;

*(vi)* $\frac{1}{a'_1} < a < \frac{1}{a_2}$,

*we define it as system (2.3.9.1). Then the Julia set of system (2.3.9.1) is*

$$\begin{aligned}\mathscr{J}(2.3.9.1) = \partial &\left\{ z_0 : ||z_0 + c|| \leq r_1,\ \tfrac{1}{a_1} \leq r_1 \leq \tfrac{1}{a'_1} \right\} \\ \bigcup \partial &\left\{ z_0 : a \leq ||z_0 + c|| \leq r_2,\ \tfrac{1}{a_2} \leq r_2 \leq \tfrac{1}{a'_2} \right\}.\end{aligned}$$

***Proof*** For system (2.3.9.1), according to Theorem 2.7, we can obtain the conclusion immediately. □

It implies that its Julia set is 3 concentric circles with the center of $-c$ and the radiuses of $r_1$, $a$ and $r_2$ respectively, where $\frac{1}{a_1} \leq r_1 \leq \frac{1}{a'_1}$, $\frac{1}{a_2} \leq r_2 \leq \frac{1}{a'_2}$.

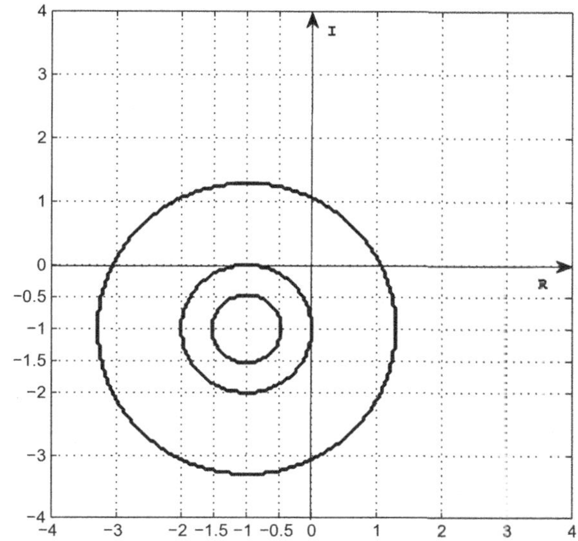


**Fig. 2.21** The Julia set of system (2.3.10)

**Example 2.5** In system (2.3.9), let $a_{1n} = \sqrt{3} + \frac{i}{n+1}$, $c_{1n} = 1 + \frac{i}{n+1}$, $a_{2n} = \frac{\sqrt{2}}{4} + \frac{\sqrt{2}i}{4(n+1)}$, $c_{2n} = \frac{1}{n+1} + i$, $c = 1 + i$ and $a = 1$, we have

$$z_{n+1} = \begin{cases} (\sqrt{3} + \frac{i}{n+1})[z_n + (1 + \frac{i}{n+1})]^2 - (1 + \frac{i}{n+2}) : & \|z_0 + (1+i)\| < 1\,; \\ (\frac{\sqrt{2}}{4} + \frac{\sqrt{2}}{4(n+1)}i)[z_n + (\frac{1}{n+1} + i)]^2 - (\frac{1}{n+2} + i) : & \|z_0 + (1+i)\| \geq 1\,. \end{cases} \tag{2.3.10}$$

Because

(i) $0 < \sqrt{3} < \|a_{1n}\| \leq \|a_{10}\| = 2$ and $0 < \frac{\sqrt{2}}{4} < \|a_{2n}\| \leq \|a_{20}\| = \frac{1}{2}$ for $\forall n$;
(ii) $\|c_{1n}\| \leq \|c_{10}\| = \sqrt{2}$ and $\|c_{2n}\| \leq \|c_{20}\| = \sqrt{2}$ for $\forall n$;
(iii) $c_{10} = c_{20} = c = 1 + i$;
(vi) $a = 1$, $\frac{1}{a_1'} = \frac{\sqrt{3}}{3}$, $\frac{1}{a_2} = 2$, so $\frac{1}{a_1'} < a < \frac{1}{a_2}$,
by Theorem 2.8, we have

$$\mathscr{J}(2.3.10) = \partial\left\{ z_0 : \|z_0 + (1+i)\| \leq r_1,\ 0.5 \leq r_1 < \frac{\sqrt{3}}{3} \right\} \bigcup \partial\left\{ z_0 : 1 \leq \|z_0 + (1+i)\| \leq r_2,\ 2 \leq r_2 < 2\sqrt{2} \right\}$$

This implies that the Julia set of system (2.3.10) is 3 concentric circles with the center of $-c = -1 - i$ and the radiuses of $r_1$, 1 and $r_2$ respectively, where $0.5 \leq r_1 < \frac{\sqrt{3}}{3}$, $2 \leq r_2 < 2\sqrt{2}$. Its Julia set is demonstrated in Fig. 2.21.

We can see that the simulation matches Theorem 2.8 well from the Fig. 2.21.

For system (2.3.9), let $a_{1n} \equiv a_1 \neq 0$, $a_{2n} \equiv a_2 \neq 0$, and $c_{1n} = c_{2n} \equiv c$, it becomes to a special system

$$z_{n+1} = \begin{cases} a_1(z_n + c)^2 - c : & ||z_0 + c|| < a\,; \\ a_2(z_n + c)^2 - c : & ||z_0 + c|| \geq a\,. \end{cases} \tag{2.3.11}$$

From Theorem 2.8, we can obtain the following corollary directly.

**Corollary 2.2** *For the complex dynamic system (2.3.11), when* $\frac{1}{||a_1||} < a < \frac{1}{||a_2||}$, *we have*

$$\mathcal{J}(11) = \partial\left\{ z_0 : ||z_0 + c|| \leq \frac{1}{||a_1||} \right\} \bigcup \partial\left\{ z_0 : a \leq ||z_0 + c|| \leq \frac{1}{||a_2||} \right\}.$$

*This implies that its Julia set is 3 concentric circles with the center of* $-c$ *and the radiuses of* $\frac{1}{||a_1||}$, $a$ *and* $\frac{1}{||a_2||}$ *respectively.*

**The Cases that Julia Sets Are Non-concentric Circles**

**Theorem 2.9** *If the piecewise system (2.3.9) satisfies the following conditions*
*(i)* $0 < a_1' \leq ||a_{1n}|| \leq a_1$ *and* $0 < a_2' \leq ||a_{2n}|| \leq a_2$ *for* $\forall n$;
*(ii)* $||c_{1n}|| \leq c_1$ *and* $||c_{2n}|| \leq c_2$ *for* $\forall n$;
*(iii)* $||c_{10} - c|| \leq a - \frac{1}{a_1'}$ *and* $||c_{20} - c|| \geq a + \frac{1}{a_2'}$,
*we define it as system (2.3.9.2). Then the Julia set of system (2.3.9.2) is*

$$\begin{aligned} \mathcal{J}(2.3.9.2) = \partial&\left\{ z_0 : ||z_0 + c_{10}|| \leq r_1,\ \frac{1}{a_1} \leq r_1 \leq \frac{1}{a'_1} \right\} \\ \bigcup \partial&\left\{ z_0 : ||z_0 + c_{20}|| \leq r_2,\ \frac{1}{a_2} \leq r_2 \leq \frac{1}{a'_2} \right\}. \end{aligned}$$

***Proof*** For system (2.3.9.2), according to Theorem 2.7, we can obtain the conclusion directly. □

In addition, from the third condition of Theorem 2.9, we have

$$\begin{aligned} ||c_{10} - c_{20}|| &= ||c_{10} - c + c - c_{20}|| \\ &= ||(c - c_{20}) - (c - c_{10})|| \\ &\geq ||c - c_{20}|| - ||c - c_{10}|| \\ &\geq (a + \tfrac{1}{a'_2}) - (a - \tfrac{1}{a'_1}) \\ &= \tfrac{1}{a'_1} + \tfrac{1}{a'_2} \geq r_1 + r_2\,. \end{aligned}$$

As a result, its Julia set is two circles that one with the center of $-c_{10}$ and the radius of $r_1$ while another with the center of $-c_{20}$ and the radius of $r_2$, where $\frac{1}{a_1} \leq r_1 \leq \frac{1}{a_1'}$, $\frac{1}{a_2} \leq r_2 \leq \frac{1}{a_2'}$. What's more, These two circles are away from each other when $||c_{10} - c_{20}|| > r_1 + r_2$, and they are externally tangent when $||c_{10} - c_{20}|| = r_1 + r_2$.

**Example 2.6** In system (2.3.9), let $a_{1n} = \frac{\sqrt{3}}{2} + \frac{i}{2(n+1)}$, $c_{1n} = \frac{3}{2} - \frac{3i}{2(n+1)}$, $a_{2n} = 1 + \frac{\sqrt{3}i}{n+1}$, $c_{2n} = -\frac{3}{5(n+1)} + \frac{4i}{5}$, $c = 1.5 - 1.5i$ and $a = 1.5$, then it leads to

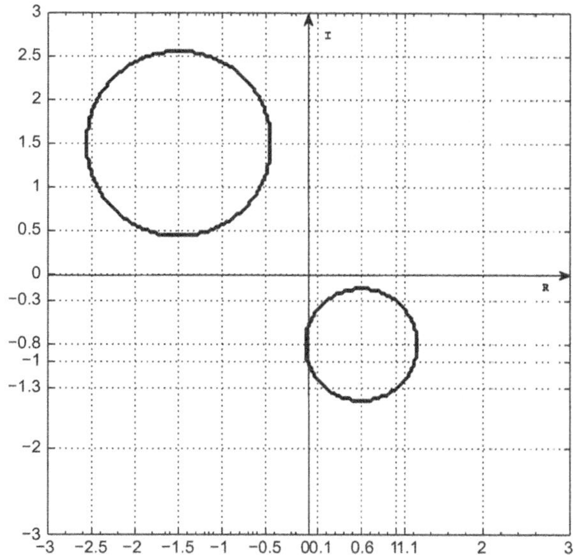


**Fig. 2.22** The Julia set of system (2.3.12)

$$z_{n+1} = \begin{cases} (\frac{\sqrt{3}}{2} + \frac{i}{2(n+1)})[z_n + (\frac{3}{2} - \frac{3i}{2(n+1)})]^2 - (\frac{3}{2} - \frac{3i}{2(n+2)}): & ||z_0 + (1.5 - 1.5i)|| < 1.5\,; \\ (1 + \frac{\sqrt{3}i}{n+1})[z_n + (-\frac{3}{5(n+1)} + \frac{4i}{5})]^2 - (-\frac{3}{5(n+2)} + \frac{4i}{5}): & ||z_0 + (1.5 - 1.5i)|| \geq 1.5\,. \end{cases} \tag{2.3.12}$$

Following the conditions

(i) $0 < \frac{\sqrt{3}}{2} < ||a_{1n}|| \leq ||a_{10}|| = 1$ and $0 < 1 < ||a_{2n}|| \leq ||a_{20}| = 2$ for $\forall n$;

(ii) $||c_{1n}|| \leq ||c_{10}|| = \frac{3\sqrt{2}}{2}$ and $||c_{2n}|| \leq ||c_{20}|| = 1$ for $\forall n$;

(iii) $a = 1.5,\ \frac{1}{a_1'} = \frac{2\sqrt{3}}{3},\ \frac{1}{a_2'} = 1,\ c_{10} = 1.5 - 1.5i,\ c_{20} = -0.6 + 0.8i,\ c = 1.5 - 1.5i$, so $||c_{10} - c|| \leq a - \frac{1}{a_1'}$ and $||c_{20} - c|| \geq a + \frac{1}{a_2'}$,

applying Theorem 2.9, we have

$$\begin{aligned} \mathscr{J}(16) = \partial \left\{ z_0 : ||z_0 + 1.5 - 1.5i|| \leq r_1,\ 1 \leq r_1 < \frac{2\sqrt{3}}{3} \right\} \\ \bigcup \partial \{ z_0 : ||z_0 - 0.6 + 0.8i|| \leq r_2,\ 0.5 \leq r_2 < 1 \}\,. \end{aligned}$$

Note that

$$||c_{10} - c_{20}|| > \frac{1}{a_1'} + \frac{1}{a_2'} \geq r_1 + r_2\,.$$

The Julia set of system (2.3.12) is two circles away from each other, one with the center of $-c_{10} = -1.5 + 1.5i$ and the radius of $r_1$ while another with the center of $-c_{20} = 0.6 - 0.8i$ and the radius of $r_2$, where $1 \leq r_1 < \frac{2\sqrt{3}}{3}$, $0.5 \leq r_2 < 1$. Its Julia set is demonstrated in Fig. 2.22.

### The Cases that Julia Sets Are Not Circles

**Theorem 2.10** *If piecewise system* (2.3.9) *satisfies the following conditions*

*(i)* $0 < a_1' \le ||a_{1n}|| \le a_1$ *and* $0 < a_2' \le ||a_{2n}|| \le a_2$ *for* $\forall n$;

*(ii)* $||c_{1n}|| \le c_1$ *and* $||c_{2n}|| \le c_2$ *for* $\forall n$;

*(iii)* $||c_{10} - c|| \ge a + \frac{1}{a_1'}$ *and* $a + \frac{1}{a_2} > ||c_{20} - c||$,

*(vi)* $||c_{20} - c|| > a - \frac{1}{a_2} \ge 0$ *or* $||c_{20} - c|| > \frac{1}{a_2'} - a \ge 0$,

*we define it as system (2.3.9.3). Then the Julia set of system (2.3.9.3) is*

$$\mathscr{J}(2.3.9.3) = \partial\Big(\Big\{z_0 : ||z_0 + c_{20}|| \le r_2,\ \frac{1}{a_2} \le r_2 \le \frac{1}{a_2'}\Big\}$$
$$-\Big\{z_0 : ||z_0 + c_{20}|| \le r_2,\ \frac{1}{a_2} \le r_2 \le \frac{1}{a_2'}\Big\}\bigcap\{z_0 : ||z_0 + c|| < a\}\Big).$$

***Proof*** For system (2.3.9.3), according to Theorem 2.7, we can obtain the conclusion directly. □

It implies that its Julia set is formed by two intersected circles, one with the center of $-c_{20}$ and the radius of $r_2$ while another with the center of $-c$ and the radius of $a$, where $\frac{1}{a_2} \le r_2 \le \frac{1}{a_2'}$.

**Example 2.7** (i) In system (2.3.9), let $a_{1n} = 10 + \frac{10i}{n+1}$, $c_{1n} = \frac{16}{n+1}$, $a_{2n} = \frac{1}{2} + \frac{\sqrt{3}i}{2(n+1)}$, $c_{2n} = -2 - \frac{2i}{n+1}$, $c = -2 - 3i$ and $a = 1$, then it leads to

$$z_{n+1} = \begin{cases} (10 + \frac{10i}{n+1})(z_n + \frac{16}{n+1})^2 - \frac{16}{n+2} : & ||z_0 - 2 - 3i|| < 1\,; \\ (\frac{1}{2} + \frac{\sqrt{3}i}{2(n+1)})[z_n + (-2 - \frac{2i}{n+1})]^2 - (-2 - \frac{2i}{n+2}) : & ||z_0 - 2 - 3i|| \ge 1\,. \end{cases} \quad (2.3.13)$$

For system (2.3.13) satisfies all the conditions of Theorem 2.10, we have

$$\mathscr{J}(18) = \partial(\{z_0 : ||z_0 - 2 - 2i|| \le r_2,\ 1 \le r_2 < 2\}$$
$$-\{z_0 : ||z_0 - 2 - 2i|| \le r_2,\ 1 \le r_2 < 2\}\bigcap\{z_0 : ||z_0 - 2 - 3i|| < 1\}).$$

It implies the Julia set of system (2.3.13), the boundary with the shape of crescent moon, is formed by the circle with the center of $-c_{20} = 2 + 2i$ and the radius of $r_2$ removing the intersecting part of it and the circle with the center of $-c = 2 + 3i$ and the radius of 1, where $1 \le r_2 < 2$. Its Julia set is demonstrated in Fig. 2.23 by the boundary of the shadow region.

(ii) In system (2.3.9), let $a_{1n} = \frac{3}{n+1} + 4i$, $c_{1n} = \frac{9i}{n+3}$, $a_{2n} = \frac{\sqrt{2}}{2} + \frac{\sqrt{2}i}{2(n+1)}$, $c_{2n} = \frac{1}{n+1} - i$, $c = 0.5 - 0.5i$ and $a = 1.5$, it becomes to

$$z_{n+1} = \begin{cases} (\frac{3}{n+1} + 4i)(z_n + \frac{9i}{n+3})^2 - \frac{9i}{n+4} : & ||z_0 + 0.5 - 0.5i|| < 1.5\,; \\ (\frac{\sqrt{2}}{2} + \frac{\sqrt{2}i}{2(n+1)})[z_n + (\frac{1}{n+1} - i)]^2 - (\frac{1}{n+2} - i) & ||z_0 - 0.5 - 0.5i|| \ge 1.5\,. \end{cases} \quad (2.3.14)$$

For system (2.3.14) satisfies all the conditions of Theorem 2.10, we have

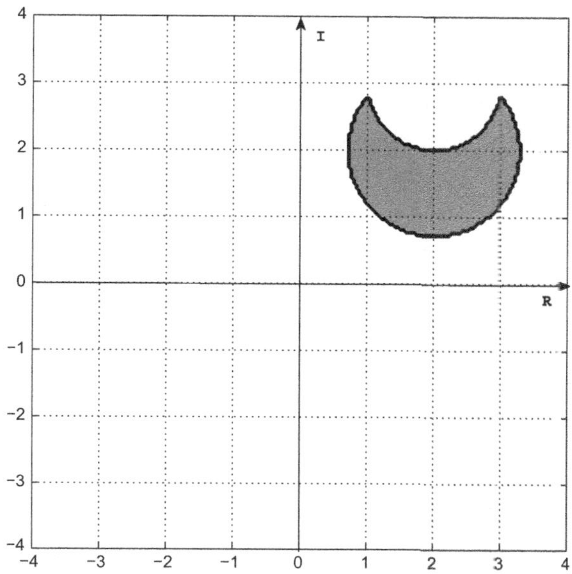


**Fig. 2.23** The Julia set of system (2.3.13)

$$\begin{aligned}\mathscr{J}(19) = \partial( &\{ z_0 : ||z_0 + 1 - i|| \le r_2,\ 1 \le r_2 < \sqrt{2}\,\} \\ &-\{ z_0 : ||z_0 + 1 - i|| \le r_2,\ 1 \le r_2 < \sqrt{2}\,\} \\ &\bigcap \{ z_0 : ||z_0 + 0.5 - 0.5i|| < 1.5 \}\,)\,,\end{aligned}$$

It implies the Julia set of system (2.3.14), the boundary with the shape of crescent moon, is formed by the circle with the center of $-c_{20} = -1 + i$ and the radius of $r_2$ removing the intersecting part of it and the circle with the center of $-c = -0.5 + 0.5i$ and the radius of 1.5, where $1 \le r_2 < \sqrt{2}$. Its Julia set is demonstrated in Fig. 2.24 by the boundary of the shadow region.

(iii) In system (2.3.9), let $a_{1n} = 8 + \frac{6i}{n+1}$, $c_{1n} = \frac{12}{n+2}$, $a_{2n} = \frac{1}{3(n+1)} + \frac{\sqrt{3}i}{3}$, $c_{2n} = 2 + \frac{9i}{5(n+1)}$, $c = 1 + 2i$ and $a = 1$, then it leads to

$$z_{n+1} = \begin{cases} (8 + \frac{6i}{n+1})(z_n + \frac{12}{n+2})^2 - \frac{12}{n+3} : & ||z_0 + 1 + 2i|| < 1\,; \\ (\frac{1}{3(n+1)} + \frac{\sqrt{3}i}{3})[z_n + (2 + \frac{9i}{5(n+1)})]^2 - (2 + \frac{9i}{5(n+2)}) : & ||z_0 + 1 + 2i|| \ge 1\,. \end{cases} \tag{2.3.15}$$

For system (2.3.15) satisfies all the conditions of Theorem 2.10, we have

$$\begin{aligned}\mathscr{J}(20) = \partial( &\{ z_0 : ||z_0 + 2 + 1.8i|| \le r_2,\ 1.5 \le r_2 < \sqrt{3}\,\} \\ &-\{ z_0 : ||z_0 + 2 + 1.8i|| \le r_2,\ 1.5 \le r_2 < \sqrt{3}\,\} \\ &\bigcap \{ z_0 : ||z_0 + 1 + 2i|| < 1 \}\,)\,,\end{aligned}$$

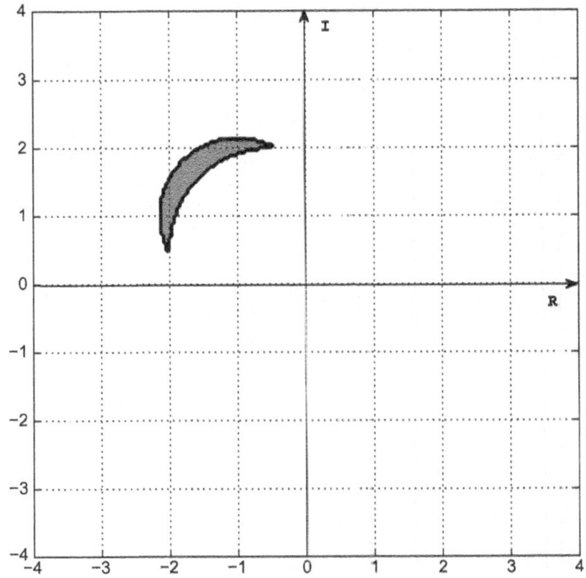


Fig. 2.24 The Julia set of system (2.3.14)

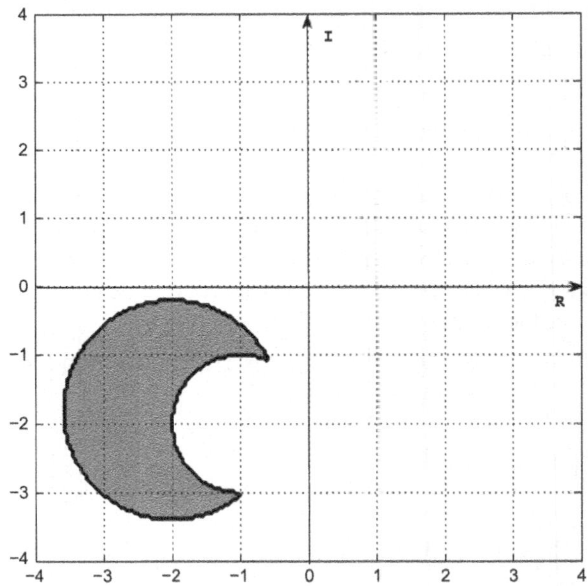


Fig. 2.25 The Julia set of system (2.3.15)

It implies the Julia set of system (2.3.15), the boundary with the shape of crescent moon, is formed by the circle with the center of $-c_{20} = -2 - 1.8i$ and the radius of $r_2$ removing the intersecting part of it and the circle with the center of $-c = -1 - 2i$ and the radius of 1, where $1.5 \le r_2 < \sqrt{3}$. Its Julia set is demonstrated in Fig. 2.25 by the boundary of the shadow region.

(vi) In system (2.3.9), let $a_{1n} = \frac{2}{n+1} + \sqrt{5}i$, $c_{1n} = \frac{7i}{n+1}$, $a_{2n} = 1 + \frac{\sqrt{3}i}{3(n+1)}$, $c_{2n} = -\frac{2}{n+1} + 2i$, $c = -1.5 + 1.5i$ and $a = 1$, then it becomes

$$z_{n+1} = \begin{cases} (\frac{2}{n+1} + \sqrt{5}i)(z_n + \frac{7i}{n+1})^2 - \frac{7i}{n+2} : & ||z_0 - 1.5 + 1.5i|| < 1 ; \\ (1 + \frac{\sqrt{3}i}{3(n+1)})[z_n + (-\frac{2}{n+1} + 2i)]^2 - (-\frac{2}{n+2} + 2i) : & ||z_0 - 1.5 + 1.5i|| \ge 1 . \end{cases} \quad (2.3.16)$$

Because system (2.3.16) satisfies all the conditions of Theorem 2.10, we have

$$\begin{aligned} \mathscr{J}(21) = \partial(\{\, z_0 : ||z_0 - 2 + 2i|| \le r_2, \ \frac{\sqrt{3}}{2} \le r_2 < 1 \,\} \\ - \left\{ z_0 : ||z_0 - 2 + 2i|| \le r_2, \ \frac{\sqrt{3}}{2} \le r_2 < 1 \right\} \\ \bigcap \{\, z_0 : ||z_0 - 1.5 + 1.5i|| < 1 \,\}\,) , \end{aligned}$$

It implies the Julia set of system (2.3.16), the boundary with the shape of crescent moon, is formed by the circle with the center of $-c_{20} = 2 - 2i$ and the radius of $r_2$

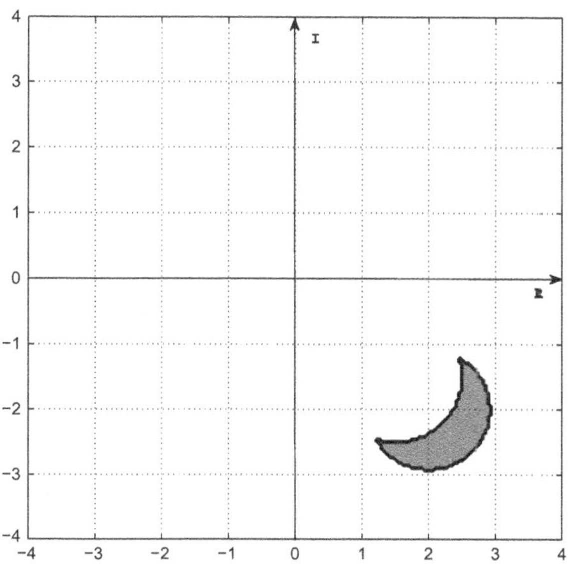


**Fig. 2.26** The Julia set of system (2.3.16)

removing the intersecting part of it and the circle with the center of $-c = 1.5 - 1.5i$ and the radius of 1, where $\frac{\sqrt{3}}{2} \leq r_2 < 1$. Its Julia set is demonstrated in Fig 2.26 by the boundary of the shadow region.

We can see that all the simulated results match Theorem 2.10 well from these figures above. What's more, here the Julia sets of these four systems are located in four different quadrants respectively. This means that we can control the locations and sizes of their Julia sets.

# Chapter 3
# Fractal Control and Synchronization of Classical Model

Based on the study of the typical Langevin problem and the dynamics of the particle's movement, the physical applications of M-J sets are discussed [174, 175]. They show that the fractal structure of the M-J sets may visually reflect the change rule of the particle's velocity. And the classical Mandelbrot set of the complex system $z_{n+1} = z_n^2 + c$ has a concrete physical meaning in the classical particle movement. For all kick strengths $c$ being in Mandelbrot set, the velocity of the particle remains bounded, whereas for $c$ being not in Mandelbrot set the velocity asymptotically diverges—the particle 'falls of' the potential well. Julia sets also have the physical meaning. For each $c$, they describe the closure of the union of all possible unstable periodic orbits of the particle.

However, the particle's movement needs to show different behaviors or similar behaviors for different systems according to the practical requirement. Recently, the control and synchronization of Julia sets are considered [196, 207, 208], which provides a valid method to analyze and control the movement of the particle. In this part, the feedback control is taken on the Mandelbrot set of (1). We show that Mandelbrot set of the controlled system changes with the changing of values of the control parameter, such as rotating, expanding, shrinking and so on. Subsequently, this control method is extended to control the generalized Mandelbrot set of the complex system

$$z_{n+1} = z_n^r + c, \tag{3.0.1}$$

where $r \geq 2$ is an arbitrary integer. By taking the reference item to be the trajectory of another system, synchronization of Mandelbrot sets of different systems is achieved.



S. T. Liu et al., *Fractal Control and Its Applications*,
https://doi.org/10.1007/978-981-15-5459-9_3

## 3.1 Feedback Control and Synchronization of Mandelbrot Sets of Complex Quadratic Polynomial Map

### 3.1.1 Feedback Control of Mandelbrot Set

According to the fractal character of M-set, we introduce the auxiliary reference feedback control [209] into the system (3.0.1). Then we have the following controlled system

$$z_{n+1} = z_n^2 + c + g_n(z_n - z_{ref}), \tag{3.1.1}$$

where $z_{ref}$ is the reference item and $g_n$ is the feedback item. Take

$$g_n = kz_{ref} + \sum_{j=0}^{l} a_j z_{n-j}, \tag{3.1.2}$$

where $l$ is a non-negative integer depending on the dimension of the system. $k$ and $a_j$ depend on the stability of the target. To simplify our illustration, we choose the simplest form, i.e., $l = 0$ , then (3.1.2) is

$$g_n = kz_{ref} + a_0 z_n. \tag{3.1.3}$$

Substituting (3.1.3) into (3.1.1), we have

$$z_{n+1} = (1 + a_0)z_n^2 + (k - a_0)z_{ref} z_n + c - kz_{ref}^2. \tag{3.1.4}$$

Let $k = a_0$, we have

$$z_{n+1} = (1 + k)z_n^2 + c - kz_{ref}^2. \tag{3.1.5}$$

To discuss the Mandelbrot set of the controlled system (3.1.5), we discuss the structure of the Mandelbrot sets of the following systems. Consider

$$z_{n+1} = \frac{1}{d} z_n^2 - dc, \tag{3.1.6}$$

$$z_{n+1} = z_n^2 + dc, \tag{3.1.7}$$

$$z_{n+1} = z_n^2 + c - d, \tag{3.1.8}$$

$$z_{n+1} = z_n^2 + (\cos\theta + i \sin\theta)c. \tag{3.1.9}$$

Let $f(z) = z^2 + c$, $f_5(z) = (1 + k)z^2 + c - kz_{ref}^2$, $f_6(z) = \frac{1}{d}z^2 + dc$, $f_7(z) = z^2 + dc$, $f_8(z) = z^2 + c - d$ and $f_9(z) = z^2 + (\cos\theta + i\sin\theta)c$. When the parameter $d$ is changed in systems (3.1.6)–(3.1.9), we obtain the information about different

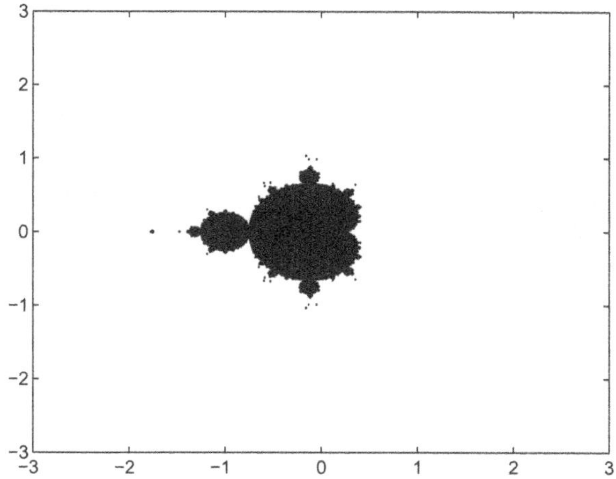


**Fig. 3.1** Mandelbrot set $M_1$ of system (3.0.1)

Mandelbrot sets, such as the position, size, shape and so on. In consequence, we can adjust the values of $d$ according to different requirements. Recall that the Mandelbrot set $M$ is defined to be the parameter set $c$ that enables the Julia set of $f$ to be connected,

$$M = \{c \in C : J(f) \text{ is connected}\}. \tag{3.1.10}$$

In addition, the recursive form of $f$ could be used to illustrate another character of the Mandelbrot set,

$$M = \{c \in C : \{f^n(0)\}_{n \geq 1} \text{ is bounded}\}. \tag{3.1.11}$$

Denote the Mandelbrot sets of system (3.0.1), (3.1.5)–(3.1.9) by $M_1$, $M_5 - M_9$ and the $M_1$ is illustrated in Fig. 3.1.

Let $w_{n+1} = \frac{1}{d}w_n^2 + dc$ for system (3.1.6), then by induction, we can get $w_n = d * z_n$. Thus, for any constant $d$, the iteration of system (3.1.6) at 0 is bounded if and only if the iterations of system (3.0.1) at 0 is bounded. From the definition of Mandelbrot set, we have the following conclusion.

**Proposition 3.1** *The Mandelbrot sets of system* (3.1.6) *and* (3.0.1) *are identical.*

This conclusion shows that the Mandelbrot sets of the controlled system (3.1.6) will not be changed with different values of the control parameter $d$, see Fig. 3.2.

**Proposition 3.2** *The Mandelbrot set of system* (3.1.7) $M_7 = \{\frac{1}{d}c : c \in M_1\}$.

***Proof*** Let $c' = dc$, then the Mandelbrot set $M'$ of $f_8(z) = z^2 + c'$ is same to $M_1$ about the system parameter $c'$. Since $c = \frac{1}{d}c'$, we have $M_8 = \{\frac{1}{d}c : c \in M_1\}$. $M_8$ increases as $d$ decreases, which means that smaller control values can get bigger Mandelbrot sets and bigger control value can get smaller Mandelbrot sets, see Fig. 3.3.

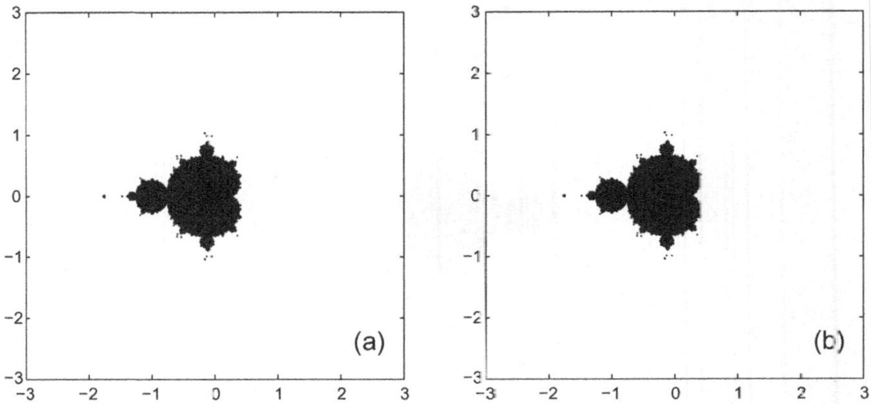


**Fig. 3.2** Control of Mandelbrot sets of system (3.1.6). **a** $d = 0.49$; **b** $d = 21$

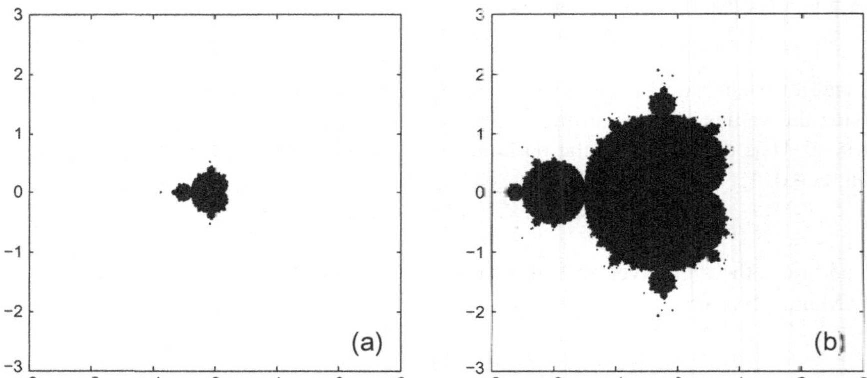


**Fig. 3.3** Control of Mandelbrot sets of system (3.1.7). **a** $d = 19$; **b** $d = 0.51$

With the similar analysis to that of Proposition 3.2, we have (see Fig. 3.4)

**Proposition 3.3** *(1) The Mandelbrot set* $M_8$ *of system* (3.1.8) *can be obtained by moving* $M_1$ $d$ *units.*
*(2) The Mandelbrot set* $M_9$ *of system* (3.1.9) *can be obtained by rotating* $M_1$ $\theta$ *angle.*

We now discuss the Mandelbrot set of the controlled system (3.1.5). From (3.1.5) we have

$$z_{n+1} = (1+k)z_n^2 + \frac{1}{1+k}(1-k)(c - kz_{ref}^2). \tag{3.1.12}$$

By Proposition 3.2, we know that the Mandelbrot sets of system (3.1.12) or system(3.1.5) is same to the Mandelbrot sets of the following system

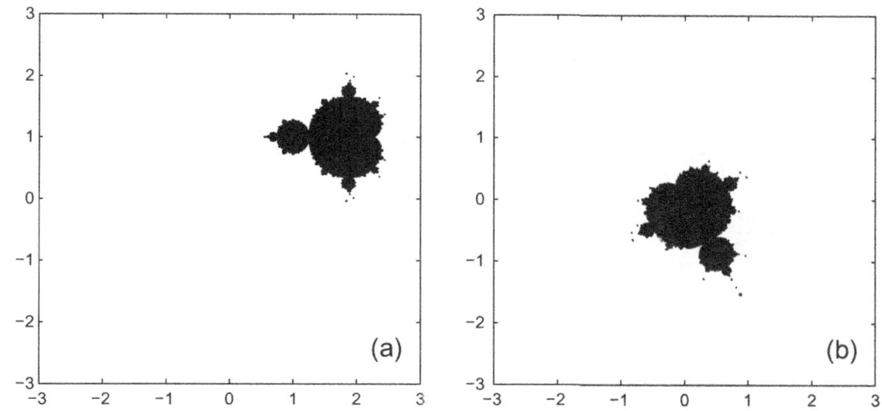

**Fig. 3.4** Control of Mandelbrot sets of system (3.1.8) and (3.1.10). **a** $d = 2 + i$; **b** $\theta = -\frac{2\pi}{3}$

$$\begin{aligned} z_{n+1} &= z_n^2 + (1+k)(c - kz_{ref}^2) \\ &= z_n^2 + (1+k)c - (1+k)kz_{ref}^2. \end{aligned} \tag{3.1.13}$$

From Propositions 3.2 and 3.3, we can get that the Mandelbrot set of system (3.1.12) is also that of the controlled system (3.1.5). According to our expectation, we could select appropriate reference item $z_{ref}$ in the controlled system (3.1.5) to control the Mandelbrot sets at any position on the complex plane, and control the size of Mandelbrot set by changing $k$, see Fig. 3.5.

Figure 3.5 shows that the Mandelbrot sets rotate $\theta$, or expand, or shrink once the auxiliary reference control item $z_{ref}$ and the feedback coefficient $k$ are changed.

In fact, by amending the feedback item in (3.1.2), the auxiliary reference feedback control method can be applied to the generalized Mandelbrot set of the complex system

$$z_{n+1} = z_n^r + c, \tag{3.1.14}$$

where $r > 2$ is an integer. For example, take $g_n = k \cdot \sum_{i=1}^{n} z_n^{n-i} z_{ref}^{i-1}$, then the controlled system of (3.1.14) is

$$z_{n+1} = z_n^r + c + k(z_n^r - z_{ref}^r). \tag{3.1.15}$$

The changing of Mandelbrot sets of (3.1.15) can be analyzed in a similar way. Thus we omit the discussion here.

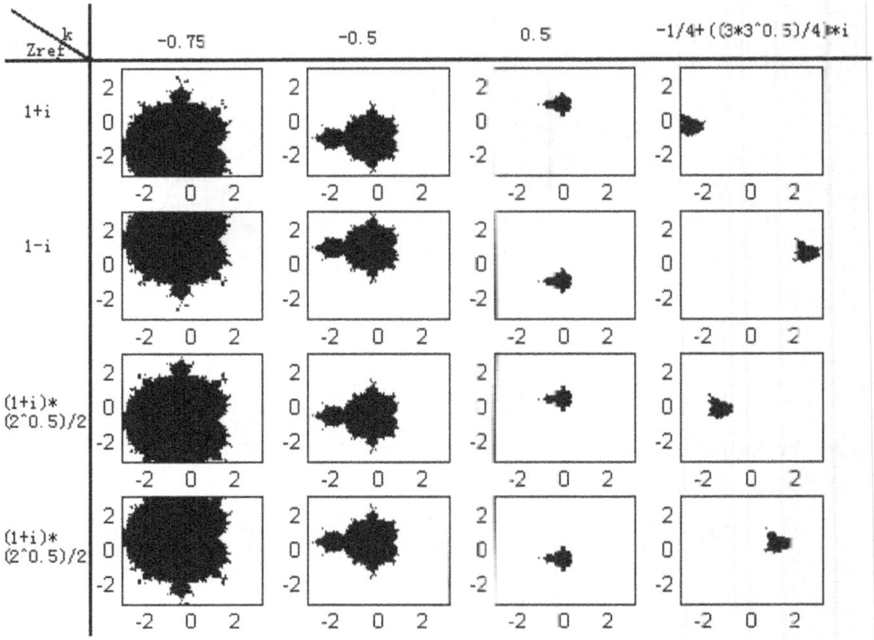


**Fig. 3.5** Control of Mandelbrot sets of system (3.1.5) with different parameters

### 3.1.2 *Synchronization of Mandelbrot Sets*

Generally speaking, chaos synchronization has been studied extensively and there are many methods to realize chaos synchronization. In this section, we will consider the synchronization of generalized Mandelbrot sets of (3.1.14) and the following complex system

$$w_{n+1} = w_n^s + c', \tag{3.1.16}$$

where $s$ is not equal to $r$ in (3.1.14). Therefore the generalized Mandelbrot sets of (3.1.14) and (3.1.16) are different. We consider to control the generalized Mandelbrot set of (3.1.14) changing to be the generalized Mandelbrot set of (3.1.16). Since $z_{ref}$ in the auxiliary reference feedback control item usually is chosen to be the objective, we take $z_{ref} = (w_n)^{\frac{s}{r}}$. Then

$$z_{n+1} = z_n^r + c + k(z_n^r - w_n^s), \tag{3.1.17}$$

and so

$$\begin{aligned} z_{n+1} - w_{n+1} &= z_n^r + c + k(z_n^r - w_n^r) - w_n^s - c' \\ &= (1+k)(z_n^r - w_n^s) + c - c' \\ &= (1+k)(z_n - w_n)(z_n^{r-1} + z_n^{r-2} w_n + \cdots + z_n w_n^{r-2} + w_n^{r-1}) \\ &+ (1+k)(w_n^r - w_n^s) + c - c'. \end{aligned}$$

Considering only the iterations which are bounded since $c$ is not in the generalized Mandelbrot set if for some $n_0$, $f^{n_0}(0; c)$ is not bounded from the definition of Mandelbrot set. That is to say that there exists a positive real number $T$ and we consider only the iterations included in the set $\{z : |z| \le T\}$. Let $M = \max\{T^r, T^s\}$. From the definition of Mandelbrot set, we take the initial values $z_1 = w_1 = 0$ and $c = c'$ in the computing process. Then

$$|z_3 - w_3| = |(1+k)(c^r - c^s)| \le 2M|1+k|,$$

and so

$$\begin{aligned} |z_{n+1} - w_{n+1}| &\le rM|1+k||z_n - w_n| + 2M|1+k| \\ &\le rM|1+k|(rM|1+k||z_{n-1} - w_{n-1}| + 2M|1+k|) + 2M|1+k| \\ &= (rM|1+k|)^2|z_{n-1} - w_{n-1}| + (rM|1+k| + 1)(2M|1+k|) \\ &\le \cdots \\ &\le (rM|1+k|)^{n-3}|z_3 - w_3| + [(rM|1+k|)^{n-4} + \cdots + rM|1+k| + 1](2M|1+k|) \\ &\le [(rM|1+k|)^{n-3} + (rM|1+k|)^{n-4} + \cdots + rM|1+k| + 1](2M|1+k|) \\ &= 2M|1+k|\frac{1 - (rM|1+k|)^{n-2}}{1 - rM|1+k|}. \end{aligned}$$

Take $rM|1+k| < 1$, then the right side of the above equation goes to $\frac{2M|1+k|}{1-rM|1+k|}$ as $n \to \infty$. If $|1+k| \to 0$ in this case, that is $k \to -1$, we have $|z_{n-1} - w_{n+1}| \to 0$. Thus the synchronization of trajectories of systems (3.1.16) and (3.1.17) is achieved, that is to say that, Mandelbrot set of (3.1.17) will change to be Mandelbrot set of (3.1.16) with $k \to -1$.

Take $s = 2$ in (3.1.16) and $r = 4.99$ in (3.1.14), then Mandelbrot sets of them are illustrated in Figs. 3.1 and 3.6 respectively. From Fig. 3.7, obviously, Mandelbrot sets synchronization has not been achieved when the values of $k$ is small. However, Mandelbrot set of (3.1.17) is changing to be the Mandelbrot set of (3.1.14) with the values of $k$ going to $-1$.

## 3.2 Coupling of Mandelbrot Sets

In this section, we introduce the definitions of synchronization and coupling of two different Mandelbrot sets and consider the changing and coupling between two different Mandelbrot sets. Then, by the nonlinear coupling method, the synchronization and coupling of two different Mandelbrot sets are achieved, which make one

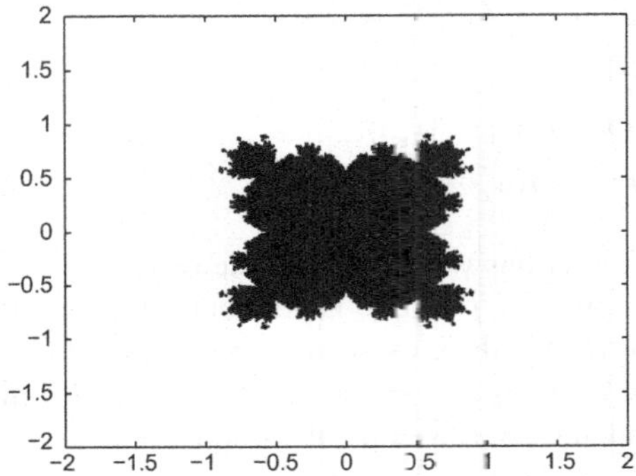


**Fig. 3.6** Mandelbrot set of system (3.1.14) when $r = 4.99$

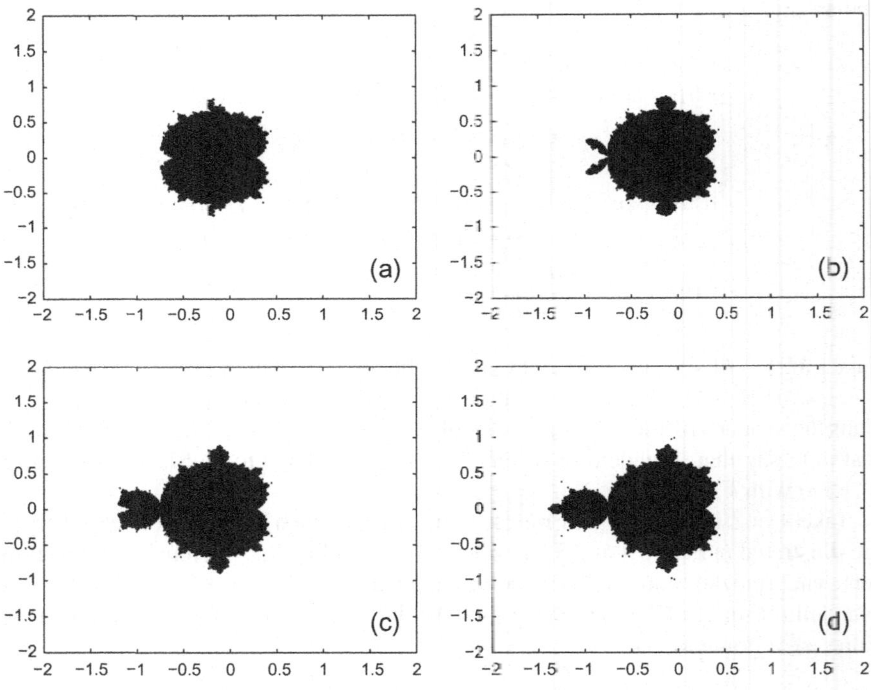


**Fig. 3.7** Mandelbrot set of system (3.1.15) when $r = 4.99$. **a** $k = -0.11$; **b** $k = -0.39$; **c** $k = -0.69$; **d** $k = -0.84$

Mandelbrot set change to be another one and also make two different Mandelbrot sets change to be the same one.

Consider two systems in the complex domain $\mathbb{C}$

$$z_{n+1} = f(z_n; c), \tag{3.2.1}$$

and

$$w_{n+1} = g(w_n; d). \tag{3.2.2}$$

Obviously, the Mandelbrot sets of (3.2.1) and (3.2.2) are determinate and have not any relations between them. Denote them by $M^c$ and $M^d$ respectively.

To connect $M^c$ with $M^d$, we couple the systems (3.2.1) and (3.2.2). Introduce the coupled items into (3.2.1) and (3.2.2), then

$$z_{n+1} = f(z_n; c) + p(z_n, w_n; k), \tag{3.2.3}$$

$$w_{n+1} = g(w_n; d) + q(z_n, w_n; k), \tag{3.2.4}$$

where $p$ and $q$ are the coupling items and are the continuous maps about $z_n$ and $w_n$, and $k$ is the coupling strength. It is obvious that one Mandelbrot set is corresponding with a value of $k$ and denote by $M_k^c$ and $M_k^d$ the Mandelbrot sets of systems (3.2.3) and (3.2.4) respectively. Coupling of Mandelbrot sets of (3.2.1) and (3.2.2) is taken to occur if

$$\lim_{k\to k_0} (M_k^c \cup M_k^d - M_k^c \cap M_k^d) = \emptyset \tag{3.2.5}$$

for some $k_0$ ($k_0$ may be $\infty$). Specifically, when $q(z_n, w_n; k) = 0$, (3.2.5) becomes

$$\lim_{k\to k_0} (M_k^c \cup M^d - M_k^c \cap M^d) = \emptyset. \tag{3.2.6}$$

Then the synchronization of Mandelbrot sets of (3.2.1) and (3.2.2) is achieved. That is, the Mandelbrot set of (3.2.3) will change to be the Mandelbrot set of (3.2.2) when $k \to k_0$.

From (3.2.5), it is not difficult to see that the Mandelbrot sets of (3.2.3) and (3.2.4) will change to be the same one when $k \to k_0$ and from (3.2.6), the Mandelbrot set $M_k^c$ will change to be the Mandelbrot set $M^d$ when $k \to k_0$.

Now, we will introduce a nonlinear coupling method to achieve the synchronization and coupling of two different Mandelbrot sets.

Considering only the iterations which are bounded since $c \notin M$ if for some $n_0$, $f^{n_0}(0; c)$ is not bounded from the definition of Mandelbrot set. That is to say that there exists a positive real number $T$, we consider only the iterations included in the set $\{z : |z| \le T\}$. If the iterations of the coupled systems (3.2.3) and (3.2.4) are asymptotic, the boundedness of the iterations will be same and then the Mandelbrot sets of (3.2.3) and (3.2.4) will be same.

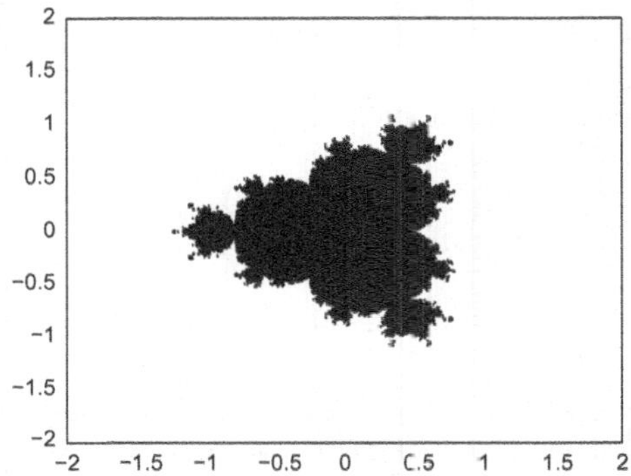


**Fig. 3.8** Mandelbrot set $M^d$ of $g(w_n; d) = w_n^4 + d$

Take $p(z_n, w_n; k) = k[g(w_n; d) - f(z_n; c)]$ and $q(z_n, w_n; k) = 0$, then system (3.2.3) becomes

$$z_{n+1} = f(z_n; c) + k[g(w_n; d) - f(z_n; c)], \tag{3.2.7}$$

which with (3.2.2) derives

$$z_{n+1} - w_{n+1} = (1 - k)[f(z_n; c) - g(w_n; d)].$$

Therefore

$$\begin{aligned} |z_{n+1} - w_{n+1}| &= |1 - k||f(z_n; c) - g(w_n; d)| \\ &\le |1 - k|(2T). \end{aligned} \tag{3.2.8}$$

If $|1 - k| \to 0$ in the right side of (3.2.8), $|z_{n+1} - w_{n+1}| \to 0$. From the previous discussion, we know that the synchronization of the iterations is achieved and so the synchronization of Mandelbrot sets is achieved.

Take $f(z_n; c) = z_n^2 + c$, $g(w_n; d) = w_n^4 + d$, then Figs. 3.1 and 3.8 are their Mandelbrot sets respectively.

It is easy to see from Fig. 3.9 that the Mandelbrot set $M^c$ is changing to $M^d$ with $k$ increasing. When $k \to 1$ is big enough, it will turn to be $M^d$ which illustrates that the synchronization of Mandelbrot sets $M^c$ and $M^d$ is achieved.

Take $p(z_n, w_n; k) = k[g(w_n; d) - f(z_n; c)]$ and $q(z_n, w_n; k) = k[f(z_n; c) - g(w_n; d)]$, then from (3.2.3) and (3.2.4) we have

$$z_{n+1} - w_{n+1} = (1 - 2k)[f(z_n; c) - g(w_n; d)].$$

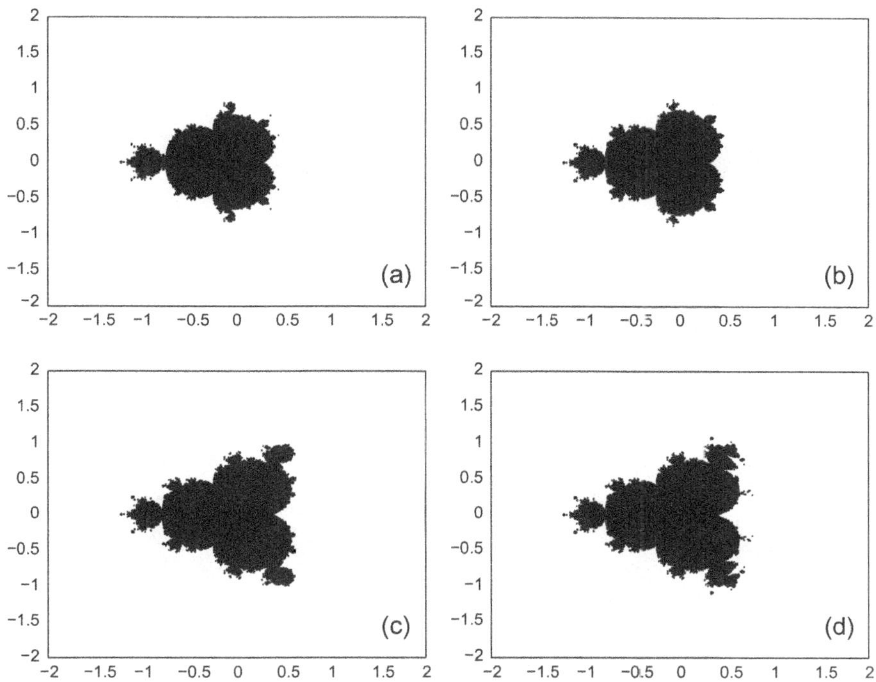

Fig. 3.9 Synchronization of Mandelbrot sets $M^c$ and $M^d$. **a** $k = 0.00011$; **b** $k = 0.11$; **c** $k = 0.29$; **d** $k = 0.36$

Therefore

$$\begin{aligned} |z_{n+1} - w_{n+1}| &= |1 - 2k||f(z_n; c) - g(w_n; d)| \\ &\leq |1 - 2k|(2T'). \end{aligned} \tag{3.2.9}$$

If $|1 - 2k| \to 0$ in the right side of (3.2.9), $|z_{n+1} - w_{n+1}| \to 0$. From the previous discussion, we know that coupling of the iterations is achieved and so the coupling of Mandelbrot sets is achieved and they will turn to be the same one. Take $f(z_n; c) = z_n^2 + c$, $g(w_n; d) = w_n^4 + d$ and the following Fig. 3.10 illustrates the coupling of their Mandelbrot sets.

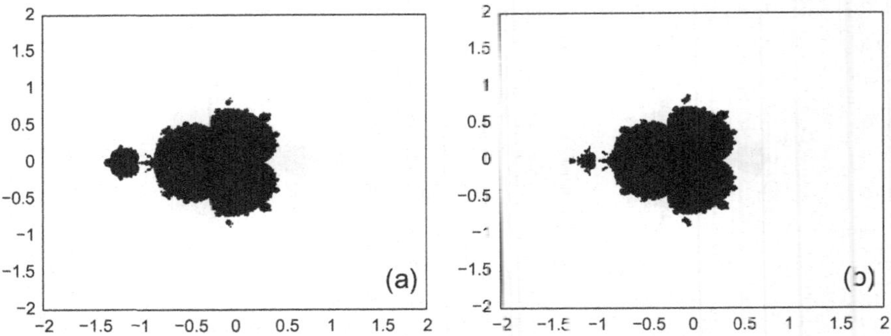


**(a)** Coupling of Mandelbrot sets $M^c$ and $M^d$ when $k = 0.11$

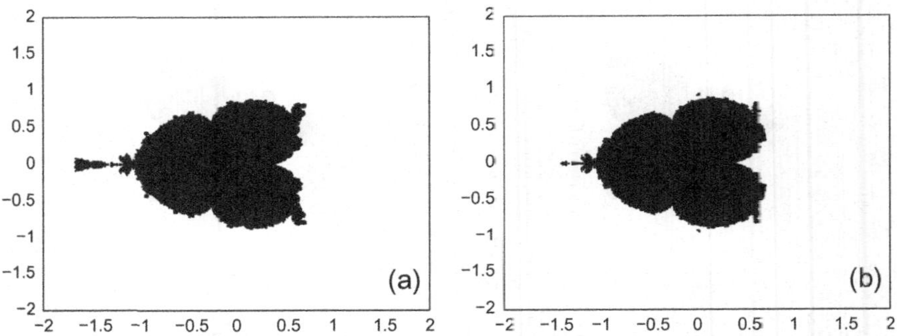


**(b)** Coupling of Mandelbrot sets $M^c$ and $M^d$ when $k = 0.31$.

**Fig. 3.10** Coupling of Mandelbrot sets $M^c$ and $M^d$ with $k$. **a** Mandelbrot set of system (3.2.3); **b** Mandelbrot set of system (3.2.4)

From Fig. 3.10, the Mandelbrot sets $M^c$ and $M^d$ are changing and turning to be the same when $k \to \frac{1}{2}$, which illustrates that the coupling of Mandelbrot sets $M^c$ and $M^d$ is achieved.

# Chapter 4
# Control and Synchronization of Julia Sets Generated by a Class of Complex Time-Delay Rational MAP

Recently, the properties and applications of Julia sets from more general complex maps were investigated. Saitoh et al. [210] discussed a complex system extended from Logistic map, and got the condition under which the Julia set is Cantor set. Fornaess [211] studied the complex Henon map and pointed out that the complex Henon map's maximum entropy set equals to its Julia set. By extending the complex map from integral expression into rational expression, Blanchard et al. [212, 213] investigated the fractal behavior of the following system

$$z_{n+1} = \frac{z_n^{p+q} + \lambda}{z_n^q} + c, \tag{4.0.1}$$

where $\lambda$ and $c$ are complex numbers, $p$ and $q$ are positive integers. Because of the existence of zero and pole points, system (4.0.1) revealed a richer variety of fractal properties (Blanchard et al. [212, 213]; Qiu et al. [214]; Morabito and Devaney [215]).

In [216], Zhang proposed a optimal control item, which does not need to calculate the fixed points, to research the control of the Julia sets from a class of complex perturbed rational map $z_{n+1} = \frac{1}{2}\left(z_n + \frac{\lambda}{z_n^q}\right)$. In this chapter, we modify the general complex rational map (4.0.1) into a new system by adding time-delay and a complex constant $\lambda_2$ into the denominator and deals with the control and synchronization of its corresponding Julia sets. The new system is denoted as follows:

$$z_{n+1} = \frac{z_n^{p+q} + \lambda_1}{z_{n-t}^q + \lambda_2} + c, \tag{4.0.2}$$

where $p, q, t \in N^+$, $p \geq 2$, $q > 0$, $\lambda_1$, $\lambda_2$, $c$ are complex constants. Obviously, If $t = 0$ and $\lambda_2 = 0$, system (4.0.2) reduces to system (4.0.1). Furthermore, if $p = 2$ and $\lambda_1 = 0$, system (4.0.2) changes into system $z_{n+1} = z_n^2 + c$. For convenience, the Julia set of system (4.0.2) is denoted as CTRM-Julia set in this chapter.


S. T. Liu et al., *Fractal Control and Its Applications*,
https://doi.org/10.1007/978-981-15-5459-9_4

## 4.1 Definition and Properties of the Julia Set from Complex Time-Delay Rational Map

For convenience, system (4.0.2) is redefined as

$$R(z_{n-t}, z_{n-t+1}, \ldots, z_n) = z_{n+1} = \frac{P(z_n)}{Q(z_{n-t})} + c, \tag{4.1.1}$$

where $P(z_n) = z_n^{p+q} + \lambda_1$ and $Q(z_{n-t}) = z_{n-t}^q + \lambda_2$. From Definition 2.1, it is clear that the initial point is a single value $z_0$. While for system (4.1.1), the number of initial points relies on $t$ because of the existence of delay. That is, when $n = 0$, we get the initial points of system (4.1.1), $z_{-t}, z_{-t+1}, \ldots, z_{-1}, z_0$, which is denoted as:

$$\psi = (z_{-t}, z_{-t+1}, \ldots, z_{-1}, z_0).$$

Then, the definition of Julia set from system (4.1.1) is given as follows.

**Definition 4.1** Set $z_{-t}, z_{-t+1}, \ldots, z_{-1}$ as the fixed variables and $z_0 = x_0 + iy_0$ as the free variable. The filled Julia set of system (4.1.1) is defined as $K_R$ such that

$$K_R = \{z_0 | R^n(z_{-t}, z_{-t+1}, \ldots, z_0) \nrightarrow \infty, n \to \infty\}.$$

While Julia set of system (4.1.1) is the boundary of $K_R$, *i.e.s* $J_R = \partial K_R$.

In consideration of the circumstance that the denominator of (4.1.1) may equal to 0 in the process of iteration, the following assumption is given.

**Assumption 4.1** *If there is an initial value* $\psi^* = (z_{-t}^*, z_{-t+1}^*, \ldots, z_0^*)$ *and* $k \in \mathbb{N}$ *satisfy that the denominator of (4.1.1) equal to 0 in the kth iteration of* $\psi^*$, *we regard it as* $R^k(\psi^*) \to \infty$, *which illustrates that* $\psi^* \notin K_R$.

From the above analysis, it is clear that the trajectories of the points in the filled Julia set $K_R$ have different boundness properties from the points out of it. In other words, the Julia set $J_R$ is closely related to the trajectories of points in $\mathbb{C}$. In order to determine the points whose trajectories tend to infinity under function $R$, an escape criterion is needed (Julia [189]; Getz and Helmstedt [217]). Thus, the following lemmas and theorems are given which are helpful for the rest of the research.

**Lemma 4.1** ([218]) *If a complex system f has attractive fixed points, the filled Julia set* $K_f$ *is the combined basin of attraction, the Julia set* $J_f$ *is the boundary of the attractive domain.*

**Theorem 4.1** *The escape criterion of the CTRM-Julia set is* $S_R$ *which is denoted as*

$$S_R = \max_{1 \le i \le t} \left\{ |z_{-i}|, \frac{1 + \sqrt{1 + 4a_i b_i}}{2a_i} \right\},$$

*where* $a_i = \frac{|z_{-i}|^q}{|z_{-i}|^q + |\lambda_2|}$, $b_i = \frac{|\lambda_1|}{|z_{-i}|^q + |\lambda_2|} + |c|$, $\varepsilon \in \mathbb{R}$ *is an arbitrary real number.*

***Proof*** $\forall \varepsilon \in \mathbb{R}$, set $S_R^{\varepsilon} = \max\limits_{1 \leq i \leq t} \left\{ |z_{-i}|, \frac{1+\varepsilon+\sqrt{(1+\varepsilon)^2+4a_i b_i}}{2a_i} \right\}$. When the initial value $z_{-t}, z_{-t+1}, \ldots, z_{-1}$ is given, the value of $S_R$ is fixed. If $|z_0| > S_R^{\varepsilon}$, we have

$$
\begin{aligned}
|z_1| &> |\frac{|z_0|^{q+2}-|\lambda_1|}{|z_{-t}|^q+|\lambda_2|} - |c|| \\
&> \frac{|z_{-t}|^q}{|z_{-t}|^q+|\lambda_2|}|z_0|^2 - \frac{|\lambda_1|}{|z_{-t}|^q+|\lambda_2|} - |c| \\
&= a_t|z_0|^2 - b_t \\
&> (1+\varepsilon)|z_0|.
\end{aligned}
$$

Obviously $|z_1| > |z_0| > S_R^{\varepsilon}$, similarly,

$$
\begin{aligned}
|z_2| &> \frac{|z_1|^{q+2}-|\lambda_1|}{|z_{-t+1}|^q+|\lambda_2|} - |c| \\
&> \frac{|z_{-t+1}|^q}{|z_{-t+1}|^q+|\lambda_2|}|z_1|^2 - (\frac{|\lambda_1|}{|z_{-t+1}|^q+|\lambda_2|} + |c|) \\
&= a_{t-1}|z_1|^2 - b_{t-1} \\
&> (1+\varepsilon)|z_1| \\
&> (1+\varepsilon)^2|z_0|.
\end{aligned}
$$

Through mathematical induction, we have

$$
|z_n| = |R^n(z_{-t}, z_{-t+1}, \ldots, z_0)| > (1+\varepsilon)^n|z_0| \to \infty, (n \to \infty).
$$

So only when $|z_0| < S_R^{\varepsilon}$, $\lim\limits_{n\to\infty} |z_n|$ would be bounded. It is noted that it holds for arbitrary $\varepsilon > 0$ here, let $\varepsilon \to 0$, we get $S_R^{\varepsilon} \to S_R$. In other words, The escape criterion of the Julia set from system (4.1.1) is $S_R$. □

**Theorem 4.2** *For system (4.1.1), $J_R$ and $K_R$ are closed sets.*

***Proof*** From Assumption 4.1, it is known that there exits a region $K_1'$ which contains the pole point $\psi^*$ and its neighbourhood. If $z_0 \in K_1'$, $\lim\limits_{n\to\infty} R^n(z_{-t}, z_{-t+1}, \ldots, z_0) \to \infty$. Besides, system (4.1.1) is continuous, so for arbitrary large $T > 0$, there exists $\delta_1 > 0$, $m > 1$, such that to every $z \in O(z_0, \delta_1)$, $|R^m(z_{-t}, z_{-t+1}, \ldots, z_0)| > T$.

Denote $K = K_R \cup K_1'$ and $K^{\mathbb{C}} = K_2'$. If $z_0 \in K_2'$, then there exists $k$, such that $R^k(z_{-t}, z_{-t+1}, \ldots, z_0) = \infty$. $R^k(z_{-t}, z_{-t+1}, \ldots, z_0)$ is continuous on coordinate $z$, so there exists $\delta_2 > 0$, for arbitrary $z \in O(z_0, \delta_2)$, we have $|R^k(z_{-t}, z_{-t+1}, \ldots, z_0)| > T$. Thus, we get that if $z_0 \in K_1' \cup K_2'$, there exists neighborhood whose points belong to $K_1' \cup K_2'$. So $K_R^{\mathbb{C}} = K_1' \cup K_2'$ is open set, and $K_R$ is closed set, namely $K_f$ is closed set.($K^{\mathbb{C}}$ means the complementary set of $K$ in complex plane.) □

**Corollary 4.1** *For system (4.1.1), $J_R$ and $K_R$ are compact sets.*

***Proof*** From Theorem 4.1, it is known that $K_R$ is bounded. From Theorem 4.2, it is known that $K_R$ is closed set. Thus $K_R$ is compact set. Since $J_R = \partial^+ K_R$, $J_{\bar{j}}$ is also compact set. □

**Corollary 4.2** *Infinity is an attracting fixed point of system (4.1.1), thus the Julia set $J_R$ is the boundary of the attractive domain of infinity.*

***Proof*** From Theorem 4.1, it is known that when $z_0 > S_R$, $R^n(z_0) \to \infty$ as $n \to \infty$, then Infinity is an attracting fixed point of system (4.1.1). From Lemma 4.1, it is known that the Julia set $J_R$ is the boundary of the attractive domain of infinity. □

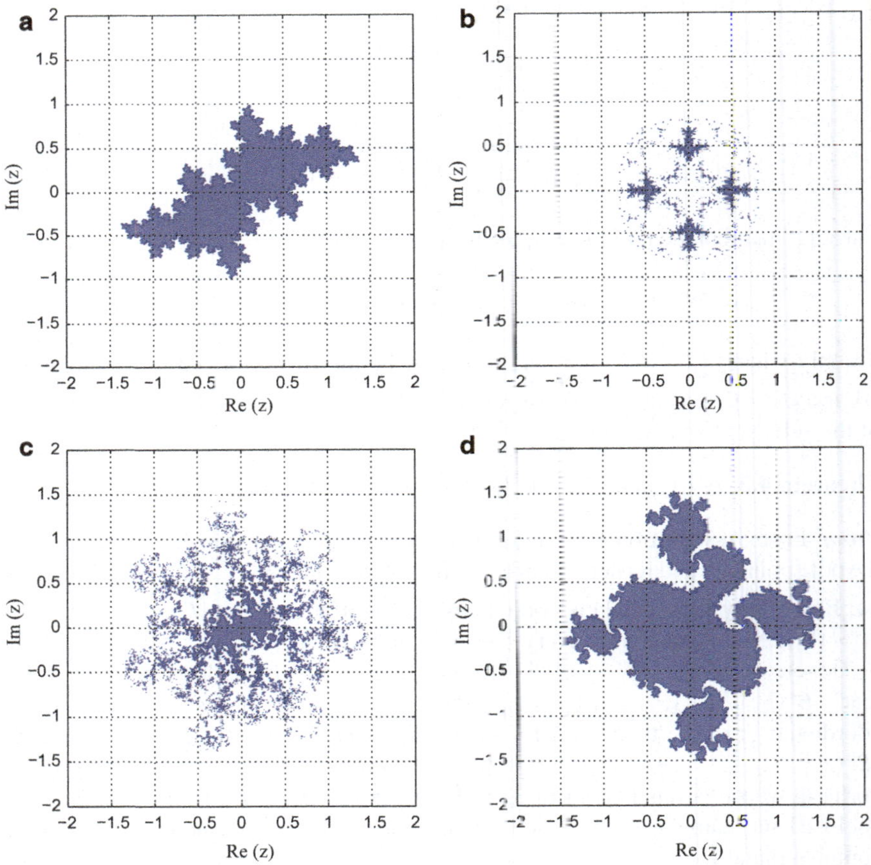


**Fig. 4.1** **a** $p = 2, q > 0, \lambda_1 = 0, \lambda_2 = 0, c = -0.3905 - 0.58768i$ *and* $t = 0$, **b** $p = 3, q = 1, \lambda_1 = 0.1, \lambda_2 = 0, c = 0, t = 1$ *and* $z_{-1} = 0.61$, **c** $p = 5, q = 2, \lambda_1 = 0.001, \lambda_2 = 2.5 - 1.2i, c = -0.9 + 0.5i, t = 1$ *and* $z_{-1} = 0.61$, **d** $p = 3, q = 1, \lambda_1 = 0, \lambda_2 = 2 + 0.5i, c = -1, t = 1$ *and* $z_{-1} = 0.31i$

Some simulations are represented in Fig. 4.1. It is observed from Fig. 4.1 that the Julia sets from complex time-delay rational map have more complex fractal structure than the general complex systems without time delay. In particular, when $p = 2$, $q = 0$, $\lambda_1 = 0$, $\lambda_2 = 0$, $c = -0.3905 - 0.58768i$ and $t = 0$, the result is the Julia set of $z_{n+1} = z_n^2 + c$ (Andreadis and Karakasidis [219] ).

In the next section, our solution of the control of CTRM-Julia set is to reconstruct the attractive domain of infinity. Moreover, the attractive domain of infinity is closely related to the trajectories of the points in complex plane, the synchronization of CTRM-Julia sets can be obtained by changing the trajectories of points in complex plane.

## 4.2 Control of the Julia Set from Complex Time-Delay Rational Map

As mentioned in above section, we consider the problem of reconstructing the attractive domain of infinity of system (4.1.1) to realize the control of $J_R$. In other words, the controlling item $u(n)$ we designed is needed to ensure that the fixed point is unchanged. Furthermore, it must ensure that the fixed point is still attractive for the following controlled system

$$z_{n+1} = \frac{P(z_n)}{Q(z_{n-t})} + c + u(n). \tag{4.2.1}$$

To make sure that the fixed point is unchanged, we apply the hybrid control method (Luo et al. [220]) which contains both parameter perturbation and state feedback to system (4.1.1).

$$u(n) = -\alpha(z_{n+1} - z_n), \tag{4.2.2}$$

where $0 \le \alpha \le 1$. Then we get the controlled system

$$R^u = z_{n+1} = (1 - \alpha)(\frac{P(z_n)}{Q(z_{n-t})} + c) + \alpha z_n. \tag{4.2.3}$$

Obviously, the controlled system (4.2.3) becomes the original system (4.1.1) when $\alpha = 0$. It has been proved that the fixed point is unchanged by adding the hybrid controlling item (Luo et al. [220]). In other words, infinity is still the fixed point of the controlled system (4.2.3). In the following, we will discuss the structure of Julia set of the controlled system (4.2.3). The following theorem is given:

**Theorem 4.3** *The escape criterion of the Julia set from the controlled system (4.2.3) is $S_{R^u}$ which is denoted as*

$$S_{R^u} = \max_{1 \le i \le t} \{|z_{-i}|, \frac{l + \sqrt{l^2 + 4a_i b_i}}{2a_i}\}.$$

*where $l = \frac{1+\alpha}{1-\alpha}$. $a_i$ and $b_i$ are the same as Theorem 4.1.*

***Proof*** For an arbitrary real number $\varepsilon$, set $S^{\varepsilon}_{R^u} = \max_{1 \le i \le t} \{|z_{-i}|, \frac{l^{\varepsilon} + \sqrt{(l^{\varepsilon})^2 + 4a_i b_i}}{2a_i}\}$, where $l^{\varepsilon} = \frac{1+\alpha+\varepsilon}{1-\alpha}$. For the controlled system (4.2.3), if $|z_0| > S^{\varepsilon}_{R^u}$, we have

$$\begin{aligned} |z_1| &> |(1-\alpha)(\tfrac{|z_0|^{q+2} - |\lambda_1|}{|z_{-t}|^q + |\lambda_2|} - |c|) - \alpha|z_0|| \\ &> (1-\alpha)a_t|z_0|^2 - \alpha|z_0| - (1-\alpha)b_t \\ &> (1+\varepsilon)|z_0|. \end{aligned}$$

Obviously $|z_1| > |z_0| > S^{\varepsilon}_{R^u}$, similarly,

$$\begin{aligned} |z_2| &> (1-\alpha)(\tfrac{|z_1|^{q+2} - |\lambda_1|}{|z_{-t+1}|^q + |\lambda_2|} - |c|) - \alpha|z_1| \\ &= (1-\alpha)a_{t-1}|z_1|^2 - \alpha|z_1| - (1-\alpha)b_{t-1} \\ &> (1+\varepsilon)|z_1| \\ &> (1+\varepsilon)^2|z_0|. \end{aligned}$$

Through mathematical induction, we have

$$|z_n| = |(R^u)^n(z_{-t}, z_{-t+1}, \ldots, z_0)| > (1+\varepsilon)^n|z_0| \to \infty.$$

So only when $|z_0| < S^{\varepsilon}_{R^u}$, $\lim_{n\to\infty} |z_n|$ would be bounded. It is noted that it holds for arbitrary $\varepsilon > 0$ here, let $\varepsilon \to 0$, we get $S^{\varepsilon}_{R^u} \to S_{R^u}$. In other words, the escape criterion of the Julia set from system (4.2.3) is $S_{R^u}$. □

From Theorem 4.3, it is known that the infinity is still the attractive fixed point. That is to say, the way to depict Julia set by constructing the attractive domain of infinity is still valid. When $\alpha$ increases, it is clear that $l = \frac{1+\alpha}{1-\alpha}$ increases too, namely $S_{R^u}$ increases. When $\alpha \to 1$, we have $S_{R^u} \to \infty$. Actually, when $\alpha \to 1$, the controlled system becomes $z_{n+1} = z_n$ which means that the values are unchanged for all the points in complex plane. It also means that the Julia set of controlled system covers the whole complex plane.

For example, we take $p = 3,\ q = 1,\ \lambda_1 = 0,\ \lambda_2 = 2 + 0.5i,\ c = -1$. The value of the delay $t = 1$ and $z_{-1} = 0.3i$. The origin Julia set with these parameters is shown in Fig. 4.2d and the controlled Julia sets is shown in Fig. 4.2. It is clear from the simulations in Fig. 4.2 that the controlled Julia set expands outward with the increase of the controlled parameter $\alpha$.

In the practical applications, we can take suitable values of $\alpha$ to have the desired fractal behaviors of the system.

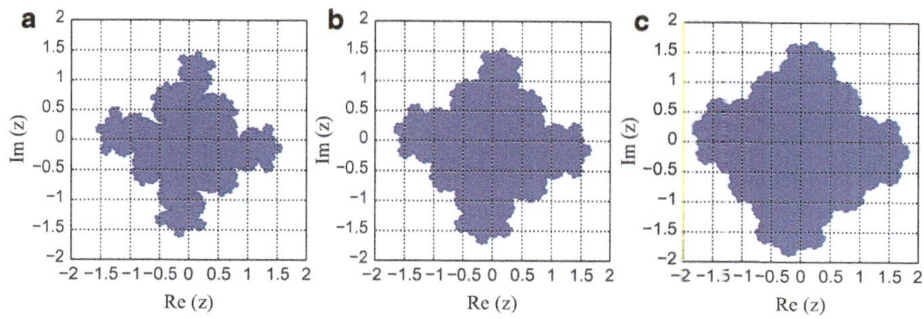


**Fig. 4.2** **a** The controlled Julia set of (4.2.3) $J_{R^u}$ with $p = 3$, $q = 1$, $\lambda_1 = 0$, $\lambda_2 = 2 + 0.5i$, $c = -1$, $t = 1$, $z_{-1} = 0.3i$, $\alpha = 0.11$. **b** $J_{R^u}$ with the same parameters as (**a**) except $\alpha = 0.31$. **c** $J_{R^u}$ with the same parameters as (**a**) except $\alpha = 0.51$

## 4.3 Synchronization of the Julia Sets from Complex Time-Delay Rational Map

Consider two different complex time delay rational maps as follows

$$R_1 = z_{n+1} = \frac{z_n^{p+q} + \lambda_1}{z_{n-t}^q + \lambda_2} + c, \tag{4.3.1}$$

and

$$R_2 = w_{n+1} = \frac{w_n^{p^*+q^*} + \lambda_1^*}{w_{n-t^*}^{q^*} + \lambda_2^*} + c^*. \tag{4.3.2}$$

where $p^*, q^*, t^* \in N^+$ , $p^* \geq 2$ , $q^* > 0$ , $\lambda_1^*$ , $\lambda_2^*$ , $c^*$ are complex constants. Define the Julia sets of system (4.3.1) and (4.3.2) as $J_{R_1}$ and $J_{R_2}$. A coupling term $O(\cdot)$ is added to system (4.3.1) to associate $J_{R_1}$ with $J_{R_2}$, we have

$$R_1^O = R_1 + O(z_n, w_n, \lambda_1, \lambda_2, t, c, \lambda_1^*, \lambda_2^*, t^*. c^*, k). \tag{4.3.3}$$

where $k$ represents the uncertain coupling parameter or coupling strength. It is clear that the change of $k$ will be accompanied by the change of Julia set of coupled system (4.3.3). Thus the following definition is given.

**Definition 4.2** (Wang and Liu [221]) The synchronization between the Julia sets of (4.3.1) and (4.3.2) occurs if

$$\lim_{k \to k_0} \left( J_{R_1^O} \cup J_{R_2} - J_{R_1^O} \cap J_{R_2} \right) = \emptyset,$$

or

$$\left(J_{R_1^O} \cup J_{R_2} - J_{R_1^O} \cap J_{R_2}\right)\Big|_{k=k_0} = \emptyset,$$

for some $k_0$.

It is known that the trajectories of the points in the filled Julia set have different boundness properties from the points out of it. In other words, the Julia sets of system (4.3.1) and (4.3.2) are closely related to the trajectories of the points in $\mathbb{C}$. Thus, our solution of the synchronization between $J_{R_1}$ and $J_{R_2}$ is to synchronize the trajectories.

**Theorem 4.4** (Wang and Liu [221]) *The synchronization between the Julia sets of (4.3.1) and (4.3.2) is realized if* $\forall \upsilon_0 \in \mathbb{C}$, *such that*

$$\lim_{n\to\infty}\lim_{k\to k_0}\left|(R_1^O)^n(\upsilon_0) - R_2^n(\upsilon_0)\right| = 0,$$

*or*

$$\lim_{n\to\infty}\left|(R_1^O)^n(\upsilon_0)|_{k=k_0} - R_2^n(\upsilon_0)\right| = 0,$$

*for some* $k_0$.

In addition, the following Lemma and Theorem are given which are helpful for the following research.

**Lemma 4.2** ([218]) *Denote* $K_f$ *as the filled Julia set of any complex system* $f$, *for the initial point* $\alpha_0 \in K_f$, *we have* $f^k(\alpha_0) \in K_f$, $k \in \mathbb{N}$.

**Theorem 4.5** *For the Julia set* $K_f$ *of any complex system* $f$, *if* $\exists \alpha_0, n_0$ *satisfy that* $f^{n_0}(\alpha_0) \notin K_f$, *then* $\alpha_0 \notin K_f$ *is also true.*

***Proof*** Set $\alpha^* = f^{n_0}(\alpha_0)$. Since $\alpha^* \notin K_f$, we have $\lim_{n\to\infty} f^n(\alpha^*) \to \infty$ which indicates that $\lim_{n\to\infty} f^{n+n_0}(\alpha_0) \to \infty$. Set $n + n_0 = N$, we have $\lim_{N\to\infty} f^N(\alpha_0) \to \infty$. From the definition of Julia set, it is clear that $\alpha_0 \notin K_f$. □

Based on Theorem 4.4, the nonlinear coupling term ([222], Liu and Liu)

$$O(z_n, w_n, \lambda_1, \lambda_2, t, c, \lambda_1^*, \lambda_2^*, t^*, c^*, k)$$

is designed as:

$$O_1(\cdot) = k[R_2 - R_1].$$

Add it to system (4.3.1), we get the coupled system

$$R_1^O = R_1 + k[R_2 - R_1] \tag{4.3.4}$$

It is known from Theorem 4.1 that the Julia set of complex time delay system is bounded which means that it can be obtained by calculating the iterations of the points in a bounded space. That is, we only need to consider those initial points in $D$ such that $J_{R_1} \cap J_{R_2} \subset D$. Then from Theorem 4.5, the scope can be further narrowed to those points whose trajectories are still in $D$. For $D$ is a bounded space, there exists $T > 0$ that

$$max(|(R_1)^n(\upsilon_0)|, |(R_2)^n(\upsilon_0)|) < T. \tag{4.3.5}$$

From Theorem 4.4 and (4.3.5), we obtain:

$$\begin{aligned} &|(R_1^O)^n(\upsilon_0) - (R_2)^n(\upsilon_0)| \\ &= |1-k||(R_1)^n(\upsilon_0) - (R_2)^n(\upsilon_0)| \\ &\leq |1-k|(2T). \end{aligned} \tag{4.3.6}$$

If $k \to 1$ in the right-hand of the above inequality (4.3.6), we have

$$|(R_1^O)^n(\upsilon_0) - (R_2)^n(\upsilon_0)| \to 0.$$

It is evident that the synchronization of the trajectories of the systems (4.3.2) and (4.3.4) is achieved. Based on Theorem 4.4, the synchronization between the Julia sets of system (4.3.1) and system (4.3.2) is achieved.

For example, the parameters of system (4.3.1) are chosen as $p = 3, q = 1$, $\lambda_1 = 0.1, \lambda_2 = 0, c = 0$, and $t = 1, z_{-1} = 0.6$. The Julia set of system (4.3.1) is shown in Fig. 4.1b. The parameters of system (4.3.2) are chosen as the same as the previous section, its Julia set is shown in Fig. 4.1d.

In Fig. 4.3, the Julia sets of coupled system (4.3.4) are present. It is apparent from Fig. 4.3 that when $k$ increases, the symmetric structure of $J_{R_1}$ disappears along with the appearance of the outside part of $J_{R_2}$. When $k \to 1$, $J_{R_1}$ becomes more and more similar with $J_{R_2}$. Therefore the synchronization between Julia sets of the system (4.3.1) and (4.3.2) is realized.

In the following, we apply the **synchronization index** method to quantify the synchronization process of Julia sets. Let us firstly recall the details of the method which is proposed by Wang and Liu [221].

(1) Denote $D : [a, b] \times [c, d]$ as the initial space which satisfies that $J_{R_1^O} \cup J_{R_2} \subset D$, where $a, b, c, d \in \mathbb{N},\ a \leq x_0 \leq b,\ c \leq y_0 \leq d\ and\ z_0 = x_0 + iy_0$. By dividing the two intervals, $[a, b]$ and $[c, d]$, into $M - 1$ subintervals, $M^2$ points with the following coordinates are obtained.

$$\begin{cases} x_0 = a + (i-1)(\frac{b-a}{M-1}), \\ y_0 = c + (j-1)(\frac{d-c}{M-1}). \end{cases}$$

Each point $z_0$ has a unique serial number $(i, j)$ with $1 \leq i, j \leq M$.

(ii) Set $n = 200$ as the escape time limit and $S = max(S_{R_1}, S_{R_2})$ as the escape radius. A criterion $\phi(z_0)$ is designed for each point as follows:

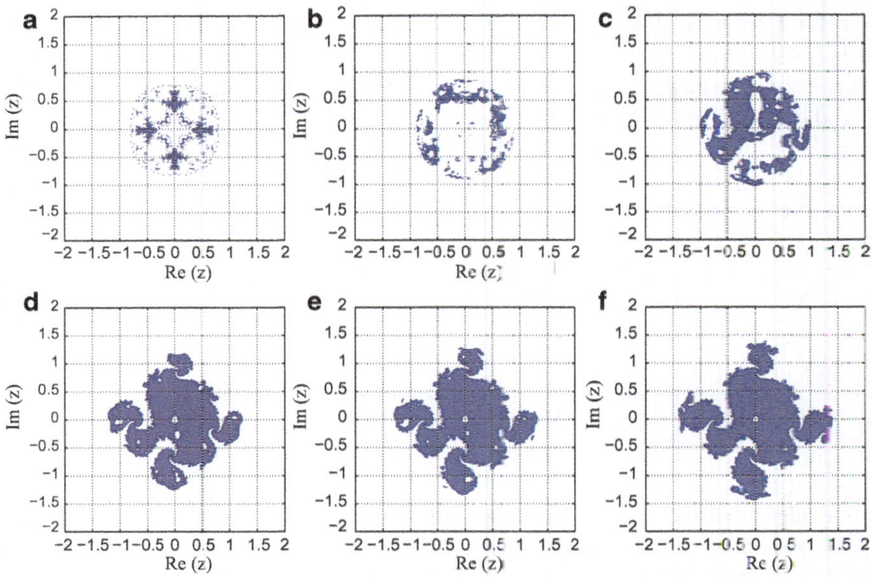


**Fig. 4.3** The parameters of system $R_1$ are chosen as $p = 3$, $q = 1$, $\lambda_1 = 0.1$, $\lambda_2 = 0$, $c = 0$, $t = 1$ *and* $z_{-1} = 0.6$ (see Fig. 4.1b). The parameters of system $R_2$ are chosen as $p = 3$, $q = 1$, $\lambda_1 = 0$, $\lambda_2 = 2 + 0.5i$, $c = -1$, $t = 1$ *and* $z_{-1} = 0.3i$ (see Fig. 4.1d). Then the changes of Julia sets from the coupled system (4.3.4) with different coupling strengths are shown in: (**a**):$k = 0.01$, (**b**):$k = 0.11$, (**c**):$k = 0.31$, (**d**):$k = 0.49$, (**e**):$k = 0.71$, (**f**):$k = 0.79$

$$\phi(z_0)=\begin{cases} 1, if\ |(R_1^o)^n(z_0)|<S\ and\ |(R_2)^n(z_0)|>S, \\ \quad or\ |(R_1^o)^n(z_0)|>S\ and\ |(R_2)^n(z_0)|<S, \\ 0, if\ |(R_1^o)^n(z_0)|<S\ and\ |(R_2)^n(z_0)|<S, \\ \quad or\ |(R_1^o)^n(z_0)|>S\ and\ |(R_2)^n(z_0)|>S. \end{cases}$$

(iii) The **synchronization index** of $J_{R_1^o}$ and $J_{R_2}$, which is defined as $\Phi(z_0)$, is calculated by:

$$\begin{aligned} \Phi(z_0) &= \frac{1}{M^2}\sum_D(\phi(z_0)) \\ &= \frac{1}{M^2}[num(J_{R_1^o}) + num(J_{R_2}) - 2num(J_{R_1^o} \cap J_{R_2})], \end{aligned} \tag{4.3.7}$$

where $num(A)$ means the number of points in a set $A$ under the background of the escape time algorithm. If the $J_{R_1^o}$ and $J_{R_2}$ are separately plotted with red and black and their shared part is recoloured with gray (see Fig. 4.4), then

$$\Phi(z_0) = \frac{1}{M^2}(num(black) + num(red)). \tag{4.3.8}$$

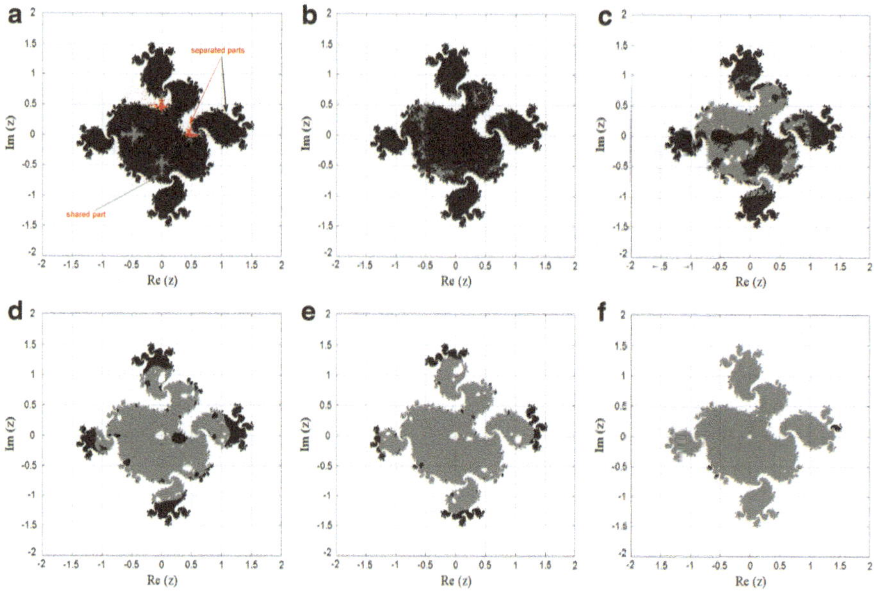


**Fig. 4.4** The parameters of systems $R_1$ and $R_2$ are same as Fig. 4.3. Then the shared part and the separated parts of $J_{R_1^O}$ and $J_{R_2}$ with different coupling strengthes are shown in: (**a**):$k = 0.01$, (**b**):$k = 0.11$, (**c**):$k = 0.29$, (**d**):$k = 0.51$, (**e**):$k = 0.71$, (**f**):$k = 0.79$

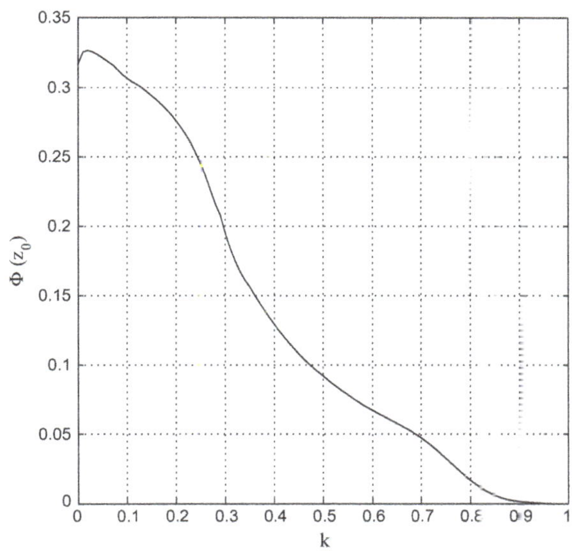


**Fig. 4.5** The change curve of the synchronization process with $k \in [0, 1]$ (The parameters of systems $R_1$ and $R_2$ are the same as Fig. 4.3)

From (4.3.7) and (4.3.8), it is known that there is an inverse proportion relationship between the synchronization degree and $\Phi(z_0)$. That is, $J_{R_1^o}$ and $J_{R_2}$ are more similar if $\Phi(z_0)$ is smaller and the synchronization is realized if the black and red parts disappear (see Fig. 4.4). *i.e.s* $\Phi(z_0) = 0$.

(vi) After calculating all $\Phi(z_0)$ for $k \in [0, 1]$ at the interval of 0.01, the change curve of the synchronization process is illustrated in Fig. 4.5.

As shown in Fig. 4.5, as the coupling strength $k$ increases, $\Phi(z_0)$ becomes smaller which indicates that the synchronization degree becomes higher. When $k \rightarrow 1$ the representation of $\Phi(z_0) \rightarrow 0$ indicates that $J_{R_1^o}$ totally changes to $J_{R_2}$. By using this method, the relationship between the degree of synchronization and the coupling strength can be observed visually.

# Chapter 5
# Control and Synchronization of Spatial Fractals

## 5.1 Synchronization Between the Spatial Julia Sets of Complex Lorenz System and Complex Henon Map

The spatial Julia set, which is the natural extension of planar Julia set, has also attracted significant interest in recent years. Early results focused mainly on the topics of drawing the graphics [223–225] and property analysis [226]. In recent years, scholars devoted more attention to its application. In [227–229], Shudo et al. studied three classical quantum maps and made the conjecture that Julia sets play a fundamental role in the complex semiclassical theory of quantum tunneling.

It is practical that in applications, the spatial Julia sets of different systems have different fractals behaviours, sometimes we need the spatial fractals behaviours of two different systems to show similar or same properties, this is the problem of synchronization of spatial Julia sets which has been addressed by a few of researchers. The optimal control method was applied in [208] when the author discussed synchronization of the Julia sets of Henon map. Liu [222, 230] achieved the synchronization of Julia sets in coupled map lattice via gradient control method and nonlinear coupling method. Nevertheless, it is noted that the previous work has focused only on the synchronization between Julia sets of a same system with just different parameters. There still remains a need for an efficient method which can solve the problem of different-structure synchronization between Julia sets of different systems. Therefore, this chapter focuses on a kind of Lorenz system [231] which was known as the most basic chaotic system, applies two kinds of nonlinear coupling methods to realize the synchronization between Julia sets of complex Lorenz system and complex Henon map [225].


S. T. Liu et al., *Fractal Control and Its Applications*,
https://doi.org/10.1007/978-981-15-5459-9_5

### 5.1.1 Spatial Julia Set of Complex Lorenz Model

Consider the real discrete reduced Lorenz system which was initiated in [231]: $L(a, \tau) : \mathbb{R}^2 \to \mathbb{R}^2$

$$\begin{cases} x_{n+1} = (1 + a\tau)x_n - \tau x_n y_n, \\ y_{n+1} = (1 - \tau)y_n + \tau x_n^2. \end{cases} \tag{5.1.1}$$

where $a$ and $\tau$ are real parameters.

The chaotic and bifurcation behaviors of system (5.1.1) have been shown in [232–234]. In order to investigate the fractals behavior, system (5.1.1) is generalized to complex domain as follows:

$$\begin{cases} z_{n+1} = (1 + a\tau)z_n - \tau z_n w_n, \\ w_{n+1} = (1 - \tau)w_n + \tau z_n^2, \end{cases} \tag{5.1.2}$$

where $a, \tau \in \mathbb{C}$, $z_n$ and $w_n$ are denoted as $z_n = x_n + iy_n$, $w_n = s_n + it_n$. Define system (5.1.2) as $L_{\mathbb{C}}(a, \tau)$ and its Julia set as:

**Definition 5.1** The filled Julia set of the map $L_{\mathbb{C}}$ denoted by $J_F(L_{\mathbb{C}})$ is the set of all the values of initial conditions $(x_0, y_0, s_0, t_0)$ such that

$$\lim_{k\to\infty} |L_{\mathbb{C}}^k(x_0, y_0, s_0, t_0)| \nrightarrow \infty.$$

where $L_{\mathbb{C}}^k(\cdot)$ represents the kth iteration of the initial point.

**Definition 5.2** The Julia set of $L_{\mathbb{C}}$ which is the boundary of the filled Julia set is denoted as

$$J(L_{\mathbb{C}}) = \partial(J_F(L_{\mathbb{C}})). \tag{5.1.3}$$

As defined above, $J(L_{\mathbb{C}})$ belongs to a four-dimensional space. The simplest approach to visualize $J(L_{\mathbb{C}})$ is to take a three-dimensional slice by fixing one of the four real coordinates. The exiting primary methods boundary tracing [223] and ray tracing [224] can be used to solve this problem. However, their resulting image can not adequately represent the complicated fractals structure. It is noted that the widespread escape-time algorithm can calculate the Julia set very accurately in the complex plane [235]. In our implementation we extend the escape-time algorithm to address the numerical calculation of $J(L_{\mathbb{C}})$. Details are as follows:

(1) For eight real numbers $a, b, c, d, e, f, g, h$ and a natural number $M$, four intervals $[a, b]$, $[c, d]$, $[e, f]$ and $[g, h]$ are constructed as the space of initial points $(x_0, y_0, s_0, t_0)$, i.e. $a \le x_0 \le b$, $c \le y_0 \le d$, $e \le s_0 \le f$, and $g \le t_0 \le h$.
(2) Each of the intervals is divided into $M - 1$ subintervals, hence, $M^4$ points are obtained in a space lattice with the following coordinates

$$\begin{cases} x_0 = a + (i-1)(\frac{b-a}{M-1}), \\ y_0 = c + (j-1)(\frac{d-c}{M-1}), \\ s_0 = e + (p-1)(\frac{f-e}{M-1}), \\ t_0 = g + (q-1)(\frac{h-g}{M-1}). \end{cases}$$

Each point has a unique serial number $(i, j, p, q)$ with $1 \le i, j, p, q \le M$. Define $S$ as the set which contains the $M^4$ points.

(3) Set $N$ as the escape time limit and $R$ as the escape radius, the value of the distance $r$ of the $N$th iteration from the original point in $S$ is calculated, i.e.

$$r = \sqrt{x_N{}^2 + y_N{}^2 + s_N{}^2 + t_N{}^2}.$$

If $r \le R$, $(x_0, y_0, s_0.t_0) \in K$, otherwise, the point is abandoned. Finally, a new set $K$ is obtained.

(4) By fixing one of the four real coordinates of the points in $K$, for instance $t_0 = t^*$, the points in $K$ of the form $(x_0, y_0, s_0, t^*)$ are sought out and recombined to set $K_{t^*}$. Thus, the 3-D slice of Julia set of $L_{\mathbb{C}}$, defined as $J_{t^*}(L_{\mathbb{C}})$, can be simulated as $\partial K_{t^*}$ in $(x, y, s)$ space.

- Changing $t^*$ gives a four-dimensional representation of $J(L_{\mathbb{C}})$.
- Fixing one of the other three coordinates of the points in $K_{t^*}$ gives a 2-D slice of $J(L_{\mathbb{C}})$.

**Remark 5.1** Under the background of this algorithm, we define $S(A)$ as the number of the points in set $A$, which is helpful for the research in the next section.

By taking the four intervals as $[-2, 2]$, $[-2, 2]$, $[-3, 3]$, $t_0 = 0$ and fixing $M = N = 400$, $R = 10^5$, the 3-D slice of $J(L_{\mathbb{C}})$ with $a = 0.361$, $\tau = 1.5$ considered in [231] is presented in Fig. 5.2. In Fig. 5.1, the 2-D slices by $y_0 = 0$ and by $s_0 = 0$ are illustrated. Actually, the 2-D slice by $y_0 = 0$ is the combined basin of attraction when $L_{\mathbb{C}} : \mathbb{C}^2 \to \mathbb{C}^2$ turn into $L(a, \tau) : \mathbb{R}^2 \to \mathbb{R}^2$(see Fig. 10a in [231]).

### *5.1.2 Synchronization Between Spatial Julia Sets of the Complex Lorenz System and Henon Map*

Let us firstly recall the definition of synchronization between Julia sets and apply it to our study. Consider the following two different systems: the complex Lorenz system

$$L_{\mathbb{C}} = \begin{cases} z_{n+1} = (1 + a_1\tau_1)z_n - \tau_1 z_n w_n, \\ w_{n+1} = (1 - \tau_1)w_n + \tau_1 z_n^2, \end{cases} \tag{5.1.4}$$

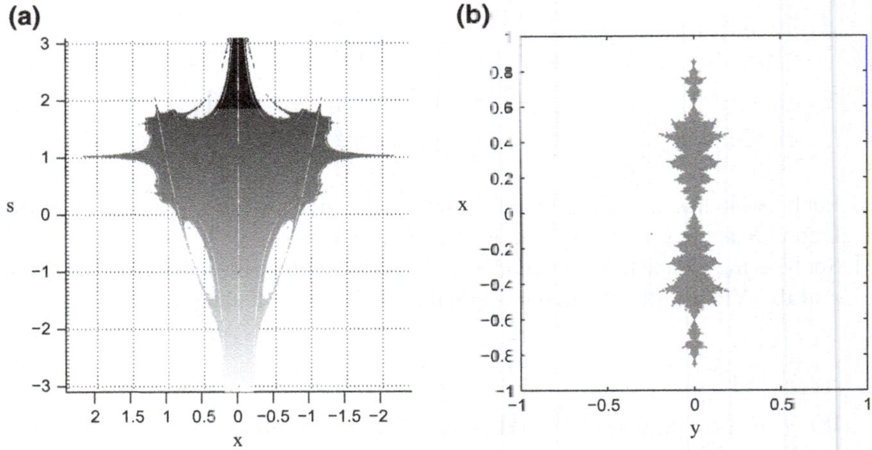


**Fig. 5.1** **a** The filled Julia set of the 2-D slice of $J_{t^*=0}(L_{\mathbb{C}})$ with $y_0 = 0$, **b** The filled Julia set of the 2-D slice of $J_{t^*=0}(L_{\mathbb{C}})$ with $s_0 = 0$

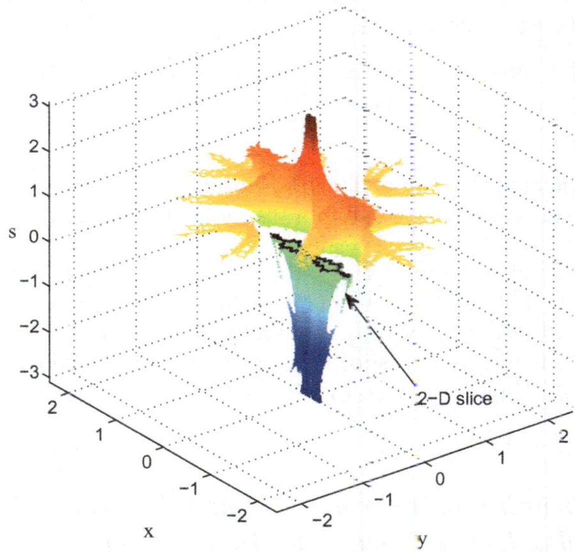


**Fig. 5.2** The 3-D slice of the Julia set of complex Lorenz map with $a = 0.361$, $\tau = 1.5$, in which we cut it through by $s_0 = 0$ to clearly see the 2-D slice

and the complex Henon map [225]:

$$H_{\mathbb{C}} = \begin{cases} p_{n+1} = {p_n}^2 + c - bq_n, \\ q_{n+1} = p_n, \end{cases} \tag{5.1.5}$$

where $b, c \in \mathbb{C}$, $p_n$ and $q_n$ are denoted as $p_n = x'_n + iy'_n$, $q_n = s'_n + it'_n$. Define the Julia sets of system (5.1.4) and (5.1.5) as $J(L_{\mathbb{C}})$ and $J(H_{\mathbb{C}})$. A coupling term $O(\cdot)$ is added to system (5.1.4) to associate $J(L_{\mathbb{C}})$ with $J(H_{\mathbb{C}})$, we have

$$L^*_{\mathbb{C}} = L_{\mathbb{C}} + O(z_n, w_n, p_n, q_n, a, \tau, b, c, k). \tag{5.1.6}$$

where $k$ represents the uncertain coupling parameters. It is clear that the change of $k$ will be accompanied by the change of Julia set of coupled system (5.1.6). Thus the following definition is given.

**Definition 5.3** The synchronization between the Julia sets of (5.1.4) and (5.1.5) occurs if

$$\lim_{k\to k_0} \left(J(L^*_{\mathbb{C}}) \cup J(H_{\mathbb{C}}) - J(L^*_{\mathbb{C}}) \cap J(H_{\mathbb{C}})\right) = \emptyset,^{[11,20]}$$

or

$$\left(J(L^*_{\mathbb{C}}) \cup J(H_{\mathbb{C}}) - J(L^*_{\mathbb{C}}) \cap J(H_{\mathbb{C}})\right)\big|_{k=k_0} = \emptyset,$$

for some $k_0$.

Obviously, the trajectories of the points in the filled Julia set have different boundness property from the points out of it. In other words, the Julia sets of system (5.1.4) and (5.1.5) are closely related to the trajectories of the points in $\mathbb{C}^2$. Thus, our solution of the synchronization between $J(L_{\mathbb{C}})$ and $J(H_{\mathbb{C}})$ is to synchronize the trajectories.

**Theorem 5.1** *The synchronization between the Julia sets of (5.1.4) and (5.1.5) is realized if* $\forall \upsilon_0 \in \mathbb{C}^2$, *such that*

$$\lim_{n\to\infty} \lim_{k\to k_0} \left(L^*_{\mathbb{C}})^n(\upsilon_0) - (H_{\mathbb{C}})^n(\upsilon_0)\right| = 0.$$

*or*

$$\lim_{n\to\infty} \left|(L^*_{\mathbb{C}})^n(\upsilon_0)|_{k=k_0} - (H_{\mathbb{C}})^n(\upsilon_0)\right| = 0,$$

*for some* $k_0$.

***Proof*** If the condition of the theorem is satisfied, the iteration of the same initial point under (5.1.5) and (5.1.6) have the same boundness property, which means that

$$J(L^*_{\mathbb{C}}) = J(H_{\mathbb{C}}),$$

and then the synchronization of Julia sets of (5.1.4) and (5.1.5) is achieved. □

Based on Theorem 5.1, the synchronization between Julia sets of the complex Lorenz system and Henon map is realized via two methods in the following parts.

**The Nonlinear Coupling Method**
The nonlinear coupling term $O_1(z_n, w_n, p_n, q_n, a, \tau, b, c, k)$ is designed as:

$$O_1(\cdot) = k[H_{\mathbb{C}} - L_{\mathbb{C}}].$$

Add it to system (5.1.4), we get

$$L_{\mathbb{C}}^* = L_{\mathbb{C}} + k[H_{\mathbb{C}} - L_{\mathbb{C}}]. \tag{5.1.7}$$

It is apparent from above analysis that the spatial Julia set can be obtained by calculating the iteration of the points in a bounded space. That is, we only need to consider those initial points in $D$ such that $J(L_{\mathbb{C}}^*) \cap J(H_{\mathbb{C}}) \subset D$. The scope can be further narrowed to those points whose trajectories are still in $D$ since $\alpha_0 \notin J(f)$ if $\exists n_0$ such that $f^{n_0}(\alpha_0) \notin D$ ($f$ means $L_{\mathbb{C}}^*$ or $H_{\mathbb{C}}$, $\alpha_0$ means an arbitrary initial value which belongs to $\mathbb{C}^2$). For $D$ is a bounded space, there exists $T > 0$ that

$$max(|(L_{\mathbb{C}}^*)^n(\upsilon_0)|, |(H_{\mathbb{C}})^n(\upsilon_0))| < T. \tag{5.1.8}$$

From Theorem 5.1 and (5.1.8), we obtain:

$$\begin{aligned} |(L_{\mathbb{C}}^*)^n(\upsilon_0) - (H_{\mathbb{C}})^n(\upsilon_0)| &= |1-k|\left|(L_{\mathbb{C}})^n(\upsilon_0) - (H_{\mathbb{C}})^n(\upsilon_0)\right| \\ &\leq |1-k|(2T). \end{aligned} \tag{5.1.9}$$

If $k \to 1$ in the right-hand of the above inequality (5.1.9), we have

$$|(L_{\mathbb{C}}^*)^n(\upsilon_0) - (H_{\mathbb{C}})^n(\upsilon_0)| \to 0.$$

Evidently, the synchronization of the trajectories of the systems (5.1.5) and (5.1.6) is achieved. Based on Theorem 5.1, the synchronization between the Julia sets of system (5.1.4) and system (5.1.5) is achieved.

For example, set $a = 0.361, \tau = 1.5$ and $b = 0.31, c = -1.18$ considered in [225]. $J_{t^*=0}(H_{\mathbb{C}})$ and the changing process of $J_{t^*=0}(L_{\mathbb{C}}^*)$ are respectively shown in Figs. 5.3 and 5.4. In Fig. 5.4, $J_{t^*=0}(L_{\mathbb{C}}^*)$ is fixed in the same space lattice as $J_{t^*=0}(H_{\mathbb{C}})$. We can see easily that when $k$ increases, the symmetric wing structure of $J_{t^*=0}(L_{\mathbb{C}}^*)$ disappears along with the appearance of the main joint part of $J_{t^*=0}(H_{\mathbb{C}})$. When $k \to 1$, $J_{t^*=0}(L_{\mathbb{C}}^*)$ becomes the same as $J_{t^*=0}(H_{\mathbb{C}})$. Therefore the synchronization between Julia sets of the complex Lorenz system and Henon map is realized.

According to the research on chaotic synchronization [236, 237], it is easy to quantify the synchronization process by plotting the trajectory of the error equation. However, the synchronization of Julia sets focuses on two point sets instead of two initial points. To quantify the synchronization process of Julia sets, a mathematical tool inspired by the results in [172] is proposed. Details are as follows:

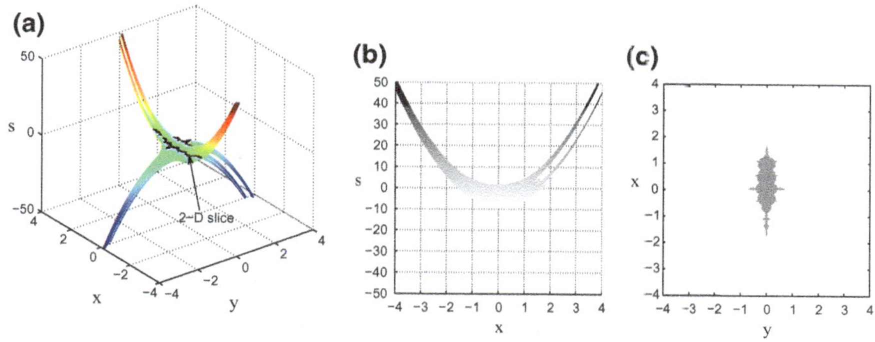


**Fig. 5.3** **a** 3-D slice of the Julia set of complex Henon map with $b = 0.31$, $c = -1.18$, **b** The filled Julia set of 2-D slice in $\mathbb{R}^2$, **c** The filled Julia set of 2-D slice with $s_0 = 0$

(1) Fix $J_{t^*=0}(L^*_{\mathbb{C}})$ and $J_{t^*=0}(H_{\mathbb{C}})$ in the same space lattice

$$D = [-4, 4] \times [-4, 4] \times [-50, 50].$$

$M,\ N,\ R$ are the same as the previous description. A criterion is designed for each of the points $p_0 \in D$ as follows:

$$\phi(p_0) = \begin{cases} 1, if\ |(L^*_{\mathbb{C}})^N(p_0)| < R\ and\ |(H_{\mathbb{C}})^N(p_0)| > R, \\ \quad or\ |(L^*_{\mathbb{C}})^N(p_0)| > R\ and\ |(H_{\mathbb{C}})^N(p_0)| < R, \\ 0, if\ |(L^*_{\mathbb{C}})^N(p_0)| < R\ and\ |(H_{\mathbb{C}})^N(p_0)| < R, \\ \quad or\ |(L^*_{\mathbb{C}})^N(p_0)| > R\ and\ |(H_{\mathbb{C}})^N(p_0)| > R. \end{cases}$$

(2) The **synchronization index** of $J_{t^*=0}(L^*_{\mathbb{C}})$ and $J_{t^*=0}(H_{\mathbb{C}})$, which is defined as $\Phi(p)$, is calculated by:

$$\begin{aligned} \Phi(p) &= \frac{1}{M^3}\sum_D(\phi(p)) \\ &= \frac{1}{M^3}[S(J_{t^*=0}(L^*_{\mathbb{C}})) + S(J_{t^*=0}(H_{\mathbb{C}})) - 2S(J_{t^*=0}(L^*_{\mathbb{C}}) \cap J_{t^*=0}(H_{\mathbb{C}}))]. \end{aligned}$$

It is the fact that if $J_{t^*=0}(L^*_{\mathbb{C}})$ and $J_{t^*=0}(H_{\mathbb{C}})$ are plotted respectively with two different colors, red and blue, then the connected part of them is recoloured with green (see Fig. 5.5). It is apparent that $\sum_D(\phi(p))$ represents the number of points which belong to the blue and red parts. As $k$ increases, the green part grows whereas the blue and red parts go against this tendency. This means that $J_{t^*=0}(L^*_{\mathbb{C}})$ and $J_{t^*=0}(H_{\mathbb{C}})$ are more similar if $\Phi(p)$ is smaller and the synchronization is realized if $\Phi(p) \to 0$.

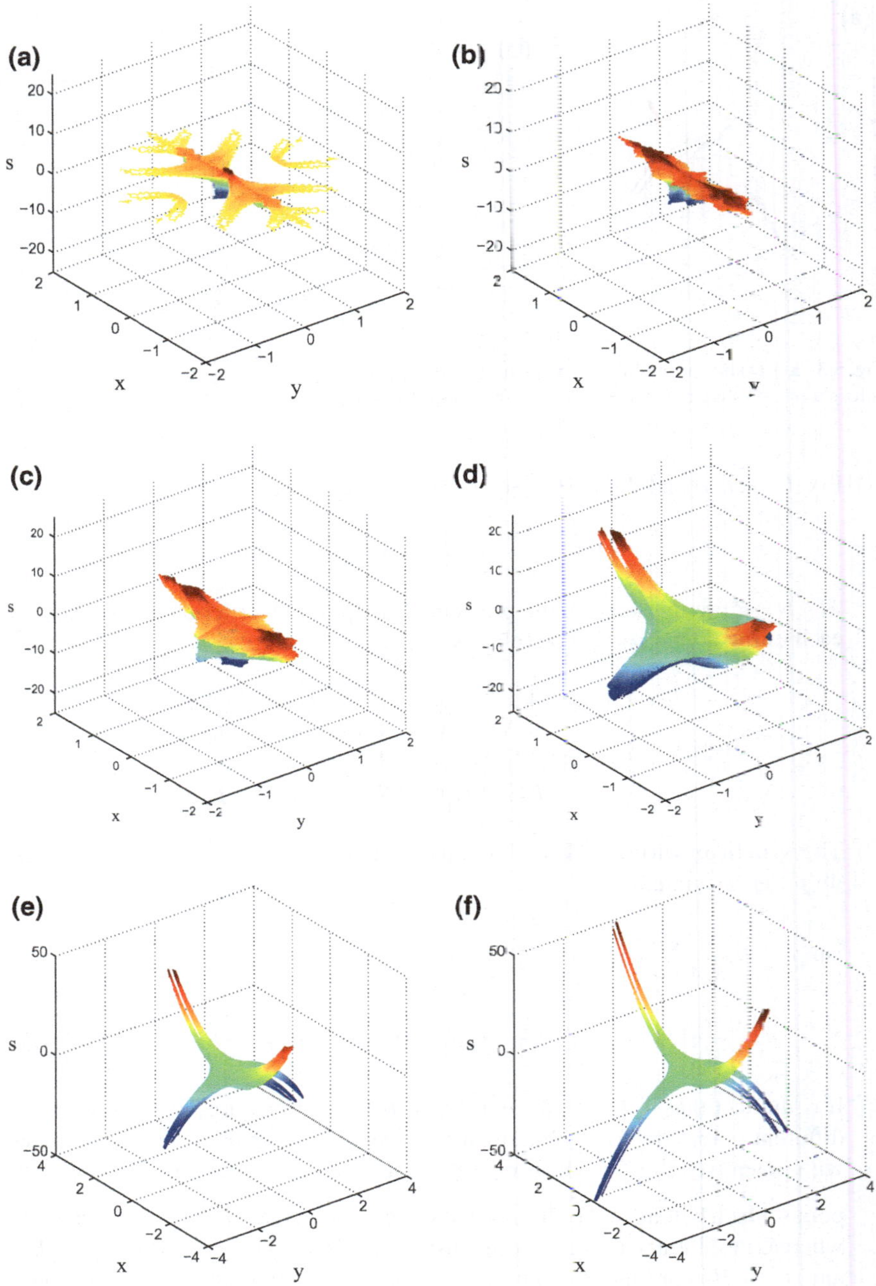


**Fig. 5.4** The changing process of $J_{t^*=0}(L_{\mathbb{C}}^*)$ as (**a**) $k = 0.001$, (**b**) $k = 0.11$, (**c**) $k = 0.31$, (**d**) $k = 0.49$, (**e**) $k = 0.71$, (**f**) $k = 0.89$

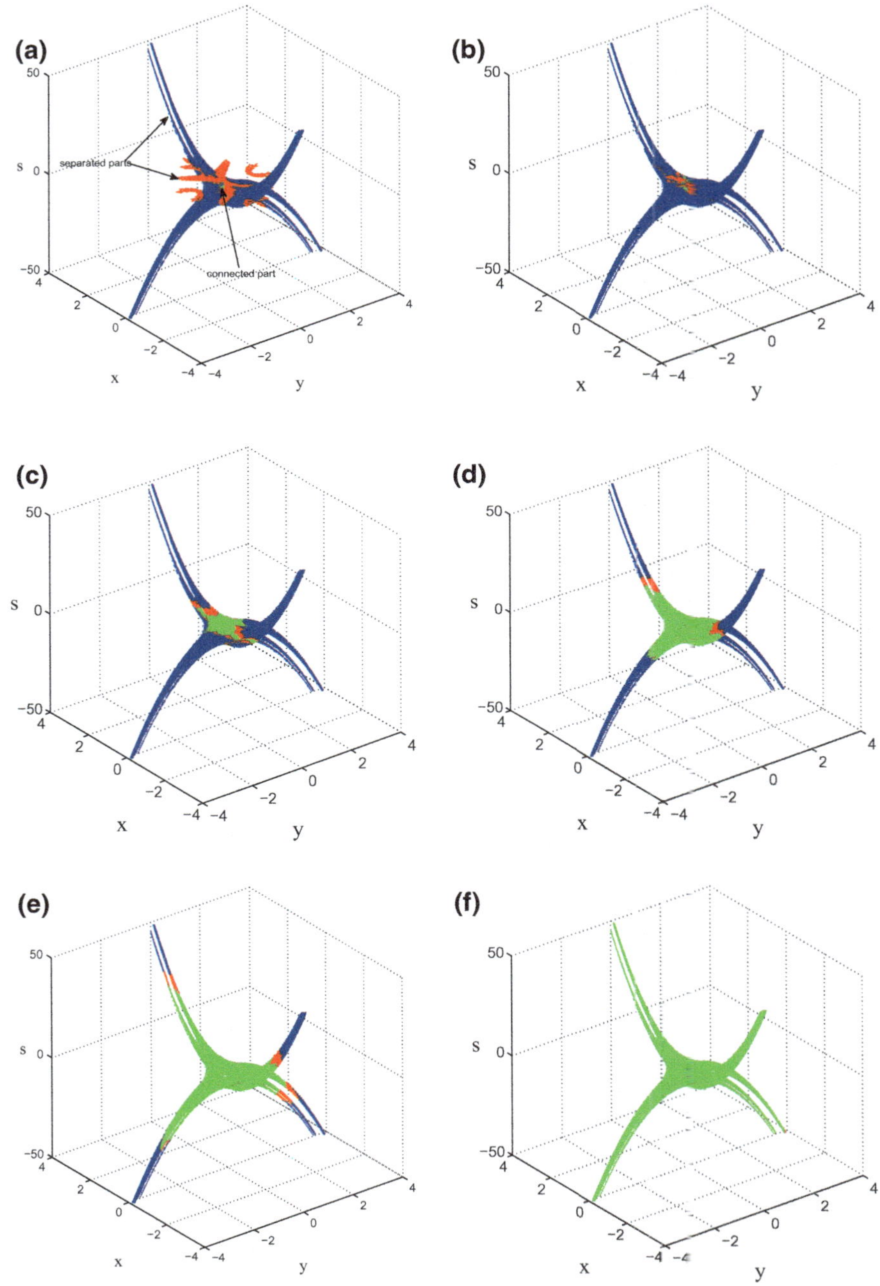


**Fig. 5.5** The changes of the connected part and separated parts of $J_{t^*=0}(L_{\mathbb{C}}^*)$ and $J_{t^*=0}(H_{\mathbb{C}})$ with (**a**) $k = 0.001$, (**b**) $k = 0.11$, (**c**) $k = 0.31$, (**d**) $k = 0.49$, (**e**) $k = 0.71$, (**f**) $k = 0.89$

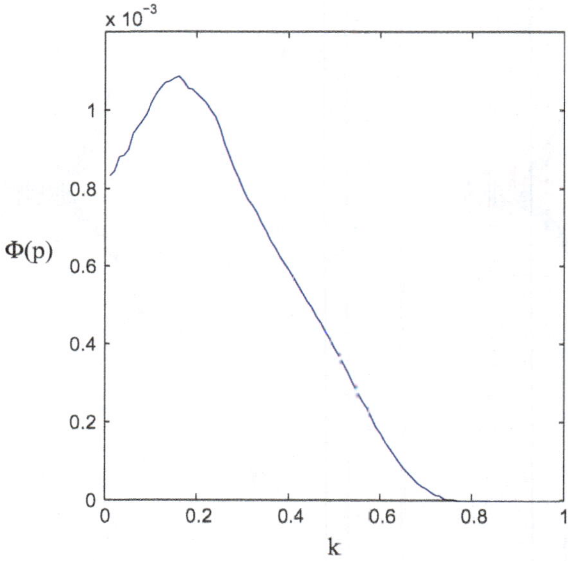


**Fig. 5.6** The change curve of the synchronization process

(3) After calculating all $\Phi(p)$ for $k \in [0, 1]$ at the interval of 0.01, the change curve of the synchronization process is illustrated in Fig. 5.6. As shown in Fig. 5.6, when $k > 0.18$, $\Phi(p)$ becomes smaller as $k$ increases. When $k \to 1$, the representation of $\Phi(p) \to 0$ indicates that $J_{t^*=\mathbb{C}}(L_{\mathbb{C}}^*)$ totaly changes to $J_{t^*=0}(H_{\mathbb{C}})$.

**The Nonlinear Coupling Matrix Method**

In this part, the nonlinear coupling term is designed in a novel form in which the coupling parameter $k$ is changed into a coupling matrix $K$. The implementation of this coupling term is presented in the followings

For convenience, system (5.1.4) and (5.1.5) are redefined as:

$$\alpha_{n+1} = \begin{bmatrix} z_{n+1} \\ w_{n+1} \end{bmatrix} = A\alpha_n + F^{'}(\alpha_n), \tag{5.1.10}$$

$$\beta_{n+1} = \begin{bmatrix} p_{n+1} \\ q_{n+1} \end{bmatrix} = B\beta_n + G^{'}(\beta_n), \tag{5.1.11}$$

in which

$$A = \begin{bmatrix} 1 + a_1\tau_1, & 0 \\ 0, & 1 - \tau_1 \end{bmatrix}, B = \begin{bmatrix} 0, & -b \\ 1, & 0 \end{bmatrix},$$

$$F^{'}(\alpha_n) = \begin{bmatrix} -\tau_1 z_n w_n \\ \tau_1 z_n^2 \end{bmatrix}, G^{'}(\beta_n) = \begin{bmatrix} {p_n}^2 + c \\ 0 \end{bmatrix}.$$

The nonlinear coupling term is designed as:

$$O_2(\cdot) = w(\alpha_n) - v(\beta_n), \tag{5.1.12}$$

where $w(\alpha_n)$ and $v(\beta_n)$ are undetermined items about $\alpha_n$ and $\beta_n$ in the following forms:

$$w(\beta_n) = G^{'}(\beta_n) + (B - A + K)\beta_n,$$

$$v(\alpha_n) = F^{'}(\alpha_n) + K\alpha_n,$$

where

$$K = \begin{bmatrix} k_{11} & k_{12} \\ k_{21} & k_{22} \end{bmatrix}$$

is the coupling matrix.

Add (5.1.12) into system (5.1.10), we have

$$\alpha^*_{n+1} = A\alpha^*_n + F^{'}(\alpha^*_n) + O_2(\cdot). \tag{5.1.13}$$

As the same as the method mentioned in the previous description, the synchronization of the Julia sets of (5.1.10) and (5.1.11) is achieved by the trajectories synchronization of system (5.1.11) and system (5.1.13), so the error equation is written as:

$$\begin{aligned} e_{n+1} &= \alpha^*_{n+1} - \beta_{n+1} \\ &= A\alpha^*_n + F^{'}(\alpha^*_n) + G^{'}(\beta_n) + (B - A + K)\beta_n - F^{'}(\alpha_n) - \\ &\quad K\alpha_n - B\beta_n - G^{'}(\beta_n) \\ &= (A - K)(\alpha^*_n - \beta_n) \\ &= (A - K)e_n \end{aligned} \tag{5.1.14}$$

Thus, the problem of the trajectories synchronization can be solved by taking the stability of the solution of (5.1.14) into consideration.

From Eq. (5.1.14), it is known that the solution of (5.1.14) is stable if

$$-1 < \lambda\{A - K\} < 1, \tag{5.1.15}$$

where $\lambda\{A - K\}$ are the eigenvalues of $A - K$.

Then the trajectories synchronization is achieved which suggests that the synchronization between Julia sets of (5.1.10) and (5.1.11) is realized.

For an instance, take the same parameters as the previous subsection, we have $A - K = \begin{bmatrix} 1 + a_1\tau_1 - k_{11}, & -k_{12} \\ -k_{21}, & 1 - \tau_1 - k_{22} \end{bmatrix}$. To meet the conclusion of (5.1.15), a proper

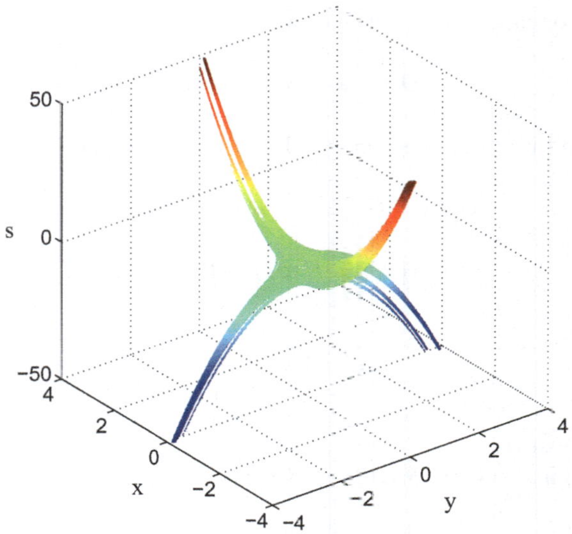


**Fig. 5.7** The Julia set of coupled system (5.1.13)

$K$ is chosen as $\begin{bmatrix} 2, & 0.2 \\ 0.3, & 0.5 \end{bmatrix}$. The Julia set of the controlled system (5.1.13) is shown in Fig. 5.7 which verifies the effectiveness of the synchronization method.

The conclusion can be drawn that both the two methods can be effectively employed to realize the synchronization. Although the former coupling term is simple to implement, the coupling parameter $k$ need to be adjusted. On the the contrary, the synchronization can be realized via the later complicated coupling term through choosing a proper coupling matrix $K$.

## 5.2 Identification Control of Space Fractal Julia Set

### *5.2.1 Design of Synchronization Controller and Parameter Identifier for Spatial Julia Sets*

Consider a complex dynamical system with two independent variables:

$$z_{m+1,n} + \omega z_{m,n+1} = f(c, z_{m,n}), \tag{5.2.1}$$

where $f(.)$ is a nonlinear function, $c$ is a plural, $\omega$ is a real parameter and $\omega \neq 0$. If

$$f(c, z_{m,n}) = z_{m,n}^2 + c, \tag{5.2.2}$$

then the complex dynamic system (5.2.1) is

$$z_{m+1,n} + \omega z_{m,n+1} = z_{m,n}^2 + c. \tag{5.2.3}$$

When $\omega = 0$, $n = n_0$,

$$z_{m+1,n_0} = z_{m,n_0}^2 + c. \tag{5.2.4}$$

The complex dynamical system (5.2.4) is the most basic Julia set

It's worth noting that the continuous form of the complex dynamical system (5.2.1) is

$$\frac{\partial \phi(z,v)}{\partial z} + \omega \frac{\partial \phi(z,v)}{\partial v} = f(\phi(z,v)), \tag{5.2.5}$$

Equation (5.2.5) is a one-dimensional convection equation in physics, so studying complex dynamical system (5.2.1) can provide useful information for studying one-dimensional complex convection system. System (5.2.1) contains two independent variables, so its motion state is spatial motion behavior, so system (5.2.1) is called a multivariable complex dynamic system. Introduce the following definition:

**Definition 5.4** ([238]) For system (5.2.1), if a state $z_{m,n}$ presents a periodicity feature in the iterative process, and the points with slight differences with it all leave it after the iteration, which means that they show repulsive characteristics thus the state $z_{m,n}$ is called the spatial exclusivity periodic point of system (5.2.1).

**Definition 5.5** ([239]) A closure of a space repulsive periodic point is called a space Julia set. Make

$$N_r = \{r, r+1, r+2, \cdots | r \text{ is a nonnegative integer}\}$$

.

Consider two space Julia sets with the same structure in complex domain

$$x_{m+1,n} + \omega x_{m,n+1} = f(x_{m,n}, c), \tag{5.2.6}$$

$$y_{m+1,n} + \omega y_{m,n+1} = g(y_{m,n}, c_{m,n}). \tag{5.2.7}$$

where $x_{m,n}$ is the state of the driving system, $y_{m,n}$ is the state of the responding system, and $f(.)$, $g(.)$ are nonlinear functions, $c$ and $c_{m,n}$ are plurals, $\omega$ is a real parameter and $\omega \neq 0$.

To associate the spatial Julia set of system (5.2.6) with system (5.2.7), couple system (5.2.6) and (5.2.7) and introduce a controller into system (5.2.7).

$$y_{m+1,n} + \omega y_{m,n+1} = g(y_{m,n}, c_{m,n}) + u_{m,n}(x_{m,n}, y_{m,n}, \omega). \tag{5.2.8}$$

where system (5.2.6) is the driving system and system (5.2.8) is the responding system. Controller $u_{m,n}$ enables the driving system and the responding system to achieve asymptotic synchronization. Moreover, $\omega$ is a given real number and $c$ is the complex number that the driving system needs to identify.

When the system parameters $\omega$, $c$, $c_{m,n}$ are given, its corresponding space Julia set is also determined, set them as $J^*$, $J_n$ respectively, if

$$\lim_{n\to\infty} (J_n \cup J^* - J_n \cap J^*) = \emptyset, \tag{5.2.9}$$

then we call the synchronization of the space Julia set of system (5.2.6) and system (5.2.8) is realized.

It is a fact that from the definition of space Julia set, the Julia set $J(f)$ is closely related to the orbit of $f$. Therefore, we utilize orbit synchronization to realize the synchronization of space Julia set and design a controller. The initial iteration value of system (5.2.6) and (5.2.8) are identical, when $m \to \infty$, $n \to \infty$, the system (5.2.6) and (5.2.8) realize asymptotic synchronization, $c_{m,n} \to c$, the unknown parameter $c$ of the driving system is identified.

**Design of Synchronous Controller and Parameter Identifier for Spatial Julia Sets**

Consider a class of extensive complex system spatial Julia set $z_{m-1,n} + \omega z_{m,n-1} = z_{m,n}^2 + c$, where $c$ is a plural, *omega* is a real parameter and $\omega \neq 0$, discuss the synchronization control of the system and the identification of the parameter $c$. As a result, the driving system (5.2.6) and the responding system (5.2.8) are respectively as follows:

$$x_{m+1,n} + \omega x_{m,n+1} = x_{m,n}^2 + c, \tag{5.2.10}$$

$$y_{m+1,n} + \omega y_{m,n+1} = y_{m,n}^2 + c_{m,n} + u_{m,n}. \tag{5.2.11}$$

Let the error between the driving system (5.2.10) and the responding system (5.2.11) be

$$e_{m,n} = x_{m,n} - y_{m,n}. \tag{5.2.12}$$

Assume the driving system is a fractal set of the unknown parameter $c$. The purpose is to make the driving system (5.2.10) and the responding system (5.2.11) achieve asymptotic synchronization and identify the unknown parameter $c$ by designing an appropriate synchronization controller in the case of unknown driving system parameters.

Therefore, the synchronization controller of responding system (5.2.11) is designed as

$$u_{m,n} = x_{m,n}^2 - y_{m,n}^2, \tag{5.2.13}$$

and the parameter identifier is devised as

$$c_{m+1,n} - \omega c_{m,n+1} - (1-\omega)c_{m,n} = k e_{m,n}, \tag{5.2.14}$$

where

$$\begin{cases} 2\omega^2 - 2 < k < 0 \quad , \quad 0 < \omega < 1, \\ \frac{1-2\omega}{4} < k < \frac{3+2\omega-\omega^2}{4} \quad , \quad 1 \le \omega < 3. \end{cases}$$

and $\omega$, $k$ are real parameters.

**Theorem 5.2** *For the system (5.2.10) and (5.2.11), when Eq. (5.2.13) of the synchronous controller and (5.2.14) of the parameter identifier are adopted, the trajectories of the driving system (5.2.10) and the responding system (5.2.11) starting from any initial value can achieve global asymptotic synchronization, and the parameter c of the space Julia set of the driving system can be identified.*

***Proof*** In terms of the above analysis, if the driving system (5.2.10) and the responding system (5.2.11) can realize orbit synchronization, the space Julia set of the system (5.2.10) and (5.2.11) starting from any initial value can achieve global asymptotic synchronization. And in order to prove the orbit synchronization between the system (5.2.10) and (5.2.11), which means to prove $|x_{m,n} - y_{m,n}| \to 0$ ($n \to \infty, n \to \infty$), that is, $|e_{m,n}| \to 0$ ($m \to \infty, n \to \infty$).

Obviously, from Eqs. (5.2.10), (5.2.11) and (5.2.13),

$$x_{m+1,n} - y_{m+1,n} + \omega(x_{m,n+1} - y_{m,n+1}) = c - c_{m,n}. \tag{5.2.15}$$

And according to Eq. (5.2.12), it is known that

$$x_{m+1,n} - y_{m+1,n} = e_{m+1,n}, \tag{5.2.16}$$

and

$$x_{m,n+1} - y_{m,n+1} = e_{m,n+1}. \tag{5.2.17}$$

Substitute (5.2.16) and (5.2.17) into (5.2.15), then

$$e_{m+1,n} + \omega e_{m,n+1} = c - c_{m,n}. \tag{5.2.18}$$

Thus,

$$e_{m+2,n} + \omega e_{m+1,n+1} = c - c_{m+1,n}, \tag{5.2.19}$$

$$\omega e_{m+1,n+1} + \omega^2 e_{m,n+2} = \omega c - \omega c_{m,n+1}. \tag{5.2.20}$$

Substitute (5.2.14) into (5.2.18)–(5.2.20), then

$$\begin{aligned} &\omega^2 e_{m,n+2} - e_{m+2,n} + (1-\omega)e_{m+1,n} - \\ &(1-\omega)\omega e_{m,n+1} - k e_{m,n} = 0. \end{aligned} \tag{5.2.21}$$

In order to simplify the proof process, only consider a special condition, let $k = k_1 + k_2$, then

$$\begin{cases} \omega^2 e_{m,n+2} + (1-\omega)\omega e_{m,n-1} - k_1 e_{m,n} = 0, \\ -e_{m+2,n} + (1-\omega) e_{m+1,n} - k_2 e_{m,n} = 0. \end{cases} \tag{5.2.22}$$

The characteristic equation of (5.2.22) is

$$\begin{cases} \omega^2\lambda^2 + (1-\omega)\omega\lambda - k_1 = 0, \\ \mu^2 - (1-\omega)\mu + k_2 = 0. \end{cases} \tag{5.2.23}$$

According to the characteristic root of Eq. (5.2.23), the discussion can be divided into two situations:

(1) when $2\omega^2 - 2 < k \le 0$ and $0 < \omega < 1$, make $2\omega^2 - \omega < k_1 < \omega$, $\omega - 2 < k_2 < -\omega$. Then

$$\begin{cases} \Delta_1 = \omega^2(1-\omega)^2 + 4\omega^2 k_1 \ge 0, \\ \Delta_2 = (1-\omega)^2 - 4k_2 \ge 0. \end{cases} \tag{5.2.24}$$

Therefore, the solution of Eq. (5.2.22) is

$$\begin{cases} e_{m,n} = t_1\lambda_1^n + t_2\lambda_2^n, \\ e_{m,n} = t_3\mu_1^n + t_4\mu_2^n. \end{cases} \tag{5.2.25}$$

Where $t_1, t_2, t_3, t_4$ are the undetermined constants, and through the characteristic root of Eq. (5.2.23), we can be obtained

$$\begin{cases} |\lambda_{1,2}| = \left| \frac{-(1-\omega)\omega \pm \sqrt{\omega^2(1-\omega)^2 + 4\omega^2 k_1}}{2\omega^2} \right| < 1, \\ |\mu_{1,2}| = \left| \frac{1-\omega \pm \sqrt{(1-\omega)^2 - 4k_2}}{2} \right| < 1. \end{cases} \tag{5.2.26}$$

From Eq. (5.2.26),

$$\frac{|\lambda_{1,2}|}{|\mu_{1,2}|} = \left| \frac{(\omega-1) \pm \sqrt{(1-\omega)^2 + 4k_1}}{\omega(1-\omega) \pm \omega\sqrt{(1-\omega)^2 - 4k_2}} \right|, \tag{5.2.27}$$

when $\omega$, $k_1$ and $k_2$ are within the allowable range, $\frac{\lambda_{1,2}|}{|\mu_{1,2}|} = t$. $t$ ia a real parameter, then $|\lambda_{1,2}| = t|\mu_{1,2}|$. Let $t_3 = t_1 t^n$, the stable solution of Eq. (5.2.22) can be obtained,

$$e_{m,n} = t_3\mu_1^n + t_4\mu_2^n, \tag{5.2.28}$$

that is, $|e_{m,n}| \to 0 (m \to \infty, n \to \infty)$. The solution of Eq. (5.2.22) is also the solution of Eq. (5.2.21), so the solution of Eq. (5.2.21) is also stable, $|e_{m,n}| \to 0 (m \to \infty, n \to \infty)$.

(2) when $\frac{1-2\omega}{4} < k < \frac{3+2\omega-\omega^2}{4}$, and $1 \le \omega < 3$, make $-\frac{\omega^2}{4} < k_1 < -\frac{(1-\omega)^2}{4}$, $\frac{(1-\omega)^2}{4} < k_2 < 1$, so

$$\begin{cases} \Delta_1 < 0, \\ \Delta_2 < 0. \end{cases} \tag{5.2.29}$$

Therefore, the solution of Eq. (5.2.22) is

$$\begin{cases} e_{m,n} = t_5\lambda_3^n \cos(n\theta) + t_6\lambda_3^n \sin(n\theta), \\ e_{m,n} = t_7\mu_3^n \cos(n\theta) + t_8\mu_3^n \sin(n\theta). \end{cases} \tag{5.2.30}$$

where $t_5, t_6, t_7, t_8$ are undetermined constants.

From $-\frac{\omega^2}{4} < k_1 < -\frac{(1-\omega)^2}{4}$, $\frac{(1-\omega)^2}{4} < k_2 < 1$, and $1 \le \omega < 3$, thus

$$\begin{cases} \lambda_3 = \frac{\sqrt{2(1-\omega)^2+4k_1}}{2\omega} < 1, \\ \mu_3 = k_2 < 1. \end{cases} \tag{5.2.31}$$

From Eq. (5.2.31),

$$\frac{\lambda_3}{\mu_3} = \frac{\sqrt{2(1-\omega)^2+4k_1}}{2\omega k_2}. \tag{5.2.32}$$

when $\omega, k_1$ and $k_2$ are within the allowable range, $\frac{\lambda_3}{\mu_3} = t$. $t$ is a real parameter, $\lambda_3 = t\mu_3$. Make $t_7 = t_5 t^n$, $t_8 = t_6 t^n$, then the solution of Eq. (5.2.22) is

$$e_{m,n} = t_7\mu_3^n \cos(n\theta) + t_8\mu_3^n \sin(n\theta). \tag{5.2.33}$$

This solution is stable, which means $|e_{m,n}| \to 0$ $(m \to \infty, n \to \infty)$. The solution of Eq. (5.2.22) is also the solution of Eq. (5.2.21), then the solution of Eq. (5.2.21) is stable, i.e $|e_{m,n}| \to 0(m \to \infty, n \to \infty)$.

In summary, the trajectories of the driving system (5.2.10) and the responding system (5.2.11) starting from any initial value can achieve global asymptotic synchronization.

**Example 5.1** System (5.2.10) $c = 0.75 + 0.15i$, $1 < \omega < 3$, make $k = -0.41$, the initial parameter $c_{0,0} = -0.55 - 0.55i$, MATLAB calculation accuracy is 16 bits, here we take 4 decimal places. When iterating 12 steps, i.e. $m = n = 12$, the spatial Julia set of driving system (5.2.10) and responding system (5.2.11) are shown in Fig. 5.8a, b. What's more, when 20 steps are iterated, i.e. $m = n = 20$, the spatial Julia set of the driving system (5.2.10) and the responding system (5.2.11) are exhibited in Fig. 5.8c, d. When iterating 80 steps, the spatial Julia set of the driving system (5.2.10) and the responding system (5.2.11) are demonstrated in Fig. 5.8e, f.

As can be seen from Fig. 5.8, with the increase of iterative steps $m$ and $n$, the spatial Julia set of the driving system (5.2.10) and the responding system (5.2.11) are asymptotically synchronized.

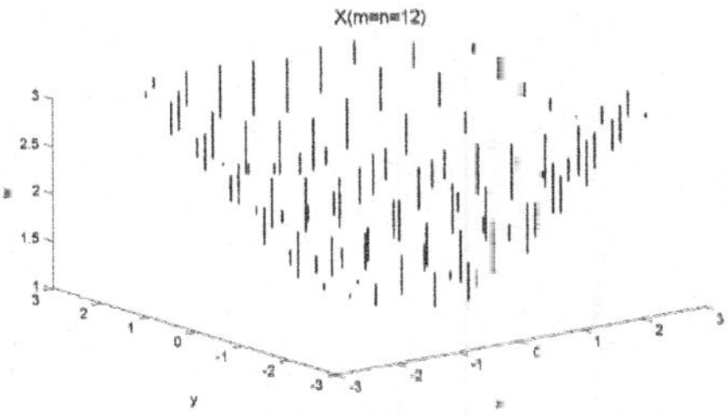


(a) When $m = n = 12$, the space Julia set of system (5.2.10)

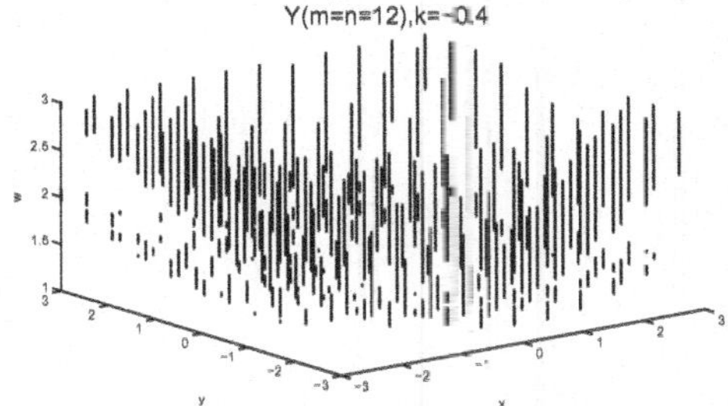


(b) When $m = n = 12$, the space Julia set of system (5.2.11)

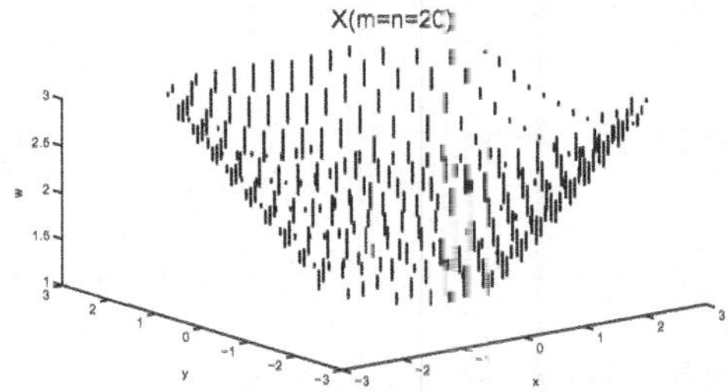


(c) When $m = n = 20$, the space Julia set of system (5.2.10)

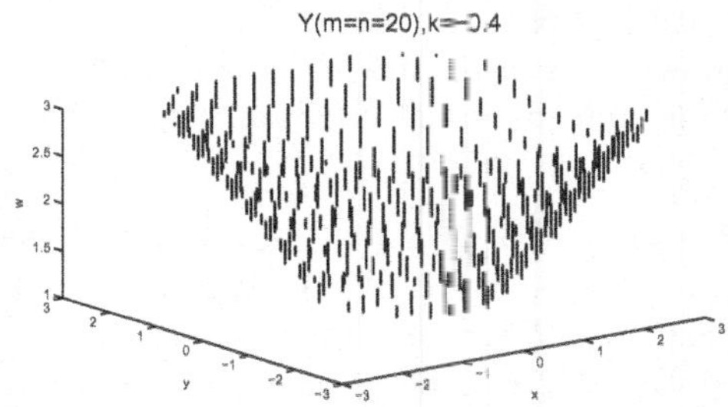


(d) When $m = n = 20$, the space Julia set of system (5.2.11)

**Fig. 5.8** The spatial Julia set of the target system (5.2.10) and the response system (5.2.11) at $k = -0.41$

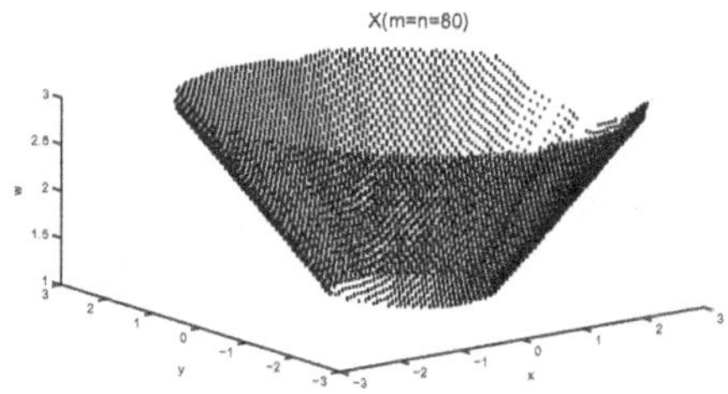


(e) When $m = n = 80$, the space Julia set of system (5.2.10)

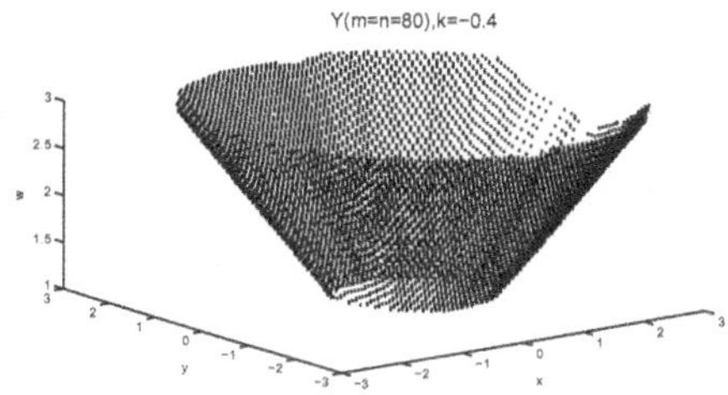


(f) When $m = n = 80$, the space Julia set of system (5.2.11)

**Fig. 5.8** (continued)

When $k = -0.41$, the real part identification process and imaginary part identification process of the parameter $c_{m,n}$ in the responding system (5.2.11) are shown in Fig. 5.9. The changing process of the error $|e_{m,n}|$ between driving system (5.2.10) and responding system (5.2.11) is shown in Fig. 5.10.

According to Figs. 5.9 and 5.10, with the increase of the number of iterations, the real and imaginary parts of the parameter $c_{m,n}$ are stable at 0.7231 and 0.1443 respectively, and the real and imaginary parts of the error $e_{m,n}$ asymptotically approach zero. Therefore, the responding system (5.2.11) and the driving system (5.2.10) are asymptotically synchronized, and the unknown parameters $c$ of the driving system (5.2.10) are identified.

When the value of $k$ gradually decreases, the changing process of error $|e_{m,n}|$ between the driving system (5.2.10) and the responding system (5.2.11) is shown in Fig. 5.11.

According to Fig. 5.11, when the value of $k$ gets smaller and smaller, the error $|e_{m,n}|$ tends to 0 faster and faster, and the identification effect becomes better and better.

When $k = 5.01$, iterating 80 steps, the spatial Julia set of driving system (5.2.10) and responding system (5.2.11) are shown in Fig. 5.12a, b.

When $k = 5$, the real part identification process and imaginary part identification process of the parameter $c_{m,n}$ in the responding system (5.2.11) are shown in Fig. 5.13, and the changing process of the error $e_{m,n}$ between the driving system (5.2.10) and responding system (5.2.11) is shown in Fig. 5.14.

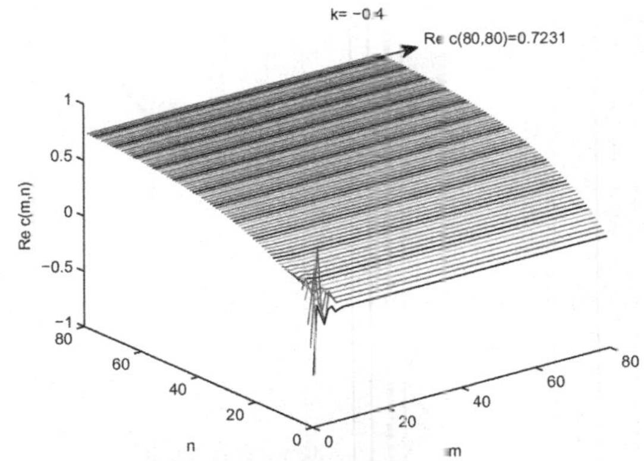


(a) the real part identification process of the parameter $c_{m,n}$ in the responding system (5.2.11)

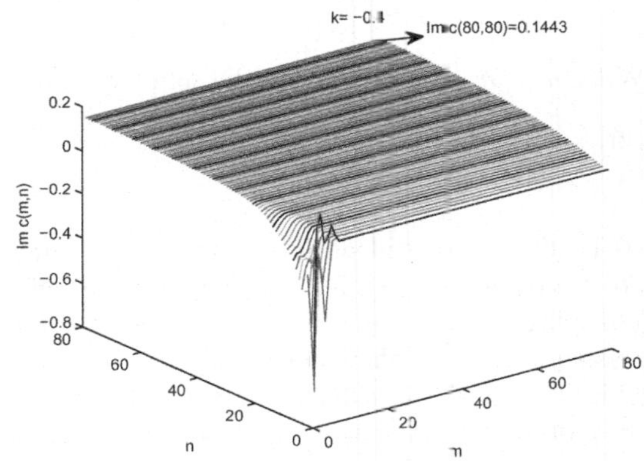


(b) the imaginary part identification process of the parameter $c_{m,n}$ in the responding system (5.2.11)

**Fig. 5.9** The identification process of $c_{m,n}$ in the responding system (5.2.11) at $k = -0.41$

According to the simulation results, with the increase of the number of iterations, the real and imaginary part of the parameter $c_{m,n}$ of the responding system (5.2.11) tend to be infinite, thus the driving system (5.2.10) and the responding system (5.2.11) cannot achieve synchronization, also, the parameters $c$ cannot be identified. And the real and imaginary part of the error $e_{m,n}$ don't go to 0.

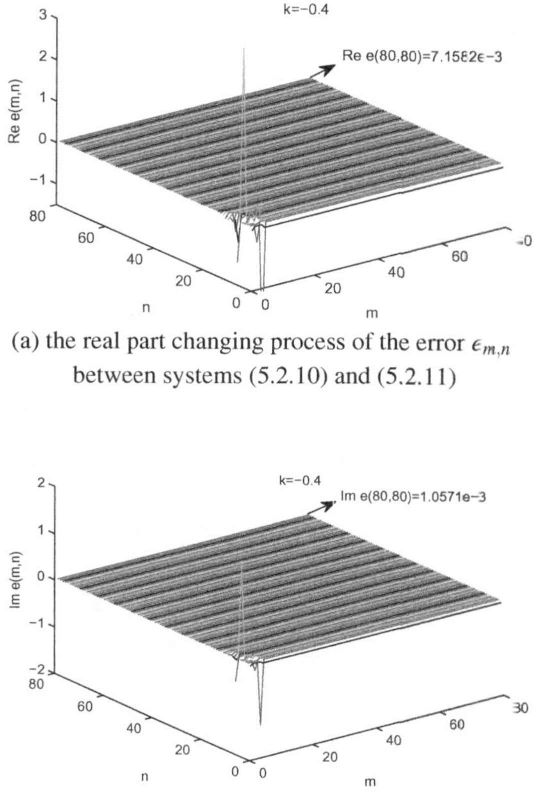


(a) the real part changing process of the error $e_{m,n}$ between systems (5.2.10) and (5.2.11)



(b) the imaginary part changing process of the error $e_{m,n}$ between systems (5.2.10) and (5.2.11)

**Fig. 5.10** The $e_{m,n}$ between systems (5.2.10) and (5.2.11) changing with $k = -0.41$

## 5.2.2 *Design of Synchronization Controller and Parameter Identifier for the Basic Julia Set*

In particular, in the spatial fractal Julia set driving system (5.2.10) and the responding system (5.2.11), make $\omega = 0$, $n = n_0$, discuss parameter identification problems of its basic Julia set $z_{m+1,n_0} = z_{m,n_0}^2 + c$ . At this time, the forms of the driving system and the responding system are as follows:

$$x_{m+1,n_0} = x_{m,n_0}^2 + c, \tag{5.2.34}$$

$$y_{m+1,n_0} = y_{m,n_0}^2 + c_{m,n_0} + u_{m,n_0}. \tag{5.2.35}$$

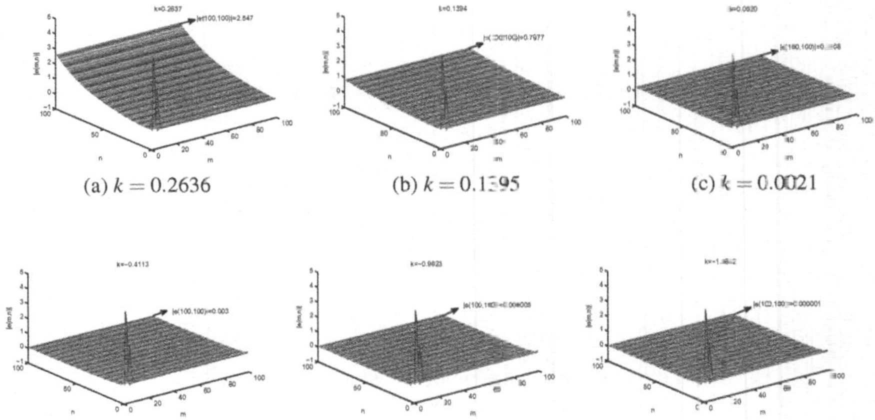

(a) $k = 0.2636$ (b) $k = 0.1395$ (c) $k = 0.0021$

(d) $k = -1.0861$ (e) $k = -0.9624$ (f) $k = -0.4112$

**Fig. 5.11** The $\mid e_{m,n} \mid$ between systems (5.2.10) and (5.2.11) changing with $k$ when $k$ becomes smaller

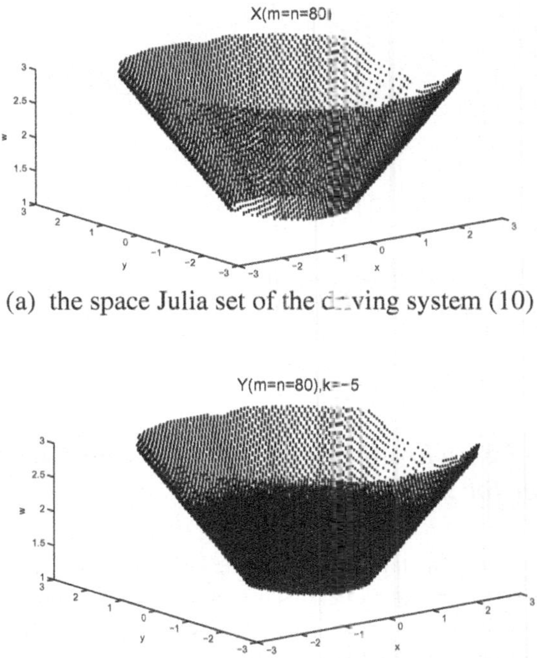


(a) the space Julia set of the driving system (10)

(b) the space Julia set of the responding system (11)

**Fig. 5.12** The spatial Julia sets of systems (5.2.10) and (5.2.11) at $k = 5.01$ and $m = n = 80$

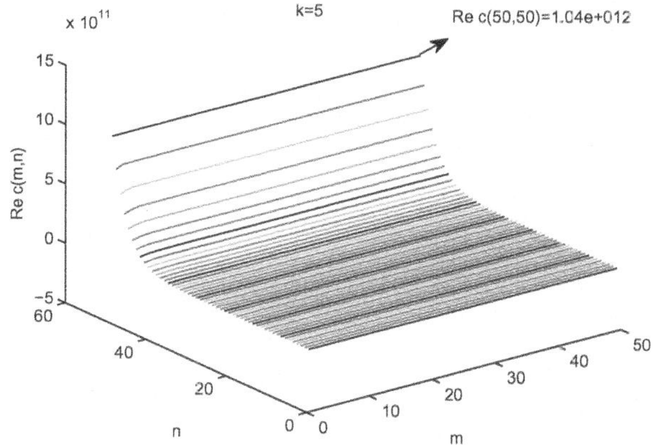


(a) The real identification process of $c_{m,n}$ of the response system (5.2.11)

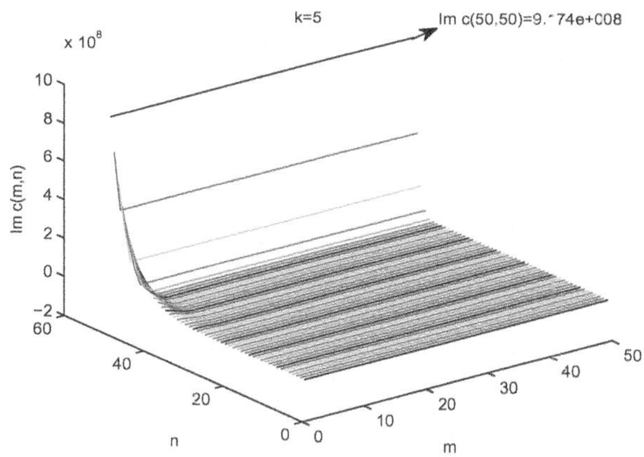


(b) The imaginary identification process of $c_{m,n}$ of the response system (5.2.11)

**Fig. 5.13** The identification process of $c_{m,n}$ of the response system (5.2.11) at $k = 5.01$

Let the error variable between the driving and responding system be

$$e_{m,n_0} = x_{m,n_0} - y_{m,n_0}. \tag{5.2.36}$$

Take the structure of synchronization controller $u_{m,n_0}$ as

$$u_{m,n_0} = x^2_{m,n_0} - y^2_{m,n_0}. \tag{5.2.37}$$

And take the structure of the parameter identifier as

$$c_{m+1,n_0} - c_{m,n_0} = ke_{m,n_0}. \tag{5.2.38}$$

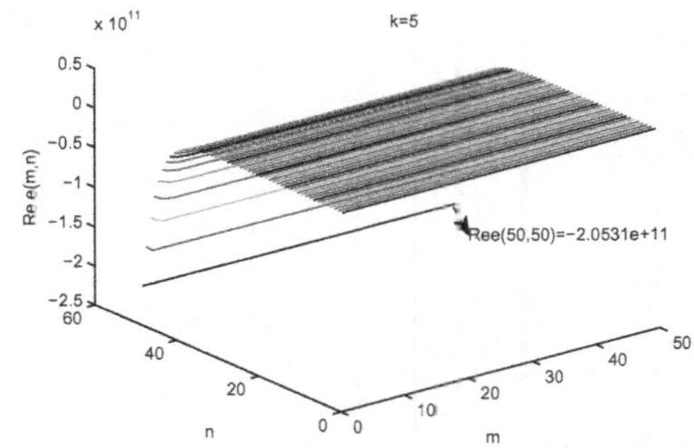


(a) the real part changing process of the error $e_{m,n}$ between driving system (5.2.10) and responding system (5.2.11)

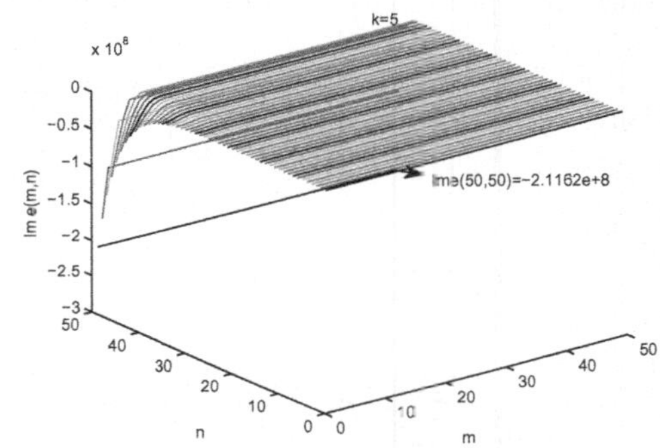


(b) the imaginary part changing process of the error $e_{m,n}$ between driving system (5.2.10) and responding system (5.2.11)

**Fig. 5.14** The $e_{m,n}$ between systems (5.2.10) and (5.2.11) changing with $k = 5.01$

where $k$ is a constant which satifies $0 < k < 1$, and the driving system (5.2.34) and the responding system (5.2.35) starting from any initial values can achieve global asymptotic synchronization [240, 241].

## 5.3 Control of the Spatial Mandelbrot Set Generated in Coupled Map Lattice

To begin with, the French mathematician Gaston Julia investigated the simple complex map

$$z_{n+1} = z_n^\alpha + c \ (c \in \mathbb{C}, \alpha = 2). \tag{5.3.1}$$

Following the footsteps of Julia, Mandelbrot [242] proposed another classical fractal set named Mandelbrot set which is formed with all the values of $z_0$ that the orbit at $z_0 = 0$ is bounded. The closely relationship between Julia set and Mandelbrot set is that Julia sets are connected for the value of $c$ which belongs to Mandelbrot set.

Generally speaking, Mandelbrot set has drawn increasing attentions on its structure research [243], properties analysis [244] and other theoretic features [167, 245]. It has also derived out some new research topics including cross disciplines about the electronic system [246], particle movement [177], biology [247] and so forth. (These researches on theory and application include the generalized M-set with $\alpha > 2$ or $\alpha < 0$).

It's worth noting that the interesting fractal structure and good utility of the Mandelbrot set have stimulated scholars′ enthusiasm for seeking higher dimensional Mandelbrot sets. The higher dimensional Mandelbrot set, denoted as spatial Mandelbrot set, can be constructed in two cases: hypercomplex system with one variable [181, 248, 249] and general complex system with more than one variable [250]. In the former case, Garant-Pelletier and Rochon [249] gave a multicomplex version of the so-called Fatou–Julia theorem and pointed out the symmetries of the principal 3-D slices of the generalized Mandelbrot set for tricomplex numbers. Wang and Sun [181] proved the connection of the general quaternionic Mandelbrot set and calculated the boundary of the stability region of the fixed point. In the later case, Danca [186] unveiled the graphical results of the spatial Mandelbrot-like sets generated from a class of alternated complex map (with a single system variable and two parameter variables), and left some visually observations to be proved about the boundedness and symmetry properties of these sets. Liu [250] presented the fractal structure of the spatial Mandelbrot set generated in coupled map lattice (CML M-Set for short) and achieved its control.

Although there are many achievements about the spatial Mandelbrot set, few research has focused on the control problem [250]. That is because that Mandelbrot set is formed with the values $c$ rather than the system state variables $z$. To sum up, we need proper mathematical transform to realize the control goal of $c$ by controlling the system state variable $z$. In [250], Liu achieved the size change of the CML Mandelbrot set without changing its structure properties. This section is to achieve better control effect which includes both symmetry change, position change and size change.

### 5.3.1 The Mandelbrot Set in Coupled Map Lattice

Consider the complex coupled map lattice system as follows:

$$z_{n+1}(i) = (1-\varepsilon)f(z_n(i)) + \frac{\varepsilon}{2}[f(z_n(i-1)) + f(z_n(i+1))], \tag{5.3.2}$$

where $n$ is the discrete time index, $\varepsilon \in [0, 1]$ is a real coupling strength, $i = 1, 2, \ldots, N$ is the lattice site. In general, the periodic boundary conditions $z_n(0) = z_n(N)$ and $z_n(N+1) = z_n(1)$ are assumed.

Take $f(z_n(i)) = z_n^2(i) + c(i)$ and $N = 2$, we have:

$$\begin{cases} z_{n+1}(1) = (1-\varepsilon)z_n^2(1) + \varepsilon z_n^2(2) + (1-\epsilon)c(1) + \epsilon c(2), \\ z_{n+1}(2) = (1-\varepsilon)z_n^2(2) + \varepsilon z_n^2(1) + (1-\epsilon)c(2) + \epsilon c(1), \end{cases} \tag{5.3.3}$$

where $Z_n = (z_n(1), z_n(2))^T$, $C = (c(1), c(2))^T$ ($T$ denotes transpose). For convenience, we denote

$$C_\epsilon = (c_\epsilon(1), c_\epsilon(2))^T = ((1-\epsilon)c(1) + \epsilon c(2), (1-\epsilon)c(2) + \epsilon c(1))^T.$$

Thus, system (5.3.3) can be reduced into

$$Z_{n+1} = F(Z_n, \varepsilon, C) = G(Z_n, \varepsilon) + C_\epsilon.$$

Besides, denote $F(Z_n, \varepsilon, C) = F(Z_n) = (F_{(1)}, F_{(2)})^T$ and $G(Z_n, \varepsilon) = G(Z_n) = (G_{(1)}, G_{(2)})^T$. By setting $Z_0 = \mathbf{0} = (0, 0)^T$, a sequence of $Z_n$ is gotten as follows:

$$\begin{aligned} Z_0 &= \mathbf{0}, \\ Z_1 &= F(\mathbf{0}) = G(\mathbf{0}) + C_\epsilon = C_\epsilon, \\ Z_2 &= F^2(\mathbf{0}) = G(Z_1) + C_\epsilon = G(C_\epsilon) + C_\epsilon, \\ &\cdots \\ Z_n &= F^n(\mathbf{0}) = \underbrace{G(G\cdots G(G}_{n-1}(C_\epsilon) + C_\epsilon) + C_\epsilon) + C_\epsilon. \end{aligned}$$

It is obvious that the trajectories of $Z_n$ are closely related to the initial constant $C_0$. Thus, the Mandelbrot set in coupled map lattice is defined as follows:

**Definition 5.6** The Mandelbrot set in coupled map lattice, denoted as $M(F)$, is the spatial set of all the initial constant $C_0 = (c_0(1), c_0(2))^T$ such that

$$\lim_{k\to\infty} \|F^k(\mathbf{0}, \varepsilon, C_0)\| \nrightarrow \infty,$$

in which we denote $z_n(1) = x_n + jy_n$, $z_n(2) = s_n + jt_n$, $c(1) = u + jv$ and $c(2) = p + jq$ ($j^2 = -1$).

As shown in the above definition, $M(F)$ is formed by the initial constant $C_0 = (c_0(1), c_0(2))^T$, which illustrates that $M(F)$ belongs to a four-dimensional space.

It is known that two common methods can be applied to visualize $M(F)$ in three-dimensional space. By taking $p_0$=$q_0$=$r_0$ and then finding all the initial points $(u_0, v_0, r_0, r_0)$ which satisfies Definition 5.6, literature [250] depicted $M(F)$ in $(u, v, r)$ space. In [197, 221], $M(F)$ was depicted in $(u, v, p)$ space by fixing one of the four real coordinates, such as $q_0 = 0$ and then finding all the initial points $(u_0, v_0, p_0, 0)$ which satisfies Definition 5.6. The $M(F)$ generated by the former method can be regarded as compressing the four-dimensional space into three-dimensional. Whereas, the later method can be regarded as taking a three-dimensional slice. In this contribution, $M(F)$ is visualized via the second method by taking $q_0 = 0$. The results are presented in Fig. 5.15 which fully display the beautiful and complex fractal structure of $M(F)$.

### 5.3.2 Symmetry of the Mandelbrot Set in Coupled Map Lattice

Symmetry is an important property of Mandelbrot set which has been investigated in the planar case [243] and the spatial case of multicomplex system [249]. For the 3-D slice of Mandelbrot set in coupled map lattice in this part, the following theorem is given.

**Theorem 5.3** *The 3-D slice of the Mandelbrot set in coupled map lattice*

$$M(F)_{q_0=0}$$

*is symmetric around the plane u_0_p.*

***Proof*** Divide system (5.3.3) into real part and imaginary part, one gets

$$\begin{cases} x_{n+1} = (1-\varepsilon)(x_n^2 - y_n^2) + \varepsilon(s_n^2 - t_n^2) + (1-\epsilon)u + \epsilon p, \\ y_{n+1} = 2(1-\varepsilon)x_n y_n + 2\varepsilon t_n s_n + (1-\epsilon)v + \epsilon q, \\ s_{n+1} = (1-\varepsilon)(s_n^2 - t_n^2) + \varepsilon(x_n^2 - y_n^2) + (1-\epsilon)p + \epsilon u, \\ t_{n+1} = 2(1-\varepsilon)s_n t_n + 2\varepsilon x_n y_n + (1-\epsilon)q + \epsilon v. \end{cases} \tag{5.3.4}$$

Suppose any two points $\overset{*}{C_0}, \bar{C_0} \in M(F)_{q_0=0}$ in the $(u, v, p)$ space, which are symmetric around the plane $u\_0\_p$, that is,

$$\overset{*}{C_0} = (u_0, v_0, p_0, 0)^T, \quad \bar{C_0} = (u_0, -v_0, p_0, 0)^T.$$

For the initial point $Z_0 = (0, 0, 0, 0)$, the $n$-th iterations of $Z_0$ with different parameters $\overset{*}{C_0}, \bar{C_0} \in M(F)_{q_0=0}$ are denoted as $\overset{*}{Z_n} = (\overset{*}{x}_n, \overset{*}{y}_n, \overset{*}{s}_n, \overset{*}{t}_n)^T$, $\bar{Z_n} = (\bar{x}_n, \bar{y}_n, \bar{s}_n, \bar{t}_n)^T$.

(1) For $n = 1$, one can get:

$$\overset{*}{Z_1} = \begin{cases} \overset{*}{x}_1 = (1-\varepsilon)u_0 + \varepsilon p_0, \\ \overset{*}{y}_1 = (1-\varepsilon)v_0, \\ \overset{*}{s}_1 = (1-\varepsilon)p_0 + \varepsilon u_0, \\ \overset{*}{t}_1 = \varepsilon v_0 \end{cases} \quad and \quad \bar{Z_1} = \begin{cases} \bar{x}_1 = (1-\varepsilon)u_0 + \varepsilon p_0, \\ \bar{y}_1 = (\varepsilon - 1)v_0, \\ \bar{s}_1 = (1-\varepsilon)p_0 + \varepsilon u_0, \\ \bar{t}_1 = -\varepsilon v_0 \end{cases}$$

After the first iteration, the first and third coordinates of the two points remain the same, while the second and forth coordinates become opposite numbers.

(2) Assume the truth for $n = k > 1$, that is,

$$\overset{*}{Z}_k = (\overset{*}{x}_k, \overset{*}{y}_k, \overset{*}{s}_k, \overset{*}{t}_k)^T, \quad \bar{Z}_k = (\bar{x}_k, \bar{y}_k, \bar{s}_k, \bar{t}_k)^T = (\overset{*}{x}_k, -\overset{*}{y}_k, \overset{*}{s}_k, -\overset{*}{t}_k)^T. \quad (5.3.5)$$

Then this assumption leads us to the following by substituting (5.3.5) into (5.3.4)

$$\begin{cases} \overset{*}{x}_{k+1} = (1-\varepsilon)\left(\overset{*2}{x_k} - \overset{*2}{y_k}\right) + \varepsilon\left(\overset{*2}{s_k} - \overset{*2}{t_k}\right) + (1-\epsilon)u_0 + \epsilon p_0, \\ \overset{*}{y}_{k+1} = 2(1-\varepsilon)\,\overset{*}{x}_k\overset{*}{y}_k + 2\varepsilon\overset{*}{t}_k\overset{*}{s}_k + (1-\epsilon)v_0, \\ \overset{*}{s}_{k+1} = (1-\varepsilon)(\overset{*}{s}_k - \overset{*}{t}_k) + \varepsilon\left(\overset{*2}{x_k} - \overset{*2}{y_k}\right) + (1-\epsilon)p_0 + \epsilon u_0, \\ \overset{*}{t}_{k+1} = 2(1-\varepsilon)\,\overset{*}{s}_k\overset{*}{t}_k + 2\varepsilon\,\overset{*}{x}_k\overset{*}{y}_k + \epsilon v_0 \end{cases}$$

and

$$\begin{cases} \bar{x}_{k+1} = (1-\varepsilon)\left(\overset{*2}{x_k} - \overset{*2}{y_k}\right) + \varepsilon\left(\overset{*2}{s_k} - \overset{*2}{t_k}\right) + (1-\epsilon)u_0 + \epsilon p_0, \\ \bar{y}_{k+1} = -2(1-\varepsilon)\,\overset{*}{x}_k\overset{*}{y}_k - 2\varepsilon\overset{*}{t}_k\overset{*}{s}_k - (1-\epsilon)v_0, \\ \bar{s}_{k+1} = (1-\varepsilon)(\overset{*}{s}_k - \overset{*}{t}_k) + \varepsilon\left(\overset{*2}{x_k} - \overset{*2}{y_k}\right) + (1-\epsilon)p_0 + \epsilon u_0, \\ \bar{t}_{k+1} = -2(1-\varepsilon)\,\overset{*}{s}_k\overset{*}{t}_k - 2\varepsilon\,\overset{*}{x}_k\overset{*}{y}_k - \epsilon v_0. \end{cases}$$

Thus, $\overset{*}{x}_{k+1} = \bar{x}_{k+1}$, $\overset{*}{s}_{k+1} = \bar{s}_{k+1}$ and $\overset{*}{y}_{k+1} = -\bar{y}_{k+1}$, $\overset{*}{t}_{k+1} = -\bar{t}_{k+1}$ still stand.

(3) Consequently, we now know from the principle of the mathematical inductive method that $\overset{*}{x}_n = \bar{x}_n$, $\overset{*}{s}_n = \bar{s}_n$ and $\overset{*}{y}_n = -\bar{y}_n$, $\overset{*}{t}_n = -\bar{t}_n$ are true for all positive integers $n$.

Then, we arrive at

$$\lim_{n\to\infty} \| \overset{*}{Z}_n \| = \lim_{n\to\infty} \| \bar{Z}_n \|.$$

Thus, for any two points $\overset{*}{C}_0$ and $\bar{C}_0$ which are symmetric around the plane $u\_0\_p$, the iterations behave the same boundedness property or not, which ensure that they simultaneously belong to $M_F$ or not. Then, from Definition 5.6, the two parts of $M(F)_{q_0=0}$ divided by the plane $u\_0\_p$ is symmetric.

The examples in Figs. 5.15b, c and 5.16 illustrate the validity of Theorem 5.3.

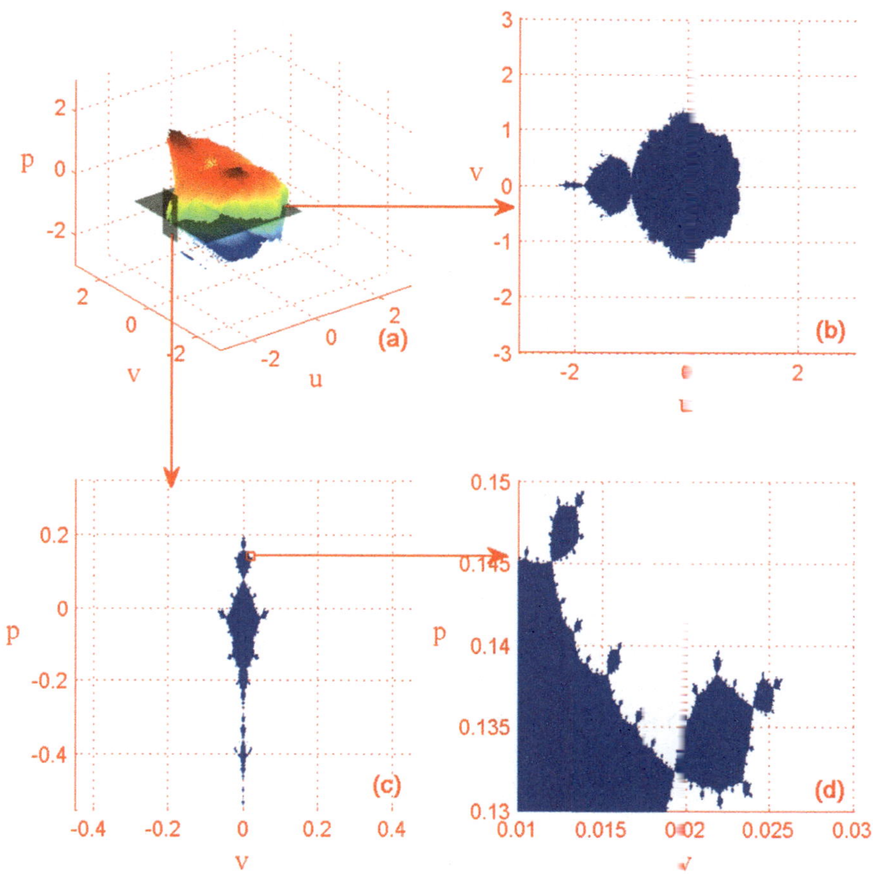


**Fig. 5.15** **a** The Mandelbrot set in coupled map lattice $M(F)_{q0=0}$ with $\epsilon = 0.41$, **b** The 2-D slice of $M(F)_{q0=0}$ with $p_0 = -0.51$, **c** The 2-D slice of $M(F)_{q0=0}$ with $u_0 = -2.31$, **d** The partial enlarged views of $M(F)_{q0=0,u0=-2.31}$

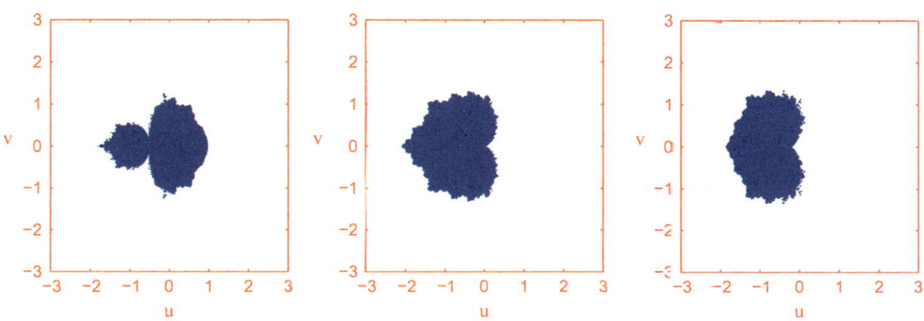


**Fig. 5.16** The 2-D slices of $M(F)_{q0=0}$ with (**a**): $p_0 = -1.001$, (**b**): $p_0 = 0.501$, (**c**): $p_0 = 0.701$

### 5.3.3 Control of the Mandelbrot Set in Coupled Map Lattice

In this section, a proper controller is designed for the 3-D slice of CML M-set to achieve its symmetry change, position change and size change. Let us next propose four lemmas which will play an important role in the subsequent development

**Lemma 5.1** *For the following system:*

$$\begin{cases} z_{n+1}(1) = d_1[(1-\varepsilon)\left(\frac{1}{d_1}z_n(1)\right)^2 + \varepsilon\left(\frac{1}{d_1}z_n(2)\right)^2 + c_\varepsilon(1)]. \\ z_{n+1}(2) = d_2[(1-\varepsilon)\left(\frac{1}{d_2}z_n(2)\right)^2 + \varepsilon\left(\frac{1}{d_2}z_n(1)\right)^2 + c_\varepsilon(2)]. \end{cases} \tag{5.3.6}$$

*where $d_1, d_2$ belong to a bounded set of complex numbers and $d_1, d_2 \neq 0$. Then the Mandelbrot set of system (5.3.6), denoted as $M(\bar{F})$, is identical with the Mandelbrot set of original system (5.3.3).*

***Proof*** Rewrite system (5.3.6) as:

$$\begin{cases} z_{n+1}(1) = \bar{F}_{(1)}\,(z_n(1), z_n(2)) = d_1 F_{(1)}\left(\frac{1}{d_1}z_n(1), \frac{1}{d_1}z_n(2)\right), \\ z_{n+1}(2) = \bar{F}_{(2)}\,(z_n(1), z_n(2)) = d_2 F_{(2)}\left(\frac{1}{d_2}z_n(1), \frac{1}{d_2}z_n(2)\right). \end{cases}$$

Mandelbrot set is generated by iterations of the initial point $Z_0 = \mathbf{0}$, it follows that

$$\frac{1}{d_1}\bar{F}_{(1)}(\mathbf{0}) = F_{(1)}(\mathbf{0}).$$

Then the n-th iteration of system (5.3.6) can be shown as

$$\begin{aligned} F^n_{(1)}(\mathbf{0}) &= F^{n-1}_{(1)}(F_{(1)}(\mathbf{0})) = F^{n-1}_{(1)}\left(\frac{1}{d_1}\bar{F}_{(1)}(\mathbf{0})\right) \\ &= F^{n-2}_{(1)}\left(F_{(1)}\left(\frac{1}{d_1}\bar{F}_{(1)}(\mathbf{0})\right)\right) = F^{n-2}_{(1)}\left(\frac{1}{d_1}\bar{F}^2_{(1)}(\mathbf{0})\right) \\ &= \cdots\cdots \\ &= F_{(1)}\left(\frac{1}{d_1}\bar{F}^{n-1}_{(1)}(\mathbf{0})\right) \\ &= \frac{1}{d_1}\bar{F}^n_{(1)}(\mathbf{0}). \end{aligned} \tag{5.3.7}$$

In the meanwhile, one can get

$$F^n_{(2)}(\mathbf{0}) = \frac{1}{d_2}\bar{F}^n_{(2)}(\mathbf{0}). \tag{5.3.8}$$

Since $d_1, d_2$ are bounded, from (5.3.7) and (5.3.8), one can see that $F^n(\mathbf{0})$ and $\overline{F}^n(\mathbf{0})$ behave the same boundedness property or not. Therefore, according to Definition 5.6, the Mandelbrot set of system (5.3.9) $M(\overline{F})$ is the same with the Mandelbrot set of original system (5.3.3). □

**Lemma 5.2** *For the following system:*

$$\begin{cases} z_{n+1}(1) = G_{(1)}(Z_n) + (1-\epsilon)l_1 c(1) + \epsilon l_2 c(2), \\ z_{n+1}(2) = G_{(2)}(Z_n) + (1-\epsilon)l_2 c(2) + \epsilon l_1 c(1), \end{cases} \tag{5.3.9}$$

*where $L = (l_1, l_2)$ and $l_1, l_2 \in \mathbb{R}^+$. Then, the Mandelbrot set of system (5.3.9) can be expressed as $M(5.3.9) = \left\{\left(\frac{c(1)}{l_1}, \frac{c(2)}{l_2}\right)^T : C = (c(1), c(2))^T \in M(F)\right\}$.*

***Proof*** Let $C' = (c(1)', c(2)')^T = (l_1 c(1), l_2 c(2))^T$, then the Mandelbrot set $M'$ of $Z_{n+1} = F(Z_n, \epsilon, C')$ is the same as $M(F)$ about the system parameter $C'$. Since $C = \left(\frac{c(1)'}{l_1}, \frac{c(2)'}{l_2}\right)^T$, we have $M(5.3.9) = \left\{\left(\frac{c(1)}{l_1}, \frac{c(2)}{l_2}\right)^T : C = (c(1), c(2))^T \in M(F)\right\}$.
□

Similarly, one can get the following two lemmas:

**Lemma 5.3** *For the following system:*

$$\begin{cases} z_{n+1}(1) = G_{(1)}(Z_n) + (1-\epsilon)(c(1)+m_1) + \epsilon(c(2)+m_2), \\ z_{n+1}(2) = G_{(2)}(Z_n) + (1-\epsilon)(c(2)+m_2) + \epsilon(c(1)+m_1), \end{cases} \tag{5.3.10}$$

*where $m_1 = a + bj$, $m_2 = c + dj$. Then, the Mandelbrot set of system (5.3.10) can be obtained by translating $M(F)$ $a$ units, $b$ units, $c$ units and $d$ units respectively along the negative direction of $u$-axis, $v$-axis, $p$-axis and $q$-axis.*

**Lemma 5.4** *For the following system:*

$$\begin{cases} z_{n+1}(1) = G_{(1)}(Z_n) + (1-\epsilon)(cos\theta + jsin\theta)c(1) + \epsilon c(2), \\ z_{n+1}(2) = G_{(2)}(Z_n) + (1-\epsilon)c(2) + \epsilon(cos\theta + jsin\theta)c(1), \end{cases} \tag{5.3.11}$$

*where $\theta \in (0, 2\pi)$. Then, the Mandelbrot set of system (5.3.11) can be achieved by rotating $M(F)$ with $\theta$ angle clockwise in the $u$-$v$ plane.*

Based on the above lemmas, the controller $U_n$ is designed as follows:

$$U_n = \begin{cases} u_n(1) = k_1[(1-\varepsilon)z_n^2(1) + \varepsilon z_n^2(2)] - C_{ref}(1), \\ u_n(2) = k_2[(1-\varepsilon)z_n^2(2) + \varepsilon z_n^2(1)] - C_{ref}(2). \end{cases} \tag{5.3.12}$$

The auxiliary reference item $C_{ref} = (C_{ref}(1), C_{ref}(2))^T$ is denoted as:

$$\begin{cases} C_{ref}(1) = (1-\epsilon)c_{ref}(1) - \frac{\epsilon}{1+k_1}[(k_2-k_1)c(2) - (1+k_2)c_{ref}(2)], \\ C_{ref}(2) = (1-\epsilon)c_{ref}(2) - \frac{\epsilon}{1+k_2}[(k_1-k_2)c(1) - (1+k_1)c_{ref}(1)], \end{cases}$$

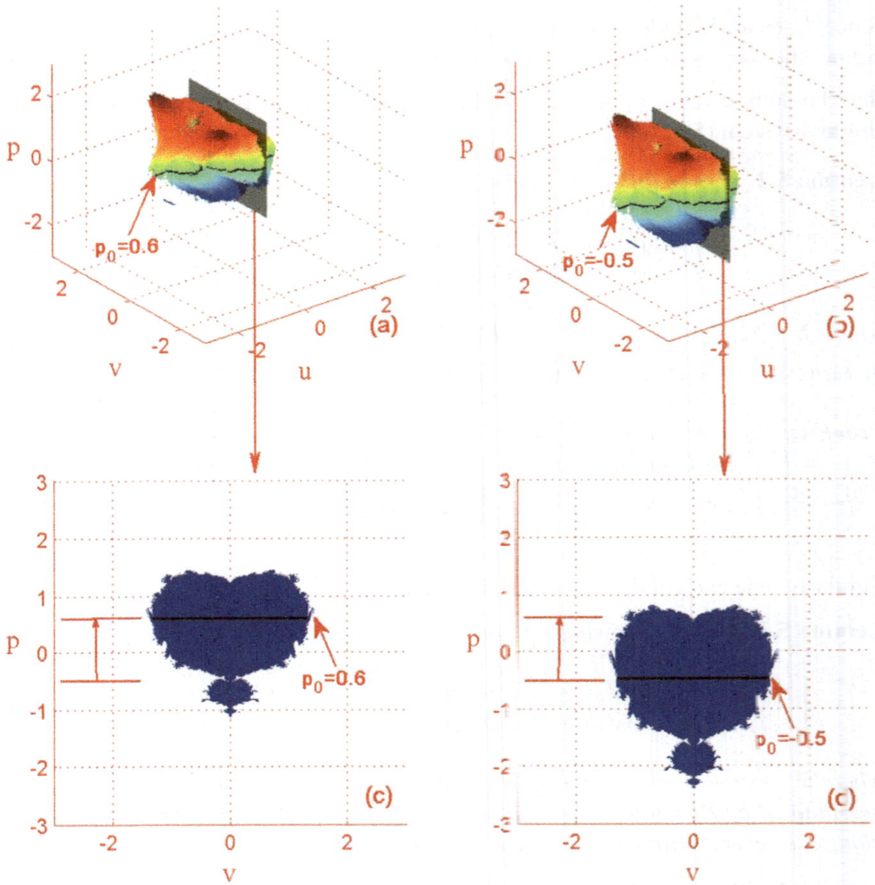


**Fig. 5.17** **a** The controlled CML M-Set with $k_1, c_{ref}(1) = 0$ and $k_2 = 0.25, c_{ref}(2) = 1$. **b** The original CML M-Set without control. **c** The slice of **(a)** with $u_0 = 0$. **d** The slice of **(b)** with $u_0 = 0$.

where $k_1, c_{ref}(1) \in \mathbb{C},\ k_2 > -1, c_{ref}(2) \in \mathbb{R}$. Substitute the above controller into (5.3.3), the controlled system can be shown as:

$$F_c = \begin{cases} F_c(1): & \frac{1}{1+k_1}[(1-\varepsilon)((1+k_1)z_n(1))^2 + \varepsilon((1+k_1)z_n(2))^2] \\ & + (1-\epsilon)(c(1) - c_{ref}(1)) + \frac{(1+k_2)\epsilon(c(2)-c_{ref}(2))}{1+k_1}, \\ F_c(2): & \frac{1}{1+k_2}[(1-\varepsilon)((1+k_2)z_n(2))^2 + \varepsilon((1+k_2)z_n(1))^2] \\ & + (1-\epsilon)(c(2) - c_{ref}(2)) + \frac{(1+k_1)\epsilon(c(1)-c_{ref}(1))}{1+k_2}. \end{cases}$$

By means of Lemma 5.1, the controlled system can be converted to the following form:

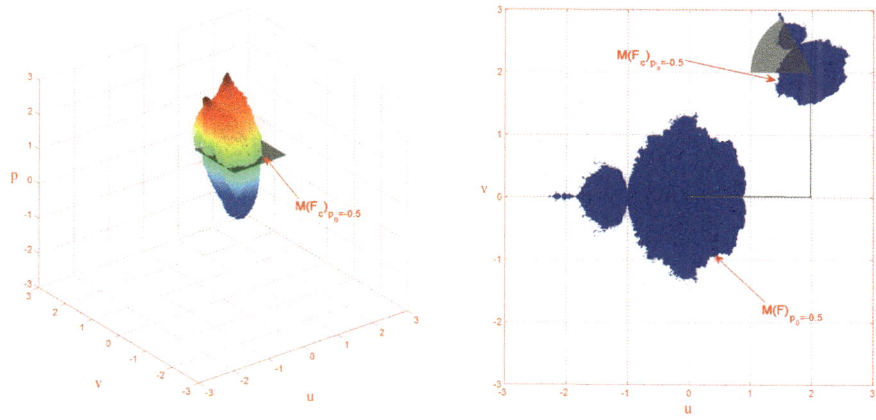


**Fig. 5.18** **a** The controlled CML M-Set with $k_2, c_{ref}(2) = 0$ and $k_1 = \sqrt{3}j, c_{ref}(1) = 2 + 2j$, **b** $M(F)_{p0=-0.5}$ and $M(F_c)_{p0=-0.5}$

$$Z_{n+1} = \begin{cases} (1-\varepsilon)z_n^2(1) + \varepsilon z_n^2(2) + (1-\epsilon)(1+k_1)(c(1) - c_{ref}(1)) \\ \qquad + \epsilon(1+k_2)(c(2) - c_{ref}(2)), \\ (1-\varepsilon)z_n^2(2) + \varepsilon z_n^2(1) + (1-\epsilon)(1+k_2)(c(2) - c_{ref}(2)) \\ \qquad + \epsilon(1+k_1)(c(1) - c_{ref}(1)). \end{cases} \tag{5.3.13}$$

Let $1 + k_1 = |1 + k_1|(cos\vartheta + jsin\vartheta)$, where $\vartheta$ is the argument of complex number $1 + k_1$ and $c_{ref}(1) = m(1) + jm(2)$. Then, applying the Lemmas 5.2–5.4, one gets Theorem 5.4.

**Theorem 5.4** *The Mandelbrot set of controlled system (5.3.13), denoted as $M(F_c)$, can be obtained through the following five steps,*

*(1) Transform $c(2)$ to $\frac{c(2)}{1+k_2}$, then let $M_I = \left\{(c(1), \frac{c(2)}{1+k_2})^T\right\}$.*
*(2) Move $M_I$ $c_{ref}(2)$ units along the positive direction of p-axis, one gets $M_{II}$. Then let $M_{II} = \{(c_{III}(1), c_{II}(2))^T\}$.*
*(3) Transform $c_{II}(1)$ to $\frac{c_{II}(1)}{|1+k_1|}$, then let $M_{III} = \left\{(\frac{c_{II}(1)}{|1+k_1|}, c_{II}(2))^T\right\}$.*
*(4) Rotate $M_{III}$ $\vartheta$ angle clockwise in the u-v plane, one gets $M_{IV}$.*
*(5) Move $M_{IV}$ $m(1)$ units and $m(2)$ units respectively along the positive direction of u-axis and v-axis, one can get $M(F_c)$.*

For comprehensive testify the control strategy, three examples with different parameters $K, C_{ref}$ are respectively presented in Figs. 5.17–5.19.

In Fig. 5.17, the control parameters are taken as $k_1, c_{ref}(1) = 0$ and $k_2 = 0.25$, $c_{ref}(2) = 1$. By means of Theorem 5.4(1) and Theorem 5.4(2), the controlled CML M-Set $M(F_c)$ becomes four-fifths of the original uncontrolled longitudinal size, and then moves 1 unit along the positive direction of $p$-axis.

To show the control effect more clearly, the slice of $M(F)$ with $p_0 = -0.5$ is taken for example. As presented in Fig. 5.17a, b, the original slice $p_0 = -0.5$ moves

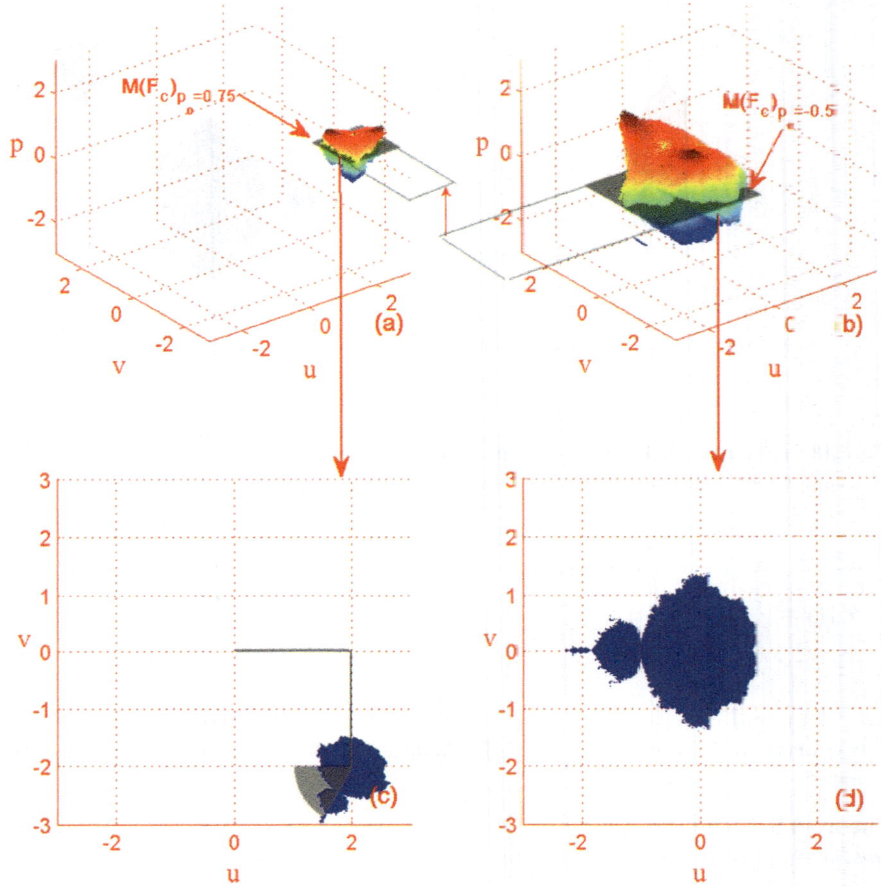


**Fig. 5.19** **a** The controlled CML M-Set with $k_2 = c_{ref}(2) = 1$ and $k_1 = -\sqrt{3}j$, $c_{ref}(1) = 2 - 2j$. **b** The original CML M-Set without control. **c** The slice of (**a**) with $p_0 = 0.75$. **d** The slice of (**b**) with $p_0 = -0.5$

to the target location $p_0 = 0.6$ by the following two steps:

$$p_0 = -0.5 \xrightarrow[Theorem 5.4(1)]{} p_0 = \frac{-0.5}{1.25} = -0.4 \xrightarrow[Theorem 5.4(2)]{} p_0 = -0.4 + 1 = 0.6.$$

From the above Fig. 5.18, the controller acts on the $u$-$v$ plane by taking $k_2, c_{ref}(2) = 0, k_1 = \sqrt{3}j$ and $c_{ref}(1) = 2 + 2j$. Let $1 + k_1 = 2\left(cos\frac{\pi}{3} + jsin\frac{\pi}{3}\right)$. From Theorem 5.4(3), Theorem 5.4(4) and Theorem 5.4(5), the controlled CML M-Set $M(F_c)$ is achieved by shrinking the transverse size of $M(F)$ to half, rotating $\frac{\pi}{3}$

angle clockwise and then translating 2 units respectively along the positive direction of $u$-axis, $v$-axis.

What's more, choose $k_2 = c_{ref}(2) = 1$, $k_1 = -\sqrt{3}j$ and $c_{ref}(1) = 2 + 2j$, the CML M-Set $M(F_c)$ is shown in Fig. 5.19. Let $1 + k_1 = 2\left(\cos\frac{5\pi}{3} + j\sin\frac{5\pi}{3}\right)$. According to Theorem 5.4, by shrinking both the transverse size and longitudinal size of $M(F)$ to half, rotating $\frac{\pi}{3}$ angle counterclockwise and then moving 1 unit, 2 units and 2 units respectively along the positive direction of $p$-axis, $u$-axis, $v$-axis, the CML M-Set $M(F_c)$ is finally obtained.

The slice of $M(F)$ with $p_0 = -0.5$ is taken for example to show the control effect more clearly. As presented in Fig. 5.19a, b, the original slice $p_0 = -0.5$ moves to the target location $p_0 = 0.75$ by the following two steps:

$$p_0 = -0.5 \xrightarrow[Theorem 5.4(1)]{} p_0 = \frac{-0.5}{2} = -0.25 \xrightarrow[Theorem 5.4(2)]{} p_0 = -0.25 + 1 = 0.75.$$

# Chapter 6
# Fractal Phenomena and Control in Economical Models

Recently, fractal curves have been studied widely in different fields [251–253], such as microwave, optical technology and tumor growths. Weierstrass–Mandelbrot function (WMF), a classical function to generate fractal curves, has attracted huge interest from mathematicians and physicists. Now there exist many works [254–264] focusing on the fractal dimension, different graphical representations, and physical applications of WMF.

One important problem is that, we usually have the requirements for controlling fractal behaviors of curves described by WMF according to some practical requirements. Here "control" refers to constrain the fluctuations of time series in WMF model as conditions have micro variations. So it is an interesting topic to study control of WMF. Consequently, a great deal of attention [216, 230, 265, 266] has been focused on the control of fractal systems theoretically and simulatively. However, the control problem of WMF has not been addressed yet.

On the other hand, theory of wavelets has recently achieved a tremendous success [267–273] in many areas of nonlinear science. Here we have introduced a control method using wavelet basis for WMF and investigated its control effects.

## 6.1 Control Effects of Morlet Wavelet Term on Weierstrass–Mandelbrot Function Model

### 6.1.1 WMF Model with Morlet Wavelet Term

Weierstrass [274] has proposed the original function that is a continuous everywhere but differentiable nowhere as follows:


S. T. Liu et al., *Fractal Control and Its Applications*,
https://doi.org/10.1007/978-981-15-5459-9_6

$$w(t) = \sum_{n=0}^{\infty} a^n \cos(b^n \pi t), \tag{6.1.1}$$

where $0 < a < 1$, $b$ is a positive odd integer and $ab > 1 + 3\pi/2$. Mandelbrot [164] has generalized the definition of Weierstrass function as following form named by WMF:

$$w(t) = \sum_{n=-\infty}^{+\infty} \frac{(1 - e^{i\gamma^n t})e^{i\phi_n}}{\gamma^{(2-D)n}}, \tag{6.1.2}$$

where $1 < D < 2$, $\gamma > 1$ and $\phi_n$ is an arbitrary phase.

Because of importance of Morlet wavelet term in nonlinear science and significance of parameter $\gamma$ in variation of WMF [254], we introduce a Morlet wavelet term in WMF as a controller and obtain the following controlled WMF model:

$$w(t, u) = \sum_{n=-\infty}^{+\infty} \frac{(1 - e^{i(\gamma+u)^n t})e^{i\phi_n}}{(\gamma + u)^{(2-D)n}}, \tag{6.1.3}$$

where $u$ is controller as $u = Be^{-\beta^2|w-\xi|/\eta^2}\cos(\pi|w - \xi|/\eta)$, $\beta > 0$ is parameter controlling form of wavelet, $\eta > 0$ denotes parameter which controls its length, $\xi > 0$ means controlling of its situation and $B > 0$ stands for weight coefficient.

We recall the following conclusions [275, 276], which will be helpful for the main results in next sections.

**Lemma 6.1** *There exists*

$$\sum_{n=-\infty}^{\infty} f(n) = \sum_{m=-\infty}^{\infty} \hat{f}(k),$$

*for appropriate functions $f$, where $\hat{f}$ is the Fourier transform as*

$$\hat{f}(v) = F\{f(x)\} = \int_{-\infty}^{+\infty} f(n)e^{i2\pi mn}dn.$$

**Lemma 6.2**

$$\Gamma(x + iy) = \int_0^{\infty} t^{x+iy-1}e^{-t}dt,$$

*and*

$$|\Gamma(x + iy)| \sim \sqrt{2\pi}|y|^{x-0.5}e^{-\pi|y|/2}.$$

Based on Lemmas 6.1 and 6.2, we can easily prove the convergence of series of Eq. (6.1.3) as Lemma 6.3, which is necessary to obtain its control relationships. In this part, for convenience, we assume $\phi_n = 0$.

**Lemma 6.3** *The series $w(t, u)$ satisfying Eq. (6.1.3) is convergent, if*

$$e^{-2m\pi^2/\ln(\gamma+u)}$$

*and*

$$e^{-i2\pi m \ln t/\ln(\gamma+u)}$$

*decay as $m \to +\infty$ and the term*

$$|2\pi m|^{D-2.5}$$

*decays as $m \to -\infty$.*

### 6.1.2 Control Relationships

Then we study its control effects as $u$ changes.

Without loss of generality, we first give cosine series as one of the two different forms of Eq. (6.1.3) to consider its control effects as follows,

$$C(t, u) = Real(w(t, u)) = \sum_{n=-\infty}^{+\infty} \frac{(1 - \cos[(\gamma + u)^n t])}{(\gamma + u)^{(2-D)n}}. \tag{6.1.4}$$

From the above basic lemmas in Sect. 6.1.1, we obtain the following corollary to define $C \doteq C(t, u)$ for simplicity.

**Corollary 6.1** *When $D > 1.5$, we have*

$$|C| \leq \frac{M_2\eta^2 + M_1B\eta^2 + M_1B\beta^2\xi}{\eta^2 + M_1B\beta^2}, \tag{6.1.5}$$

*where*

$$M_1 = (1.5 - D)M_R^{\frac{1}{1.5-D}},$$

$$M_2 = D - 0.5 + (\gamma - 1)M_1,$$

$$M_R = \lim_{p\to+\infty} \frac{\sqrt{2\pi}}{t^{D-2}} \sum_{m=-p}^{+p} e^{\frac{-2\pi m^2}{\ln(\gamma+u)}} |2\pi m|^{D-2.5},$$

*and*

$$\eta^2 + B\beta^2 M_1 > 0.$$

According to Corollary 6.1, we obtain the following theorem.

**Theorem 6.1** *The scope of C satisfying Eq. (6.1.4) will increase as B increases monotonously when*

$$D > 1.5,$$

$$\frac{0.5 - D}{\gamma - 1} < M_1 < 0,$$

*and*

$$\eta^4 + \beta^2\eta^2\xi < B\beta^2\eta^2 M_1 + \beta^2\eta^2 M_2 - B\beta^2 M_1.$$

***Proof*** After taking

$$G(B, \beta, \eta, \xi) = \frac{M_2\eta^2 + M_1 B\eta^2 + M_1 B\beta^2\xi}{\eta^2 + M_1 B\beta^2}$$

from inequality (6.1.5), we have the following relationship:

$$\frac{\partial G}{\partial B} = \frac{M_1\eta^4 + B\beta^2 M_1^2 + \beta^2\eta^2\xi M_1 - \beta^2\eta^2 M_1 M_2 - M_1^2 B\beta^2\eta^2}{(\eta^2 + M_1 B\beta^2)^2}.$$

We get easily that

$$\frac{\partial G}{\partial B} > 0,$$

if $D > 1.5$, $\frac{0.5-D}{\gamma-1} < M_1 < 0$ and

$$\eta^4 + \beta^2\eta^2\xi < B\beta^2\eta^2 M_1 + \beta^2\eta^2 M_2 - B\beta^2 M_1.$$

The proof is thus completed.

The other three Theorems 6.2–6.4 for the relationships between the fluctuations of the series described by Eq. (6.1.4) and the variation of parameters $\beta$, $\eta$ and $\xi$ in the control term $u = Be^{-\beta^2|w-\xi|/\eta^2}\cos(\pi|w-\xi|/\eta)$ are shown as following:

**Theorem 6.2** *The scope of $C(t, u)$ satisfying Eq. (6.1.4) will decrease as $\beta$ increases when*

$$\beta > max\left\{\frac{D - 0.5}{\xi}, \sqrt[3]{\frac{(\gamma - 1)\eta^2}{B\xi}}\right\}.$$

**Theorem 6.3** *The scope of $C(t, u)$ satisfying Eq. (6.1.4) will increase with the monotonously increase of $\eta$ when $\xi > D - 0.5 + (\gamma + B - 1)M_1$.*

**Theorem 6.4** *The scope of $C(t, u)$ satisfying Eq. (6.1.4) will decrease as $\xi$ increases when $0 < \eta < \beta\sqrt{-BM_1}$.*

Based on Theorems 6.1–6.4, we can get different control effects.

### 6.1.3 Results and Discussion

In order to verify effectiveness of our control method, we give some simulation examples as follows. Because many real phenomena have fractal dimensions $D$ when $D > 1.5$, we choose parameters as $\gamma = 1.6$, $D = 1.8$, $t \in (0, 1)$ and then consider the control effects of Morlet wavelet term on $C(t, u)$ of Eq. (6.1.4).

Using Theorem 6.1, we get that $\partial G/\partial B > 0$ when $\beta = 0.1$, $\eta = 1$ and $\xi = 1$, meaning that the scope of $C(t, u)$ increases as $B$ increases. This is illustrated in Fig. 6.1.

When $\beta$ changes as $B = 1$, $\eta = 1$ and $\xi = 1$, we obtain $\partial G/\partial \beta < 0$ applying Theorem 6.2, meaning that scope of $C(t, u)$ decreases as $\beta$ increases. This is shown in Fig. 6.2.

Based on Theorem 6.3, we have $\partial G/\partial \eta > 0$ as $\beta = 0.1$, $B = 1$ and $\xi = 1$, meaning that scope of $C(t, u)$ increases as $\eta$ increases. This is observed in Fig. 6.3.

Furthermore, based on Theorem 6.4, we get $\partial G/\partial \xi < 0$ when $\xi = 0.1$, $B = 1$ and $\eta = 1$, meaning that the scope of $C(t, u)$ decreases as $\xi$ increases. This is described in Fig. 6.4.

From Figs. 6.1, 6.2, 6.3 and 6.4, we find that scope of $C(t, u)$ increases or decreases as parameters $B$, $\beta$, $\eta$ or $\xi$ change. However, according to simulation results above, variation of $C(t, u)$ is complex although its fractal dimension described by the box counting method is always 1.8 [277]. So we can not get useful knowledge from the one fractal dimension. Then we consider its two other statistical characteristics: multifractality and Hurst exponent.

As multifractality [278–280], we calculate

$$P_i(\varepsilon) \sim \varepsilon^{\alpha},$$

firstly, where $P_i$ means the probability of this mass at $i$ which is relative to total mass for a box size $\varepsilon$, and $\alpha$ is named as singular exponent. Then we get relationships between $N_\alpha$, total number of the boxes that have same value of $\alpha$ and box size $\varepsilon$ as

$$N_\alpha(\varepsilon) \sim \varepsilon^{-f(\alpha)},$$

where $f(\alpha)$ stands for fractal dimension of the set that has same value of $\alpha$. $\Delta\alpha$ means the variation range of exponent and $\Delta f(\alpha) > 0$ represents that probability in the max position of $C(t, u)$ is larger than that in min position of $C(t, u)$. For its different forms of $C(t, u)$ shown in Figs. 6.1, 6.2, 6.3, 6.4 and 6.5a, the multifractality has similar form that $f(\alpha)$ increases at first shortly and then decreases when $\alpha$ increases. Figure 6.5b shows the form by using an example. In addition, it is well known that generalized dimension $D_q$ generates different dimensions such as fractal dimension, correlation dimension, etc., where $q$ means the weight. So it is useful to study the relationship between $D_q$ of series of controlled WMF model and $\alpha$ for further understanding complex fractal behavior of WMF model. Because $D_q$ has the following scaling relationship:

$$(q - 1)D_q = q\alpha - f(\alpha),$$

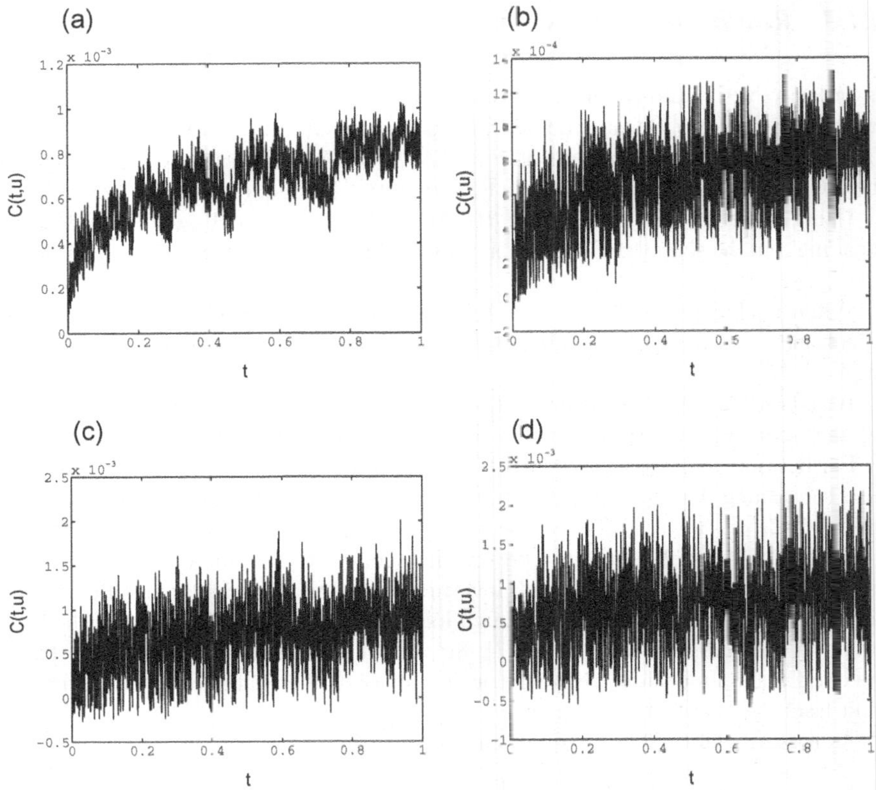

Fig. 6.1 Variation of $C(t, u)$ with $t$ for different values of $B$: **a** $B = 0.11$, **b** $B = 0.31$, **c** $B = 0.49$ and **d** $B = 0.69$. In each case $\beta = 0.1$, $\xi = 1$ and $\eta = 1$

we have variation between $D_q$ and $\alpha$. Figure 6.5c shows the increasing relationship clearly.

In order to verify its different variations of series as $u$ changes, we calculate their $\Delta\alpha$ and $\Delta f(\alpha)$ shown in Table 6.1. Then we get that $\Delta f(\alpha)$ and $\Delta\alpha$ increase at first and then decrease as $B$ increases. Moreover, $\Delta f(\alpha)$ and $\Delta\alpha$ increase as $\eta$ increases monotonously. In addition, as $\xi$ or $\beta$ increases, $\Delta f(\alpha)$ and $\Delta\alpha$ decrease.

Hurst exponent $(H)$ is then calculated. First, we obtain the cumulative deviate series

$$X(t) = \sum_{i=1}^{t}(\xi_i - m),$$

where

$$m = \frac{1}{n}\sum_{i=1}^{n}\xi_i,$$

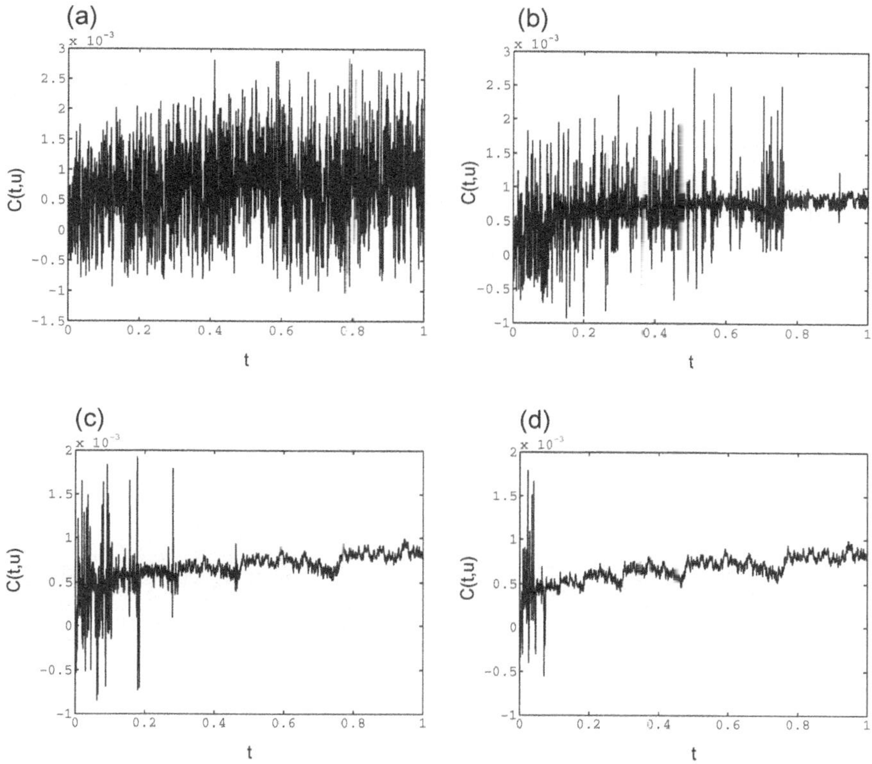

Fig. 6.2 Variation of $C(t, u)$ with $t$ as $\xi = 1$, $\eta = 1$, $B = 1$ and different values of $\beta$: **a** $\beta = 0.09$, **b** $\beta = 0.21$, **c** $\beta = 0.31$ and **d** $\beta = 0.39$

for $t = 1, 2, \ldots, n$. Then we get

$$\frac{R(n)}{S(n)} \sim n^H,$$

where the range

$$R(n) = max(X(t)) - min(X(t))$$

and the standard deviation

$$S(n) = \sqrt{\frac{1}{n}\sum_{i=1}^{n}(X_i - m)^2}.$$

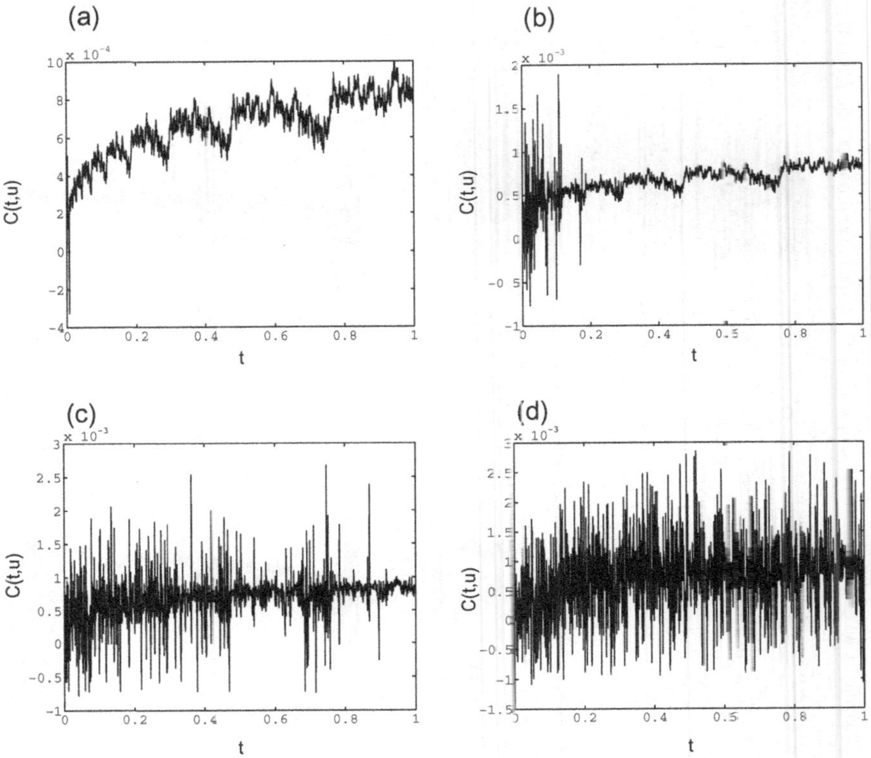


**Fig. 6.3** Variation of $C(t, u)$ with $t$ for $\beta = 0.1, \xi = 1, B = 1$ and different values of $\eta$: **a** $\eta = 0.09$, **b** $\eta = 0.31$, **c** $\eta = 0.49$ and **d** $\eta = 0.71$

So we calculate its variation of $H$ using the R/S method [281] as shown in Table 6.1 for Eq. (6.1.4). Then we get that when $\beta$ and $\xi$ increase, respectively, $H$ increases, while $H$ decreases as $B$ and $\eta$ increase, respectively.

## 6.2 Generalized Weierstrass–Mandelbrot Function Model for Actual Stocks Markets Indexes with Nonlinear Characteristics

The search for effective methods of understanding the dynamical behavior of financial markets is always a very rich discipline [282–285]. One of the most interesting topics is the study of stock market indexes. Actual stock market indexes have many nonlinear characteristics, such as fractal [286–288] or multifractal [279, 280, 289, 290], Hurst

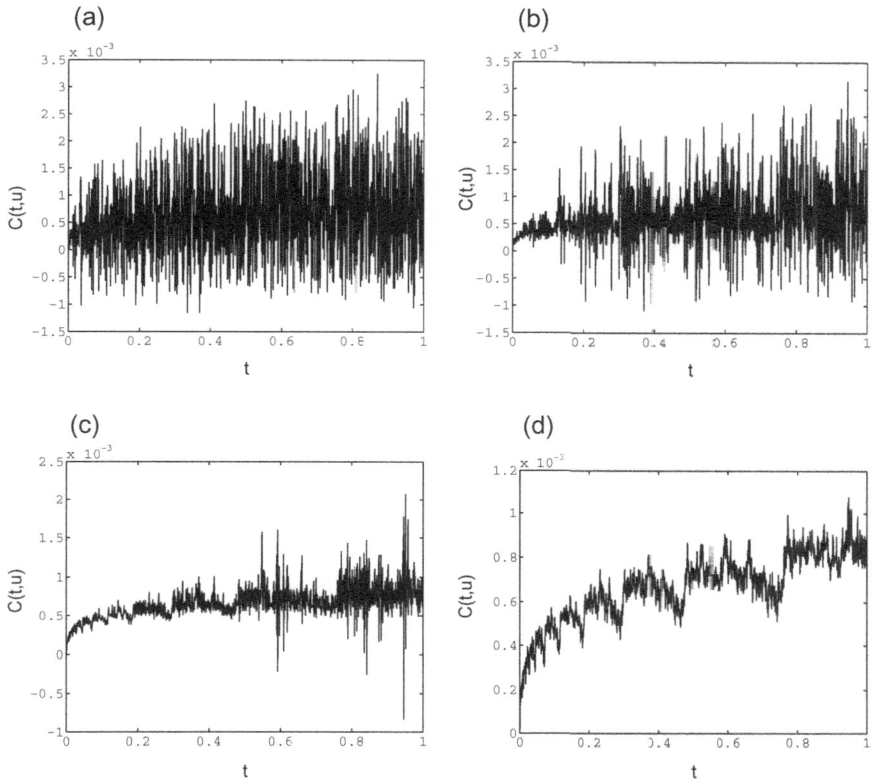


**Fig. 6.4** Variation of $C(t, u)$ with $t$ as $\xi$ changes when $\beta = 0.1$, $\eta = 1$, $B = 1$: **a** for $\xi = 15.01$, **b** for $\xi = 19.99$, **c** for $\xi = 24.99$ and **d** for $\xi = 30.01$

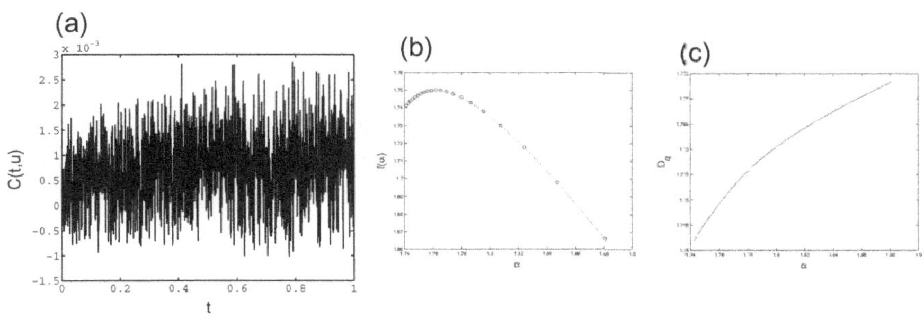


**Fig. 6.5** **a** Variation of $C(t, u)$ with $t$ for $\beta = 0.11$, $\xi = 1$, $\eta = 1$, $B = 1$. **b** Relation between spectrum $f(\alpha)$ and exponent $\alpha$. **c** Relation between generalized dimension $D_q$ and $\alpha$ for $C(t, u)$ shown in Fig. 6.5a

**Table 6.1** Different effects of $u$ on $C(t, u)$ as parameters change

| Parameters | $\Delta f(\alpha)$ | $\Delta\alpha$ | $H$ |
|---|---|---|---|
| $\beta = 0.1$ | 0.1545 | 0.2454 | 0.5827 |
| $\beta = 0.2$ | 0.0909 | 0.1394 | 0.6619 |
| $\beta = 0.3$ | 0.0294 | 0.0475 | 0.8924 |
| $\beta = 0.4$ | 0.0079 | 0.0144 | 0.9355 |
| $B = 0.1$ | 0.0548 | 0.0890 | 0.9579 |
| $B = 0.3$ | 0.1413 | 0.2209 | 0.7708 |
| $B = 0.5$ | 0.2239 | 0.3344 | 0.6377 |
| $B = 0.7$ | 0.2139 | 0.3199 | 0.5938 |
| $\eta = 0.1$ | 0.0056 | 0.0119 | 0.9956 |
| $\eta = 0.3$ | 0.0264 | 0.0415 | 0.9393 |
| $\eta = 0.5$ | 0.0766 | 0.1154 | 0.7279 |
| $\eta = 0.7$ | 0.0925 | 0.1533 | 0.6139 |
| $\xi = 15$ | 0.1375 | 0.2134 | 0.6456 |
| $\xi = 20$ | 0.0839 | 0.1262 | 0.6562 |
| $\xi = 25$ | 0.0136 | 0.0247 | 0.9227 |
| $\xi = 30$ | 0.0081 | 0.0166 | 0.9984 |

exponent ($H$) [291, 292] from 0.5 to 1 which means the time sequence is not random walker, chaos [293–296] and other nonlinear stylized facts [297–300].

For these nonlinear characteristics or nonlinear stylized facts of actual stock market indexes, there are many interesting works. For example, different methods were used to calculate the fractal/multifractal dimension $D$ [286, 288, 289] and the Hurst exponent $H$ [292, 301–304]. Moreover, chaotic models were proposed to forecast the nonlinearities of bifurcation and long-term motion for the stock price index [305–307]. McMillan [308] extended a nonlinear functional form to predict the nonlinear relationship between stock market returns and a variety of financial and macroeconomic variables described by a nonparametric regression method. Then he studied the negative forecasting effect of volume on nonlinear future returns which supported the logistic smooth-transition model [309]. Models based on hierarchical/neural networks were introduced to understand the nonlinearity of the complex financial systems [310–314].

However, the above works did not focus on the nonlinear mathematical model with two nonlinearities: fractal and Hurst exponent, which are important to explain and understand the complexity of the actual stock market indexes dynamically. In other words, these approaches focus on the calculation of fractal dimension and Hurst exponent of actual stock market indexes and on the simulation of other nonlinear stylized facts. They did not find a mathematical model to study deeply the further relationship between the dynamical behavior of actual stock market indexes and the variation of fractal dimension and Hurst exponent. We think the relationship is very important to comprehend the dynamical behavior of actual stock market indexes.

But few consider the relationship. So our contribution is to find the relationship and use it to predict the dynamical behavior of actual stock market indexes. So it is important to propose a mathematical model to describe the dynamical behavior of actual stock market indexes with two nonlinear characteristics: fractal dimension and Hurst exponent.

Because of the importance of fractal dimension and Hurst exponent for the actual stock market indexes and a mathematical model named Weierstrass–Mandelbrot function (WMF) which has fixed fractal dimension and constant Hurst exponent, we want to find a relationship between the WMF and the dynamical behavior of actual stock market indexes to understand the complex variation of actual markets indexes deeply. Hence we study the variation of WMF as parameters change in 3-dimensional space and the relationship between the parameters and the actual stock market indexes in order to find an effective simulating method for the actual stock market indexes.

Our model is generalized by WMF which has been studied for many years. Weierstrass [274] proposed the original function as a continuous everywhere but differentiable nowhere function (6.1.1). In 1977, Mandelbrot [164], as the fractal ancestor, generalized the definition of Weierstrass function (6.1.2). He discussed the variation of WMF and calculated its Hausdorff–Besicovitch dimension [315] which is larger than 1, meaning that it is fractal. WMF has been used to simulate different real phenomena in many fields [255, 259, 260, 262–264, 316], such as climate, turbulent and engineering science.

### 6.2.1 Generalized Model of WMF

According to the value of $\phi_n$ in system (6.1.2), there are two different forms of WMF.

When $\phi_n = 0$, we have the cosine series:

$$C(t) = ReW(t) = \sum_{n=-\infty}^{+\infty} \frac{1 - \cos(\gamma^n t)}{\gamma^{(2-D)n}}.$$

If $\phi_n = n\pi$, we get the sine series:

$$A(t) = -ImW(t) = \sum_{n=-\infty}^{+\infty} \frac{(-1)^n \sin(\gamma^n t)}{\gamma^{(2-D)n}}.$$

Without loss of generality, we generalize the cosine series as an example to consider its dynamical behavior as $D$ and $\gamma$ change in higher dimensional space and its simulations for actual stocks markets indexes.

When we assume that parameters $D$ and $\gamma$ change, we have the following generalized WMF model:

$$C(t,\gamma,D)=ReW(t)=\sum_{n=-\infty}^{+\infty}\frac{1-\cos(\gamma^n t)}{\gamma^{(2-D)n}}. \tag{6.2.1}$$

This model is in 3-dimensional space and its graph will vary with different parameters $\gamma$ and $D$.

For the reason that $C \doteq C(t,\gamma,D)$ of system (6.2.1) has the fractal scaling property [242] when its fractal dimension $D \in (1.5, 2)$ and $\gamma > 1$, we simulated the dynamical behavior of system (6.2.1) in 3-dimensional space as shown in Fig. 6.6a using Matlab when $D \in (1.5, 2)$, $t \in (0, 1)$ and $\gamma \in (1.6, 2)$. It can been seen clearly that the fluctuation frequency of $C$ as $D$ changes is much larger than that of $C$ as $\gamma$ changes comparing with Fig. 6.6b, c.

Hurst exponent ($H$) is an effective statistic to clarify the time sequences from random walk to biased random walk [281, 301]. Because the actual stock market indexes have different Hurst exponents as $1 > H > 0.5$, we study the relationships between $H$ of generalized WMF model and its parameters $\gamma$ and $D$ dynamically.

The process of calculating $H$ named rescaled range analysis (R/S) is listed as follows: we first calculate $X(t,N)$ from a time sequence $\xi(t)$, $t = 1, 2, \ldots$

$$X_{t,N}=\sum_{u=1}^{t}(\xi(u)-M_N),$$

where $X_{t,N}$ is the accumulative deviation in $N$ period, $\xi_u$ is the inflow of the year $u$ and

$$M_N=\frac{1}{N}\sum_{t=1}^{N}\xi(t)$$

is the mean value of $\xi_t$ in $N$ period. Then we calculate the range $R$:

$$R=Max(X_{t,N})-Min(X_{t,N}),\ \ 1\le t\le N,$$

where $Max(X)$ is the maximum value of $X$ and $Min(X)$ is the minimum value of $X$. $S$ is the standard deviation defined as follows:

$$S=\left(\frac{1}{N}\sum_{t=1}^{N}(\xi_t-M_N)^2\right)^{1/2}.$$

In order to compare different time sequences, the relationship is proposed by

$$\frac{R}{S}\propto(N)^H,$$

where $R/S$ is the rescaled range, $N$ is the observation frequency and $H$ is the Hurst exponent.

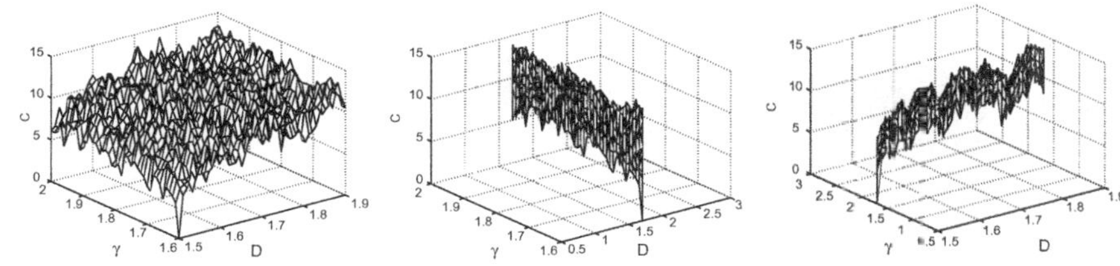


**Fig. 6.6** The dynamical behavior and its cross-section of $C$ as the graph of generalized WMF model, where $\gamma$ is the Weierstrass spectrum and $D$ is taken to be the Hausdorff–Besicovitch dimension for system (6.2.1)

If $H = 0.5$, the time sequence is random walker. When $0 \le H < 0.5$, it is anti-persistence or Ergodic. As $0.5 < H < 1$, the time sequence has persistent and even increasing trend, meaning that the time sequence is biased random walk or fractional brownian motion.

Because of the definition of $H$, we take the actual time sequence of stock market indexes as

$$\xi(t) = \sum_{n=-t}^{+t} \frac{1 - \cos(\gamma^n t)}{\gamma^{(2-D)n}},$$

where $D$ is the fractal dimension of the time sequence and the parameter $\gamma$ means the spectrum of the time sequence. Then we have the following results.

### 6.2.2 Simulations and Results

Based on the results shown in Fig. 6.6, it is clear that the fluctuation frequency of $C$ changes complexly as $D$ and $\gamma$ change. In order to clarify the relationship between WMF and the actual stock market indexes, we should consider common features of them. Because WMF has fixed fractal dimension $D$ and Hurst exponent $H$ which are easy to be calculated in reality for the actual stock market indexes, we first simulate the dynamical behavior of $H$ for WMF as $\gamma$ and $D$ change, respectively shown in Fig. 6.7. According to Fig. 6.7a, we can see clearly that the dynamical behavior of $H$ is complex. Figure 6.7b shows that $H$ will decrease at first and then increase as $D$ increases when $\gamma$ is fixed. In addition, the variation of $H$ as $\gamma$ changes becomes more complicated than that of $H$ as $D$ changes shown in Fig. 6.7c.

**Remark 6.1** Furthermore, it is clear that $\gamma$ and $D$ play important roles in the complexity of the behavior of WMF dynamically [317]. And then the same $H$ can be obtained as $\gamma$ is different and $D$ is fixed which is illustrated in Fig. 6.7.

So in order to find the relationship between WMF and the actual stock market indexes, we next simulate the actual stock market indexes choosing $\gamma = 1.6$ randomly for WMF. We use the boxcount method and R/S method to calculate the

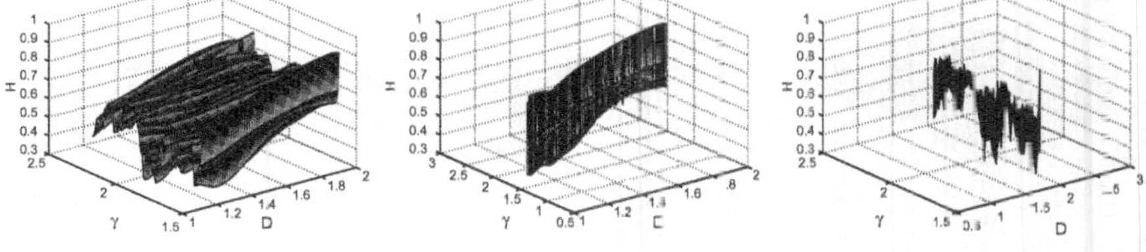


**Fig. 6.7** The dynamical behavior and its cross-section of $H$ as Hurst exponent for system (6.2.1)

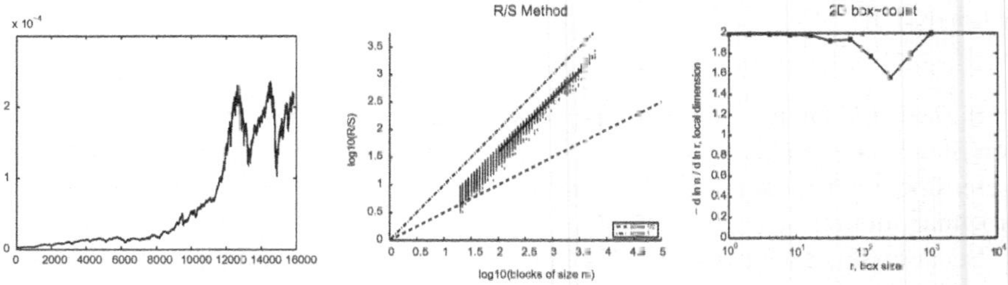


**Fig. 6.8** The variation of actual stocks market indexes

fractal dimension $D$ and Hurst exponent $H$ of actual stock market indexes: S&P 500 from 1950.01.02–2010.12.30 daily, respectively, in which $D = 1.9$, $H = 0.8$ as mean values of simulated values according to Fig. 6.3. Then we simulate the actual S&P 500 indexes daily using the generalized WMF model as the easy process as follows:

(1) Calculate the Hurst exponent $H$ of actual stocks markets indexes.
(2) Calculate the fractal dimension $D$ of actual stocks markets indexes using sandbox method [277].
(3) Choose $\gamma$ for WMF as the values of $D$ and $H$ according to the relationship of Fig. 6.7.
(4) Find the initial value of actual stocks market indexes $v_0$.
(5) Take $\gamma$, $D$ and $v_0$ into generalized WMF model (5.2.1).
(6) Plot the graph of generalized WMF model and the actual graph of stocks market indexes.
(7) Compare the fit line between the graph of generalized WMF model and the graph of actual stocks markets indexes.

For comparing the difference between our simulated results and the actual stock market indexes, we normalize these values in range (0, 1).

Figure 6.9 shows the simulation of WMF for actual S&P 500 indexes in different time ranges (see Fig. 6.9a, c, e) and the fitting lines (see Fig. 6.9b, d, f). Obviously, the fitting lines have different forms. It means that we can not get clarified simulations for different time ranges of actual time series according to the same values $D$ and $\gamma$. For example, Fig. 6.9a, b show that our method can simulate effectively at the first

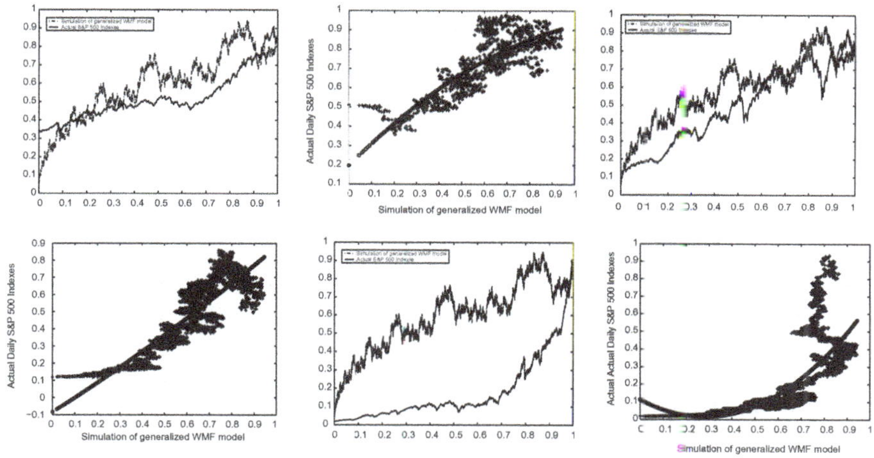


**Fig. 6.9** The variation of actual daily S& P 500 index and simulation of generalized WMF model

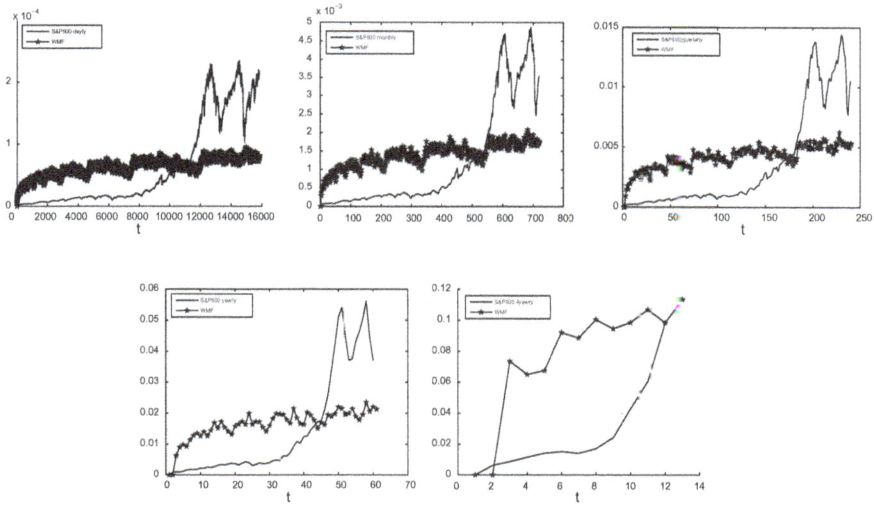


**Fig. 6.10** The variation of actual daily S&P 500 index and simulation of generalized WMF model at different time ranges from 1950.01.03 to 2010.12.30

one-third roughly. Figure 6.9c, d illustrate that our method can simulate effectively at the last one-third roughly. Figure 6.9e, f show that our method cannot simulate effectively for the whole interval. For further finding the relationship between WMF and S&P 500 indexes, we also calculate the correlation coefficient $r$ between WMF and S&P 500 shown in Fig. 6.10, such as $r = 0.6506$ daily, $r = 0.6555$ monthly (21 days = 1 month), $r = 0.6403$ quarterly (3 monthes = 1 quarter), $r = 0.5780$ yearly

(4 quarters = 1 year) and $r = 0.5471$ four yearly. The obtained value of correlation coefficient $r$ represents the different linear effectiveness of our method for simulating the actual stocks markets indexes.

**Remark 6.2** Based on Fig. 6.9 and the different values of correlation coefficient $r$ shown in Fig. 6.10, it is clear that our easy process has better effectiveness for the higher time frequency such as daily and monthly data than that for the lower time frequency such as quarterly, yearly and four yearly data.

Obviously, the correlation coefficient is small, meaning that the easy process must be modified furtherly. The reason that our easy process can not get better effectiveness may be the value of $\gamma$ is chosen randomly and $D$ is fixed for all different time ranges. So in order to get more effectiveness, we should find a way to decide the value of $\gamma$ and should calculate different $D$ according to different time ranges. Because the correlation coefficient $r$ of our easy method is maximal for the monthly time series, we next consider the other vital characteristics of S&P 500 indexes monthly to find exact value of $\gamma$. Furthermore, there are many characteristics of time series in which skewness and kurtosis are the two main characteristics. So we calculate the distribution of the skewness and kurtosis of S&P 500 monthly using box plot shown in Fig. 6.11. It is shown that the discrete range of skewness is larger than that of kurtosis. Moreover, the skewness is more symmetry than kurtosis. In order to find the relationship between $\gamma$ and $D$ and the values of skewness and kurtosis, we plot the variation of these relationship shown in Fig. 6.12. Obviously, as $D$ increases and $\gamma$ is fixed, the skewness of WMF decreases monotonously which is shown in Fig. 6.12a; however, the kurtosis of WMF increases monotonously as $D$ increases and $\gamma$ is fixed, which is shown in Fig. 6.12b. When $\gamma$ increases and $D$ is fixed, the skewness of WMF decreases at first monotonously and then decreases with wide fluctuation shown in Fig. 6.12c. Meanwhile, the kurtosis of WMF increases at first monotonously and then increases with wide fluctuation shown in Fig. 6.12d when $\gamma$ increases and $D$ is fixed. So we can choose appropriate value of $\gamma$ using the actual values of skewness and kurtosis of the actual stocks markets indexes shown in Fig. 6.12 and calculate different $D$ for different time ranges.

**Remark 6.3** Actually, we can calculate the relationship between skewness or kurtosis and any $D$ or $\gamma$ for any different time range. It means that we can first calculate the actual value of $D$ for the aimed actual stocks markets indexes. Then we calculate the skewness and kurtosis of the aimed actual stocks markets indexes to find the appropriate value of $\gamma$.

Then our simulating process is modified as follows:

(1) Calculate the value of $t$ for the lillietest method which can decide the Gauss distribution when $t = 0$.
(2) Choose the series as $t = 1$ which means the series is not Gauss distribution.
(3) Calculate the skewness and kurtosis of the actual stocks markets index.
(4) Calculate the Hurst exponent $H$ of actual stocks markets index.

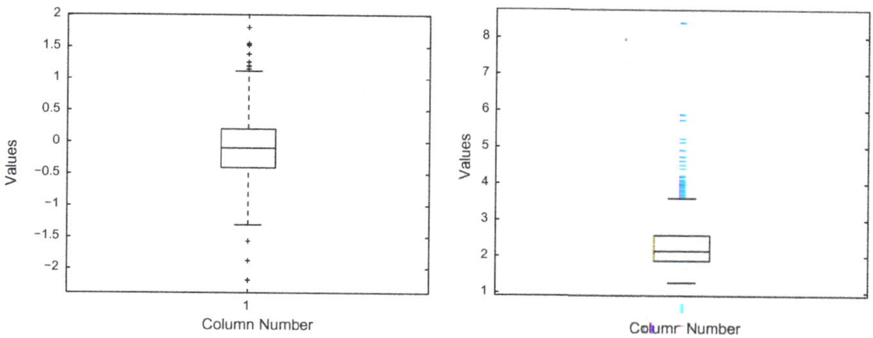


**Fig. 6.11** Distribution of skewness and kurtosis of for S& P 500 index monthly

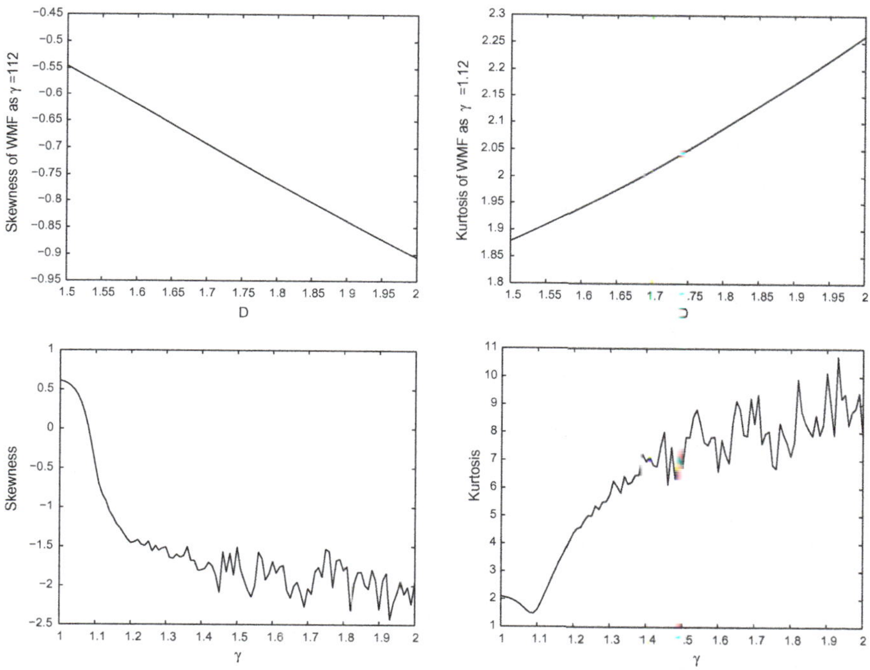


**Fig. 6.12** Variations of skewness and kurtosis for WMF as $\gamma$ and $D$ change

(5) Calculate the fractal dimension $D$ of actual stocks markets index using sandbox method and choose the range of $\gamma$ for WMF according to the relationship of Fig. 6.7.
(6) Choose exact value of $\gamma$ as results shown in Fig. 6.12.
(7) Find the initial value of actual stocks market indexes $v_0$.
(8) Take $\gamma$, $D$ and $v_0$ into generalized WMF model.

(9) Plot the graph of generalized WMF model and the actual graph of stocks market index.
(10) Compare the fit line between the graph of generalized WMF model and the graph of actual stocks markets index.

For example, we simulate different time ranges of S&P 500 indexes by our method when we choose appropriate $\gamma$ and $D$. One is the daily time series and the other is the yearly time series. At first we look for the time ranges with fractal characteristic. We use the lillietest method to find the nonrandom time series of S&P 500 at different time ranges shown in Table 6.2, in which $t = 0$ means the time series is random or belongs to Gauss distribution and $t = 1$ means it is not random. Then we use the box method to find the distribution of the skewness and kurtosis of the non-random time series shown in Fig. 6.11. We then calculate the value of $D$ using the boxcount method and then obtain the definite value of $\gamma$ according to the relationship monotonously shown in Fig. 6.12c, d. Based on our results, we choose $\gamma = 1.02$ for WMF to simulate the actual S&P 500 indexes from 1950.01.03–1950.12.29 daily shown in Fig. 6.13a and choose $\gamma = 1.05$ for WMF to simulate the actual S&P 500 indexes from 1950.01.03–1957.12.23 yearly illustrated in Fig. 6.13b. It is clear that the simulation shown in Fig. 6.13 is more effective than that of shown in Figs. 6.9 and 6.10, in which WMF and S&P 500 indexes vary at the similar behavior dynamically and the correlation coefficients of WMF and S&P 500 indexes are $r = 0.8729$ daily and $r = 0.9828$ yearly, respectively.

In addition, Fig. 6.14 illustrates the fit box results. It is shown that fit effect like diverse range is small and numbers of approximating is much more as time ranges become larger, respectively.

### 6.2.3 Discussions and Conclusions

Prediction models for actual financial markets indexes have been proposed by different methods, such as by using a multivariate time-series analysis [283, 318] and by neural networks [313, 319, 320]. However, these models did not consider the two main nonlinear characteristics: fractal dimension $D > 1.5$ and the Hurst exponent $1 > H > 0.5$. In addition, there are few models to simulate the actual financial markets indexes based on the two nonlinear characteristics dynamically. So we propose a new simulation model based on a WMF model which has been used widely in many real fields, such as turbulence, tissue study and climate study.

Because of the complexity of actual financial markets, many terms have effects on the real systems. Hence we firstly considered its dynamical behavior of generalized WMF model as parameters changed in 3-dimensional space. Then we simulated actual S&P 500 index using WMF effectively, in which parameters $D$ and $\gamma$ were chosen by the distribution of the skewness and kurtosis of actual Daily S&P 500 index, respectively. In addition, our results show that we can simulate different actual

**Table 6.2** Values of lillietest, skewness and kurtosis of actual S&P 500 from 1950 to 2012 monthly (21 days = 1 month)

| Time range | t | Skewness | Kurtosis |
|---|---|---|---|
| 1953.01.02–1953.01.30 | 1 | −1.0208 | 2.6703 |
| 1953.02.02–1953.02.27 | 0 | −0.7919 | 2.7002 |
| 1953.03.02–1953.03.31 | 0 | −0.7020 | 2.4557 |
| 1953.04.01–1953.04.30 | 0 | −0.2439 | 1.9439 |
| 1953.05.01–1953.05.29 | 1 | 0.4910 | 1.5611 |
| 1953.06.01–1953.06.30 | 0 | 0.2161 | 1.5822 |
| 1953.07.01–1953.07.31 | 0 | −0.4548 | 2.1017 |
| 1953.08.03–1953.08.31 | 0 | −0.8765 | 3.6901 |
| 1953.09.01–1953.09.30 | 0 | −0.4997 | 1.9551 |
| 1953.10.01–1953.10.30 | 0 | 0.8910 | 3.0951 |
| 1953.11.02–1953.11.30 | 0 | 0.1519 | 2.1609 |
| 1962.01.02–1962.01.31 | 0 | 0.0888 | 2.5576 |
| 1962.02.01–1962.02.28 | 1 | 0.3818 | 1.9282 |
| 1962.03.01–1962.03.30 | 0 | −0.0149 | 2.1357 |
| 1962.04.02–1962.04.30 | 1 | −0.8889 | 2.5001 |
| 1962.05.01–1962.05.31 | 1 | −0.0230 | 1.4150 |
| 1962.06.01–1962.06.29 | 1 | −0.7963 | 2.8259 |
| 1962.07.02–1962.07.31 | 0 | −0.8251 | 2.8258 |
| 1962.08.01–1962.08.31 | 0 | −0.2862 | 1.5748 |
| 1962.09.04–1962.09.28 | 0 | −0.3315 | 2.3269 |
| 1962.10.01–1962.10.31 | 0 | −0.1798 | 3.0893 |
| 1962.11.01–1962.11.30 | 0 | −0.0829 | 1.8208 |
| 1994.01.03–1994.01.31 | 1 | −0.7680 | 2.3752 |
| 1994.02.01–1994.02.28 | 0 | −0.1685 | 2.2338 |
| 1994.03.01–1994.03.31 | 1 | −1.2885 | 3.8923 |
| 1994.04.04–1994.04.29 | 0 | −0.5634 | 2.4535 |
| 1994.05.02–1994.05.31 | 0 | 0.4330 | 1.6788 |
| 1994.06.01–1994.06.30 | 1 | −0.7890 | 2.6376 |
| 1994.07.01–1994.07.29 | 0 | 0.0862 | 2.6368 |
| 1994.08.01–1994.08.31 | 1 | 0.9818 | 2.4771 |
| 1994.09.01–1994.09.30 | 0 | −0.1197 | 2.9368 |
| 1994.10.03–1994.10.31 | 0 | −0.5228 | 2.9238 |
| 1994.11.01–1994.11.30 | 1 | −0.7151 | 1.8208 |

stock market indexes in different time ranges when we chose different parameters of generalized WMF model effectively.

Our method can simulate time series with fractal dimension $D$ and Hurst exponent $H$ which may be helpful for improving the understanding of the underlying dynam-

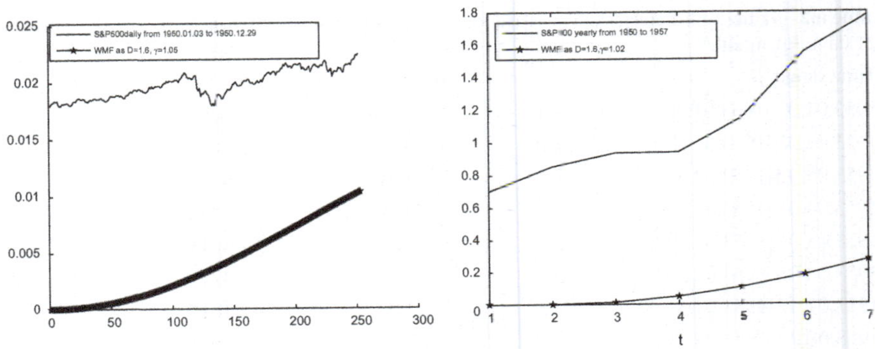


**Fig. 6.13** Simulation effectiveness of WMF for different time ranges of S& P 500 indexes

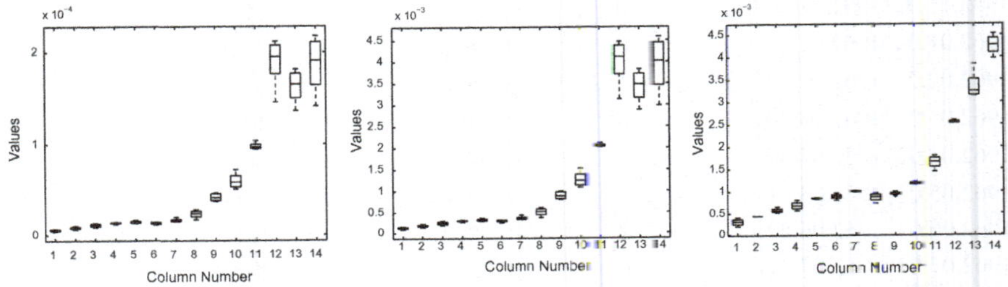


**Fig. 6.14** Variation of polyfit distribution between WMF and S& P 500 indexes at different time ranges

ical behavior of actual stock market indexes, especially for the market researchers and analysts as follows:

(1) Our method can simulate different time ranges dynamically, especially for high frequency of time series, which may be helpful for understanding the real dynamical behavior of time series.
(2) Fractal and Hurst exponent can be calculated easily, but we can not get more details from them and especially can not predict the tendency of the time series dynamically. Our methods present a simulation process to simulate the dynamical behavior of the real time series, which may be useful for the market researchers and analysts to predict the tendency of the stock markets nonlinearly.
(3) In addition, our method can be useful not only for the stock market indexes, but also for the other time series, such as time series in biology and engineering.

## 6.3 Fractal Dimension of the Controlled Julia Sets of the Output Duopoly Competing Evolution Model

An analysis of stock market was presented by the use of a simple fractal function. What's more, based on a detailed analysis of financial index behavior, the authors proposed a method to identify the stage of the current financial growth and estimated the time in which the index value was going to reach the maximum [288]. In [321], the existence of fractal basin boundaries has been examined, which is important to nonlinear economic systems. In [322], the authors proposed a coordination index of spatial distribution in the regional transportation economy by the use of the fractal theory. Combined the coordination index and other indicators, the coordination degree of regional transportation economy has been reflected more comprehensively. The spatial structure of China's vegetation has been analyzed quantitatively based on fractal theory in [323].

Generally speaking, dimension is an important feature of geometric objects, and it is the required number of independent coordinates to determine a position of geometric object. People have been accustomed to the integer dimension in Euclidean space. But in the research of clouds and some natural phenomena such as the coastline, researchers found that the dimension of these cases can not use the traditional Euclidean dimension to describe. And fractal dimension is one of the most basic variables to quantitatively represent the random shapes and phenomena with self-similarity. Fractal dimension is a very important concept in fractal theory, and it is different from the Euclidean geometry dimension that shows integer degree of freedom. Dimension can be a fraction in fractal theory, and this kind of dimension is the important concept which is introduced by the physicists when they study the theory of chaotic attractor. In order to quantitatively describe the irregularity of objective things, Hausdorff, a mathematician, studied the fractal dimension and extended the dimension from integer to fraction at the beginning of the 20th century, and it broke through the limit that topological sets' dimension were integer. And now, the measures and dimension of Hausdorff have become the core concepts and mathematical foundations in the fractal theory [324]. Fractal dimension has many different definitions for different applications, such as Hausdorff dimension, box-counting dimension, similarity dimension and information dimension, etc. In practical applications, the calculation of Hausdorff dimension is very difficult, so the box-counting dimension is often used to describe fractal. Box-counting dimension is not only convenient for practical applications, but also has important theoretical significance and it is easy to use box-counting dimension to describe the 'irregularity' of fractal [315, 325]. And in this section, we utilize the box-counting dimension to discuss.

First of all, Cournot model which was put forward in 1838 by the French economist Anthony Augustine Library, was the early oligopoly model, and lots of research has been promoted with this model. Based on cournot model, the output duopoly competing evolution model was introduced [326]. In order to get the biggest profit, sellers need to make the output decisions. Under the assumption that sellers don't have entire acquaintance about the market, they adjust their output just according to the local margin profit and the competitor's output decision. In [327], the different results were showed by the output duopoly competing evolution model from the economic chaos control which was introduced in game theory. In this part, the basic theory and method of fractal are used into the output duopoly competing evolution model to analyze this model from the fractal viewpoint.

Furthermore, in numerous fractal dimension, the definition of Hausdorff based on the Caratheodory is the oldest and it may be one of the most important definitions. For any set, the Hausdorff dimension can be defined, but it is hard to calculate or estimate its value in practice. On the contrary, box-counting dimension has become one of the most widely used dimensions because it is easy to calculate and estimate.

**Definition 6.1** ([315]) Assuming that $F$ is any non-empty bounded set on $R^d$. $N_\delta(F)$ is the minimum number of sets that can cover $F$ with the maximum diameter $\delta$. Then, the upper and lower box-counting dimension of $F$ can be defined as follows

$$\underline{\dim}_B F = \underline{\lim}_{\delta \to 0} \left( \frac{\log N_\delta(F)}{-\log \delta} \right)$$

$$\overline{\dim}_B F = \overline{\lim}_{\delta \to 0} \left( \frac{\log N_\delta(F)}{-\log \delta} \right)$$

If these two values are equal, this common value is called the box-counting dimension. And it can be written as

$$\dim_B F = \lim_{\delta \to 0} \left( \frac{\log N_\delta(F)}{-\log \delta} \right).$$

### 6.3.1 Julia Set of the Output Duopoly Competing Evolution Model

Consider the output duopoly competing evolution model as follows:

$$\begin{cases} x(t+1) = x(t) + \alpha_1 x(t)[a - b(x(t) + y(t)) - (b + 2e_1)x(t) - d_1](1 - s), \\ y(t+1) = y(t) + \alpha_2 y(t)[a - b(x(t) + y(t)) - (b + 2e_2)y(t) - d_2](1 - s). \end{cases} \tag{6.3.1}$$

where $x(t)$ and $y(t)$ are the output of two companies at time $t$ respectively, $i = 1, 2, \ldots$; $a, b$ are positive constants, and $a$ is the highest price in the market; assume that the cost functions are $C(x) = c_1 + d_1 x + e_1 x^2$ and $C(y) = c_2 + d_2 y + e_2 y^2$, and the $1st$ derivative are positive and the $2nd$ derivative are negative, so we can get $d_i, e_i > 0$; since companies' marginal profits are demanded to be less than the price of the same product in the market, we have $d_1 + 2e_1 x < a$, $d_2 + 2e_2 y < a$; $\alpha_i$ are positive parameter, which show the companies' adjustment speed. Julia set is a notation about the complex dynamical system but the model (6.3.1) is a real system and it is not irrelevant to consider Julia set. However, if we take $z(t) =$

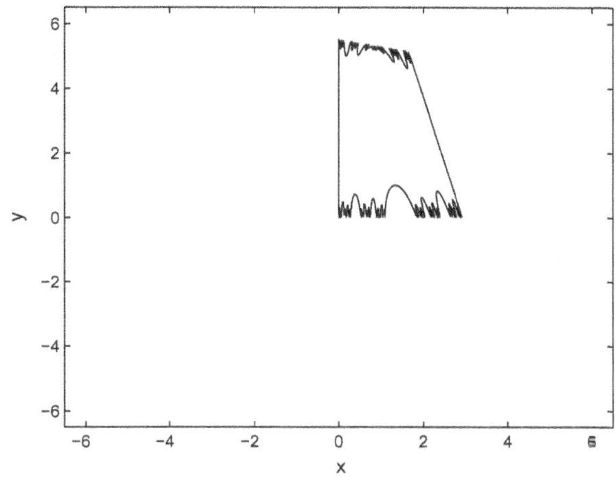


**Fig. 6.15** Original Julia sets of system

$x(t) + iy(t)$, the (6.3.1) can be seen as a map on the complex plane, and then we build the Julia set. Let $H(x, y) = (x + \alpha_1 x[a - b(x + y) - (b + 2e_1)x - d_1](1 - s), y + \alpha_2 y[a - b(x + y) - (b + 2e_2)y - d_2](1 - s))$.

**Definition 6.2** Let $K$ be the set of $(x, y)$ in $R^2$ whose trajectory is limited, i.e.

$$K = \{(x, y) \in R^2 | \{H^n(x, y)\}_{n=1}^{\infty} \text{ is bounded}\}.$$

Then the set $K$ is called the filled Julia set corresponding the map $H(x, y)$. The boundary of $K$ is called the Julia set of the map $H(x, y)$, which is denoted by $J_H$, i.e. $J_H = \partial K$.

For example, the parameters of model (6.3.1) are taken to be $a = 10, b = 1, d_1 = 1, d_2 = 1, e_1 = 1, e_2 = 1.1, s = 0.3, \alpha_1 = 0.52, \alpha_2 = 0.1$, then the corresponding Julia set is shown in Fig. 6.15.

The curve in Fig. 6.15 represents the initial output of two competitors. When we take the values inside of the curve, the two competitors' output will be stable and bounded. Otherwise, one of competitors or both competitors' output will tend to infinity, that is to say, the total supply will tend to infinity.

According to the definition of the fixed point, we can get the fixed points of system (6.3.1) as follows:

(1) $(x^*, y^*) = (0, 0)$;
(2) $x^* = 0, y^* \neq 0$. Then we have $a - by^* - (b + 2e_2)y^* - d_2 = 0$ and $(x^*, y^*) = (0, \frac{a-d_2}{2(b+e_2)})$;
(3) $x^* \neq 0, y^* = 0$. Then we have $a - bx^* - (b + 2e_1)x^* - d_1 = 0$ and $(x^*, y^*) = (\frac{a-d_1}{2(b+e_1)}, 0)$;
(4) $x^* \neq 0, y^* \neq 0$. Then we have

$$(x^*, y^*) = \left( \frac{(a-d_1)(2b+2e_2) - b(a-d_2)}{3b^2+4be_1+4be_2+4e_1e_2}, \frac{(a-d_2)(2b+2e_1) - b(a-d_1)}{3b^2+4be_1+4be_2+4e_1e_2} \right)$$

Here, we just consider the fixed point that both $x^*$ and $y^*$ are not zero. So for the example above, we can get the fixed point $(x^*, y^*) = (1.8228, 1.7089)$.

### 6.3.2 *Julia Set's Control and the Change of Its Box-Counting Dimension*

Chaos control has attracted attentions and many control methods have been introduced [328, 329]. In recent years, control of Julia sets is discussed. In this section we will consider the control of Julia sets of model (6.3.1) and compute the box-counting dimension of the controlled Julia sets to describe the complexity of Julia set and the system.

Rewrite the model (6.3.1)

$$\begin{cases} x(n+1) = x(n) + \alpha_1 x(n)[a - b(x(n) + y(n)) - (b+2e_1)x(n) - d_1](1-s), \\ y(n+1) = y(n) + \alpha_2 y(n)[a - b(x(n) + y(n)) - (b+2e_2)y(n) - d_2](1-s). \end{cases} \tag{6.3.2}$$

In this section, two control methods will be introduced to control Julia set of the model (6.3.2).

Let

$$\begin{cases} F(x_n, y_n) = \alpha_1 x_n[a - b(x_n + y_n) - (b+2e_1)x_n - d_1](1-s), \\ G(x_n, y_n) = \alpha_2 y_n[a - b(x_n + y_n) - (b+2e_2)y_n - d_2](1-s). \end{cases}$$

**Step Hysteresis Control Method**

Introduce the control items [330] into (6.3.2), we can get the following controlled system

$$\begin{cases} x_{n+1} = x_n + F(x_n, y_n) + k(x_n - x_{n-1}), n \geqslant 3, \\ y_{n+1} = y_n + G(x_n, y_n) + k(y_n - y_{n-1}), n \geqslant 3, \end{cases} \tag{6.3.3}$$

where $k$ is the control parameter. This control method starts from the third step, and the second step is

$$\begin{cases} x_2 = x_1 + \alpha_1 x_1[a - b(x_1 + y_1) - (b+2e_1)x_1 - d_1](1-s), \\ y_2 = y_1 + \alpha_2 y_1[a - b(x_1 + y_1) - (b+2e_2)y_1 - d_2](1-s). \end{cases}$$

It is clear that the fixed point of the original system is also the fixed point of the controlled system. To simplify the Jacobi matrix of the controlled system, the controlled system can be rewritten as the following form

$$\begin{cases} x_{n+1} = x_n + F(x_n, y_n) + k(x_n - s_n), n \geqslant 3, \\ y_{n+1} = y_n + G(x_n, y_n) + k(y_n - t_n), n \geqslant 3, \\ s_{n+1} = x_n, n \geqslant 3, \\ t_{n+1} = y_n, n \geqslant 3. \end{cases} \tag{6.3.4}$$

Let $f(x, y, s, t) = x + F(x, y) + k(x - s)$, $g(x, y, s, t) = y + G(x, y) + k(y - t)$, $p(x, y, s, t) = x$, $q(x, y, s, t) = y$. And the fixed point of (6.3.4) is $(x^*, y^*, s^*, t^*)$, where $s^* = x^*, t^* = y^*$. Therefore the Jacobi matrix of the controlled model (6.3.4) at this fixed point is

$$J = \frac{\partial(f, g, p, q)}{\partial(x, y, s, t)}|_{(x^*, y^*, s^*, t^*)} = \begin{bmatrix} a_1 & a_2 & -k & 0 \\ a_3 & a_4 & 0 & -k \\ 1 & 0 & 0 & 0 \\ 0 & 1 & 0 & 0 \end{bmatrix},$$

where $a_1 = (1 + k) + \alpha_1(1 - s)a - \alpha_1 d_1(1 - s) - 4\alpha_1(1 - s)(b + e_1)x^* - \alpha_1 b(1 - s) y^*$, $a_2 = -\alpha_1 b(1 - s)x^*$, $a_3 = -\alpha_2 b(1 - s)y^*$, $a_4 = (1 + k) + \alpha_2(1 - s)a - \alpha_2 d_2 (1 - s) - 4\alpha_2(1 - s)(b + e_2)y^* - \alpha_2 b(1 - s)x^*$.

And the corresponding characteristic equation is

$$|\lambda E - J| = \begin{vmatrix} \lambda - a_1 & -a_2 & k & 0 \\ -a_3 & \lambda - a_4 & 0 & k \\ -1 & 0 & \lambda & 0 \\ 0 & -1 & 0 & \lambda \end{vmatrix} = 0.$$

That is

$$\lambda^4 + c\lambda^3 + d\lambda^2 + e\lambda + h = 0,$$

where $c = -(a_1 + a_4)$, $d = (a_1 a_4 - a_2 a_3)$, $e = k(a_1 - a_4)$, $h = -k^2$.

Let $p_1 = -(3c^2 - 8d)$, $q_1 = 3c^4 + 16d^2 - 16c^2 d + 16ce - 64h$, $r = -(c^3 - 4cd + e)^2$, $A = p_1^2 - 3q_1$, $B = p_1 q_1 - 3r$, $C = q_1^2 - 3p_1 r$. According to the discriminant, the results have the following several cases.

Case 1 $A = B = 0$

Let

$$\mu_1 = -\frac{p_1}{3}, \mu_2 = -\frac{q_1}{p_1}, \mu_3 = -\frac{3r}{q_1},$$

and we have $\mu_1 = \mu_2 = \mu_3$. The solutions are

$$\lambda_1 = \frac{1}{4}(-c + \sqrt{\mu_1} + \sqrt{\mu_2} + \sqrt{\mu_3}),$$

$$\lambda_2 = \frac{1}{4}(-c + \sqrt{\mu_1} - \sqrt{\mu_2} - \sqrt{\mu_3}),$$

$$\lambda_3 = \frac{1}{4}(-c - \sqrt{\mu_1} + \sqrt{\mu_2} - \sqrt{\mu_3}),$$

$$\lambda_4 = \frac{1}{4}(-c - \sqrt{\mu_1} - \sqrt{\mu_2} - \sqrt{\mu_3}).$$

Case 2 $B^2 - 4AC = 0$
Let

$$\mu_1 = -p_1 + \nu, \mu_2 = \mu_3 = -\frac{\nu}{2},$$

where $\nu = \frac{B}{A}$, and the solutions are

$$\lambda_1 = \frac{1}{4}(-c + \sqrt{\mu_1} + \sqrt{\mu_2} + \sqrt{\mu_3}),$$

$$\lambda_2 = \frac{1}{4}(-c + \sqrt{\mu_1} - \sqrt{\mu_2} - \sqrt{\mu_3}),$$

$$\lambda_3 = \frac{1}{4}(-c - \sqrt{\mu_1} + \sqrt{\mu_2} - \sqrt{\mu_3}),$$

$$\lambda_4 = \frac{1}{4}(-c - \sqrt{\mu_1} - \sqrt{\mu_2} + \sqrt{\mu_3}).$$

Case 3 $B^2 - 4AC < 0$
Let

$$\mu_1 = -\frac{1}{3}\left(p_1 + 2A\cos\left(\frac{1}{3}\arccos T\right)\right),$$

$$\mu_2 = -\frac{1}{3}\left(p_1 + 2A\cos\left(\frac{1}{3}\arccos T + \frac{2\pi}{3}\right)\right),$$

$$\mu_3 = -\frac{1}{3}\left(p_1 + 2A\cos\left(\frac{1}{3}\arccos T - \frac{2\pi}{3}\right)\right),$$

where $T = \frac{2Ap_1 - 3B}{2A^1.5}$, and the solutions are

$$\lambda_1 = \frac{1}{4}(-c + \sqrt{\mu_1} + \sqrt{\mu_2} + \sqrt{\mu_3}),$$

$$\lambda_2 = \frac{1}{4}(-c + \sqrt{\mu_1} - \sqrt{\mu_2} - \sqrt{\mu_3}),$$

$$\lambda_3 = \frac{1}{4}(-c - \sqrt{\mu_1} + \sqrt{\mu_2} - \sqrt{\mu_3}),$$

$$\lambda_4 = \frac{1}{4}(-c - \sqrt{\mu_1} - \sqrt{\mu_2} + \sqrt{\mu_3}).$$

Case 4 $B^2 - 4AC > 0$

Let

$$\Delta_1 = d^2 - ce + 12eh,\ \Delta_2 = 2d^3 - 9cde + 27e^2 + 27c^2h - 72dh,$$

$$\Delta = \frac{\sqrt[3]{2}\Delta_1}{3\sqrt[3]{\Delta_2 + \sqrt{-4\Delta_1^3 + \Delta_2^2}}} + \frac{3\sqrt[3]{\Delta_2 + \sqrt{-4\Delta_1^3 + \Delta_2^2}}}{3\sqrt[3]{2}a}.$$

Then the solutions are

$$\lambda_1 = -\frac{c}{4} - \frac{1}{2}\sqrt{\frac{c^2}{4} - \frac{2d}{3} + \Delta} - \frac{1}{2}\sqrt{\frac{c^2}{2} - \frac{4d}{3} - \Delta - \frac{-c^3 + 4cd - 8e}{4\sqrt{\frac{c^2}{4} - \frac{2d}{3} + \Delta}}},$$

$$\lambda_2 = -\frac{c}{4} - \frac{1}{2}\sqrt{\frac{c^2}{4} - \frac{2d}{3} + \Delta} + \frac{1}{2}\sqrt{\frac{c^2}{2} - \frac{4d}{3} - \Delta - \frac{-c^3 + 4cd - 8e}{4\sqrt{\frac{c^2}{4} - \frac{2d}{3} + \Delta}}},$$

$$\lambda_3 = -\frac{c}{4} + \frac{1}{2}\sqrt{\frac{c^2}{4} - \frac{2d}{3} + \Delta} - \frac{1}{2}\sqrt{\frac{c^2}{2} - \frac{4d}{3} - \Delta - \frac{-c^3 + 4cd - 8e}{4\sqrt{\frac{c^2}{4} - \frac{2d}{3} + \Delta}}},$$

$$\lambda_4 = -\frac{c}{4} + \frac{1}{2}\sqrt{\frac{c^2}{4} - \frac{2d}{3} + \Delta} + \frac{1}{2}\sqrt{\frac{c^2}{2} - \frac{4d}{3} - \Delta - \frac{-c^3 + 4cd - 8e}{4\sqrt{\frac{c^2}{4} - \frac{2d}{3} + \Delta}}}.$$

Above all, no matter what the solution of this quantic equation is, the fixed point $(x^*, y^*, s^*, t^*)$ is attractive as long as the value of $k$ can satisfy the inequality: $|\lambda_1| < 1$, $|\lambda_2| < 1$, $|\lambda_3| < 1$ and $|\lambda_4| < 1$. Then we say that the control of the system is accomplished.

For example, we take the parameters in (6.3.2) as $a = 10$, $b = 1$, $d_1 = 1$, $d_2 = 1$, $e_1 = 1$, $e_2 = 1.1$, $s = 0.3$, $\alpha_1 = 0.52$, $\alpha_2 = 0.1$, and the corresponding original Julia set is shown in Fig. 6.15. The fixed point of the system is $(x^*, y^*, s^*, t^*) = (1.8228, 1.7089, 1.8228, 1.7089)$, so according to the previous discussion, the Jacobi matrix of the control model (6.3.4) at the fixed point is

$$J = \begin{bmatrix} -1.654 + k & -0.6635 & -k & 0 \\ -0.1196 & 0.4976 + k & 0 & -k \\ 1 & 0 & 0 & 0 \\ 0 & 1 & 0 & 0 \end{bmatrix}.$$

Then, the corresponding characteristic equation is

$$|\lambda E - J| = \begin{vmatrix} \lambda - k + 1.654 & 0.6635 & k & 0 \\ 0.1196 & \lambda - k - 0.4976 & 0 & k \\ -1 & 0 & \lambda & 0 \\ 0 & -1 & 0 & \lambda \end{vmatrix} = 0.$$

That is

$$\lambda^4 + c^*\lambda^3 + d^*\lambda^2 + e^*\lambda + h^* = 0,$$

where $c^* = -(2k - 1.1564), d^* = k^2 - 1.1564k - 0.9024, e^* = -2.1516k^2, h^* = -k^2$. Let $p_1^* = 3c^2 - 8d), q_1^* = 3c^4 + 16d^2 - 16c^2d + 16ce - 64h, r^* = -(c^3 - 4cd + e)^2, A_1 = p_1^2 - 3q_1, B_1 = p_1q_1 - 3r, C_1 = q_1^2 - 3p_1r$.

Let $D = B_1^2 - 4A_1C_1$. And with the changes of the values of the control parameter $k$, we can get solutions by the four cases above.

The corresponding Julia sets are shown in Fig. 6.16.

The changes of box-counting dimension of the controlled Julia set with the changes of the values of the controlled parameter $k$ are shown in Table 6.3 or Fig. 6.17.

**The Optimal Function Control**

Introduce the control items [331] into system (6.3.2), we can get the following system

$$\begin{cases} x_{n+1} = x_n + F(x_n, y_n) + \theta(x_n + F(x_n, y_n) - x_n), \\ y_{n+1} = y_n + G(x_n, y_n) + \theta(y_n - G(x_n, y_n) - y_n), \end{cases}$$

where $\theta$ is the control parameter.

Let $h(x, y) = x + F(x, y) + \theta F(x, y), l(x, y) = y + G(x, y) - \theta G(x, y)$.

The fixed point does not need to know in this control method. But in order to analyze the stability of the fixed point, we also give the value of the fixed point denoted by $(x^*, y^*)$. So, the Jacobi matrix of the control model at the fixed point is

$$J = \begin{bmatrix} \frac{\partial h}{\partial x} & \frac{\partial h}{\partial y} \\ \frac{\partial l}{\partial x} & \frac{\partial l}{\partial y} \end{bmatrix}_{(x^*, y^*)} = \begin{bmatrix} c_{11} & c_{12} \\ c_{21} & c_{22} \end{bmatrix},$$

where $c_{11} = 1 + (1 + \theta)(\alpha_1(1 - s)a - \alpha_1 d_1(1 - s) - 4\alpha_1(1 - s)(b + e_1)x^* - \alpha_1 b(1 - s)y^*)$, $c_{12} = -(1 + \theta)(\alpha_1 b(1 - s)x^*)$, $c_{21} = -(1 + \theta)(\alpha_2 b(1 - s)y^*)$, $c_{22} = 1 + (1 + \theta)(\alpha_2(1 - s)a\alpha_2 d_2(1 - s) - 4\alpha_2(1 - s)(b + e_2)y^* - \alpha_2 b(1 - s)x^*)$. The corresponding characteristic equation is

$$|\lambda E - J| = \begin{vmatrix} \lambda - c_{11} & -c_{12} \\ -c_{21} & \lambda - c_{22} \end{vmatrix} = 0,$$

that is,

$$\lambda^2 - (c_{11} + c_{22})\lambda + (c_{11}c_{22} - c_{12}c_{21}) = 0.$$

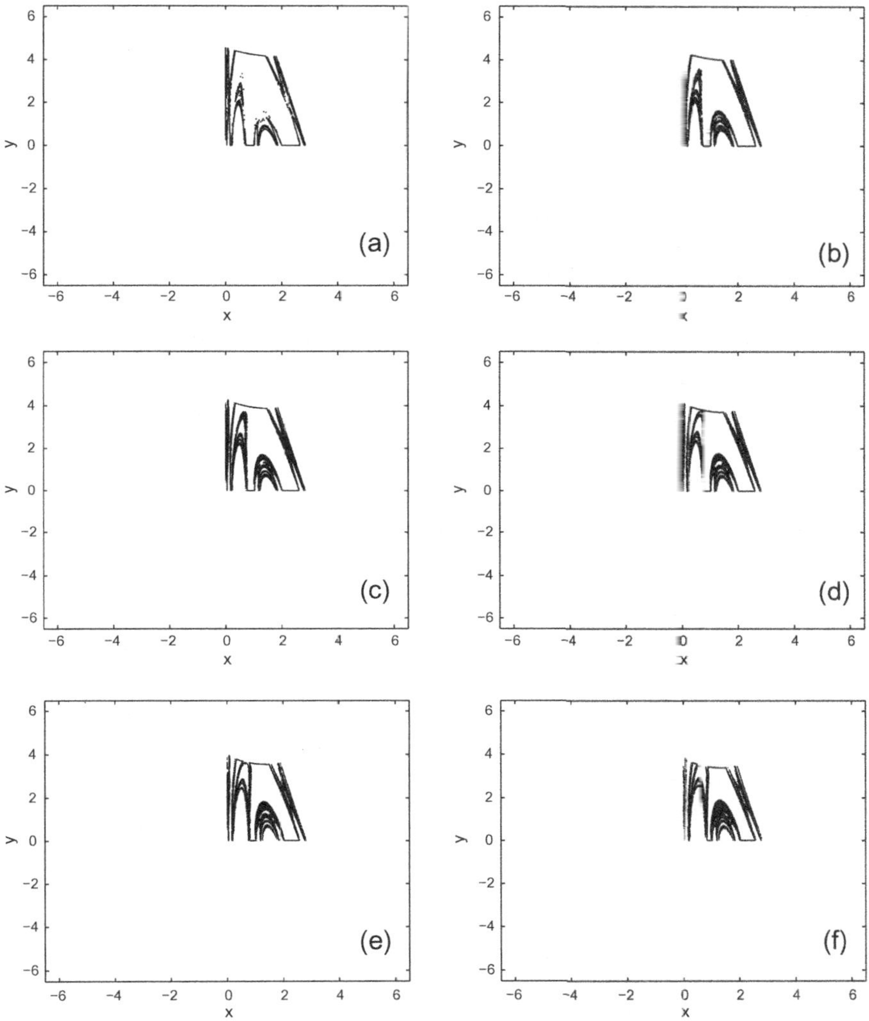


**Fig. 6.16** The changing of Julia sets of the controlled system **a** $k = 0.401$; **b** $k = 0.451$; **c** $k = 0.501$; **d** $k = 0.549$; **e** $k = 0.599$; **f** $k = 0.651$

Solving this equation, we can get $\lambda_1 = \frac{c_{11}+c_{22}+\sqrt{\Delta'}}{2}$, $\lambda_2 = \frac{c_{11}+c_{22}-\sqrt{\Delta'}}{2}$, where $\Delta' = (c_{11}+c_{22})^2 - 4(c_{11}c_{22} - c_{12}c_{21})$. When the values of $k$ satisfy the inequality $|\lambda_1| < 1$ and $|\lambda_2| < 1$, the fixed point $(x^*, y^*)$ is stable. Then we say that the control of the system is accomplished.

For example, we take the parameters in (6.3.2) as $a = 10, b = 1, d_1 = 1, d_2 = 1, e_1 = 1, e_2 = 1.1, s = 0.3, \alpha_1 = 0.52, \alpha_2 = 0.1$, and the original Julia set is shown in Fig. 6.15. The fixed point of the system can be calculated from the third part

**Table 6.3** The changes of box-counting dimension with changes of $k$

| $k$ | 0.400 | 0.405 | 0.410 | 0.415 | 0.420 | 0.425 | 0.430 | 0.435 | 0.440 | 0.445 |
|---|---|---|---|---|---|---|---|---|---|---|
| $hw$ | 1.1123 | 1.1343 | 1.3673 | 1.1766 | 1.3756 | 1.2058 | 1.3826 | 1.2061 | 1.3855 | 1.2101 |
| $k$ | 0.450 | 0.455 | 0.460 | 0.465 | 0.470 | 0.475 | 0.480 | 0.485 | 0.490 | 0.495 |
| $hw$ | 1.3891 | 1.2304 | 1.3918 | 1.2455 | 1.3921 | 1.2435 | 1.3929 | 1.2636 | 1.3935 | 1.27[illegible]2 |
| $k$ | 0.500 | 0.505 | 0.510 | 0.515 | 0.520 | 0.525 | 0.530 | 0.535 | 0.540 | 0.545 |
| $hw$ | 1.3928 | 1.2675 | 1.3928 | 1.2738 | 1.3938 | 1.2771 | 1.3939 | 1.3022 | 1.3941 | 1.2789 |
| $k$ | 0.550 | 0.555 | 0.560 | 0.565 | 0.570 | 0.575 | 0.580 | 0.585 | 0.590 | 0.595 |
| $hw$ | 1.3941 | 1.2708 | 1.3964 | 1.2882 | 1.3960 | 1.2859 | 1.3972 | 1.2865 | 1.3965 | 1.28[illegible]2 |
| $k$ | 0.60 | 0.605 | 0.610 | 0.615 | 0.620 | 0.625 | 0.630 | 0.635 | 0.640 | 0.645 |
| $hw$ | 1.3962 | 1.2759 | 1.3963 | 1.2709 | 1.3967 | 1.2762 | 1.3967 | 1.2751 | 1.3969 | 1.2759 |
| $k$ | 0.650 | 0.655 | 0.660 | 0.665 | 0.670 | 0.675 | 0.680 | 0.685 | 0.690 | 0.695 |
| $hw$ | 1.3970 | 1.2779 | 1.3967 | 1.2828 | 1.3961 | 1.2698 | 1.3953 | 1.2812 | 1.3962 | 1.26[illegible]1 |

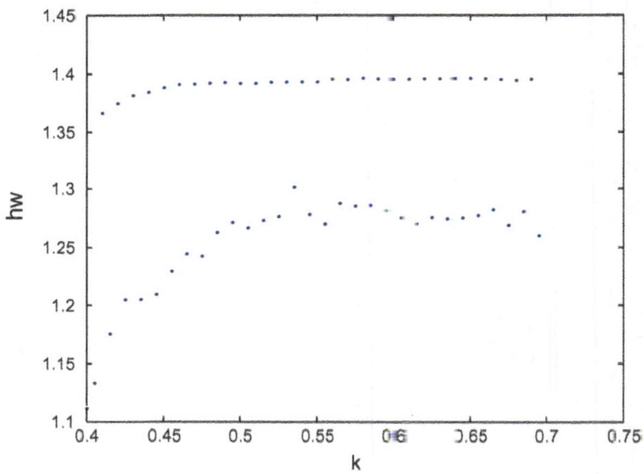


**Fig. 6.17** The scatter diagram with the change of the control parameter and box-counting dimension

$(x^*, y^*) = (1.8228, 1.7089)$. So according to the above, the Jacobi matrix of the control model at the fixed point is

$$J = \begin{bmatrix} 1 - 2.654(1+\theta) & -0.6535(1+\theta) \\ -0.1196(1+\theta) & 1 - 0.5024(1+\theta) \end{bmatrix}$$

Therefore, the characteristic equation is

$$|\lambda E - J| = \begin{vmatrix} \lambda - (1 - 2.654(1+\theta)) & 0.6535(1+\theta) \\ 0.1196(1+\theta) & \lambda - (1 - 0.5024(1+\theta)) \end{vmatrix} = 0,$$

that is,

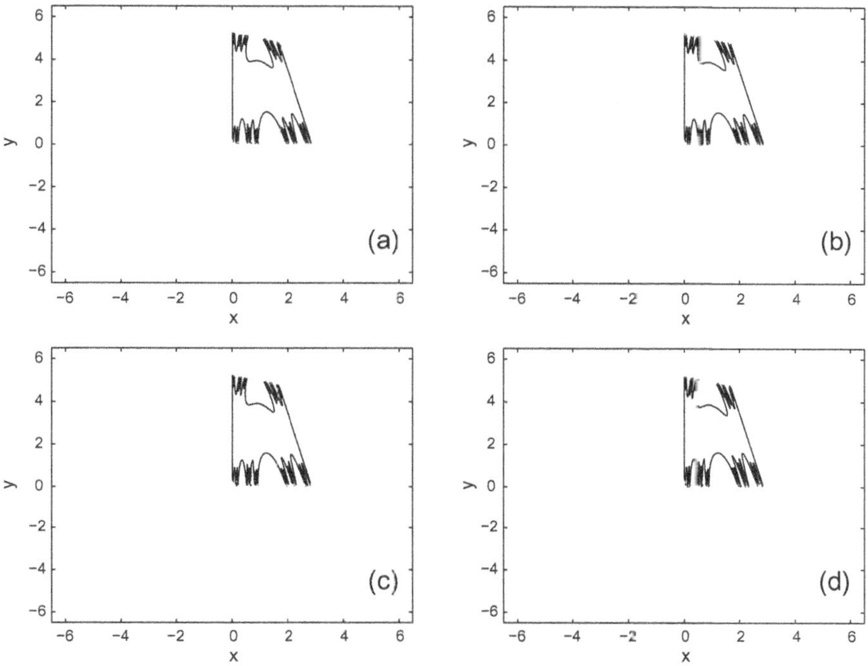

**Fig. 6.18** The changing of Julia sets of the controlled system **a** $\theta = 0.0810$; **b** $\theta = 0.0860$; **c** $\theta = 0.0910$; **d** $\theta = 0.1010$

**Table 6.4** The changes of box-counting dimension with changes of *k*

| $\theta$ | 0.0800 | 0.0805 | 0.0810 | 0.0815 | 0.0820 | 0.0825 | 0.0830 | 0.0835 | 0.0840 | 0.0845 |
|---|---|---|---|---|---|---|---|---|---|---|
| *hw* | 1.0795 | 1.0848 | 1.0815 | 1.0868 | 1.087 | 1.0655 | 1.086 | 1.0882 | 1.0864 | 1.0809 |
| $\theta$ | 0.0850 | 0.0855 | 0.0860 | 0.0865 | 0.0870 | 0.0875 | 0.0880 | 0.0885 | 0.0890 | 0.0895 |
| *hw* | 1.0858 | 1.0932 | 1.0829 | 1.0889 | 1.1113 | 1.1019 | 1.0646 | 1.0675 | 1.0688 | 1.0674 |
| $\theta$ | 0.0900 | 0.0905 | 0.0910 | 0.0915 | 0.0920 | 0.0925 | 0.0930 | 0.0935 | 0.0940 | 0.0945 |
| *hw* | 1.0604 | 1.0669 | 1.0716 | 1.0778 | 1.0831 | 1.0834 | 1.0848 | 1.1006 | 1.095 | 1.095 |
| $\theta$ | 0.0950 | 0.0955 | 0.0960 | 0.0965 | 0.0970 | 0.0975 | 0.0980 | 0.0985 | 0.0990 | 0.0995 |
| *hw* | 1.0927 | 1.0876 | 1.0946 | 1.0939 | 1.1037 | 1.1036 | 1.1153 | 1.0978 | 1.1175 | 1.1075 |

$$\lambda^2 + d_1\lambda + d_2 = 0,$$

where $d_1 = -(2 - 3.1564(1+\theta))$, $d_2 = 1 - 3.1564(1+\theta) + 1.254(1+\theta)^2$. Solving this equation, we have $\lambda_1 = \frac{-d_1+\sqrt{\Delta'}}{2}$ and $\lambda_2 = \frac{-d_1-\sqrt{\Delta'}}{2}$, where $\Delta' = d_1^2 - 4d_2$. When the values of $\theta$ satisfy the inequality $|\lambda_1| < 1$ and $|\lambda_2| < 1$, we say that the control of the system is accomplished.

According to the values of the control parameter $\theta$, the corresponding Julia sets are shown in Fig. 6.18.

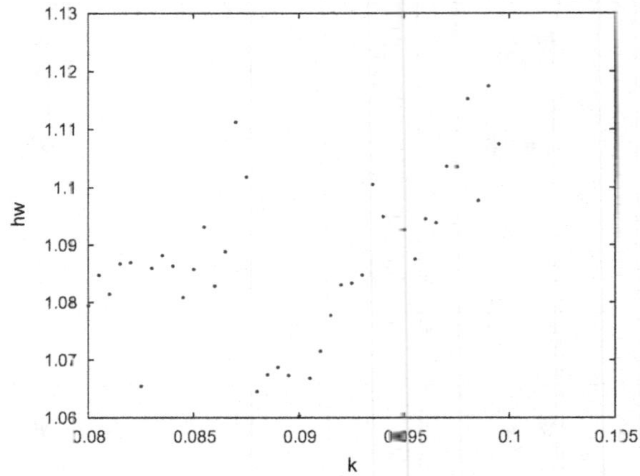


**Fig. 6.19** The scatter diagram with the change of the control parameter and box-counting dimension

The changes of the box-counting dimension of the controlled Julia set with the changes of the values of the control parameter $\theta$ are shown Table 6.4.

Though the values of $\theta$ we choose in this article are limited, but from Fig. 6.19, the change of the box-counting dimension with $\theta$ is heterogeneous, but the change tendency is that the box-counting dimension gets bigger with the value of $\theta$ gets bigger.

# Chapter 7
# Control of Julia Sets in Complex Physical Systems

Nonlinear elements are common in structural and rotating machinery applications. When such systems are subjected to periodic external excitation, they present many complex dynamic behaviors. In the past four decades, much work has been done in nonlinear dynamic systems theory. Control and analysis of the forced nonlinear systems play an important role in understanding the dynamical phenomena for mechanical, chemical and biological systems. It is a central problem in modern nonlinear dynamics theory that small external disturbances may change the various induced transitions through periodic or more complicated chaotic regimes. For example, in a forced single-degree-of-freedom weakly nonlinear system, primarily, sub- and super-harmonic resonances may occur if the linear natural frequency and the frequency of an external excitation satisfy a certain commensurate or nearly commensurate relation [332].

In nonlinear systems, the reaction-diffusion systems arise frequently in the research of chemical and biological phenomena. The dynamics of reaction-diffusion systems have been hot topics since they exhibit rich dynamic behavior including periodic and quasiperiodic solutions and chaos. Brusselator model is a famous model of chemical reactions with oscillations, which was proposed by Prigogine and Lefever in 1968. Extensive bifurcation studies of Brusselator model showed that the system exhibits rich dynamic behavior for different regions in the parameter space. It is considered to be one of the simplest reaction-diffusion models exhibiting Turing and Hopf instabilities.



S. T. Liu et al., *Fractal Control and Its Applications*,
https://doi.org/10.1007/978-981-15-5459-9_7

## 7.1 Fractal Dimension and Synchronization of the Controlled Julia Sets of a Reaction-Diffusion System

It is known that self-oscillating systems are ubiquitous in many branches of engineering and science and subjected to a periodic forcing, and they present an important class of coupled oscillators. A feature of periodically forced systems is that the frequency and amplitude of the forcing are external and can therefore be controlled. In this part, the forced Brusselator model is studied. The non-autonomous forced Brusselator chemical reaction is the classical Brusselator system with a small intensity, and the system is taken as

$$\begin{cases} \frac{du_1}{dt} = A - (B+1)u_1 + u_1^2 u_2 + \alpha \cos(\omega t), \\ \frac{du_2}{dt} = Bu_1 - u_1^2 u_2 \end{cases} \tag{7.1.1}$$

There are abundant dynamical properties about system (7.1.1) [333–335].

Julia set is one of the important sets in fractal theory [315]. In the past few years, properties and graphs of Julia set were studied. What's more, applications of Julia set are also discussed, such as the strange attractor and the boundary of the attractive domain which are related to the chaotic dynamic systems [336–338]. Based on the typical Langevin problem and dynamics of the particle's movement, the physical applications of Julia sets are given [175, 177], which shows that the fractal structure of Julia sets can visually reflect the change rule of the particle's velocity and describe the closure of the union of all possible unstable periodic orbits of the particle. And also, theories of the Julia set of Newton's method, Halley's method, and Schroder's method are studied in [192, 339, 340].

Control and synchronization of chaos are discussed extensively. Many control methods are designed. For example, activation feedback method and projective synchronization are taken for fractional order chaotic systems [341, 342]. An observer-based fuzzy neural sliding mode control scheme for interconnected unknown chaotic systems is developed [343] and the unknown nonlinearities of the interconnected chaotic systems are not restricted to the systems output only. Practically, in applications, the behaviors depicted by Julia set need to show different forms or one behavior under control becomes to be another one. Recently, Julia sets for various systems are discussed [176, 182, 183, 344] and control and synchronization of Julia sets are also studied [216, 230]. In [345], feedback control and synchronization of Julia sets of the forced Brusselator model are discussed. In addition, fractal dimension is usually one of the basic characters to depict the space availabilities and irregularities of the complex sets. It is a more similar description for the true property and station of the complex systems.

### 7.1.1 Preliminary

There are some autonomous systems which are equivalent to (7.1.1) and the following form is taken in this section since it is stable fast

$$\begin{cases} \frac{du_1}{dt} = A - (B+1)u_1 + u_1^2 u_2 + \alpha \cos(\omega t), \\ \frac{du_2}{dt} = Bu_1 - u_1^2 u_2, \\ \frac{du_3}{dt} = -\omega u_4, \\ \frac{du_4}{dt} = \omega u_3, \end{cases} \tag{7.1.2}$$

and the initial values of $u_3$ and $u_4$ are $u_3(0) = 1$ and $u_4(0) = 0$. For the derivative in time we use

$$\dot{u_i} \rightarrow \frac{u_i(t + \Delta t) - u_i(t)}{\Delta t}, i = 1, 2, 3, 4$$

Substitute these discretizations in the system (7.1.2), then the discrete version is

$$\begin{cases} x_{n+1} = a + (1 - b - r)x_n + rx_n^2 y_n + cz_n, \\ y_{n+1} = y_n + bx_n - rx_n^2 y_n, \\ z_{n+1} = z_n - d\omega_n, \\ \omega_{n+1} = \omega_n + dz_n, \end{cases} \tag{7.1.3}$$

and $z_0 = 1, \omega_0 = 0$.

Now, Julia set of the real dynamical system (7.1.3) is introduced by the use of the method in fractal theory for the complex system.

**Definition 7.1** Let $K$ be the set of $(x, y, 1, 0)$ in $R^4$ whose trajectory is limited, i.e.

$$K = \{(x, y, 1, 0) : \{f^n(x, y, 1, 0)\}_{n \geq 1} \text{ is bounded}\}.$$

Then the set $K$ is called the filled Julia set corresponding to the map $f$. The boundary of $K$ is called the Julia set of the map $f$, which is denoted by $J_f$, i.e. $J_f = \partial K$.

Evidently, from the definition of Julia set that it is a subset in the four dimensional space $R^4$. It is not convenient to show the four dimensional set in simulations. However, the initial values of the third and the forth variables are fixed. Therefore, Julia set of the system (7.1.3) can be shown in the $x - y$ plane (the initial values of the first and the second variables in (7.1.3)). It is easy to see that the boundedness of the iterative trajectory of the system plays a key role in the structure of Julia set from Definition 7.1. Therefore, Julia sets can be controlled by changing the iterative trajectory. For example, the stability of the fixed point can be considered.

Moreover, fractal dimension is one of the important tools to study systems since it can provide much information about their complexity. Hausdorff dimension is also a fractal dimension, but it is usually used in theoretical analysis. In practical applications, the box-counting dimension is frequently used since it is the one that can be calculated when working with a finite range of scales, see Definition 6.1.

In order to discuss the modulus of the eigenvalues of the Jacobi matrix of the system at the fixed point, the following lemma is given.

**Lemma 7.1** ([346]) *Let $F(\lambda) = \lambda^2 + B\lambda + C$. Suppose that $F(1) > 0$, $\lambda_1$ and $\lambda_2$ are two roots of $F(\lambda) = 0$. Then*

*(1) $|\lambda_1| < 1$ and $|\lambda_2| < 1$ if and only if $F(-1) > 0$ and $C < 1$;*
*(2) $|\lambda_1| < 1$ and $|\lambda_2| > 1$ (or $|\lambda_1| > 1$ and $|\lambda_2| < 1$) if and only if $F(-1) < 0$;*
*(3) $|\lambda_1| > 1$ and $|\lambda_2| > 1$ if and only if $F(-1) > 0$ and $C > 1$;*
*(4) $\lambda_1 = -1$ and $|\lambda_2| \neq 1$ if and only if $F(-1) = 0$ and $B \neq 0, 2$;*
*(5) $\lambda_1$ and $\lambda_2$ are complex and $|\lambda_1| = |\lambda_2| = 1$ if and only if $B^2 - 4C < 0$ and $C = 1$.*

Let $\lambda_1$ and $\lambda_2$ be two roots of the characteristic equation, which are called the eigenvalues of the fixed point $(x, y)$. $(x, y)$ is called a sink if $|\lambda_1| < 1$ and $|\lambda_2| < 1$. A sink is locally asymptotic stable. $(x, y)$ is called a source if $|\lambda_1| > 1$ and $|\lambda_2| > 1$. A source is locally unstable. $(x, y)$ is called a saddle if $|\lambda_1| > 1$ and $|\lambda_2| < 1$ (or $|\lambda_1| > 1$ and $|\lambda_2| < 1$). And $(x, y)$ is called non-hyperbolic if either $|\lambda_1| = 1$ or $|\lambda_2| = 1$.

### 7.1.2 Control of Julia Set of System (7.1.3)

In this section, we will realize the control of Julia set of system (7.1.3) by combining the parameter perturbation control method with feedback control method and compute the box-counting dimension of Julia sets of the controlled system (7.1.4) for different control parameter $k$.

First, the control of Julia set of system (7.1.3) is studied. Since Julia set is also the boundary of the attractive domain of the fixed point, the stability of the fixed point could be considered in order to control Julia set of the system.

It is easy to know that the fixed point of system (7.1.3) is $(x^*, y^*, z^*, \omega^*) = \left(\frac{a}{r}, \frac{b}{a}, 0, 0\right)$. Combining the parameter perturbation control method with feedback control to system (7.1.3), the controlled system is obtained

$$\begin{cases} x_{n+1} = k[a + (1 - b - r)x_n + r x_n^2 y_n + c z_n] + (1 - k)x_n, \\ y_{n+1} = k[y_n + b x_n - r x_n^2 y_n] + (1 - k)y_n, \\ z_{n+1} = z_n - d\omega_n + l(z_n - z^*), \\ \omega_{n+1} = \omega_n + d z_n + l(\omega_n - \omega^*), \end{cases} \tag{7.1.4}$$

where $0 < k < 1$ and $l$ are control parameters. Obviously, the controlled system (7.1.4) reduces to the original system (7.1.3) if $k = 1$ and $l = 0$.

The characteristic equation of the Jacobi matrix at the fixed point $(x^*, y^* z^*, \omega^*)$ is

$$\left[\lambda^2 - \lambda\left(2 + kb - kr - k\frac{a^2}{r}\right) + (1 + kb - kr)\left(1 - k\frac{a^2}{r}\right) + k^2\frac{a^2 b}{r}\right]\left[(\lambda - 1 - l)^2 + d^2\right] = 0.$$

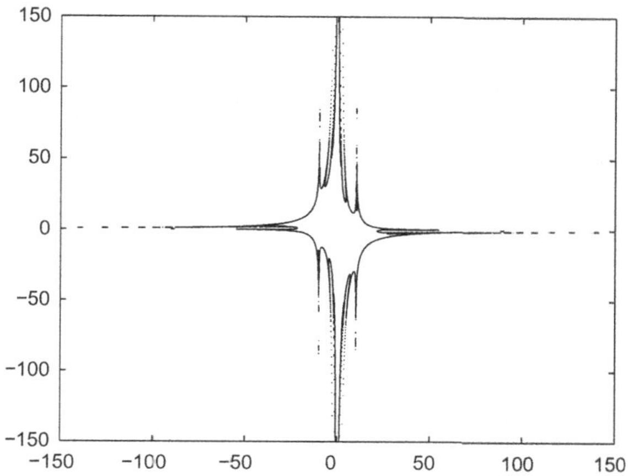


**Fig. 7.1** Julia set of system (7.1.3) when $a = 0.07, b = 0.02, r = 0.01, c = 0.2, d = 1.501$

One of the conditions to guarantee the fixed point attractive is that the modules of the eigenvalues of Jacobi matrix are less than 1. From $(\lambda - 1 - l)^2 + d^2 = 0$, we have

$$\lambda_{1,2} = (1 + l) \pm |d| i.$$

Then the region of the control parameter $l$ is $\{l : -1 - \sqrt{1 - d^2} < l < -1 + \sqrt{1 - d^2}\}$ from $|\lambda_{1,2}| < 1$. Let

$$h(\lambda) = \lambda^2 - \lambda \left(2 + kb - kr - k\frac{a^2}{r}\right) + (1 + kb - kr)\left(1 - k\frac{a^2}{r}\right) + k^2\frac{a^2 b}{r}$$

and $\lambda_{3,4}$ be the roots of $h(\lambda) = 0$. From Lemma 7.1, we know that $|\lambda_{3,4}| < 1$ if and only if $h(1) > 0$, $h(-1) > 0$ and $1 + kb - kr - k\frac{a^2}{r} + k^2 a^2 < 1$. Then the region of the control parameter $k$ is $\{k : 4 + 2(kb - br - k\frac{a^2}{r}) + k^2 a^2 > 0$ and $kb - kr - k\frac{a^2}{r} + k^2 a^2 < 0\}$.

Take $a = 0.07$, $b = 0.02$, $r = 0.01$, $c = 0.2$, $d = 1.501$ in system (7.1.3), then the fixed point is $(7, \frac{2}{7}, 0, 0)$. And Julia set is shown in Fig. 7.1.

From $|\lambda_i| < 1$ $(i = 1, 2, 3, 4)$, the ranges of the control parameters are obtained: $\{l : -1.8660 < l < -0.1340\}$ and $\{k : 0 < k < 4.2593\}$. Since $k \in (0, 1)$, the region of the control parameter $k$ is $\{k : 0 < k < 1\}$.

In order to see the changing of Julia sets of the controlled system with the changing of the control parameters, some special values of the control parameters are chosen (see Fig. 7.2).

Besides revealing changes of Julia sets of the controlled system by geometry graphs, the values of box-counting dimension of Julia sets corresponding to different control parameters $k$ are obtained (see Fig. 7.3). The following Tab illustrates the box-counting dimension of Julia sets of the controlled system (7.1.4) for different values of $k$ (Table 7.1).

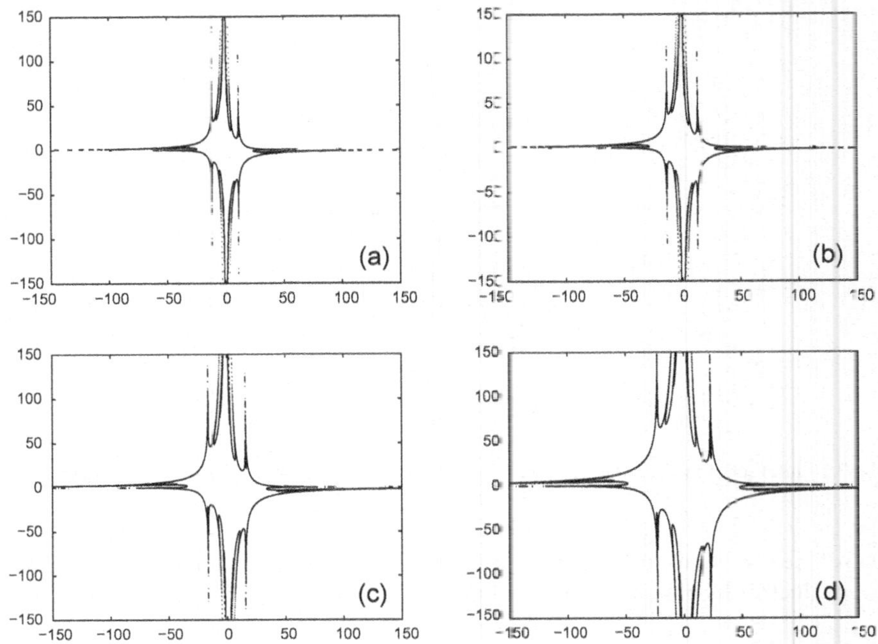


**Fig. 7.2** The changing of Julia sets of the controlled system (7.1.6) when $a = 0.07, b = 0.02, r = 0.01, c = 0.2, d = 1.501$. (**a**) $k = 0.801,\ l = -0.5$; (**b**) $k = 0.599,\ l = -0.5$; (**c**) $k = 0.401,\ l = -0.5$; (**d**) $k = 0.199,\ l = -0.5$

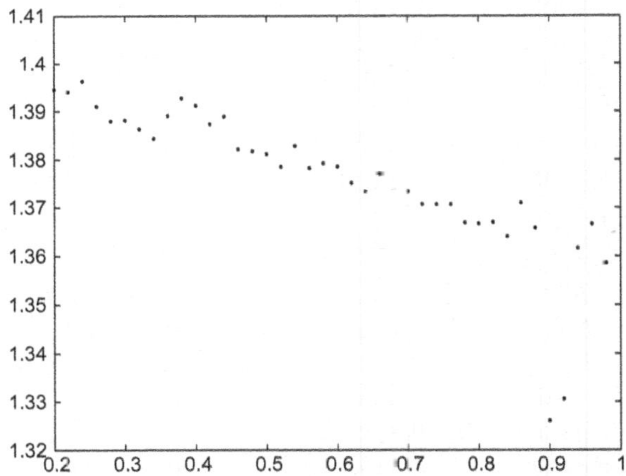

**Fig. 7.3** Box-counting dimensions of Julia sets with different control parameters $k$

**Table 7.1** Box-counting dimensions (BD) of Julia sets with different control parameters $k$

| $k$ | 0.20 | 0.22 | 0.24 | 0.26 | 0.28 | 0.30 | 0.32 | 0.34 | 0.36 | 0.38 |
|---|---|---|---|---|---|---|---|---|---|---|
| $BD$ | 1.3945 | 1.3940 | 1.3963 | 1.3910 | 1.3880 | 1.3882 | 1.3864 | 1.3344 | 1.3891 | 1.3927 |
| $k$ | 0.40 | 0.42 | 0.44 | 0.46 | 0.48 | 0.50 | 0.52 | 0.54 | 0.56 | 0.58 |
| $BD$ | 1.3912 | 1.3874 | 1.3890 | 1.3822 | 1.3818 | 1.3812 | 1.3786 | 1.3329 | 1.3783 | 1.3793 |
| $k$ | 0.60 | 0.62 | 0.64 | 0.66 | 0.68 | 0.70 | 0.72 | 0.74 | 0.76 | 0.78 |
| $BD$ | 1.3786 | 1.3752 | 1.3734 | 1.3771 | 1.3738 | 1.3734 | 1.3707 | 1.3707 | 1.3707 | 1.3669 |
| $k$ | 0.80 | 0.82 | 0.84 | 0.86 | 0.88 | 0.90 | 0.92 | 0.94 | 0.96 | 0.98 |
| $BD$ | 1.3667 | 1.3670 | 1.3641 | 1.3710 | 1.3658 | 1.3261 | 1.3306 | 1.3517 | 1.3666 | 1.3586 |

### *7.1.3 Synchronization of Julia Sets of System (7.1.3)*

Consider a system with the same form as (7.1.3) but with different parameters

$$\begin{cases} u_{n+1} = a + (1 - b - r')u_n + r'u_n^2 v_n + cs_n \\ v_{n+1} = v_n + bu_n - r'u_n^2 v_n, \\ s_{n+1} = s_n - dt_n, \\ t_{n+1} = t_n + ds_n, \end{cases} \tag{7.1.5}$$

where $r' \neq r$ and $s_0 = 1, t_0 = 0$. In order to make two different Julia sets associate with each other, four coupled items $h_i, (i = 1, 2, 3, 4)$ are introduced into (7.1.5) respectively. Then the controlled system is

$$\begin{cases} \overline{u}_{n+1} = a + (1 - b - r')\overline{u}_n + r'\overline{u}_n^2 \overline{v}_n + c\overline{s}_n + h_1, \\ \overline{v}_{n+1} = \overline{v}_n + b\overline{u}_n - r'\overline{u}_n^2 \overline{v}_n + h_2, \\ \overline{s}_{n+1} = \overline{s}_n - d\overline{t}_n + h_3, \\ \overline{t}_{n+1} = \overline{t}_n + d\overline{s}_n + h_4. \end{cases} \tag{7.1.6}$$

Let the errors of each variables in systems (7.1.3) and (7.1.6) be $e_n^1 = \overline{u}_n - x_n$, $e_n^2 = \overline{v}_n - y_n$, $e_n^3 = \overline{s}_n - z_n$, $e_n^4 = \overline{t}_n - \omega_n$ respectively. In system (7.1.6), $h_i, (i = 1, 2, 3, 4)$ are designed to be

$$h_1 = -r'\overline{u}_n^2 \overline{v}_n + rx_n^2 y_n - (r' + r)x_n + k_{11}e_n^1 + k_{12}e_n^2 + k_{13}e_n^3 + k_{14}e_n^4,$$

$$h_2 = r'\overline{u}_n^2 \overline{v}_n - rx_n^2 y_n + k_{21}e_n^1 + k_{22}e_n^2 + k_{23}e_n^3 + k_{24}e_n^4,$$

$$h_3 = k_{31}e_n^1 + k_{32}e_n^2 + k_{33}e_n^3 + k_{34}e_n^4,$$

$$h_4 = k_{41}e_n^1 + k_{42}e_n^2 + k_{43}e_n^3 + k_{44}e_n^4,$$

respectively, then

$$e_{n+1}^1 = \overline{u}_{n+1} - x_{n+1} = (k_{11} + 1 - b - r')e_n^1 + k_{12}e_n^2 + (k_{13} + c)e_n^3 - k_{14}e_n^4,$$

$$e_{n+1}^2 = (k_{21} + b)e_n^1 + (k_{22} + 1)e_n^2 + k_{23}e_n^3 + k_{24}e_n^4,$$

$$e_{n+1}^3 = k_{31}e_n^1 + k_{32}e_n^2 + (k_{33} + 1)e_n^3 + (k_{34} - d)e_n^4,$$

$$e_{n+1}^4 = k_{41}e_n^1 + k_{42}e_n^2 + (k_{43} + d)e_n^3 + (k_{44} + 1)e_n^4,$$

that is

$$\begin{bmatrix} e_{n+1}^1 \\ e_{n+1}^2 \\ e_{n+1}^3 \\ e_{n+1}^4 \end{bmatrix} = \begin{bmatrix} k_{11} + 1 - b - r' & k_{12} & k_{13} - c & k_{14} \\ k_{21} + b & k_{22} + 1 & k_{23} & k_{24} \\ k_{31} & k_{32} & k_{33} - 1 & k_{34} - d \\ k_{41} & k_{42} & k_{43} - d & k_{44} + 1 \end{bmatrix} \begin{bmatrix} e_{n+1}^1 \\ e_{n+1}^2 \\ e_{n+1}^3 \\ e_{n+1}^4 \end{bmatrix}. \tag{7.1.7}$$

Let

$$A = \begin{bmatrix} k_{11} + 1 - b - r' & k_{12} & k_{13} + c & k_{14} \\ k_{21} + b & k_{22} + 1 & k_{23} & k_{24} \\ k_{31} & k_{32} & k_{33} + 1 & k_{34} - d \\ k_{41} & k_{42} & k_{43} + d & k_{44} + 1 \end{bmatrix}.$$

It is well known that $e_n^i \to 0$ $(i = 1, 2, 3, 4)$ if the modules of the eigenvalues of the matrix $A$ is less than 1. From Lemma 7.1, we have the following theorem.

**Theorem 7.2** *$e_n^i \to 0$ $(i = 1, 2, 3, 4)$ if $k_{i,j}$ $(i, j = 1, 2, 3, 4)$ in (7.1.7) satisfy*

$$|k_{11} + 1 - b - r'| + |k_{12}| + |k_{13} + c| + |k_{14}| < 1,$$

$$|k_{21} + b| + |k_{22} + 1| + |k_{23}| + |k_{24}| < 1,$$

$$|k_{31}| + |k_{32}| + |k_{33} + 1| + |k_{34} - d| < 1,$$

$$|k_{41} + |k_{42}| + |k_{43} + d| + |k_{44} + 1| < 1.$$

For convenience, we take $k_{11} = -1 + b + r'$, $k_{13} = -c$, $k_{21} = -b$, $k_{22} = k_{33} = k_{44} = -1$, $k_{34} = d$, $k_{43} = -d$, then $|k_{12}| + |k_{14}| < 1$, $|k_{23}| + |k_{24}| < 1$, $|k_{31}| + |k_{32}| < 1$, $|k_{41}| + |k_{42}| < 1$ from Theorem 7.2. For example, we take $k_{12} = k_{14} = k_{23} = k_{24} = k_{31} = k_{32} = k_{41} = k_{42} = 0.2$, then the corresponding Julia set is shown in Fig. 7.4, which verifies the efficacy of the synchronization method.

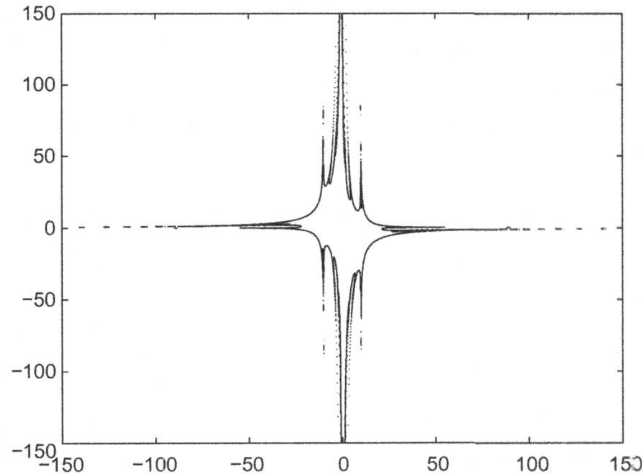


**Fig. 7.4** Julia set of system (7.1.3) when $a = 0.07, b = 0.02, r = 0.01, c = 0.2, d = 1.501$

## 7.2 Fractal Dimension Analysis of the Julia Sets of Controlled Brusselator Model

In fractal theory, fractal dimension is one of the most elemental concepts. At present, there are many definitions of fractal dimensions, including Hausdorff dimension, box-counting dimension, similarity dimension etc. In all kinds of definitions of fractal dimensions, Hausdorff dimension is the basis of the fractal theory. It can even be considered the theoretical basis of the fractal geometry. However, Hausdorff dimension is just suitable for the theoretical analysis of fractal theory and there is only a small class of fairly solid mathematical regular fractal graphics can be calculated for their Hausdorff dimension. It's hard to calculate the fractal dimension which is proposed in the practical applications. Therefore, people propose the concept of box-counting dimension. Its popularity is largely due to its relative ease of mathematical calculation and empirical estimation. In fact, in practical applications, the dimension is generally referred to as box-counting dimension see Definitin 6.1.

### *7.2.1 The Control of the Julia Set of Brusselator Model*

Brusselator model is one of the most fundamental model in nonlinear systems, and the dynamic equations are as follows

$$\begin{cases} x_{n+1} = A - (B+1)x_n + x_n^2 y_n, \\ y_{n+1} = Bx_n - x_n^2 y_n, \end{cases} \tag{7.2.1}$$

where $x, y$ denote concentration of reactant in the process of chemical reactant. $A, B > 0$ denote initial concentration of reactant.

Brusselator equations were first discovered by A.Turing in 1952 [347], and then I.Pigogine and Leefver did some systematic studies on it. They pointed out that the Brusselator equations were the most elementary and essential mathematic model which described the oscillation of biochemistry. They proved when $B > 1 + A^2$, the equations had stable and unique limit cycle. When $B \leq 1 + A^2$, there were no limit cycle [348]. It can be known that the initial concentration of reactant has an important influence on the system. In fractal theory, Julia set is a set of initial points of the system that satisfy certain conditions. With the same thought, we define the Julia set of Bursselator model. Let $F(x, y) = \left(A - (B + 1)x_n + x_n^2 y_n, Bx_n - x_n^2 y_n\right)$.

**Definition 7.2** Set $K = \left\{(x, y) \in R^2 | \{F^n(x, y)\}_{n \geq 1} \text{ is bounded}\right\}$ is called the filled Julia set of Brusselator model. The boundary of the filled Julia set is defined to be the Julia set of Brusselator model, that is $J(F) = \partial K$.

The research results demonstrate that coupled with a piecewise constant value control function, we can control the size of the limit cycle of Eq. (7.2.1) [347]. From the definition of Julia set, we found that there is a close relation between the boundedness of iterative orbits and the structure of the Julia set. So if we want to control the Julia set of Brusselator model, how to control the boundedness of the system iterative is critical. Especially, we consider the stability of the fixed point of Brusselator model, and by designing controllers to turn unstable fixed points into stable fixed points to control the boundedness of iterative orbits effectively. Then the control of the Julia set can be realized.

As was mentioned above, considering the stability of the fixed points of the system (7.2.2), we try to find controllers to make the fixed point of the system stabilize. Let the controlled Brusselator model be

$$\begin{cases} x_{n+1} = a + (1 - b - \gamma)x_n + \gamma x_n^2 y_n + u_n, \\ y_{n+1} = y_n + bx_n - \gamma x_n^2 y_n + v_n, \end{cases} \tag{7.2.2}$$

where $u_n$ and $v_n$ denote the designed controllers

**Feedback Control Method**
Take $u_n = k(x_n - x^*)$, $v_n = k(y_n - y^*)$, with the control parameter $k$, and the controlled system is

$$\begin{cases} x_{n+1} = a + (1 - b - \gamma)x_n + \gamma x_n^2 y_n + k(x_n - x^*), \\ y_{n+1} = y_n + bx_n - \gamma x_n^2 y_n + k(y_n - y^*). \end{cases} \tag{7.2.3}$$

**Theorem 7.3** *Let* $A_1 = 1 - b - \gamma + 2\gamma x^* y^* - k$, $B_1 = 1 - \gamma x^{*2} + k$, $C_1 = \gamma x^{*2}$, $D_1 = b - 2\gamma x^* y^*$, *where* $(x^*, y^*)$ *denotes the fixed point of the controlled system (7.2.3). If* $\left|(A_1 + B_1) \pm \sqrt{(A_1 - B_1)^2 + 4C_1 D_1}\right| < 2$, *the fixed point of the system is attractive.*

***Proof*** Write $\tilde{B}_1(x, y) = \big(a + (1 - b - \gamma)x + \gamma x^2 y + k(x - x^*),\ \ y + bx - \gamma x^2 y + k(y - y^*)\big)$. The Jacobi matrix of the controlled system (7.2.3) is

$$D\tilde{B}_1 = \begin{bmatrix} 1 - b - \gamma + 2\gamma x_n y_n + k & \gamma x_n^2 \\ b - 2\gamma x_n y_n & 1 - \gamma x_n^2 + k \end{bmatrix}$$

and its eigenmatrix is

$$D\tilde{B}_1 - \lambda E = \begin{bmatrix} 1 - b - \gamma + 2\gamma x_n y_n + k - \lambda & \gamma x_n^2 \\ b - 2\gamma x_n y_n & 1 - \gamma x_n^2 + k - \lambda \end{bmatrix}.$$

And the characteristic equation is

$$\lambda^2 - (2 - b - \gamma + 2\gamma x_n y_n - \gamma x_n^2 + 2k)\lambda + (1 - b - \gamma + 2\gamma x_n y_n + k)(1 - \gamma x_n^2 + k) - \gamma x_n^2(b - 2\gamma x_n y_n) = 0.$$

When the modulus of the eigenvalues $\lambda_{11}, \lambda_{12}$ of the Jacobi matrix at the fixed point is less than 1, the fixed point is attractive.

$$|\lambda_{11,12}| = \left|\frac{(A_1 + B_1) \pm \sqrt{(A_1 - B_1)^2 + 4C_1 D_1}}{2}\right| < 1,$$

that is

$$\left|(A_1 + B_1) \pm \sqrt{(A_1 - B_1)^2 + 4C_1 D_1}\right| < 2.$$

By Theorem 7.3, we know that Brusselator model can be controlled by selecting the value of $k$ which satisfies the condition, and then the control of the Julia set can be realized.

For example, take $a = 0.07$, $b = 0.02$, $\gamma = 0.01$ in system (7.2.3) and the initial Julia set is shown in Fig. 7.5, then we get $x^* = 7$, $y^* = \frac{2}{7}$. By Theorem 7.3, the range of the value of $k$ is $-1.9895 < k < 0.4695$.

Six simulation diagrams are chosen corresponding to the values of $k$ from 0.06 to 0.45 in Fig. 7.6, and we find that the trend of the change of the Julia sets is obvious. In Fig. 7.7 the box-counting dimensions of the controlled Julia sets are computed in this control method.

In the same control, six simulation diagrams are chosen corresponding to the values of $k$ from $-1$ to $-0.22$ in Fig. 7.8 to illustrate the change of the Julia sets in feedback control method. In Fig. 7.9 the box-counting dimensions of the controlled Julia sets are computed in this control method.

In feedback control method, the contraction of the left and the lower parts are faster than the right and the upper parts when the interval of control parameter $k$ between 0.06 and 0.45. In addition, the complexity of the boundary of the Julia set significantly decreases and the lower part rapidly contracts. When the interval of control parameter $k$ is between $-1$ and $-0.22$, the complexity of the boundary of

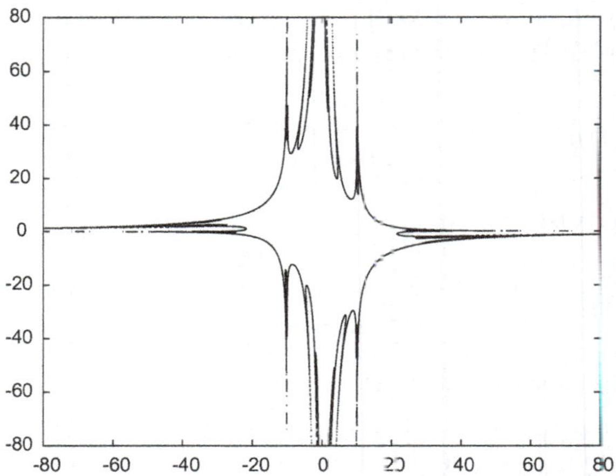


**Fig. 7.5** The original Julia set of the system

the Julia set significantly decreases, and the Julia set tends to central symmetric. In general, with the absolute value of $k$ increasing, the Julia set contracts to the center gradually.

From the perspective of the change of box-counting dimensions, with the absolute values of $k$ increasing, the change generally shows a monotonic decreasing trend. Especially when the control parameter $k$ is in the interval from 0.16 to 0.37 and $-0.9$ to $-0.4$, the monotonic change of box-counting dimensions is obvious, which indicates the effectiveness of this control method on the Julia set of Brusselator model.

**Optimal Control Method**

Take $u_n = k(f(x_n, y_n) - x_n)$, $v_n = k(g(x_n, y_n) - y_n)$, with $f(x_n, y_n) = a + (1 - b - \gamma)x_n + \gamma x_n^2 y_n$, $g(x_n, y_n) = y_n + bx_n - \gamma x_n^2 y_n$, where $k$ is the control parameter. Then the controlled system is

$$\begin{cases} x_{n+1} = a + (1-b-\gamma)x_n + \gamma x_n^2 y_n + k(f(x_n, y_n) - x_n), \\ y_{n+1} = y_n + bx_n - \gamma x_n^2 y_n + k(g(x_n, y_n) - y_n). \end{cases} \tag{7.2.4}$$

**Theorem 7.4** *Let* $A_2 = (1 - b - \gamma + 2\gamma x^* y^*)(k+1) - k$, $B_2 = (1 - \gamma x^{*2})(k+1) - k$, $C_2 = \gamma x^{*2}(k+1)$, $D_2 = (b - 2\gamma x^* y^*)(k+1)$, *where* $(x^*, y^*)$ *denotes the fixed point of the controlled system (7.2.4). If* $\left|(A_2 + B_2) \pm \sqrt{(A_2 - B_2)^2 + 4C_2 D_2}\right| < 2$, *the fixed point of the system is attractive.*

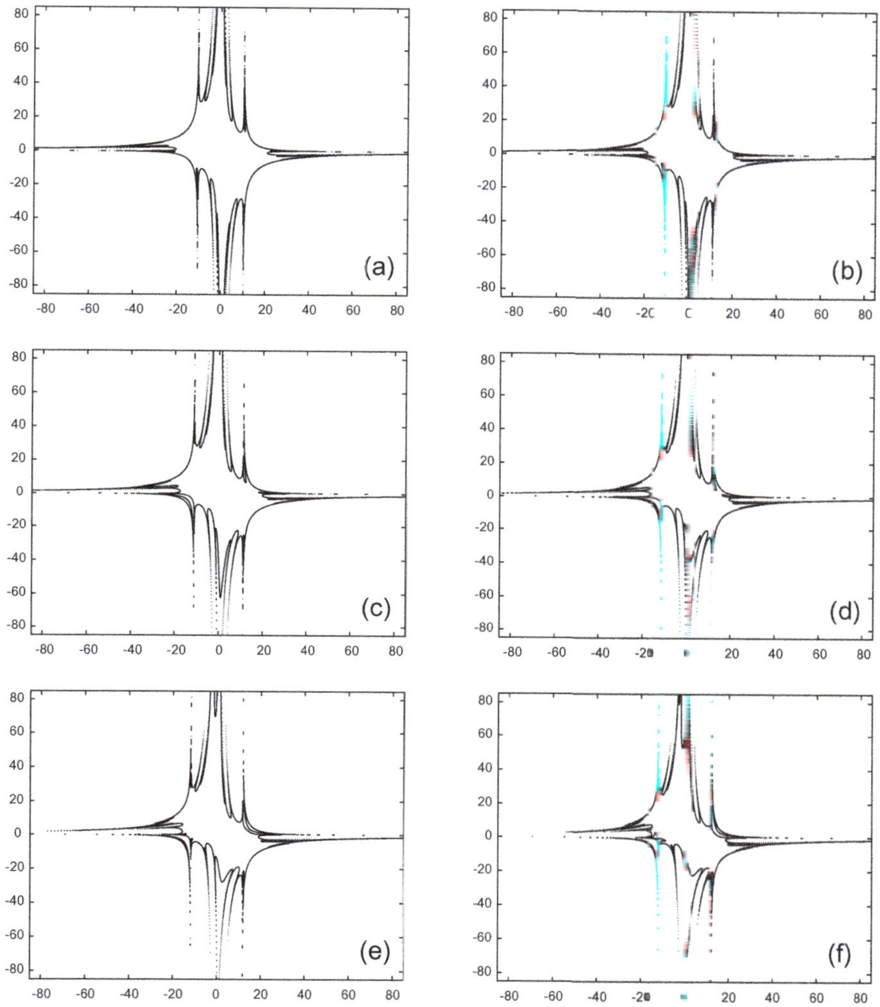


**Fig. 7.6** The change of the Julia sets of the controlled system when $a = 0.07$, $b = 0.02$ and $\gamma = 0.01$. **a** $k = 0.061$; **b** $k = 0.139$; **c** $k = 0.221$; **d** $k = 0.301$; **e** $k = 0.379$; **f** $k = 0.451$

***Proof*** Write $\tilde{B}_2(x, y) = \big(a + (1 - b - \gamma)x + \gamma x^2 y + k(f(x, y) - x),\ y + bx - \gamma x^2 y + k(g(x, y) - y)\big)$. The Jacobi matrix of the controlled system (7.2.4) is

$$D\tilde{B}_2 = \begin{bmatrix} (1 - b - \gamma + 2\gamma x_n y_n)(k + 1) - k & \gamma x_n^2 (k - 1) \\ (b - 2\gamma x_n y_n)(k + 1) & (1 - \gamma x_n^2)(k + 1) - k \end{bmatrix}$$

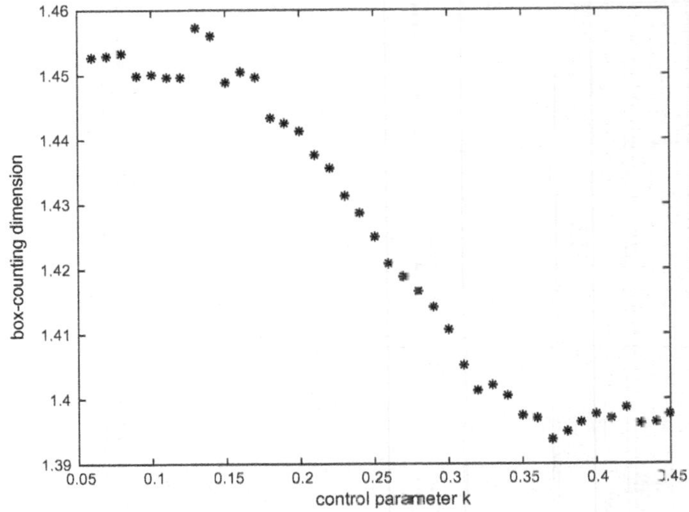


**Fig. 7.7** The change of box-counting dimensions of the Julia sets of the controlled system when $k$ is from 0.061 to 0.451

and its eigenmatrix is

$$D\tilde{B}_2 - \lambda E = \begin{bmatrix} (1 - b - \gamma + 2\gamma x_n y_n)(k+1) - k - \lambda & \gamma x_n^2 (k+1) \\ (b - 2\gamma x_n y_n)(k+1) & (1 - \gamma x_n^2)(k+1) - k - \lambda \end{bmatrix}.$$

And the characteristic equation is

$$\lambda^2 - ((2 - b - \gamma + 2\gamma x_n y_n - \gamma x_n^2)(k+1) - 2k)\lambda + ((1 - b - \gamma + 2\gamma x_n y_n)(k+1) - k)$$
$$((1 - \gamma x_n^2)(k+1) - k) - \gamma x_n^2 (b - 2\gamma x_n y_n)(k+1)^2 = 0.$$

When the modulus of the eigenvalue $\lambda_{21}, \lambda_{22}$ of the Jacobi matrix at the fixed point is less than 1, the fixed point is attractive.

$$|\lambda_{21,22}| = \left| \frac{(A_2 + B_2) \pm \sqrt{(A_2 - B_2)^2 + 4C_2 D_2}}{2} \right| < 1,$$

that is

$$\left|(A_2 + B_2) \pm \sqrt{(A_2 - B_2)^2 + 4C_2 D_2}\right| < 2.$$

By Theorem 7.4, we know that Brusselator model can be controlled by selecting the value of $k$ which satisfies the condition, and then the control of the Julia set can be realized.

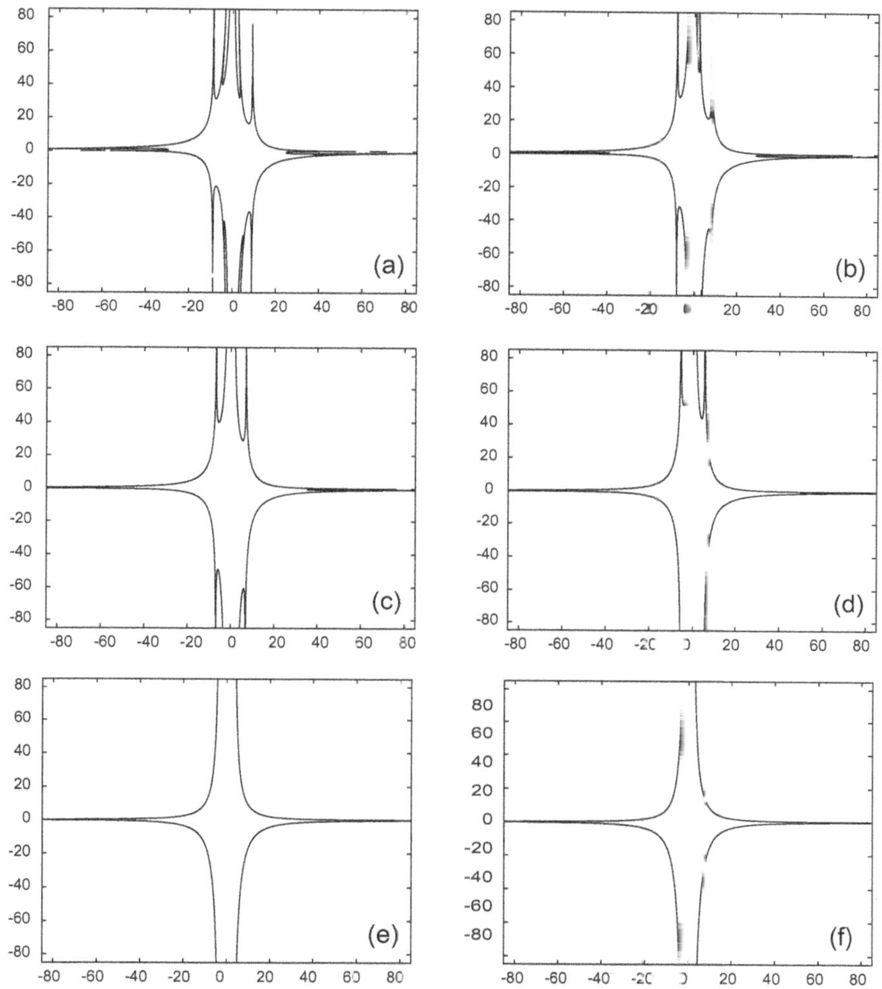

**Fig. 7.8** The change of the Julia sets of the controlled system when $a = 0.07$, $b = 0.02$ and $\gamma = 0.01$. **a** $k = -0.221$; **b** $k = -0.379$; **c** $k = -0.541$; **d** $k = -0.701$; **e** $k = -0.859$; **f** $k = -0.999$

For example, we take the values of system parameters as $a = 0.07$, $b = 0.02$, $\gamma = 0.01$, which are same to the previous description. Then the range of the value of $k$ is $-1 < k < 190.659$.

Six simulation diagrams are chosen corresponding to the values of $k$ from 0.641 to 2.601 in Fig. 7.10 to illustrate the change of the Julia sets in optimal control method. In Fig. 7.11 the box-counting dimensions of the controlled Julia sets are computed in this control method.

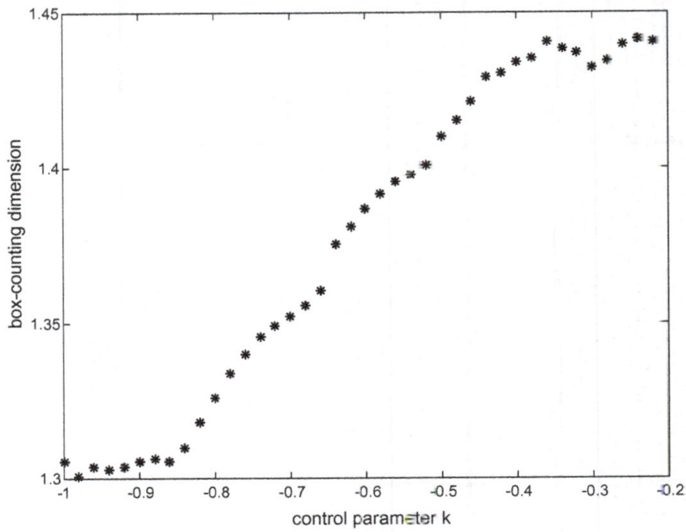


**Fig. 7.9** The change of box-counting dimensions of the Julia sets of the controlled system when $k$ is from $-0.999$ to $-0.221$

In optimal control method, the effective controlled interval of $k$ is from $-1$ to 190.6. But when the interval of $k$ is from 0.641 to 2.601, this method has the best controlled effectiveness. In this interval, with the absolute values of $k$ increasing, the Julia sets gradually contract to the center with nearly the same speed, and the overall shape of the Julia set has not changed significantly.

From the perspective of the change of box-counting dimensions, box-counting dimensions of the Julia sets generally show a monotonic decreasing trend with the absolute values of $k$ increasing. For the reason that the complexity of the Julia set can be depicted by box-counting dimension, so when the complexity of the Julia set decreases with the absolute value of $k$ increasing, it indicates that this control method has great control effectiveness.

**Gradient Control Method**

Take $u_n = k(x_n^2 - x^{*2}),\ v_n = k(y_n^2 - y^{*2})$, with the control parameter $k$. Therefore the controlled system is

$$\begin{cases} x_{n+1} = a + (1 - b - \gamma)x_n + \gamma x_n^2 y_n + k(x_n^2 - x^{*2}), \\ y_{n+1} = y_n + bx_n - \gamma x_n^2 y_n + k(y_n^2 - y^{*2}). \end{cases} \tag{7.2.5}$$

**Theorem 7.5** *Let* $A_3 = 1 - b - \gamma + 2\gamma x^* y^* + 2kx^*,\ B_3 = 1 - \gamma x^{*2} + 2ky^*,$ $C_3 = \gamma x^{*2},\ D_3 = b - 2\gamma x^* y^*$, *where* $(x^*, y^*)$ *denotes the fixed point of the controlled system (7.2.5). If* $\left|(A_3 + B_3) \pm \sqrt{(A_3 - B_3)^2 + 4C_3 D_3}\right| < 2$, *the fixed point of the system is attractive.*

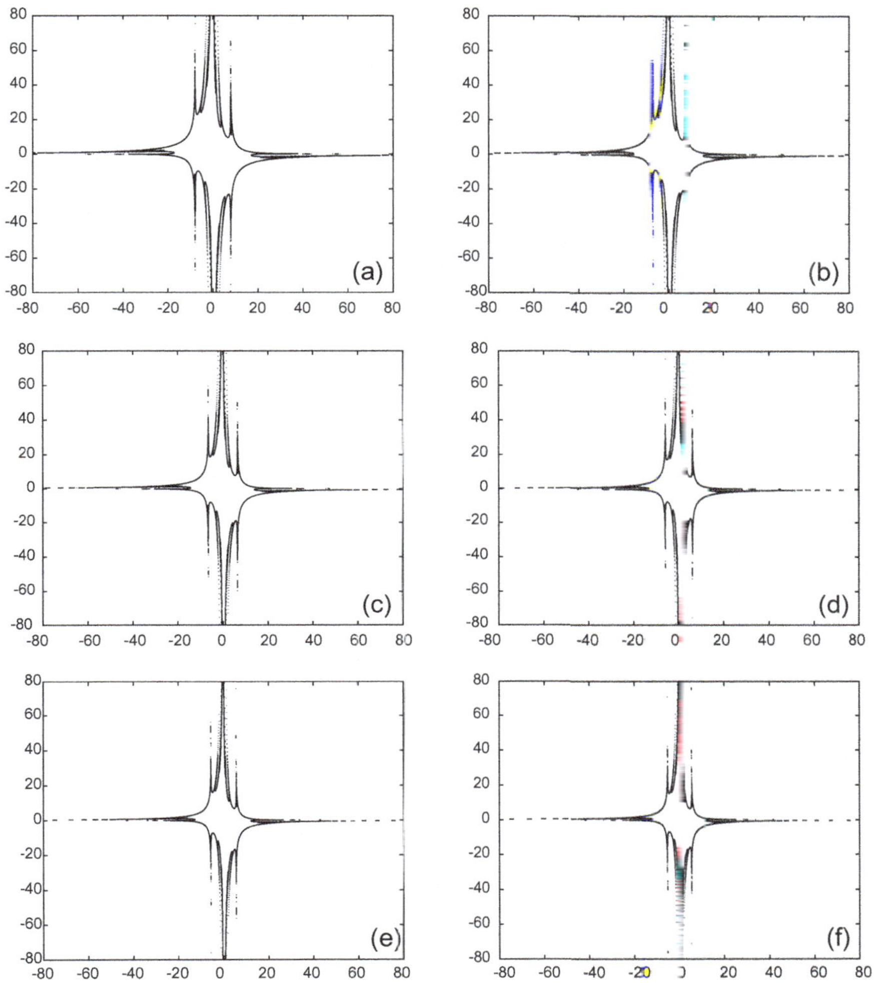

**Fig. 7.10** The change of the Julia sets of the controlled system when $a = 0.07, b = 0.02$ and $\gamma = 0.01$. **a** $k = 0.641$; **b** $k = 1.041$; **c** $k = 1.439$; **d** $k = 1.841$; **e** $k = 2.239$; **f** $k = 2.601$

***Proof*** Write $\tilde{B}_3(x, y) = \big(a + (1 - b - \gamma)x + \gamma x^2 y + k(x^2 - x^{*2}),\ \ y + bx - \gamma x^2 y + k(y^2 - y^{*2})\big)$. The Jacobi matrix of the controlled system (7.2.5) is

$$D\tilde{B}_3 = \begin{bmatrix} 1 - b - \gamma + 2\gamma x_n y_n + 2kx_n & \gamma x_n^2 \\ b - 2\gamma x_n y_n & 1 - \gamma x_n^2 + 2ky_n \end{bmatrix}$$

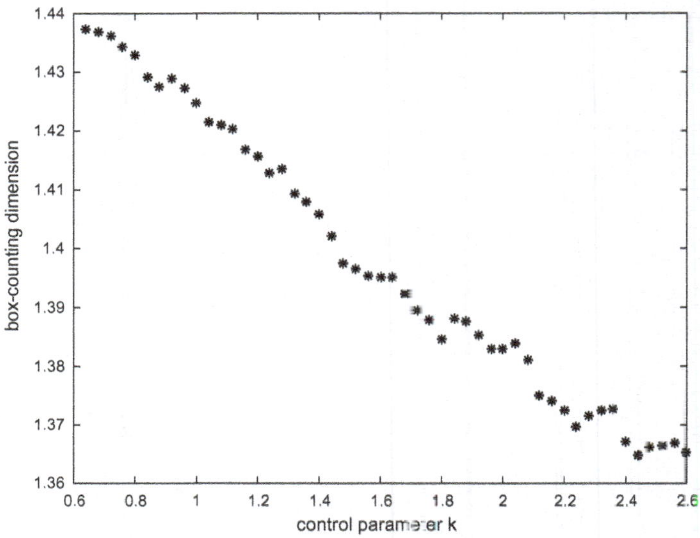


**Fig. 7.11** change of box-counting dimensions of the Julia sets of the controlled system when $k$ is from 0.641 to 2.601

and its eigenmatrix is

$$D\tilde{B}_3 - \lambda E = \begin{bmatrix} 1 - b - \gamma + 2\gamma x_n y_n + 2kx_n - \lambda & \gamma x_n^2 \\ b - 2\gamma x_n y_n & 1 - \gamma x_n^2 + 2ky_n - \lambda \end{bmatrix}$$

And the characteristic equation is

$$\lambda^2 - (2 - b - \gamma + 2\gamma x_n y_n + 2kx_n - \gamma x_n^2 + 2ky_n)\lambda + (1 - b - \gamma + 2\gamma x_n y_n + 2kx_n) \\ (1 - \gamma x_n^2 + 2ky_n) - \gamma x_n^2 (b - 2\gamma x_n y_n) = 0.$$

When the modulus of the eigenvalue $\lambda_{31}, \lambda_{32}$ of the Jacobi matrix at the fixed point is less than 1, the fixed point is attractive.

$$|\lambda_{31,32}| = \left| \frac{(A_3 + B_3) \pm \sqrt{(A_3 - B_3)^2 + 4C_3 D_3}}{2} \right| < 1,$$

that is

$$\left| (A_3 + B_3) \pm \sqrt{(A_3 - B_3)^2 + 4C_3 D_3} \right| < 2.$$

By Theorem 7.5, we know that the Brusselator model can be controlled by selecting the value of $k$ which satisfies the condition, and then the control of the Julia set can be realized.

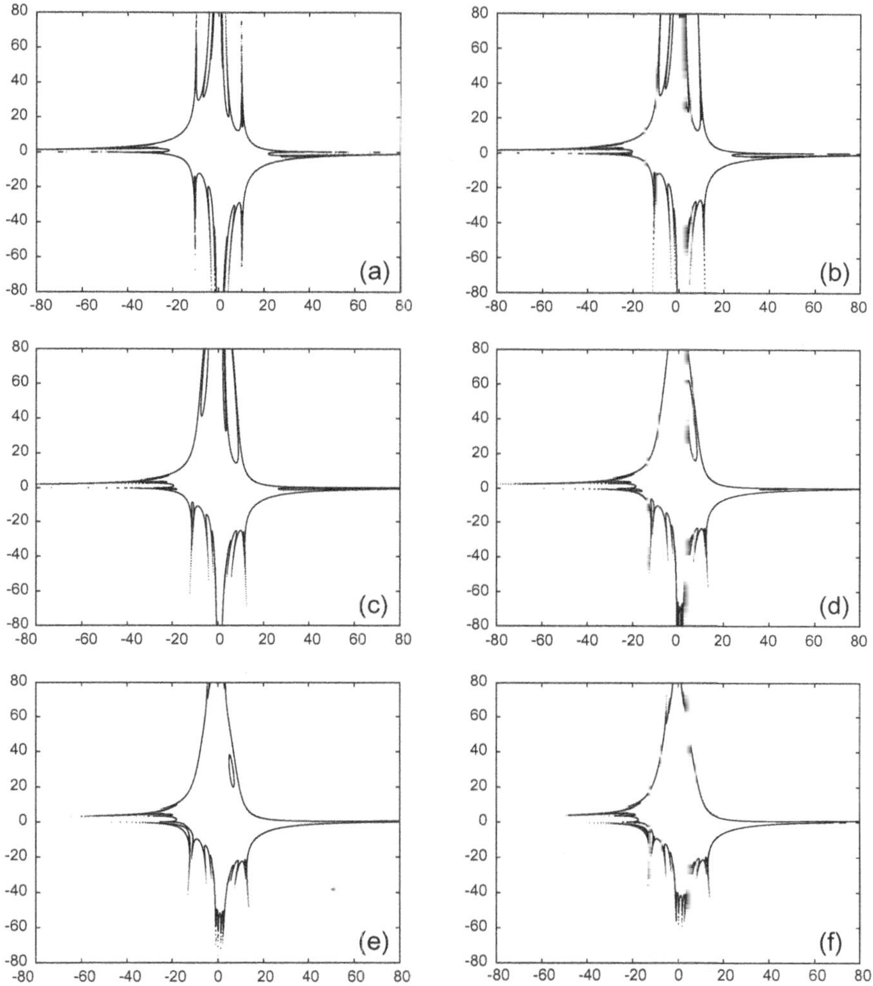


**Fig. 7.12** The change of the Julia sets of the controlled system when $a = 0.07$, $b = 0.02$ and $\gamma = 0.01$. **a** $k = -0.0006$; **b** $k = -0.0041$; **c** $k = -0.0081$; **d** $k = -0.0121$; **e** $k = -0.0159$; **f** $k = -0.0201$

For example, we take the values of system parameters as $a = 0.07$, $b = 0.02$, $\gamma = 0.01$, which are same to the previous description. Then the range of the value of $k$ is $-2.6420 < k < 0.8561$.

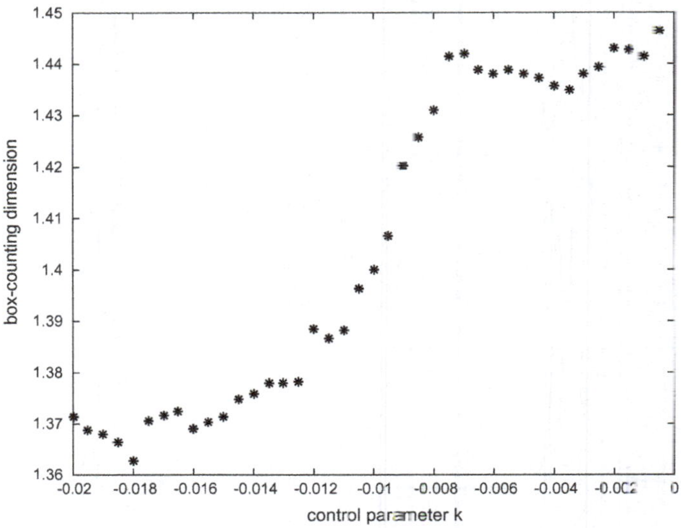


**Fig. 7.13** The change of box-counting dimensions of the Julia sets of the controlled system when $k$ is from $-0.0201$ to $-0.0006$

Six simulation diagrams are chosen corresponding to the values of $k$ from $-0.0006$ to $-0.0201$ in Fig. 7.12 to illustrate the change of the Julia sets in gradient control method. In Fig. 7.13 the box-counting dimensions of the controlled Julia sets are computed in this control method.

In the same control, six simulation diagrams are chosen corresponding to the values of $k$ from 0.0006 to 0.0201 in Fig. 7.14 to illustrate the change of the Julia sets in gradient control method. In Fig. 7.15 the box-counting dimensions of the controlled Julia sets are computed in this control method.

In gradient control method, we consider two groups of interval of the parameter $k$. It is noticed that the contraction of the left and the lower parts of the Julia set are faster than the right and the upper parts when the interval of control parameter $k$ is from $-0.0006$ to $-0.0201$, while the right and the upper parts are faster than the left and the lower parts when the interval of control parameter $k$ is from 0.0006 to 0.0201. In general, with the absolute values of $k$ increasing, the Julia sets contract to the center gradually, and the complexity of the boundary of the Julia sets decreases.

From the viewpoint of the change of box-counting dimensions, with the absolute values of $k$ increasing, the box-counting dimensions of the Julia sets generally show a monotonic decreasing trend. Especially when the control parameter $k$ is in the range from $-0.016$ to $-0.0075$ and from 0.0075 to 0.016, the change of box-counting dimensions is obviously monotonic, which indicates the effectiveness of this method on the Julia set of the Brusselator model.

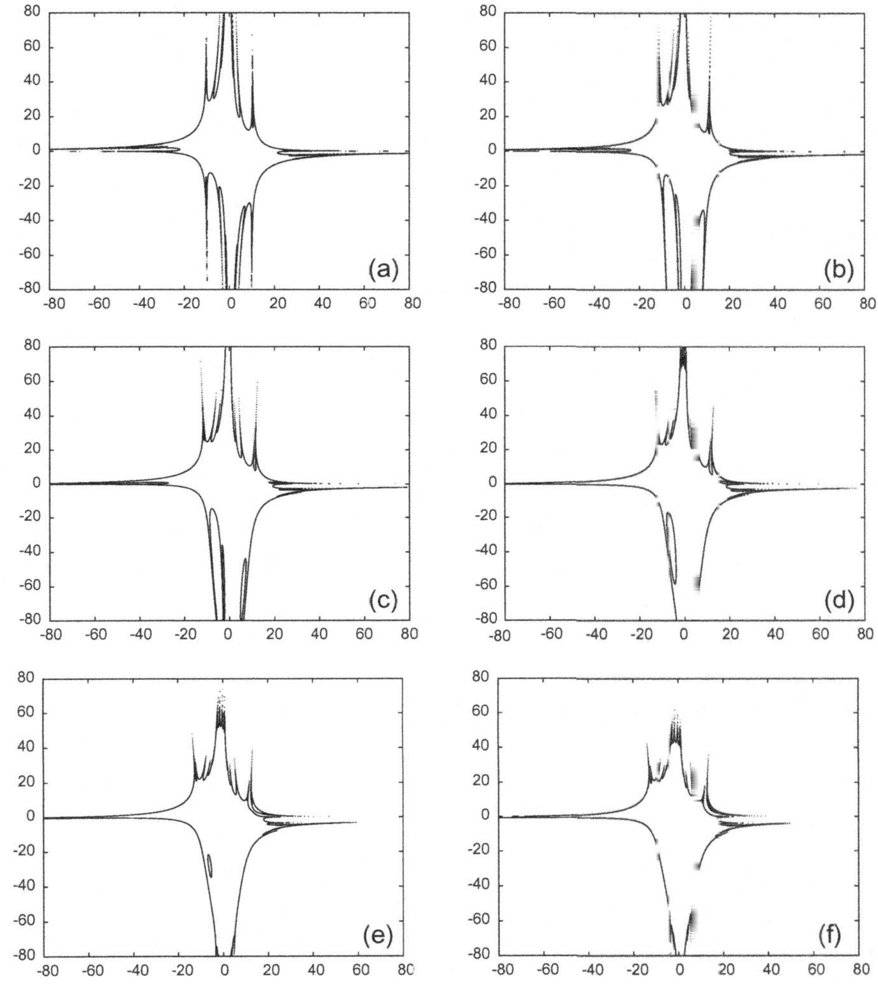


**Fig. 7.14** The change of the Julia sets of the controlled system when $a = 0.07, b = 0.02$ and $\gamma = 0.01$. **a** $k = 0.0006$; **b** $k = 0.0046$; **c** $k = 0.0084$; **d** $k = 0.0124$; **e** $k = 0.0166$; **f** $k = 0.0201$

## 7.3 Control and Synchronization of Julia Sets of Complex Standard Family

It is known that Chirikov standard map is one of the most important and extensively studied systems in nonlinear dynamics and in the theory of Hamilton and area-preserving maps [349, 350]. The model describes the motion of the kicked rotator and can be applied to many fields such as state physics, statistical mechanics, problems

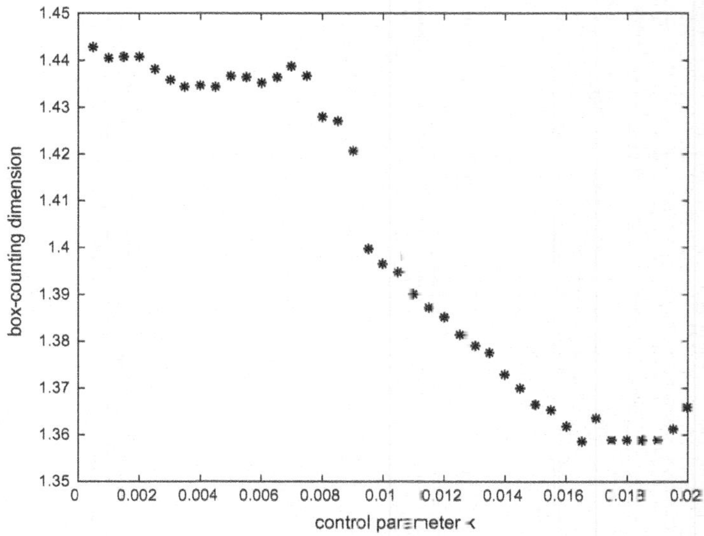


**Fig. 7.15** The change of box-counting dimensions of the Julia sets of the controlled system when $k$ is from 0.0006 to 0.0201

of quantum mechanics and quantum chaos and many others [351–353]. In general, the standard map is written as

$$\begin{cases} I_{n+1} = (1-\gamma)I_n + K\sin(\theta_n), \\ \theta_{n+1} = \theta_n + I_{n+1}, \ mod(2\pi), \end{cases}$$

where $\gamma \in [0, 1]$ is the dissipation parameter. The standard map has a straightforward physical interpretation. It is the Poincare return map for a frictionless periodically kicked pendulum. If $\gamma = 1$, the one dimensional sine-circle map is obtained. fuether-more, the two parameters family is generally given by

$$x_{n+1} = x_n + \alpha + \beta\sin(x_n), \tag{7.3.1}$$

where $\alpha$ and $\beta$ are real parameters. Equation (7.1.1) exhibits a mode-locking phase diagram which contains many infinitely Arnold tongues in which the dynamics are locked to periodic orbits, separated by gaps in which the trajectories are quasi periodic [354, 355]. Some properties of Eq. (7.1.1) are observed in experimental systems. For example, discussions in Tayleigh–Benard convection are found to be in an agreement with theoretical predictions for the properties of the transition [336, 356].

Conveniently, the one dimensional sine-circle map in the complex plane can be written in the following form

$$z_{n+1} = z_n + \alpha + \beta\sin(z_n), \tag{7.3.2}$$

where $\alpha$ and $\beta$ are real parameters. The complex dynamical properties and the existence of hairs in the Julia set of the map Eq. (7.1.1) in the complex plane have been discussed [357].

### 7.3.1 Feedback Control of Julia Set of the Complex System

The Julia set of the system (7.3.2) is the closure of the points whose orbits have imaginary part tending to $\infty$ or $-\infty$ (the two asymptotic directions) [357]. As a result, the trajectory of the imaginary part of the system (7.3.2) plays an important role in the construction of Julia set. If the trajectory could be controlled, the corresponding Julia set will be changed. In the investigation of means of controlling trajectory of systems, many methods have been introduced, such as the OGY method, the feedback control method and so on [358–363]. In order to control Julia set, we consider to control the trajectory of the imaginary part of the system (7.3.2) from the definition of Julia set with the feedback control method.

Let $z_n = x_n + iy_n$, then (7.3.2) becomes

$$\begin{cases} x_{n+1} = x_n + \alpha + \beta \sin(x_n)\cosh(y_n), \\ y_{n+1} = y_n + \beta \cos(x_n)\sinh(y_n). \end{cases} \tag{7.3.3}$$

Let $z^* = x^* + iy^*$ be the fixed point of (7.3.3). Introduce the auxiliary reference feedback control item to the second equation of Eq. (7.3.3), which is the imaginary part of the system (7.3.2), then we have the controlled system

$$\begin{cases} x_{n+1} = x_n + \alpha + \beta \sin(x_n)\cosh(y_n), \\ y_{n+1} = y_n + \beta \cos(x_n)\sinh(y_n) + k(y_n - y_{ref}), \end{cases} \tag{7.3.4}$$

where $y_{ref}$ is the reference item and $k$ is the feedback coefficient. Suppose the control objective $\overline{z} = \overline{x} + i\overline{y}$ has been achieved, then

$$\begin{cases} \overline{x} = \overline{x} + \alpha + \beta \sin(\overline{x})\cosh(\overline{y}), \\ \overline{y} = \overline{y} + \beta \cos(\overline{x})\sinh(\overline{y}) + k(\overline{y} - y_{ref}). \end{cases} \tag{7.3.5}$$

Based on the above equation, we know that $(\overline{x}, \overline{y})$ is not only the unstable fixed point of (7.3.3) but also the stable fixed point of the system (7.3.5). There are many methods to change the trajectory and we choose one of them to consider the stability of the fixed point of the system (7.3.3). Assume the control objective must be the fixed point $z^* = x^* + iy^*$, that is, $\overline{x} = x^*, \overline{y} = y^*$ when $n \to \infty$. Then

$$\begin{cases} \overline{x} = \overline{x} + \alpha + \beta \sin(\overline{x})\cosh(\overline{y}), \\ \overline{y} = \overline{y} + \beta \cos(\overline{x})\sinh(\overline{y}) \end{cases} \tag{7.3.6}$$

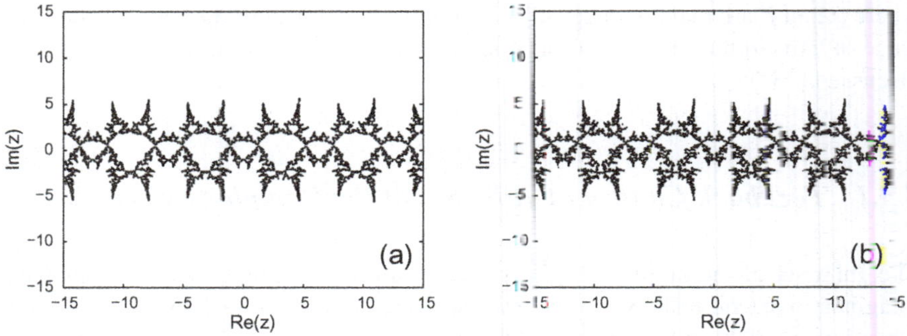


**Fig. 7.16** The changing of Julia sets of the controlled system (7.3.5) when $\alpha = 3.1$, $\beta = 0.8$: (**a**) Julia set of (7.3.5) when $k = 0.131$ and (**b**) Julia set of (7.3.5) when $k = 0.229$

and (7.3.5) must be satisfied simultaneously. Take $(x^*, y^*)$ into Eq. (7.3.5) in order to achieve the control objective $(x^*, y^*)$. Then $y_{ref}$ must be chosen to be $y^*$, which illustrates that the auxiliary reference item $y_{ref}$ must be the imaginary part of the fixed point of the system if the control objective is the fixed point.

Since Julia set of the system (7.3.2) is related to the imaginary part of the trajectory, we consider the stability of the fixed point $(x^*, y^*)$ of the controlled system (7.3.4). The first Lyapunov method is used to analyze the sufficient condition to achieve the control. The Jacobi matrix of the controlled system (7.3.4) at $(x^*, y^*)$ is

$$J = \begin{bmatrix} 1 + \beta \cos x^* \cosh y^* & \beta \sin x^* \sinh y^* \\ -\beta \sin x^* \sinh y^* & 1 + k + \beta \cos x^* \cosh y^* \end{bmatrix}.$$

It is easy to gain the eigenvalue of the above matrix is $\lambda_{1,2} = \frac{2(1+\beta \cos x^* \cosh y^*)+k \pm \sqrt{k^2 - 4(\beta \sin x^* \sinh y^*)^2}}{2}$. It is well known that, $|\lambda_{1,2}| < 1$ is one of the conditions to guarantee the stability of $(x^*, y^*)$. From the above analysis, we know that the auxiliary reference feedback could guide the trajectory of the system to the unstable fixed point and also can get a series of required output by choosing the special values of $y_{ref}$, $k$. In fact, it is not necessary to analyze the stability of the fixed point of the original system since the target to control is Julia set but not the trajectories. We make Julia set change by using the changing of the trajectories related with the fixed point, which is one of the methods to change the trajectories.

Taking $\alpha = 3.1$, $\beta = 0.8$ in (7.3.3), then we have the fixed points $z^* = x^* + iy^*$, where $x^* = \frac{(4l+3)\pi}{2}$, $y^* = \pm 2.0306$ and $l$ are integers. Then $\lambda_{1,2} = \frac{2+k \pm \sqrt{k^2 - 35.8791}}{2}$. Taking $y^* = 2.0306$, Fig. 7.16 demonstrates the changing of Julia sets of the controlled system (7.3.5) with the changing of the control parameter $k$.

### 7.3.2 Synchronization of Julia Sets of the Complex System

In this section, we will discuss the synchronization of Julia sets of the complex dissipative standard system.

Consider an identical system with the system (7.3.3) in this section but having different parameters

$$\begin{cases} p_{n+1} = p_n + \alpha' + \beta' \sin(p_n)\cosh(q_n), \\ q_{n+1} = q_n + \beta' \cos(p_n)\sinh(q_n). \end{cases} \tag{7.3.7}$$

We call the system (7.3.3) the driving system and the system (7.3.7) the response system. Julia sets of (7.3.3) or (7.3.7) are determinated once the parameters are given. From the definition of Julia set of the complex dissipative standard system, the structure of Julia set is closely related to the trajectory of the imaginary part of the system. We utilize the periodic impulsive method to implement the synchronization of the trajectory [364]. Then the synchronization of the Julia sets will be achieved. Now add the periodic pulse to the second section of (7.3.7), we have

$$\begin{cases} p_{n+1} = p_n + \alpha' + \beta' \sin(p_n)\cosh(q_n), \\ q_{n+1} = q_n + \beta' \cos(p_n)\sinh(q_n), \quad n \neq j\Delta n, \\ q_{n+1} = y_{n+1}, \quad n = j\Delta n, \end{cases} \tag{7.3.8}$$

where $j$ runs over natural numbers, implying that the kicks are applied at intervals spaced by $\Delta n$. During $\Delta n - 1$ iteration steps the response system evolves over the original system (7.3.8), while at the next step the state variable $q_i$ is replaced by the state variable $y_i$ of the driving system. Therefore, the behavior of (7.3.8) is free and independent from (7.3.3) except for the moments when $q_i$ is forced to the new value $y_i$. Figure 7.17 shows that the values of $q_i - y_i$ tend to 0, which illustrates the synchronization of the trajectory of the imaginary part of (7.3.3) and (7.3.8) occurs. Thus, Julia sets of (7.3.3) and (7.3.8) become to be the same one, which is the Julia set of (7.3.3).

Figure 7.18 is the Julia set when $\alpha' = 2$, $\beta' = 0.3$ in (7.3.7). Obviously, it is different from the Julia set in Fig. 7.16a, where $\alpha = 3.1$, $\beta = 0.8$ in (7.3.3).

From the above discussion, we know that Julia set of (7.3.8) will become to be the Julia set of (7.3.3) shown in Fig. 7.16a, which is illustrated by simulation.

In the following, we will illustrate how the synchronization work by using the conditional Lyapunov exponents of the response system. Denote (7.3.7) as

$$\begin{cases} p_{n+1} = f(p_n, q_n), \\ q_{n+1} = g(p_n, q_n). \end{cases} \tag{7.3.9}$$

By the periodic pulse method, we consider the $\Delta n$ steps as a unit. Let $F = f^{\Delta n}$, $G = g^{\Delta n}$, then

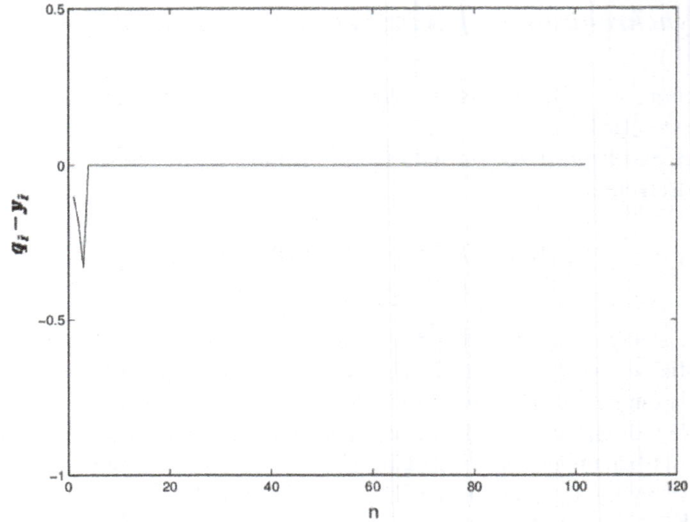


**Fig. 7.17** Synchronization of the trajectory of the imaginary part of systems (7.3.3) and (7.3.8) when $\Delta n = 2, \alpha = 3.1, \beta = 0.8$ in (7.3.3) and $\alpha' = 2, \beta' = 0.3$ in (7.3.8)

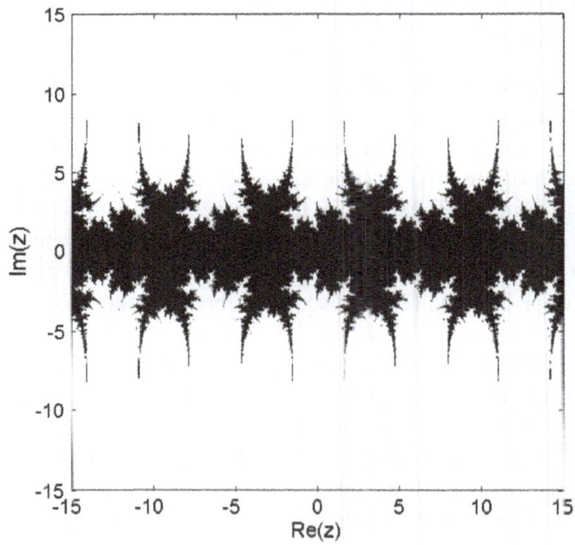


**Fig. 7.18** Julia set of Eq.(7.3.7) when $\alpha' = 2, \beta' = 0.3$

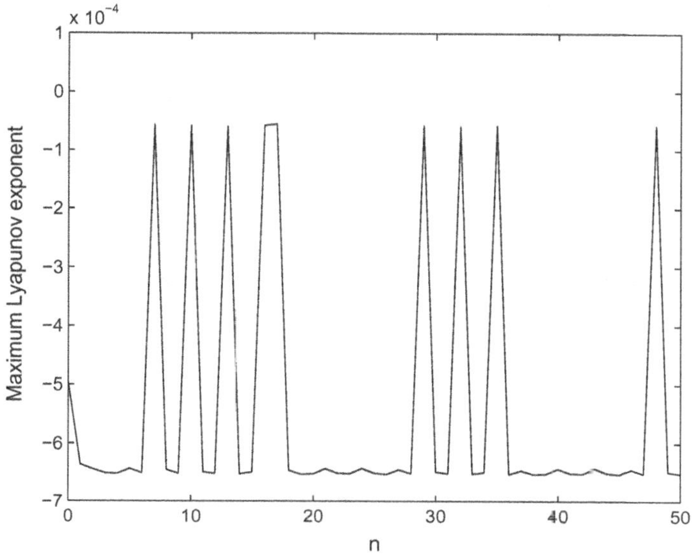


**Fig. 7.19** Maximum Lyapunov exponent spectra of the response system corresponding to the system (7.3.8)

$$\begin{cases} p_{m+1} = F(p_m, q_m), \\ q_{m+1} = G(p_m, q_m). \end{cases} \tag{7.3.10}$$

The periodic pulse added in (7.3.9) is equivalent to (7.3.10) having a drive variable. Then the conditional Lyapunov exponent is

$$\lambda = \lim_{m\to\infty} \frac{1}{m} \sum_{j=0}^{m-1} \ln \left| \frac{\partial f^{\Delta n}}{\partial p_j} \right|, \tag{7.3.11}$$

which depends on $\Delta n$. When $\lambda < 0$, synchronization can be implemented; otherwise, synchronization can not be implemented. Since $\lambda$ is related to the property of the trajectory, another form for $\lambda$ can be obtained where $\Delta n = 2$.

$$\begin{aligned} \lambda = \lim_{m\to\infty} \tfrac{1}{m} \sum_{j=0}^{m-1} \ln \big| 1 + \beta \cos p_j \cosh q_j \\ -\beta \sin p_j \sinh(q_j + \beta \cos(p_j + \alpha + \beta \sin p_j \cosh q_j) \sinh q_j).\big| \end{aligned} \tag{7.3.12}$$

To study the relation between synchronization and $\Delta n$, we calculate the maximum Lyapunov exponent of the response system. Synchronization can be implemented if the maximum Lyapunov exponent is negative for $\Delta n$. Fig. 7.19 shows the results corresponding to the system (7.3.8).

## 7.4 Identification of Julia Sets of Complex Standard Family

Based on the definition of Julia set, we know that once the system parameters are given, Julia set of the system is determinated and the number of Julia set is only one. In the practical applications, the behaviors depicted by Julia sets can be observed but the system parameters are not known. Therefore identification of Julia set is necessary.

For two systems, if the behavior of the second one is influenced by the first one but the behavior of the first one is not influenced by the second one, the first system is called the driving system and the second one is called the responsive system. Let the driving system in the complex plane be

$$z_{n+1} = v(z_n; c), \tag{7.4.1}$$

where $c$ is the system parameter and is unknown. In fact, if $c$ is known, Julia set of system is determinate. As a result, the value of $c$ is crucial to identify Julia set of system (7.4.1). Let the responsive system be

$$w_{n+1} = v(w_n; \widehat{c}_n) + u_n, \tag{7.4.2}$$

where $\widehat{c}_n$ is the parameter to design and $u_n$ is the controller.

The perpose of identification of Julia set is to design $\widehat{c}_n$ and $u_n$ to identify the unknown system parameter $c$, that is $\lim\limits_{n\to\infty} \widehat{c}_n = c$, and make the Julia set of the responsive system be the Julia set of the driving system. We chose some values of $\widehat{c}_n$ and designed them to fit the responding Julia sets in order to see the process of the identification of Julia set in this section.

### 7.4.1 Design of Controller and Parameter Laws I

In this section, the driving system and the responsive system are taken as the following form respectively

$$z_{n+1} = f(z_n) + \alpha \tag{7.4.3}$$

and

$$w_{n+1} = f(w_n) + \alpha_n + u_n, \tag{7.4.4}$$

where $u_n$ is the controller and $\alpha_n$ is the parameter to identify the parameter $\alpha$. Let $J^*$ and $J$ be the Julia sets of (7.4.3) and (7.4.4). The aim is to make the Julia set of the responsive system to be the one of the driving system and to identify the parameter $\alpha$, that is

$$J^* = J \tag{7.4.5}$$

and

$$\lim_{n\to\infty} \alpha_n = \alpha. \tag{7.4.6}$$

Let $e_n = w_n - z_n$ and $\overline{f} = -f(w_n) + f(z_n)$. Take

$$u_n = \overline{f} + Ae_n,$$

$$\alpha_{n+1} = \alpha_n - Be_n,$$

where $A$ and $B$ are real control parameters. Then we have

**Theorem 7.6** *If the control parameters A and B satisfy* $\left|\frac{(A+1)\pm\sqrt{(A+1)^2-4(A+B)}}{2}\right| < 1$, *Julia set of system (7.4.4) is changed to be the one of system (7.4.3).*

***Proof*** Let

$$\Delta\alpha_{n+1} \triangleq \alpha_{n+1} - \alpha = \Delta\alpha_n - Be_n.$$

Denote $E_n = \begin{bmatrix} e_n \\ \Delta\alpha_n \end{bmatrix}$, then $E_{n+1} = RE_n$, where $R = \begin{bmatrix} A & 1 \\ -B & 1 \end{bmatrix}$. The eigenvalues of $R$ are

$$\lambda_{1,2} = \frac{(A+1) \pm \sqrt{(A+1)^2 - 4(A+B)}}{2}.$$

It must be admitted that $|\lambda_{1,2}| < 1$ is one of the conditions to guarantee $E_n \to 0$ as $n \to \infty$. Therefore we have (7.4.6) and $e_n \to 0$ as $n \to \infty$. From the definition of Julia set, we know Julia set is closely related to the trajectories of the system. Thus Julia set of system (7.4.4) is changed to be the one of system (7.4.3).

For the complex sine-circle map (7.4.1), we take $f(z_n) = z_n + \beta \sin z_n$ and $f(w_n) = w_n + \beta \sin w_n$ in systems (7.4.3) and (7.4.4) respectively. Then

$$u_n = -w_n - \beta \sin w_n + z_n + \beta \sin z_n + A(w_n - z_n),$$

and the responsive system is

$$w_{n+1} = z_n + \beta \sin z_n + \alpha_n + A(w_n - z_n).$$

Let $\alpha = 3.1$, $\beta = 0.8$ and $A = 0.2$, $B = 0.2$ and take $u_1 = 10 + 10i$. Then $\lambda_{1,2} = 0.6 \pm 0.2i$. Therefore $|\lambda_{1,2}| < 1$. Figure 7.20a, b are the identification of the real part and imaginary part of the parameters $\alpha_n$. When the iteration numbers are 10, 15 and 20, the values of parameters are $\alpha_{10} = 3.0460 + (-0.0782)i$, $\alpha_{15} = 3.0755 + (-0.0354)i$ and $\alpha_{20} = 3.1007 + 0.0011i$. The corresponding Julia sets are Fig. 7.21b, c and d respectively, which illustrate that Julia set of system (7.4.3) is

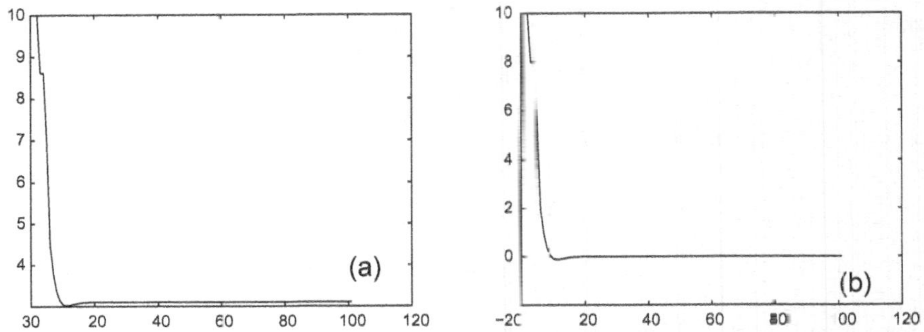


**Fig. 7.20** Identification of the parameter $\alpha$ of system (7.4.3). **a** The real part; **b** The imaginary part

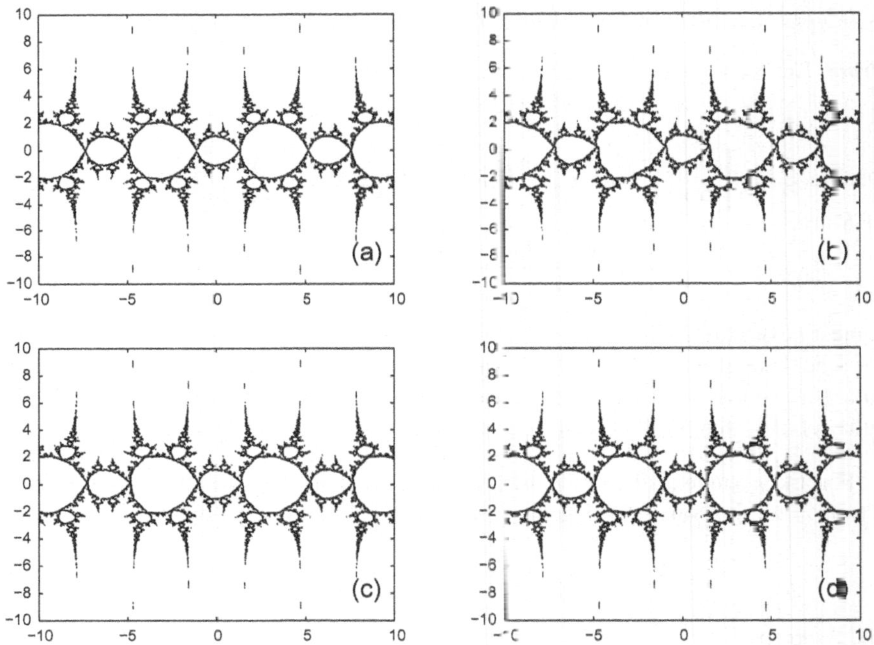


**Fig. 7.21** Identification of Julia sets. **a** Julia set of the responsive system; **b** Julia set when $n = 10$ and $\alpha = 3.0461 + (-0.0783)i$; **c** Julia set when $n = 15$ and $\alpha = 3.0754 + (-0.0355)i$; **d** Julia set when $n = 20$ and $\alpha = 3.1008 + 0.0010i$

identified with the pushing on the identification of the parameters. Fig. 7.21a is the Julia set of the responsive system.

Actually, if we take $A = 0$, the responsive system becomes $w_{n+1} = f(z_n) + \alpha_n$. In this case, the velocity to identify Julia set is quicker than $A \neq 0$.

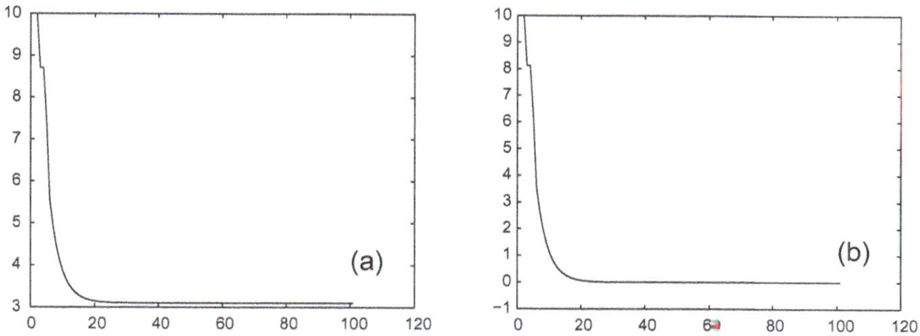


**Fig. 7.22** **a** Identification of the parameter $\alpha$ of system (7.4.8); **b** Identification of the parameter $\gamma$ of system (7.4.8)

### 7.4.2 Design of Controller and Parameter Laws II

Since system (7.3.2) is studied in the complex plane, we separate it two parts: the real part and the imaginary part. Let $z_n = x_n + i y_n$, where $x_n$ and $y_n$ are real numbers. Then system (7.3.2) becomes

$$\begin{cases} x_{n+1} = x_n + \alpha + \beta \sin x_n \cosh y_n, \\ y_{n+1} = y_n + \beta \cos x_n \sinh y_n. \end{cases} \tag{7.4.7}$$

Generally, we consider the two dimensional real system with the following form as the driving system

$$\begin{cases} x_{n+1} = x_n + \alpha + h(x_n, y_n), \\ y_{n+1} = y_n + \gamma + g(x_n, y_n), \end{cases} \tag{7.4.8}$$

where $\alpha$ and $\gamma$ are real parameters to identify. Let the responsive system be

$$\begin{cases} p_{n+1} = x_n + \alpha_n + h(x_n, y_n), \\ q_{n+1} = y_n + \gamma_n + g(x_n, y_n) \end{cases} \tag{7.4.9}$$

where $\alpha_n$ and $\gamma_n$ are parameters to design.

**Theorem 7.7** *If*

$$\begin{cases} \alpha_{n+1} = \alpha_n + \varepsilon_1 (p_n - x_n), \\ \gamma_{n+1} = \gamma_n + \varepsilon_2 (q_n - y_n), \end{cases} \tag{7.4.10}$$

*where* $\varepsilon_1$ *and* $\varepsilon_2$ *are control parameters, which satisfy* $\left|\frac{1 \pm \sqrt{1+4\varepsilon_i}}{2}\right| < 1,\ i = 1, 2,$ *then Julia set of system* (7.4.9) *is changed to be the one of system* (7.4.8).

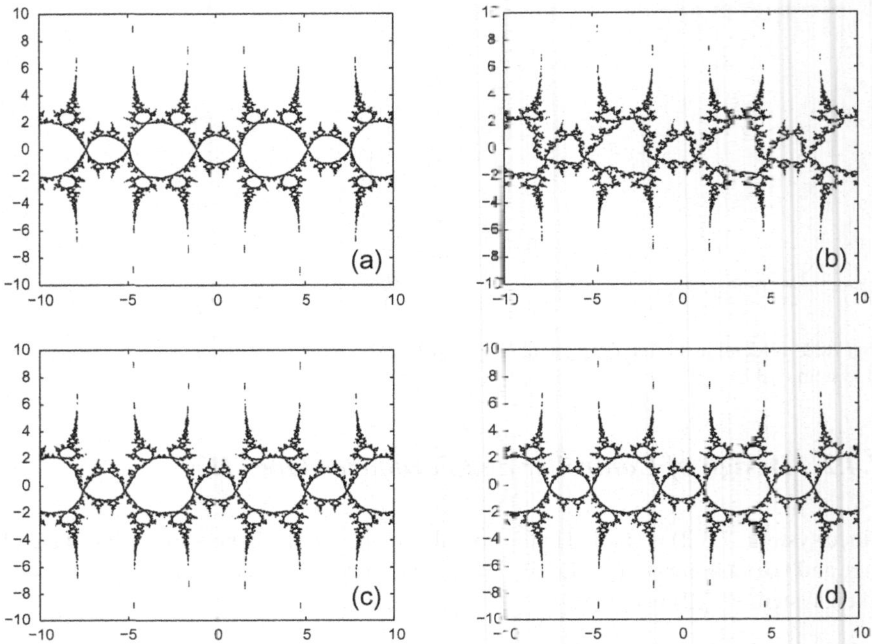

**Fig. 7.23** Identification of Julia sets. **a** Julia set of the responsive system (7.4.8); **b** Julia set when $n = 15$ and $\alpha = 3.2843, \gamma = 0.2674$; **c** Julia set when $n = 18$ and $\alpha = 3.1777, \gamma = 0.1129$; **d** Julia set when $n = 20$ and $\alpha = 3.1439, \gamma = 0.0635$

***Proof*** Let $e_n = p_n - x_n,\ d_n = q_n - y_n,\ \Delta\alpha_n = \alpha_n - \alpha,\ \Delta\gamma_n = \gamma_n - \gamma$, then

$$\begin{cases} e_{n+1} = \Delta\alpha_n, \\ d_{n+1} = \Delta\gamma_n, \\ \Delta\alpha_{n+1} = \Delta\alpha_n + \varepsilon_1 e_n, \\ \Delta\gamma_{n+1} = \Delta\gamma_n + \varepsilon_2 d_n. \end{cases} \tag{7.4.11}$$

Let

$$T = \begin{bmatrix} 0 & 0 & 1 & 0 \\ 0 & 0 & 0 & 1 \\ \varepsilon_1 & 0 & 1 & 0 \\ 0 & \varepsilon_2 & 0 & 1 \end{bmatrix},\ F_n = \begin{bmatrix} e_n \\ d_n \\ \Delta\alpha_n \\ \Delta\gamma_n \end{bmatrix},$$

then $F_{n+1} = TF_n$. The eigenvalues of $T$ are

$$\lambda_{1,2} = \frac{1 \pm \sqrt{1 + 4\varepsilon_1}}{2},\ \lambda_{3,4} = \frac{1 \pm \sqrt{1 + 4\varepsilon_2}}{2}.$$

$|\lambda_i| < 1\, (i = 1, \ldots, 4)$ are the sufficient conditions to guarantee $\lim_{n\to\infty} F_n = 0$. Therefore, $\lim_{n\to\infty} |p_n - x_n| = 0$, $\lim_{n\to\infty} |q_n - y_n| = 0$ and $\lim_{n\to\infty} \alpha_n = \alpha$, $\lim_{n\to\infty} \gamma_n = \gamma$, which illustrate that synchronization of the trajectories of the real part and the imaginary part of the system is achieved and the systems parameters are identified. Consequently, Julia set of system (7.4.9) is changed to be the one of system (7.4.8).

For the complex sine-circle map (7.4.7), we take $h(x_n, y_n) = \beta \sin x_n \cosh y_n$, $g(x_n, y_n) = \beta \cos x_n \sinh y_n$ and $\beta = 0.8,\ \alpha = 3.1, \gamma = 0$ in system (7.4.8) and $\varepsilon_1 = \varepsilon_2 = -\frac{3}{16}$ in system (7.4.10). The initial values $\alpha_1 = \gamma_1 = 10$. Then $\lambda_1 = \lambda_3 = \frac{3}{4}$ and $\lambda_2 = \lambda_4 = \frac{1}{4}$. Therefore $|\lambda_i| < 1\ (i = 1, \ldots, 4)$. Figure 7.22a, b are the procession of identification of the parameters $\alpha$ and $\gamma$ respectively. Apparently, $\lim_{n\to\infty} \alpha_n = 3.1$ and $\lim_{n\to\infty} \gamma_n = 0$. Figure 7.23a is the Julia set of system (7.4.8) when $\beta = 0.8,\ \alpha = 3.1, \gamma = 0$, which is also the one when $\beta = 0.8,\ \alpha = 3.1$ in system (7.3.1). When the iteration numbers are 15, 18 and 20, the identifications of parameters are $\alpha_{15} = 3.2844,\ \gamma_{15} = 0.2673,\ \alpha_{18} = 3.1778,\ \gamma_{18} = 0.1128$ and $\alpha_{20} = 3.1438,\ \gamma_{20} = 0.0634$. The corresponding Julia sets are Fig. 7.23b, c and d respectively, which illustrate that Julia set of system (7.4.8) is identified with the pushing on the identification of the parameters.

# Chapter 8
# Applications of Fractal Control in Biologies

In the 1920s, the American mathematician Lotka obtained the famous mathematical model [365] about the quantitative relation between hosts and parasites; in 1926, the Italian mathematician Volterra proposed the predator-prey model [366]. Since population competition model is proposed by Lotka and Volterra respectively, it is often referred to as Lotka–Volterra model. The Lotka–Volterra biomathematics model is a quantitative ecology model. The study of the model provides the theoretical foundation for the population competition relations and gives an important influence on the development of the modern ecology theory.

Based on the basic concepts and theoretical framework of this model, many mathematical and biomathematical researchers have done some improvements and practical researches. For the three-species population model with both the predator-prey relationships and the competitive relations, the unique existence condition of globally asymptotic stable periodic solutions is obtained [367] from the viewpoint of population perspective periodic coefficients of Lotka–Volterra. For the three-groups Lotka–Volterra competition model with self-diffusion and cross-diffusion, the global existence of the uniformly bounded solutions is discussed from the viewpoint of self-diffusion and cross-diffusion, and the global asymptotic stability [368] of the positive equilibrium point is proved. For non-autonomous two species competition system, one group may be diffused between the two patches, while the other population does not spread in one patch. The unique existence condition for the globally asymptotically stable positive periodic solutions is got [369] by use of the perspective of Liapunov function.

The attractive domain is a set of the initial states of the system, and the trajectories from this set converge to some attractor (which may be stable fixed points or periodic orbits or chaotic attractors, etc.). Estimation of the stable neighborhood is a very complex and interesting problem. And the boundary of the attractive domain may be very complex and may even be fractal even if the system is very simple and the attractor is a simple fixed point [370, 371]. The structure of the attractor itself is strange (probably strange attractors), therefore it is a complex work to estimate the attractive domain. But the work is very important since the system is clearly


S. T. Liu et al., *Fractal Control and Its Applications*, https://doi.org/10.1007/978-981-15-5459-9_8

understood if we know the size and structure of the attractor. Therefore, a complete analysis of a complex system is not only to analyze complexity of its attractor but also to analyze the complexity of its attractive domain from the viewpoint of the theoretical study. In addition, from the viewpoint of the application, the estimation of the attractive domain can improve the techniques of the chaos control. And the population persistence and extinction are related to the initial quantity of the population [372].

## 8.1 Fractal Analysis and Control of the Competition Model

It is common that in nature the two groups compete for the same resources from outside the system as food in competition system. The demand of resources will increase with the population increasing. However, due to the limited space and resources, it will lead to the survival of their biological environment resistance and mutual inhibition between the two groups. Finally, it will make the population change. The persistence and extinction of the population are relevant to the initial population. Assume that two competitive species in the model are populations a and b respectively. The demand for resources will be too much if the initial number of populations a or b is too large. It will lead to a shortage of resources and inadequate supply of resources, which will lead to the extinction of the two groups finally. Its competitiveness would be obviously insufficient relative to the other one if the initial number of population a or b is too small. And its number will get smaller and smaller because of the limited resources. In the course of time, the population will face extinction. However under certain conditions, as long as the initial numbers of the two competed populations are controlled in a certain range (not too big and not too small), we can ensure the survival of the two groups last forever [373].

### 8.1.1 Julia Set of Two Species Competition Model

The following two-species competition model is studied in [374]

$$\begin{cases} x(n+1) = x(n)\exp[r_1(n)(1 - \frac{x(n)}{k_1(n)} - \mu_2(n)y(n))], \\ y(n+1) = y(n)\exp[r_2(n)(1 - \mu_1(n)x(n) - \frac{y(n)}{k_2(n)})], \end{cases} \tag{8.1.1}$$

where $x(n)$ and $y(n)$ are the numbers of populations 1 and 2 respectively; $r_i(n)$ is the natural growth rate of population $i$; $k_i(n)$ is the maximum environmental capacity of population $i$, $(i = 1, 2)$. $\mu_1(n) = \frac{\sigma_1}{k_2}$, where $\sigma_1$ is the unit number of a population (relative to $k_1$) to support the amount of food consumption in two populations as the unit number of population (relative to $k_2$) to support the amount of food consumed in two populations of the $\sigma_1$ times. $\mu_2(n) = \frac{\sigma_2}{k_1}$ is defined similarly. Besides, $k_i(n)$,

$\mu_i(n)$, $r_i(n)$ are all non-negative bounded sequences for $i = 1, 2$, that is

$$0 < k_{i*} \le k_i{}^*,\ 0 < r_{i*} \le r_i{}^*,\ 0 < \mu_{i*} \le \mu_i{}^*,$$

where $a^* = \sup_{(n\in N)}\{a(n)\}$, $a_* = \inf_{(n\in N)}\{a(n)\}$ for any bounded sequence $\{a(n)\}$. The persistent of the system is studied in [374]. Moreover, they provide a mechanism, which models the oscillatory property of the species. When the coefficients are periodic, the system has a periodic solution which is globally stable under some additional conditions.

Conveniently, let $r_i(n), k_i(n), \mu_i(n)$ be constants in model (8.1.1) for $i = 1, 2$, that is,

$$\begin{cases} x(n+1) = x(n)\exp[r_1(1 - \frac{x(n)}{k_1} - \mu_2 y(n))], \\ y(n+1) = y(n)\exp[r_2(1 - \mu_1 x(n) - \frac{y(n)}{k_2})]. \end{cases} \tag{8.1.2}$$

Let $z_n = x_n + iy_n$ and regard the system (8.1.2) as a map on the complex plane to give the Julia set of system (8.1.2). Let

$$F(x, y) = x\ \exp\left[r_1\left(1 - \frac{x}{k_1} - \mu_2 y\right)\right],\ G(x, y) = y\ \exp\left[r_2\left(1 - \mu_1 x - \frac{y}{k_2}\right)\right],$$

$$H(x, y) = (F(x, y), G(x, y))$$

and

$$D = \{(x, y)|\{|H^n(x, y)|\}_{n=1}^{\infty} \text{ remains bounded}\}.$$

Define Julia set of the system (8.1.2) to be the boundary of $D$. From the property of Julia set, the equivalent definitions of model (8.1.2) $J(F) = \partial R$ and $J(f) = \partial A(\omega)$ can be obtained.

Take $r_1 = 0.8, r_2 = 1.9, k_1 = 1.6, k_2 = 1.4, \mu_1 = 0.4, \mu_2 = 0.1$ in system (8.1.2), then the corresponding Julia set of system (8.1.2) is shown in Fig. 8.1

Figure 8.1 represents the initial numbers of the two competed populations. The numbers of the two groups will be stable if the values of the initial numbers of two groups are inside the closed curve, and the numbers of one of the populations, or both the two populations will tend to infinity if the values are outside of the curve, that signifies the numbers of populations will diverge to infinity.

It must be pointed out that the module of the eigenvalues of the Jacobian matrix at the fixed point $\omega$ are smaller than 1 is one of the sufficient conditions to guarantee the attractivity of the fixed point $\omega$. Thus the corresponding range of the values of the control parameters can be obtained.

$$\begin{cases} f(x^*, y^*) = x^*, \\ g(x^*, y^*) = y^*, \end{cases}$$

that is

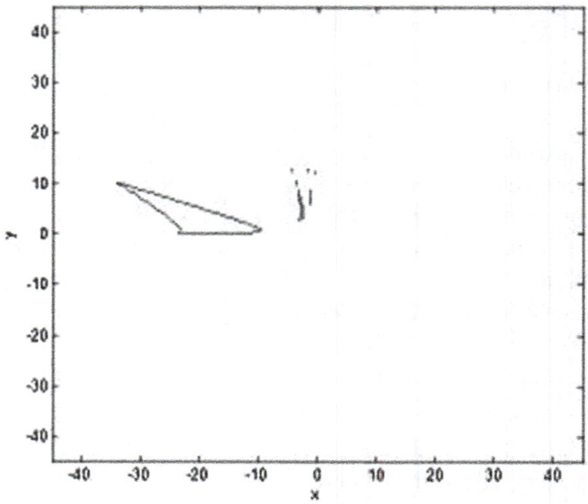


**Fig. 8.1** Julia set of system (8.1.2) when $r_1 = 0.8$, $r_2 = 1.9$, $k_1 = 1.6$, $k_2 = 1.4$, $\mu_1 = 0.4$, $\mu_2 = 0.1$

$$\begin{cases} x^* e^{r_1(1-\frac{x^*}{k_1}-\mu_2 y^*)} = x^*, \\ y^* e^{r_2(1-\mu_1 x^*-\frac{y^*}{k_2})} = y^*. \end{cases} \tag{8.1.3}$$

The solutions of the above Eq. (8.1.3) are as follows.

(1) $(x^*, y^*) = (0,0)$.

(2) If $x^* = 0$, $y^* \neq 0$, then $e^{r_2(1-\mu_1 x^*-\frac{y^*}{k_2})} = 1 = e^0$, where $r_2 > 0$ since $r_2$ is the growth rate of the population model (8.1.2). Therefore $y^* = k_2$. The fixed point of the system is $(x^*, y^*) = (0, k_2)$.

(3) If $x^* \neq 0$, $y^* = 0$, we can obtain $(x^*, y^*) = (k_1, 0)$ in the similarly way.

(4) If $x^* \neq 0$, $y^* \neq 0$, we have

$$\begin{cases} e^{r_1(1-\frac{x^*}{k_1}-\mu_2 y^*)} = 1 = e^0, \\ e^{r_2(1-\mu_1 x^*-\frac{y^*}{k_2})} = 1 = e^0. \end{cases}$$

Therefore

$$\begin{cases} x^* = \frac{k_1 k_2 \mu_2 - k_1}{k_1 k_2 \mu_1 \mu_2 - 1}, \\ y^* = \frac{k_1 k_2 \mu_1 - k_2}{k_1 k_2 \mu_1 \mu_2 - 1}. \end{cases}$$

The Jacobian of system (8.1.2) at the fixed point $(x^*, y^*)$ is

$$J = \begin{bmatrix} (1 - \frac{r_1 x^*}{k_1}) e^{r_1(1-\frac{x^*}{k_1}-\mu_2 y^*)} & -r_1 \mu_2 x^* e^{r_1(1-\frac{x^*}{k_1}-\mu_2 y^*)} \\ -r_2 \mu_1 y^* e^{r_2(1-\mu_1 x^*-\frac{y^*}{k_2})} & (1 - \frac{r_2 y^*}{k_2}) e^{r_2(1-\mu_1 x^*-\frac{y^*}{k_2})} \end{bmatrix}.$$

We hope the two competition species can exist for a long time, and neither of them becomes extinct in practice. Therefore we take the fixed point of the model as $(x^*, y^*) = (\frac{k_1k_2\mu_2-k_1}{k_1k_2\mu_1\mu_2-1}, \frac{k_1k_2\mu_1-k_2}{k_1k_2\mu_1\mu_2-1})$. Then the Jacobian at this fixed point is

$$J = \begin{bmatrix} 1-\frac{r_1x^*}{k_1} & -r_1\mu_2x^* \\ -r_2\mu_1y^* & 1-\frac{r_2y^*}{k_2} \end{bmatrix}.$$

Let

$$|\lambda E - J| = \begin{vmatrix} \lambda+\frac{r_1x^*}{k_1}-1 & r_1\mu_2x^* \\ r_2\mu_1y^* & \lambda+\frac{r_2y^*}{k_2}-1 \end{vmatrix},$$

where $E$ is the two order unit matrix. We get

$$\left(\lambda+\frac{r_1x^*}{k_1}-1\right)\left(\lambda+\frac{r_2y^*}{k_2}-1\right)-r_1r_2\mu_1\mu_2x^*y^*=0.$$

Then

$$\lambda^2+(\frac{r_1x^*}{k_1}+\frac{r_2y^*}{k_2}-2)\lambda+r_1r_2\frac{x^*}{k_1}\frac{y^*}{k_2}-r_1r_2\mu_1\mu_2x^*y^*-\frac{r_1x^*}{k_1}-\frac{r_2y^*}{k_2}+1=0.$$

Substitute

$$\begin{cases} x^* = \frac{k_1k_2\mu_2-k_1}{k_1k_2\mu_1\mu_2-1} \\ y^* = \frac{k_1k_2\mu_1-k_2}{k_1k_2\mu_1\mu_2-1} \end{cases}$$

into the above equation, then we have

$$\lambda^2+\frac{r_1(k_2\mu_2-1)+r_2(k_1\mu_1-1)-2(k_1k_2\mu_1\mu_2-1)}{k_1k_2\mu_1\mu_2-1}$$
$$\lambda+r_1r_2\frac{k_1\mu_1+k_2\mu_2-1-k_1k_2\mu_1\mu_2}{k_1k_2\mu_1\mu_2-1}-\frac{r_1(k_2\mu_2-1)+r_2(k_1\mu_1-1)}{k_1k_2\mu_1\mu_2-1}+1=0.$$

For convenience, let $a=\frac{r_1(k_2\mu_2-1)+r_2(k_1\mu_1-1)-2(k_1k_2\mu_1\mu_2-1)}{k_1k_2\mu_1\mu_2-1}$, $b=r_1r_2\frac{k_1\mu_1+k_2\mu_2-1-k_1k_2\mu_1\mu_2}{k_1k_2\mu_1\mu_2-1}-\frac{r_1(k_2\mu_2-1)+r_2(k_1\mu_1-1)}{k_1k_2\mu_1\mu_2-1}+1$. Then the above equation becomes

$$\lambda^2+a\lambda+b=0.$$

Thus the solutions of the equation can be obtained

$$\lambda_1=\frac{-a+\sqrt{a^2+4b}}{2}, \lambda_2=\frac{-a-\sqrt{a^2+4b}}{2}.$$

Therefore the fixed point $(x^*, y^*) = (\frac{k_1k_2\mu_2-k_1}{k_1k_2\mu_1\mu_2-1}, \frac{k_1k_2\mu_1-k_2}{k_1k_2\mu_1\mu_2-1})$ is attractive when $|\lambda_1| < 1$ and $|\lambda_2| < 1$, that is $|\frac{-a+\sqrt{a^2+4b}}{2}| < 1$ and $|\frac{-c-\sqrt{a^2+4b}}{2}| < 1$.

### 8.1.2 Control of Julia Set of the Competition Model

The stability of the system plays a major role in the analysis of the ecological system. A stable ecological system is very important to maintain the phase function. The ecological balance is destructed seriously with the development of the industry in recent years. We can remove some harmful control, and design some useful controls so that we can better protect the resources through the research of controling the system. Therefore, taking effective control on the system artificially is essential to maintain long-term ecosystem balance. Julia sets depicting the initial numbers of the systems can be restricted effectively according to the practical need of objective problems by controlling model (8.1.2), which meets the needs of practical problems.

Introduce the auxiliary reference feedback control items into system (8.1.2), then we have

$$\begin{cases} x(n+1) = x(n)\exp[r_1(1 - \frac{x(n)}{k_1} - \mu_2 y(n)) - p(x(n) - x_{ref}))], \\ y(n+1) = y(n)\exp[r_2(1 - \mu_1 x(n) - \frac{y(n)}{k_2} - q(y(n) - y_{ref}))], \end{cases} \tag{8.1.4}$$

where $x_{ref}$ and $y_{ref}$ are auxiliary reference items. $p$ and $q$ are the feedback control parameters.

From the above discussion, $(x^*, y^*)$ is the fixed point of system (8.1.2) which satisfies (8.1.2). Now we hope that it is attractive under the control. Thus it is also the fixed point of the controlled system (8.1.4), which means it satisfies the following equations

$$\begin{cases} x^* = x^* e^{r_1(1-\frac{x^*}{k_1}-\mu_2 y^* - p(x^*-x_{ref}))}, \\ y^* = y^* e^{r_2(1-\mu_1 x^* - \frac{y^*}{k_2} - q(y^*-y_{ref}))}. \end{cases}$$

Therefore, auxiliary reference items $x_{ref}$,$y_{ref}$ must be taken to be the fixed point $(x^*, y^*)$ of system (8.1.2). Thus the competition controlled model is as follows

$$\begin{cases} x(n+1) = x(n)\exp[r_1(1 - \frac{x(n)}{k_1} - \mu_2 y(n)) - p(x(n) - x^*))], \\ y(n+1) = y(n)\exp[r_2(1 - \mu_1 x(n) - \frac{y(n)}{k_2} - q(y(n) - y^*))]. \end{cases} \tag{8.1.5}$$

For convenience, we denote

$$\begin{cases} F(x, y) = x e^{r_1(1-\frac{x}{k_1}-\mu_2 y - p(x-x^*))}, \\ G(x, y) = y e^{r_2(1-\mu_1 x - \frac{y}{k_2} - q(y-y^*))}. \end{cases}$$

Obviously, the fixed point $(x^*, y^*) = (\frac{k_1k_2\mu_2-k_1}{k_1k_2\mu_1\mu_2-1}, \frac{k_1k_2\mu_1-k_2}{k_1k_2\mu_1\mu_2-1})$ of the original system (8.1.2) is still the fixed point of the controlled model (8.1.5).The Jacobian matrix of the controlled model (8.1.5) at the fixed point $(x^*, y^*)$ is

$$J = \begin{bmatrix} (1-\frac{r_1x^*}{k_1}-pr_1x^*)e^{r_1(1-\frac{x^*}{k_1}-\mu_2y^*)} & -r_1\mu_2x^*e^{r_1(1-\frac{x^*}{k_1}-\mu_2y^*)} \\ -r_2\mu_1y^*e^{r_2(1-\mu_1x^*-\frac{y^*}{k_2})} & (1-\frac{r_2y^*}{k_2}-qr_2y^*)e^{r_2(1-\mu_1x^*-\frac{y^*}{k_2})} \end{bmatrix}$$

$$= \begin{bmatrix} 1-\frac{r_1x^*}{k_1}-pr_1x^* & -r_1\mu_2x^* \\ -r_2\mu_1y^* & 1-\frac{r_2y^*}{k_2}-qr_2y^* \end{bmatrix}.$$

Then we have

$$|\lambda E - J| = \begin{vmatrix} \lambda+\frac{r_1x^*}{k_1}+pr_1x^*-1 & r_1\mu_2x^* \\ r_2\mu_1y^* & \lambda+\frac{r_2y^*}{k_2}+qr_2y^*-1 \end{vmatrix} = 0.$$

The following equation can be obtained.

$$\lambda^2 + \left(\frac{r_1x^*}{k_1}+\frac{r_2y^*}{k_2}+pr_1x^*+qr_2y^*-2\right)$$
$$\lambda + \left(1-\frac{r_1x^*}{k_1}-pr_1x^*\right)\left(1-\frac{r_2y^*}{k_2}-qr_2y^*\right) - r_1r_2\mu_1\mu_2x^*y^* = 0.$$

Substitute the fixed point $(x^*, y^*)$ into the above formula, then we have

$$\lambda^2 + \frac{r_1(k_2\mu_2-1+pk_1k_2\mu_2-pk_1)+r_2(k_1\mu_1-1+qk_1k_2\mu_1-qk_2)-2(k_1k_2\mu_1\mu_2-1)}{k_1k_2\mu_1\mu_2-1}\lambda +$$
$$\left[1-\frac{r_1(k_2\mu_2-1+pk_1k_2\mu_2-pk_1)}{k_1k_2\mu_1\mu_2-1}\right]\left[1-\frac{r_2(k_1\mu_1-1+qk_1k_2\mu_1-qk_2)}{k_1k_2\mu_1\mu_2-1}\right]-$$
$$\frac{r_1r_2\mu_1\mu_2(k_1k_2\mu_2-k_1)(k_1k_2\mu_1-k_2)}{(k_1k_2\mu_1\mu_2-1)^2} = 0.$$

Let

$$c = \frac{r_1(k_2\mu_2-1+pk_1k_2\mu_2-pk_1)+r_2(k_1\mu_1-1+qk_1k_2\mu_1-qk_2)-2(k_1k_2\mu_1\mu_2-1)}{k_1k_2\mu_1\mu_2-1},$$

$$d = \left[1-\frac{r_1(k_2\mu_2-1+pk_1k_2\mu_2-pk_1)}{k_1k_2\mu_1\mu_2-1}\right]\left[1-\frac{r_2(k_1\mu_1-1+qk_1k_2\mu_1-qk_2)}{k_1k_2\mu_1\mu_2-1}\right]$$
$$-\frac{r_1r_2\mu_1\mu_2(k_1k_2\mu_2-k_1)(k_1k_2\mu_1-k_2)}{(k_1k_2\mu_1\mu_2-1)^2}.$$

Thus the above equation can be rewritten as

$$\lambda^2 + c\lambda + d = 0.$$

The solutions of the equation are

$$\lambda_1' = \frac{-c+\sqrt{c^2+4d}}{2}, \lambda_2' = \frac{-c-\sqrt{c^2+4d}}{2}.$$

To guarantee the fixed point $(x^*, y^*) = (\frac{k_1k_2\mu_2-k_1}{k_1k_2\mu_1\mu_2-1}, \frac{k_1k_2\mu_1-k_2}{k_1k_2\mu_1\mu_2-1})$ attractive, we take $|\lambda_1'| < 1$ and $|\lambda_2'| < 1$, then we have $|\frac{-c+\sqrt{c^2+4d}}{2}| < 1$ and $|\frac{-c-\sqrt{c^2+4d}}{2}| < 1$. Thus the range of the control parameters $p$ and $q$ can be obtained.

For example, the system parameters of competition model (8.1.2) are taken as $r_1 = 0.8, r_2 = 1.9, k_1 = 1.6, k_2 = 1.4, \mu_1 = 0.4, \mu_2 = 0.1$. The corresponding Julia set is illustrated in Fig. 8.1. And the fixed point is $(x^*, y^*) = (1.5114, 0.5536)$. The Jacobian matrix of the controlled model at this fixed point is

$$J = \begin{bmatrix} 1-\frac{r_1x^*}{k_1} - pr_1x^* & -r_1\mu_2x^* \\ -r_2\mu_1y^* & 1-\frac{r_2y^*}{k_2} - qr_2y^* \end{bmatrix}$$

$$= \begin{bmatrix} 0.2443 - 1.2091p & -0.1209 \\ -0.4207 & 0.2487 - 1.0518q \end{bmatrix}.$$

Assume that $\lambda$ are the eigenvalues of $J$, and let

$$|\lambda E - J| = \begin{vmatrix} \lambda + 1.2091p - 0.2443 & 0.1209 \\ 0.4207 & \lambda + 1.0518q + 0.2487 \end{vmatrix} = 0.$$

We can obtain

$$(\lambda + 1.2091p - 0.2443)(\lambda + 1.0518q + 0.2487) - (0.1209)(0.4207) = 0$$

that is

$$\lambda^2 + (1.2091p + 1.0518q + 0.0044)\lambda + 1.2717pq + 0.3007p - 0.2570q - 0.1117 = 0.$$

Let $c' = 1.2091p + 1.0518q + 0.0044, d' = 1.2717pq + 0.3007p - 0.2570q - 0.1117$. Thus the above equation can be written as

$$\lambda^2 + c'\lambda + d' = 0.$$

The solutions of the equation are

$$\lambda_1'' = \frac{-c'+\sqrt{c'^2+4d'}}{2}, \lambda_2'' = \frac{-c'-\sqrt{c'^2+4d'}}{2}.$$

As the previous discussion, we can obtain the range of parameters $p$ and $q$ from $|\frac{-c'\pm\sqrt{c'^2+4d'}}{2}| < 1$.

Now we discuss the change of Julia set of the controlled model.

It is understood that Julia sets of the controlled system change as the values of the controlled parameters change (see Fig. 8.2). Julia set of the controlled model shrinks with the increasing of the controlled parameter $p$ when the controlled parameter $q$ is fixed (see Fig. 8.2a–c). Julia set of the controlled model shrinks with the increasing of the controlled parameter $q$ when the controlled parameter $p$ is fixed (see Fig. 8.2d–f).

### 8.1.3 Synchronization of Julia Sets of Model (8.1.2)

We will discuss the synchronization of Julia sets of system (8.1.2) with different parameters. The population competition system (8.1.2) is rewritten as

$$\begin{cases} x(n+1) = x(n)\exp[r_1(1-\frac{x(n)}{k_1}-\mu_2 y(n))], \\ y(n+1) = y(n)\exp[r_2(1-\mu_1 x(n)-\frac{y(n)}{k_2})], \end{cases} \tag{8.1.6}$$

and

$$\begin{cases} u(n+1) = u(n)\exp[r_1'(1-\frac{u(n)}{k_1'}-\mu_2' v(n))], \\ v(n+1) = v(n)\exp[r_2'(1-\mu_1' u(n)-\frac{v(n)}{k_2'})], \end{cases} \tag{8.1.7}$$

where parameters $r_1'$ and $r_1$, $r_2'$ and $r_2$, $k_1'$ and $k_1$, $k_2'$ and $k_2$, $\mu_1'$ and $\mu_1$, $\mu_2'$ and $\mu_2$ are not completely same, where $r_i$ and $r'_i$ are the natural growth rates of population $i$ of the system (8.1.6) and (8.1.7); $k_i$ and $k'_i$ are the maximum environmental capacity of population $i$, $(i = 1, 2)$ of the system (8.1.6) and (8.1.7); $\mu_1 = \frac{\sigma_1}{k_2}$, where $\sigma_1$ is the unit number of a population (relative to $k_1$) to support the amount of food consumption in two populations as the unit number of population (relative to $k_2$) to support the amount of food consumed in two populations of the $\sigma_1$ times, $\mu_2 = \frac{\sigma_2}{k_1}$, $\mu'_1 = \frac{\sigma'_1}{k'_2}$, $\mu'_2 = \frac{\sigma'_2}{k'_1}$ are defined similarly. Introduce the coupling item into system (8.1.6), then we have

$$\begin{cases} x(n+1) = [x(n)+k(u(n)-x(n))]\exp[r_1(1-\frac{x(n)}{k_1}-\mu_2 y(n))+ \\ \qquad k(r_1'(1-\frac{u(n)}{k_1'}-\mu_2' v(n))-r_1(1-\frac{x(n)}{k_1}-\mu_2 y(n)))], \\ y(n+1) = [y(n)+k(v(n)-y(n))]\exp[r_2(1-\mu_1 x(n)-\frac{y(n)}{k_2})+ \\ \qquad k(r_2'(1-\mu_1' u(n)-\frac{v(n)}{k_2'})-r_2(1-\mu_1 x(n)-\frac{y(n)}{k_2}))]. \end{cases} \tag{8.1.8}$$

Based on the definition of Julia set, we know Julia set can be obtained by the iteration of points within a bounded region. Assume that the bounded region is $A$, then we just need to consider the iteration of the points inside of $A$. In addition, we only need to calculate the iteration trajectory of the points inside of $A$ since $z \notin J$ if there exists $n_0$ such that $f^{n_0}(z) \notin A$ from the properties of Julia set. And because $A$ is a bounded region, there exists $M > 0$ which satisfies

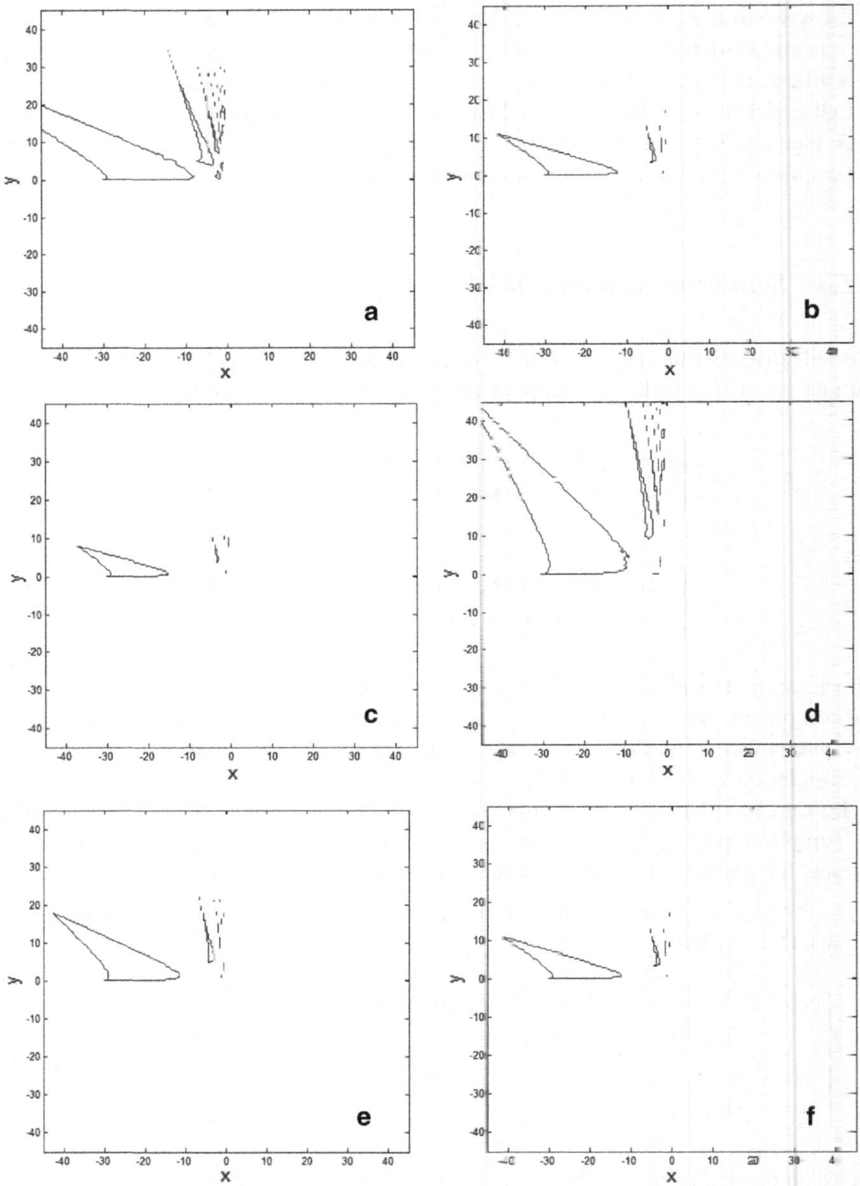


**Fig. 8.2** The changing of Julia sets of the controlled system (8.1.4) when $r_1 = 0.8$, $r_2 = 1.9$, $k_1 = 1.6$, $k_2 = 1.4$, $\mu_1 = 0.4$, $\mu_2 = 0.1$. **a** $p = -0.29, q = -0.1$; **b** $p = -0.11, q = -0.1$; **c** $p = -0.06, q = -0.1$; **d** $p = -0.11, q = -0.5$; **e** $p = -0.11, q = -0.3$; **f** $p = -0.11, q = -0$

$$\exp[r_1'(1-\tfrac{u(n)}{k_1'}-\mu_2'v(n))]<M,$$
$$\exp[r_2'(1-\mu_1'u(n)-\tfrac{v(n)}{k_2'})]<M,$$
$$\exp[r_1(1-\tfrac{x(n)}{k_1}-\mu_2y(n))+k(r_1'(1-\tfrac{u(n)}{k_1'}-\mu_2'v(n))-r_1(1-\tfrac{x(n)}{k_1}-\mu_2y(n)))]<M,$$
$$\exp[r_2(1-\mu_1x(n)-\tfrac{y(n)}{k_2})+k(r_2'(1-\mu_1'u(n)-\tfrac{v(n)}{k_2'})-r_2(1-\mu_1x(n)-\tfrac{y(n)}{k_2}))]<M$$

for all $z\in D$. Thus

$$\begin{aligned}
&|x(n+1)-u(n+1)|\\
&=|[x(n)+k(u(n)-x(n))]\exp[r_1(1-\tfrac{x(n)}{k_1}-\mu_2y(n))+k(r'_1(1-\tfrac{u(n)}{k'_1}-\mu'_2v(n))\\
&-r_1(1-\tfrac{x(n)}{k_1}-\mu_2y(n)))]-u(n)\exp[r'_1(1-\tfrac{u(n)}{k'_1}-\mu'_2v(n))]|\\
&\le |M[x(n)+k(u(n)-x(n))]-Mu(n)|\\
&=M|(1-k)(x(n)-u(n))|\\
&\le M|1-k||x(n)-u(n)|\\
&\le M|1-k|^2|x(n-1)-u(n-1)|\\
&\le\cdots\le M|1-k|^n|x(1)-u(1)|.
\end{aligned}$$

$|x(1)-u(1)|$ is bounded since the original values $x(1)$ and $u(1)$ are taken inside in a bounded set. Obviously, if $|1-k|\to 0$, that is $k\to 1$ and $n\to\infty$, we have $|x(n+1)-u(n+1)|\to 0$. In a similar way, we can get that $|y(n+1)-v(n+1)|\to 0$ if $|1-k|\to 0$, that is $k\to 1$, and $n\to\infty$. Thus the synchronization of Julia sets of systems (8.1.6) and (8.1.7) is achieved.

The coupling strength of the coupling system can be seen in the process of $k\to 1$, and it is also the degree of synchronization of the Julia sets of the coupling system, which can be seen in the following simulations.

Let the parameters be $r_1=1.35, r_2=0.09, k_1=1.7, k_2=1.2, \mu_1=0.5, \mu_2=0.1$ and $r_1'=0.8, r_2'=1.9, k_1'=1.6, k_2'=1.4, \mu_1'=0.4, \mu_2'=0.1$ in systems (8.1.6) and (8.1.7), then we have

$$\begin{cases} x(n+1)=x(n)\exp[1.35*(1-\frac{x(n)}{1.7}-0.1*y(n))],\\ y(n+1)=y(n)\exp[0.09*(1-0.5*x(n)-\frac{y(n)}{1.2})],\end{cases}$$

and

$$\begin{cases} u(n+1)=u(n)\exp[0.8*(1-\frac{u(n)}{1.6}-0.1*v(n))],\\ v(n+1)=v(n)\exp[1.9*(1-0.4*u(n)-\frac{v(n)}{1.4})].\end{cases}$$

And Julia sets are shown in Figs. 8.3 and 8.1 respectively. The coupling system is

$$\begin{cases} x(n+1)=[x(n)+k(u(n)-x(n))]\exp[1.35(1-\frac{x(n)}{1.7}-0.1y(n))+\\ \qquad k(0.8(1-\frac{u(n)}{1.6}-0.1v(n))-1.35(1-\frac{x(n)}{1.7}-0.1y(n)))],\\ y(n+1)=[y(n)+k(v(n)-y(n))]\exp[0.09(1-0.5x(n)-\frac{y(n)}{1.2})+\\ \qquad k(1.9(1-0.4u(n)-\frac{v(n)}{1.4})-0.09(1-0.5x(n)-\frac{y(n)}{1.2}))].\end{cases} \tag{8.1.9}$$

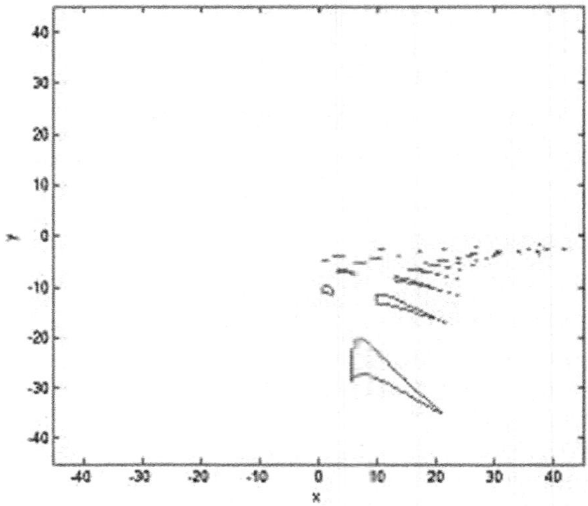


**Fig. 8.3** The original Julia set when $r_1 = 1.35, r_2 = 0.09, k_1 = 1.7, k_2 = 1.2, \mu_1 = 0.5, \mu_2 = 0.11$ in system (8.1.6)

Figure 8.3 is the initial Julia set and Fig. 8.4 is the changing of Julia sets of the coupling system (8.1.8) with different values of $k$.

## 8.2 Fractal Analysis and Control in the Predator-Prey Model

This section is concerned with the following two-dimensional predator-prey Lotka–Volterra model [375],

$$\begin{cases} \dot{x} = x(a_1 - b_1 y), \\ \dot{y} = y(-a_2 + b_2 x). \end{cases} \tag{8.2.1}$$

where $x(t), y(t)$ $(x(t) > 0, y(t) > 0)$ denote the species density of the preys and the predators at time $t$, respectively. The coefficient $a_1 > 0$ denotes the birth rate of the preys, $a_2 > 0$ denotes the birth rate of the predators, the other two coefficients $b_1$ and $b_2$ (both positive) describe interactions between the species. Stability and periodic solutions are discussed about the model [376, 377]. It must be pointed out that the discrete time models governed by difference equations are more appropriate than the continuous one when the populations have non-overlapping generations [378]. Discrete time models can also provide efficient computational models of continuous models for numerical simulations.

$$\dot{x} \to \frac{x(t + \Delta t) - x(t)}{\Delta t}, \ \dot{y} \to \frac{y(t + \Delta t) - y(t)}{\Delta t}.$$

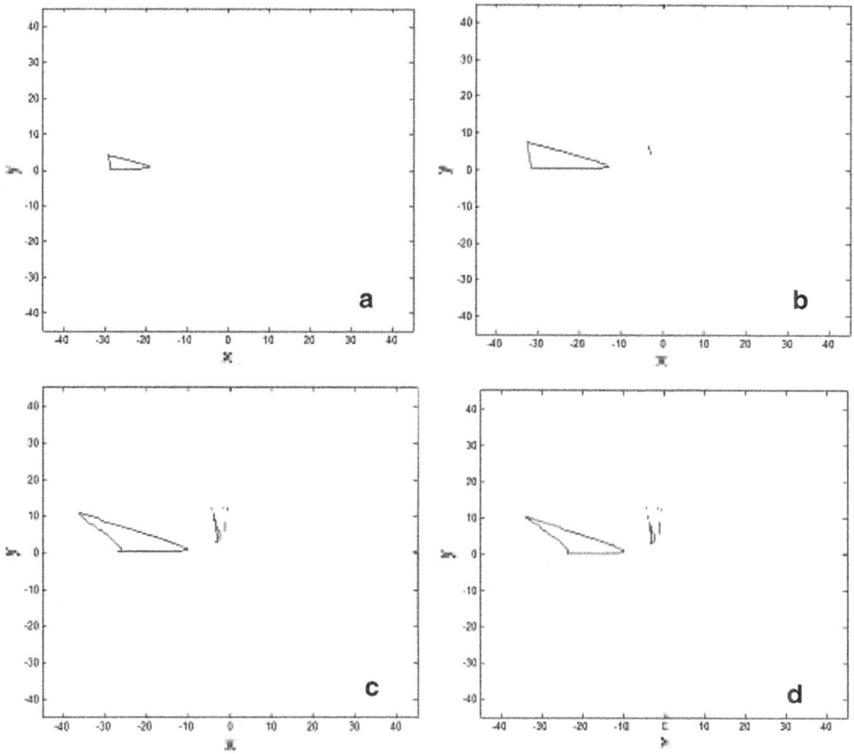

**Fig. 8.4** Julia set synchronization of system (8.1.6) and (8.1.8) with the values of $k$. **a** $k = 0.51$; **b** $k = 0.71$; **c** $k = 0.91$; **d** $k = 0.999$

By plugging these discretizations into the system (8.2.1), we have the discrete version of (8.2.1)

$$x_{n+1} = (1 + a_1\delta)x_n - b_1\delta x_n y_n,$$
$$y_{n+1} = (1 - a_2\delta)y_n + b_2\delta x_n y_n,$$

where the value of $\delta = \Delta t$ is usually small. In fact, let $\varepsilon_1 = a_1\delta, \gamma_1 = b_1\delta, \varepsilon_2 = a_2\delta, \gamma_2 = b_2\delta$, we get another discrete version of (8.2.1),

$$\begin{cases} x_{n+1} = (1 + \varepsilon_1)x_n - \gamma_1 x_n y_n, \\ y_{n+1} = (1 - \varepsilon_2)y_n + \gamma_2 x_n y_n. \end{cases} \tag{8.2.2}$$

Let $K$ be the set that every point in $K$ escapes to infinity while trajectories of points outside $K$ are bounded. The set $\mathbb{R}^2 \setminus K$ is the Julia set of the system, which is a two dimensional set. Various Julia sets can be found corresponding to different values of parameters. In fact, if we take $x$ and $y$ as the real part and the imaginary part of the complex number, Julia set of (8.2.2) can be thought as the Julia set of the complex polynomial. Thus the Julia set of (8.2.2) is nonempty, bounded and fully invariant. What's more, $\partial A(s) = J(h)$ if $s$ is an attractive fixed point of $h$, where $A(s)$ is the attractive domain of the attractive fixed point $s$ and $h$ is the complex polynomial. That is $A(s) = \{z \in \mathbb{C} : h^k(z) \to s, \text{ as } k \to \infty\}$. It is also true when $s = \infty$ [315].

From the definition and properties of Julia set, we know that the iterative trajectory of the system plays an important role in the structure of its Julia set. Therefore, we can achieve the control and synchronization of Julia sets by the iterative trajectory control. Specially, we can consider the stability of the fixed point. Thus, we can control the iterative trajectories of points to achieve the objective of the control of Julia set.

### 8.2.1 Control of Julia Set of System (8.2.2)

The fixed point of system (8.2.2) that we considered is $(x^*, y^*) = (\frac{\varepsilon_2}{\gamma_2}, \frac{\varepsilon_1}{\gamma_1})$ since there is no practical significance for the fixed point $(0, 0)$. At first we analyze the stability of the fixed point $(x^*, y^*)$ of system (8.2.2). For system (8.2.2), its Jacobian $J$ evaluated at the fixed point is

$$J = \begin{bmatrix} 1 + \varepsilon_1 - \gamma_1 y & -\gamma_1 x \\ \gamma_2 y & 1 - \varepsilon_2 + \gamma_2 x \end{bmatrix}_{(x^*, y^*)} = \begin{bmatrix} 1 & -\frac{\gamma_1 \varepsilon_2}{\gamma_2} \\ \frac{\gamma_2 \varepsilon_1}{\gamma_1} & 1 \end{bmatrix}.$$

The characteristic equation of the Jacobi matrix $J$ is

$$|\lambda I - J| = \begin{vmatrix} \lambda - 1 & \frac{\gamma_1 \varepsilon_2}{\gamma_2} \\ -\frac{\gamma_2 \varepsilon_1}{\gamma_1} & \lambda - 1 \end{vmatrix} = (\lambda - 1)^2 + \varepsilon_1 \varepsilon_2 = 0.$$

It is noted that the stable condition of this fixed point$(x^*, y^*)$ is guaranteed by the condition $|\lambda_i| < 1, i = 1, 2$. For any given parameters $\varepsilon_1, \varepsilon_2, \gamma_1$ and $\gamma_2$, the fixed point $(x^*, y^*)$ of system (8.2.2) is not stable since $|\lambda_{1,2}| > 1$, where the eigenvalues $\lambda_{1,2} = 1 \pm \sqrt{\varepsilon_1 \varepsilon_2} i$.

Introduce the linear feedback control item to (8.2.2), we obtain the controlled system

$$\begin{cases} x_{n+1} = (1 + \varepsilon_1) x_n - \gamma_1 x_n y_n + k_1 (x_n - x^*), \\ y_{n+1} = (1 - \varepsilon_2) y_n + \gamma_2 x_n y_n + k_2 (y_n - y^*). \end{cases} \tag{8.2.3}$$

Obviously, the fixed point $(x^*, y^*)$ of system (8.2.2) is also a fixed point of system (8.2.3). We hope the fixed point $(x^*, y^*)$ becomes stable in the controlled system (8.2.3) when $k_i$ $(i = 1, 2)$ are in a certain range.

Similarly, for the controlled system (8.2.3), the Jacobi matrix at the fixed point $(x^*, y^*)$ is

$$P = \begin{bmatrix} 1+\varepsilon_1-\gamma_1 y+k_1 & -\gamma_1 x \\ \gamma_2 y & 1-\varepsilon_2+\gamma_2 x+k_2 \end{bmatrix}_{(x^*,y^*)} = \begin{bmatrix} 1+k_1 & -\frac{\gamma_1\varepsilon_2}{\gamma_2} \\ \frac{\gamma_2\varepsilon_1}{\gamma_1} & 1+k_2 \end{bmatrix}.$$

The characteristic equation of the Jacobi matrix $P$ is

$$\begin{aligned} |\lambda I-P| &= \begin{vmatrix} \lambda-(1+k_1) & \frac{\gamma_1\varepsilon_2}{\gamma_2} \\ -\frac{\gamma_2\varepsilon_1}{\gamma_1} & \lambda-(1+k_2) \end{vmatrix} \\ &= \lambda^2-(2+k_1+k_2)\lambda+(1+k_1)(1+k_2)+\varepsilon_1\varepsilon_2=0. \end{aligned}$$

Then the eigenvalues are

$$\lambda_{1,2} = \frac{(2+k_1+k_2)\pm\sqrt{(k_1-k_2)^2-4\varepsilon_1\varepsilon_2}}{2}.$$

Thus we can get the range of the control parameters by the inequality $|\lambda_{1,2}|<1$.

For simplicity, take $\varepsilon_1=0.2,\ \gamma_1=0.15$ and $\varepsilon_2=0.1,\ \gamma_2=0.1$ in the system (8.2.2) and Julia set is shown in Fig. 8.5. Based on the above discussion, we know that the fixed point $(1, \frac{4}{3})$ is not stable since the eigenvalues are $\lambda'_{1,2}=1\pm\sqrt{0.02}i$. Noting that $(1, \frac{4}{3})$ is also the fixed point of the controlled system and we hope it is stable in the controlled system. Thus we have $|2+k_1+k_2\pm\sqrt{(k_1-k_2)^2-0.08}|<2$ from $|\lambda_{1,2}|<1$ $(i=1,2)$. For convenience to compute, let $k_1=k_2=k$ and then we obtain the range of the control parameter is $E=\{k: -1.9899<k<-0.0101\}$. Figure 8.6 is the changing of Julia sets of the controlled system when $k\in D$. From Fig. 8.6, we can see that Julia sets of the controlled system are expanding with the values of the control parameter $k$ decreasing. Since Julia set is also the boundary of the attractive domain of the attractive fixed point, the attractive domain is controlled which is important to the stability of the system.

### 8.2.2 Synchronization of Julia Sets of System (8.2.2)

Consider a system with the same form to (8.2.2) but with different parameters

$$\begin{cases} u_{n+1}=(1+\varepsilon_1)u_n-\gamma_1 u_n v_n, \\ v_{n+1}=(1-\varepsilon_2)v_n+a_2 u_n v_n, \end{cases} \tag{8.2.4}$$

where $\gamma_2\neq a_2$. Based on the definition of Julia set we know the Julia sets of (8.2.2) and (8.2.4) are determined and unchanged and they don't have any relations. In order to make their Julia sets associate with each other, for example, to make one Julia set change to be the other, we introduce two coupled items into (8.2.4) respectively. Then we have

$$\begin{cases} u_{n+1} = (1+\varepsilon_1)u_n - \gamma_1 u_n v_n + l_1(\gamma_1 u_n v_n - \gamma_1 x_n y_n) - l_1(u_n - x_n), \\ v_{n+1} = (1-\varepsilon_2)v_n + a_2 u_n v_n + l_2(\gamma_2 x_n y_n - a_2 u_n v_n), \end{cases} \quad (8.2.5)$$

**Theorem 8.1** *$|u_{n+1} - x_{n+1}| \to 0$ and $|v_{n+1} - y_{n+1}| \to 0$ when $n \to \infty$ and $l_i \to 1 (i = 1, 2)$, that is, the synchronization of trajectories of systems (8.2.2) and (8.2.5) is achieved.*

***Proof*** From (8.2.2) and (8.2.5), we have

$$u_{n+1} - x_{n+1} = (1 + \varepsilon_1 - l_1)(u_n - x_n) + (1 - l_1)\gamma_1(x_n y_n - u_n v_n).$$

Then

$$|u_{n+1} - x_{n+1}| \le |1 + \varepsilon_1 - l_1||u_n - x_n| + |1 - l_1||\gamma_1|(|x_n y_n| + |u_n v_n|).$$

Since Julia set is bounded, we can obtain Julia set by calculating the iteration of points in a bounded region $D$. Thus we need consider only the iterations of points in $D$. In order to get the Julia set, we consider only the points whose trajectories are also in $D$ since $z \notin J$ if there is an $n_0$ such that $h^{n_0}(z) \notin D$ from the property that Julia set is fully invariant. Since $D$ is a bounded region, there exists a $T > 0$ satisfying $|z| < T$ for all $z \in D$. Since the value of the system parameter $\gamma_1$ is bounded, there are $N > 0$ and $M > 0$ such that $|\gamma_1|(|x_n y_n| + |u_n v_n|) < N$ and $|\gamma_2||x_n||y_n| + |a_2||v_n||x_n| < M$. Next, we take the same iterative original values in systems (8.2.2) and (8.2.4). Then

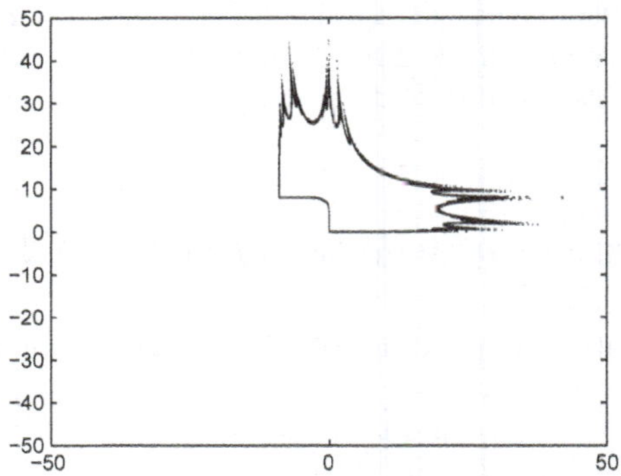


**Fig. 8.5** Original Julia sets of system (8.2.2) when $\varepsilon_1 = 0.2$, $\gamma_1 = 0.15$ and $\varepsilon_2 = 0.1$, $\gamma_2 = 0.11$

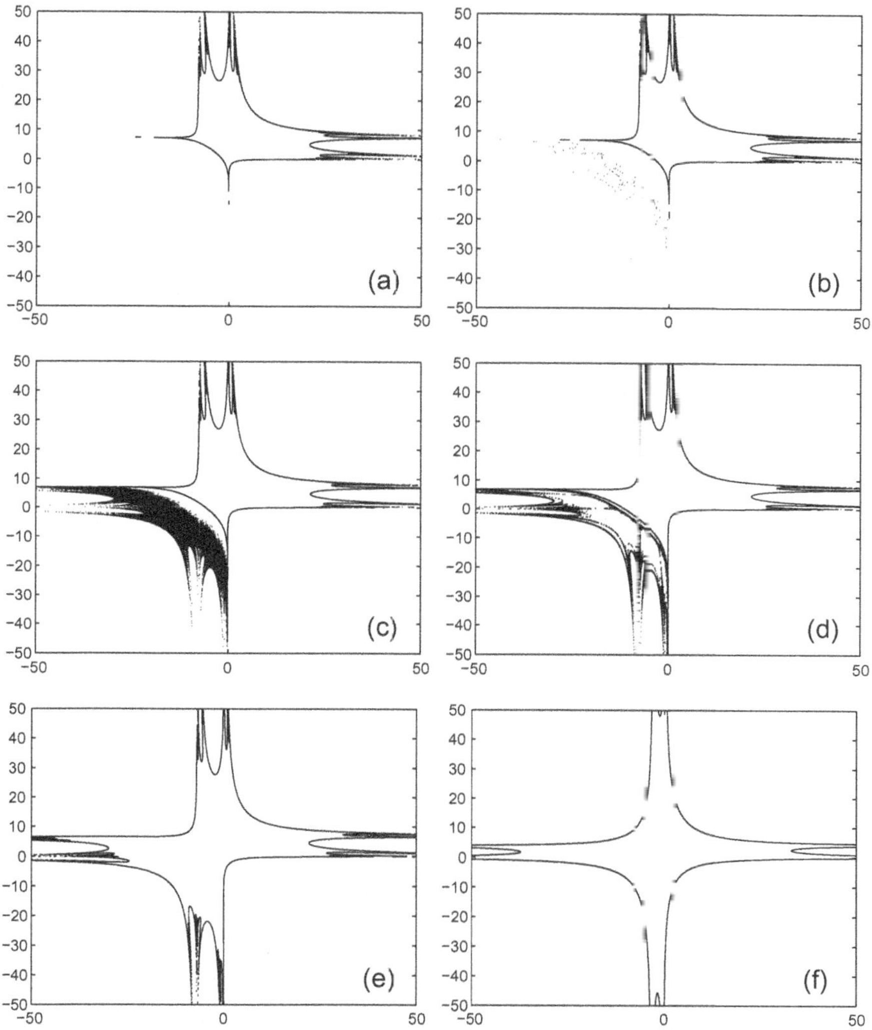


**Fig. 8.6** The changing of Julia sets of the controlled system (8.2.3) when $\varepsilon_1 = 0.2$, $\gamma_1 = 0.15$ and $\varepsilon_2 = 0.1$, $\gamma_2 = 0.11$. **a** $k = -0.11$; **b** $k = -0.13$; **c** $k = -0.15$; **d** $k = -0.17$; **e** $k = -0.21$; **f** $k = -0.61$

$$
\begin{aligned}
&|u_{n+1}-x_{n+1}| \leq |1+\varepsilon_1-l_1||u_n-x_n| + |1-l_1|N \\
&\leq |1+\varepsilon_1-l_1|(|1+\varepsilon_1-l_1||u_{n-1}-x_{n-1}| + |1-l_1|N) + |1-l_1|N \\
&= |1+\varepsilon_1-l_1|^2|u_{n-1}-x_{n-1}| + |1-l_1|N(1+|1+\varepsilon_1-l_1|) \leq \cdots \\
&\leq |1+\varepsilon_1-l_1|^n|u_1-x_1| + |1-l_1|N(1+|1+\varepsilon_1-l_1|+\cdots+|1+\varepsilon_1-l_1|^{n-1}) \\
&= |1+\varepsilon_1-l_1|^n|u_1-x_1| + |1-l_1|N\frac{1-|1+\varepsilon_1-l_1|^n}{1-|1+\varepsilon_1-l_1|}.
\end{aligned}
$$

Take $\varepsilon_1 < l_1 < 1$, then the limitation of the right side of the above equation is $|1-l_1|N\frac{1}{1-|1+\varepsilon_1-l_1|}$ as $n \to \infty$. Obviously, if $|1-l_1| \to 0$, that is $l_1 \to 1$, we have

$$
|u_{n+1}-x_{n+1}| \to 0, \ n \to \infty.
$$

Since

$$
v_{n+1}-y_{n+1} = (1-\varepsilon_2)(v_n-y_n) + (1-l_2)(\gamma_2 x_n y_n - a_2 u_n v_n),
$$

we get

$$
\begin{aligned}
&|v_{n+1}-y_{n+1}| \leq |1-\varepsilon_2||v_n-y_n| + |1-l_2||\gamma_2 x_n y_n - a_2 u_n v_n| \\
&\leq |1-\varepsilon_2||v_n-y_n| + |1-l_2|(|\gamma_2||x_n||y_n| - |a_2||v_n||y_n|)| \\
&\leq |1-\varepsilon_2||v_n-y_n| + |1-l_2|M \\
&\leq |1-\varepsilon_2|(|1-\varepsilon_2||v_{n-1}-y_{n-1}| + |1-l_2|M) + |1-l_2|M \leq \cdots \\
&\leq |1-\varepsilon_2|^n|v_1-y_1| + |1-l_2|(M+M|1-\varepsilon_2|+\cdots+M|1-\varepsilon_2|^{n-1}) \\
&= |1-\varepsilon_2|^n|v_1-y_1| + |1-l_2|M\frac{1-|1-\varepsilon_2|^n}{1-|1-\varepsilon_2|}.
\end{aligned}
$$

Since $0 < \varepsilon_2 < 1$, the limitation of the right side of the above equation is $|1-l_2|M\frac{1}{1-|1-\varepsilon_2|}$ as $n \to \infty$. Obviously, if $|1-l_2| \to 0$, that is $l_2 \to 1$, we have

$$
|v_{n+1}-y_{n+1}| \to 0, \ n \to \infty.
$$

Therefore, the synchronization of the trajectories of systems (8.2.2) and (8.2.5) is achieved. □

It follows from the definition of Julia set and the above Theorem that Julia sets of controlled system (8.2.5) are changing to be the Julia set of system (8.2.2). Therefore, the synchronization of Julia sets of the systems (8.2.2) and (8.2.5) is obtained.

Though the Julia sets are determined when the parameters are given, it is easy to see that the changing process of Julia sets by taking different coupled parameters $l_i$ $(i = 1, 2)$, which are approved by the following simulations. Figure 8.7 is the Julia set of the system (8.2.5) when $\varepsilon_1 = 0.2$, $\gamma_1 = 0.15$ and $\varepsilon_2 = 0.1$, $\alpha_2 = 0.51$. Figure 8.8 is the changing of the Julia sets of the system (8.2.5) with the changing of the control parameters. In Fig. 8.8, the black curve is the objective Julia set shown in Fig. 8.5 and the red curves are the Julia sets of the controlled system (8.2.5) with different control parameters. Obviously, Julia sets of system (8.2.5) are changing to be the Julia set of system (8.2.2) shown in Fig. 8.5.

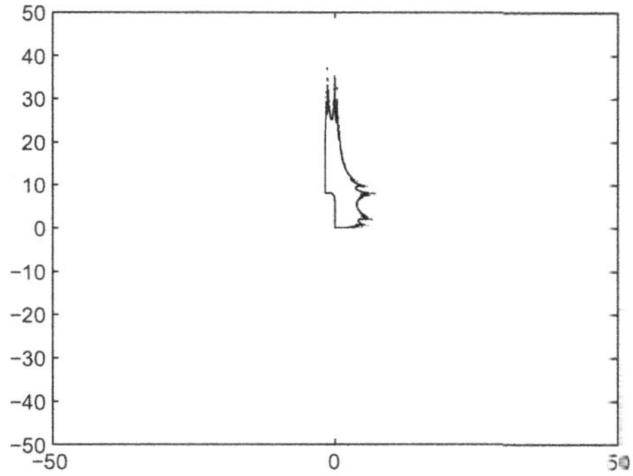


**Fig. 8.7** Original Julia sets of system (8.2.5) when $\varepsilon_1 = 0.2,\ \gamma_1 = 0.15$ and $\varepsilon_2 = 0.1,\ \alpha_2 = 0.51$

## 8.3 Fractal Analysis and Control of the Fractional Lotka–Volterra Model

As fractional differential equations are successfully applied in mathematical biology and some other interdisciplinary fields, some scholars have begun to study fractional Lotka–Volterra system [379] from 2007 and obtained some meaningful results (see e.g., [379–386]). Compared with the classic Lotka–Volterra system, the fractional one exhibits richer dynamic behaviors. Fractional operators are naturally related to memory or fractal phenomenon, so they exist in most biological systems. In 2009 and 2011, Das et al. considered fractional Lotka–Volterra system [380–382]

$$\begin{cases} {}^C D_{0+}^{\alpha} u(t) = au(t) - bu(t)v(t), \\ {}^C D_{0+}^{\beta} v(t) = cu(t)v(t) - dv(t), \\ u(0) = \delta, \quad v(0) = \gamma, \end{cases}$$

where ${}^C D_{0+}^{\alpha}$ and ${}^C D_{0+}^{\beta}$ are two Caputo fractional derivatives with $\alpha, \beta \in (0, 1]$; $a, b, c, d, \delta$ and $\gamma$ are nonnegative numbers. By using the homotopy perturbation method, the authors got some interesting results. In 2015, Elsadany et al. studied the dynamical behaviors of fractional Lotka–Volterra system [384]

$$\begin{cases} {}^C D_{0+}^{\alpha} u(t) = u(t)(r - ku(t) - v(t)), \\ {}^C D_{0+}^{\alpha} v(t) = v(t)(-1 + \beta u(t)) \end{cases}$$

and its discrete counterpart. The authors found that there are three types of bifurcations and chaos in the discrete system, but only stable (unstable) equilibria in the

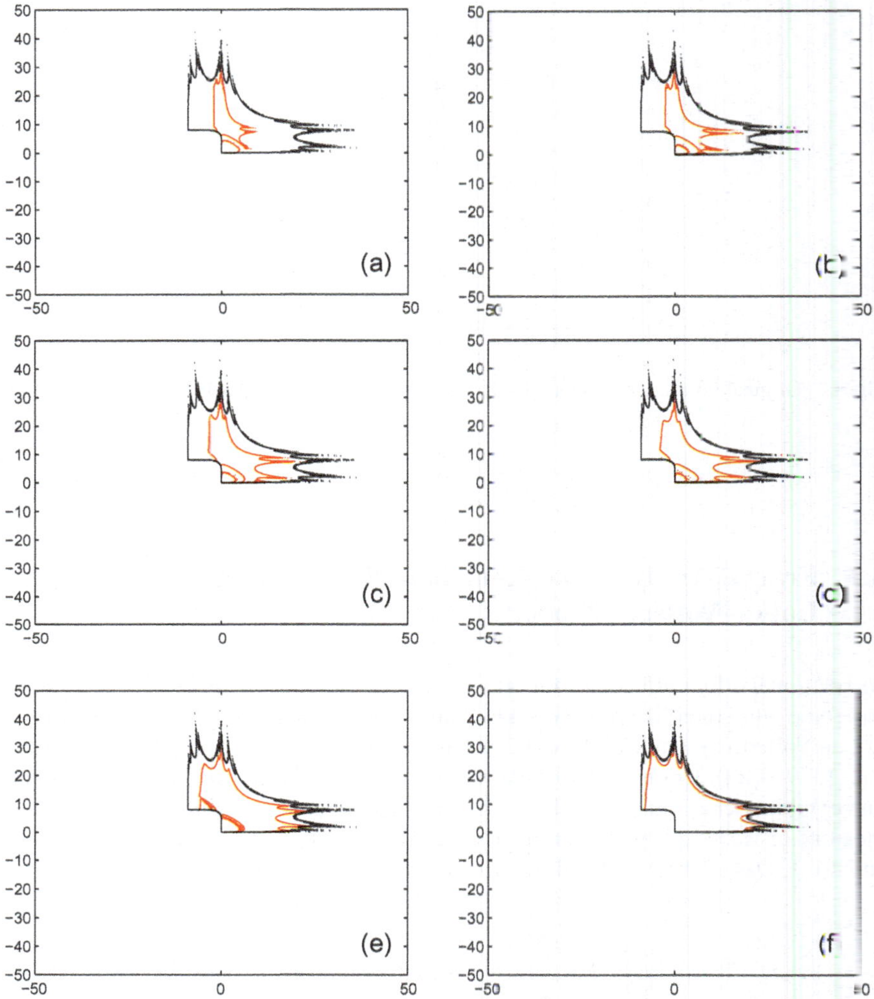


**Fig. 8.8** The changing of Julia sets of the controlled system (8.2.5) when $\varepsilon_1 = 0.2, \gamma_1 = 0.15$ and $\varepsilon_2 = 0.1, a_2 = 0.51$. **a** $l_1 = 0.2, l_2 = 0.11$; **b** $l_1 = 0.4, l_2 = 0.31$; **c** $l_1 = 0.5, l_2 = 0.51$; **d** $l_1 = 0.7, l_2 = 0.51$; **e** $l_1 = 0.8, l_2 = 0.71$; **f** $l_1 = 0.9, l_2 = 0.31$

corresponding continuous system. It is shown that the discrete system exhibits much richer dynamical behaviors than its continuous form.

It must be admitted that fractional calculus and fractal theory arise from various applications, fractional systems are worth studying from the viewpoint of fractal. Motivated by the work above, our purpose is to consider the following fractional Lotka–Volterra system by the ideas and tools of fractal:

$$\begin{cases} D_{0+}^{\alpha} u(t) = r_1 u(t) \left(1 - \frac{u(t)}{N_1} - \sigma_1 \frac{v(t)}{N_2}\right), \\ D_{0+}^{\beta} v(t) = r_2 v(t) \left(-1 + \sigma_2 \frac{u(t)}{N_1} - \frac{v(t)}{N_2}\right), \end{cases} \tag{8.3.1}$$

where $u(t)$ and $v(t)$ can be interpreted as the densities of prey and predator populations at any instant of time $t$, respectively; $r_1 > 0$ is the per capita birth rate of the prey species and $r_2 > 0$ is the per capita death rate of the predator species; $N_1 > 0$ and $N_2 > 0$ denote the maximum environmental capacity of prey and predator populations, respectively; $\sigma_1 > 0$ and $\sigma_2 > 0$ are real parameters describing the interaction of the two species; $D_{0+}^{\alpha}$ and $D_{0+}^{\beta}$ are two fractional derivatives explained later, $\alpha, \beta \in (0, 1]$.

It is well known that fractional calculus is a generalization of the classical (integer-order) differentiation and integration. The three most frequently used definitions for fractional differential operators are: the Riemann–Liouville definition, the Caputo definition and the Grünwald–Letnikov definition. More detailed information can be found in [387–389].

### 8.3.1 Fractional Derivative and Discretization

**Definition 8.1** The Riemann–Liouville fractional derivative of order $\alpha > 0$ of a continuous function $y : (0, +\infty) \to \mathbb{R}$ is given by

$$^{RL}D_{0+}^{\alpha} y(t) = \frac{1}{\Gamma(n-\alpha)} \left(\frac{d}{dt}\right)^n \int_0^t (t-s)^{n-\alpha-1} y(s) ds$$

provided that the right-hand side is pointwise defined on $(0, +\infty)$, where $\Gamma(z)$ is the Gamma function defined as $\Gamma(z) = \int_0^{\infty} t^{z-1} e^{-t} dt$, $n = \lceil \alpha \rceil$ and $\lceil \alpha \rceil$ denotes the ceiling function of the number $\alpha$.

**Definition 8.2** The Caputo fractional derivative of order $\alpha > 0$ of a continuous function $y : (0, +\infty) \to \mathbb{R}$ is given by

$$^{C}D_{0+}^{\alpha} y(t) = \frac{1}{\Gamma(n-\alpha)} \int_0^t (t-s)^{n-\alpha-1} y^{(n)}(s) ds$$

provided that the right-hand side is pointwise defined on $(0, +\infty)$, where $n = \lceil \alpha \rceil$.

**Definition 8.3** The Grünwald–Letnikov fractional derivative of order $\alpha > 0$ of a continuous function $y : (0, +\infty) \to \mathbb{R}$ is given by

$$^{GL}D_{0+}^{\alpha} y(t) = \lim_{n \to +\infty} \frac{1}{h_n^{\alpha}} \sum_{k=0}^{n} (-1)^k \binom{\alpha}{k} y(t - kh_n) \tag{8.3.2}$$

provided that the right-hand side is pointwise defined on $(0, +\infty)$, where $k \in \mathbb{N}$, $h_n = \frac{t}{n}$ is the time step of calculation and $\binom{\alpha}{k}$ is the extended binomial coefficient that is defined as

$$\binom{\alpha}{k} = \frac{\Gamma(\alpha+1)}{\Gamma(k+1)\Gamma(\alpha-k+1)} = \begin{cases} \frac{\alpha(\alpha-1)(\alpha-2)\cdots(\alpha-k+1)}{k!}, & k > 0, \\ 1, & k = 0. \end{cases}$$

**Remark 8.1** In Definitions (8.1)–(8.3), derivatives are defined by either integral or series. It is these nonlocal operations that embody the memory effect (or, nonlocality) of fractional derivatives. So a fundamental property of fractional derivatives is that they are functions taking the "past" into account.

In this section, differential operators ${}^{RL}D^{\alpha}_{0+}$, ${}^{C}D^{\alpha}_{0+}$ and ${}^{G}D^{\alpha}_{0+}$ will be denoted by $D^{\alpha}_{0+}$ uniformly without differentiating. This consideration is based on the fact that, for a wide class of functions, the Definitions (8.1)–(8.3) are equivalent under some conditions [387].

The Grünwald–Letnikov representation is adopted here for its efficiency and ease of numerical implementation. Motivated by the approximation of the numerical evaluation, the Grünwald–Letnikov fractional derivative can be evaluated through a finite sum instead of the infinite sum in Definition (8.3). So a truncation of expression (8.3.2) can be taken as:

$$ {}^{GL}D^{\alpha}_{0+}y(t) \approx \frac{1}{h_N^{\alpha}} \sum_{k=0}^{N} (-1)^k \binom{\alpha}{k} y(t - kh_N) = \frac{1}{h_N^{\alpha}} \sum_{k=0}^{N} \omega_k^{(\alpha)} y(t - kh_N), \quad (8.3.3)$$

where $N$ is called the truncation point, $h_N = \frac{t}{N}$ is the time step of truncated calculation, and

$$\omega_k^{(\alpha)} = (-1)^k \binom{\alpha}{k} = \binom{k-\alpha-1}{k} = \frac{\Gamma(k-\alpha)}{\Gamma(k+1)\Gamma(-\alpha)} \quad (8.3.4)$$

is Grünwald coefficients (or called historical correlation coefficients). For calculation of $\omega_k^{(\alpha)}$ we can use the following expression [387]:

$$\omega_0^{(\alpha)} = 1, \quad \omega_k^{(\alpha)} = \frac{k-\alpha-1}{k}\omega_{k-1}^{(\alpha)}, \quad k = 1, 2, \cdots. \quad (8.3.5)$$

Clearly at the initial time, the unknown terms in the representation (8.3.3) will be taken as equal to zero. This will be done until $N$ time steps are incorporated in the scheme.

**Remark 8.2** In particular, when $\alpha = 0$, we have

$$\omega_0^{(0)} = 1, \quad \omega_k^{(0)} = 0, \quad k = 1, 2, \cdots;$$

when $\alpha = 1$, we have

$$\omega_0^{(1)} = 1, \quad \omega_1^{(1)} = -1, \quad \omega_k^{(1)} = 0, \quad k = 2, 3, \cdots,$$

which means that the memory effect of the fractional derivative vanishes and so truncation (8.3.3) becomes a usual first-order difference scheme. For a fractional derivative defined as Definitions (8.1), (8.2) or (8.3), it is reduced to the classical one in this case [387].

From truncation (8.3.3) it can be seen that to evaluate the present value, the previous memory terms (or called hereditary biologically) must be taken into account. In order to take this effect into consideration efficiently and economically, there is the following lemma about Grünwald coefficients:

**Lemma 8.1** *The series given by Grünwald coefficients is convergent.*

***Proof*** From Eq. (8.3.4), it can be seen that

$$\omega_k^{(\alpha)} = \binom{k-\alpha-1}{k} = \frac{-\alpha(1-\alpha)(2-\alpha)\cdots(k-\alpha-1)}{k!}, \quad k = 1, 2, \cdots.$$

This shows that the series given by $\omega_k^{(\alpha)}$ is of one sign when $k$ becomes larger than the order of derivative $\alpha$. Hence for any $\alpha > 0$, we have

$$\lim_{k\to+\infty} k\left(1 - \frac{\omega_{k+1}^{(\alpha)}}{\omega_k^{(\alpha)}}\right) = \lim_{k\to+\infty} k\left(1 - \frac{k-\alpha}{k+1}\right) = \lim_{k\to+\infty} \frac{(1+\alpha)k}{k+1} = 1 + \alpha > 1.$$

According to the Raabe's Test [390], we get that $\sum_{k=0}^{\infty} \omega_k^{(\alpha)}$ is convergent. The proof is completed. □

**Remark 8.3** From Lemma 8.1, there are the conclusions in [391] immediately:

(1) The sequence given by $|\omega_k^{(\alpha)}|$ is strictly decreasing when $k$ becomes larger than the order of the derivative.
(2) Grünwald coefficients tend to zero as $k$ grows.

From the above reasons, it is known that the influence of the memory effect is fading out as time elapses, and only the newest part of the memory needs to be considered when evaluating a fractional derivative. This natural and logical feature is called the "fading memory" [391] or "short memory principle" [387, 388] that guarantees the reasonableness of truncation (8.3.3). In this regard, Spanos et al. solved a 0.5th-order example retaining coefficients of magnitude order $10^3$, which corresponds to 160 memory terms [392]; Zhu et al. took 200 memory terms in the evaluation of fractional derivative viscoelastic forces [391]. In this part, 200 memory terms are also considered, i.e., truncation point $N = 200$.

### 8.3.2 Julia Set and Fixed Point

An attractive domain is a set of the initial states of a system. Trajectories from this set converge to an attractor (fixed point, limit cycle, chaotic attractor, etc.). A Julia set can be characterized by the boundary of the attractive domain of an attractive fixed point for a polynomial system [218]. It is also the common boundary between a filled-in Julia set and an attractive domain of infinity. The desired behaviors depicted by the Julia set of a system can be achieved through the control of this set. As viewed from ecology, Julia sets provide a "map" for the stability of biological systems. People can control the initial quantity of biological populations in order to make them develop towards the expected direction (e.g. persistence or extinction).

To begin with, we discuss the discretization of system (8.3.1) before giving its Julia set. Real-world data on population is collected at discrete time points, usually every day, month, or year depending on the characteristics of different populations and the specific needs of research. So it is useful to examine the discrete dynamical version of system (8.3.1).

In order to reduce parameters, we introduce the following notations:

$$a_1 = \frac{r_1}{N_1}, \quad a_2 = \frac{r_2\sigma_2}{N_1}, \quad b_1 = \frac{r_1\sigma_1}{N_2}, \quad b_2 = \frac{r_2}{N_2}.$$

Then we obtain the following form:

$$\begin{cases} D_{0+}^{\alpha} u = u(r_1 - a_1 u - b_1 v), \\ D_{0+}^{\beta} v = v(-r_2 + a_2 u - b_2 v). \end{cases} \tag{8.3.6}$$

**Remark 8.4** In fact, fractional system (8.3.6) can be further generalized to

$$\begin{cases} D_{0+}^{\alpha} u = u(r_1' + a_1' u - b_1' v), \\ D_{0+}^{\beta} v = v(r_2' + a_2' u + b_2' v), \end{cases}$$

where $r_i', a_i', b_i' \in \mathbb{R}$ $(i = 1, 2)$. The interspecific relationship (predator-prey, symbiotic, competition, etc.) is determined by the sign of these parameters.

Now truncation (8.3.3) is used for the fractional derivatives of time, that is,

$$\begin{cases} D_{0+}^{\alpha} u(t) \to \frac{1}{h_N^{\alpha}} \sum\limits_{k=0}^{N} \omega_k^{(\alpha)} u(t - kh_N), \\ \\ D_{0+}^{\beta} v(t) \to \frac{1}{h_N^{\beta}} \sum\limits_{k=0}^{N} \omega_k^{(\beta)} v(t - kh_N). \end{cases}$$

In addition, let $N = 200$, $h_N = 1$, $u(t-k) = u_{n-1-k}$ and $v(t-k) = v_{n+1-k}$. Substitute these in system (8.3.6). Then the discrete version is

$$\begin{cases} u_{n+1} = r_1 u_n - a_1 u_n^2 - b_1 u_n v_n - \sum\limits_{k=1}^{\mathscr{N}} \omega_k^{(\alpha)} u_{n-k+1}, \\ \\ v_{n+1} = -r_2 v_n + a_2 u_n v_n - b_2 v_n^2 - \sum\limits_{k=1}^{\mathscr{N}} \omega_k^{(\beta)} v_{r-k+1}, \end{cases} \tag{8.3.7}$$

where $n \in \mathbb{N}$, $\mathscr{N} = \min\{200, n\}$ is a correction of the truncation point $N$ about the number of iterations $n$.

Let $\mathbf{F}(x, y) = [f_1(x, y), f_2(x, y)]^{\mathrm{T}}$, where $f_1(x, y) = r_1 x - a_1 x^2 - b_1 xy + \mathrm{mem}(x; \alpha, \mathscr{N})$, $f_2(x, y) = -r_2 y + a_2 xy - b_2 y^2 + \mathrm{mem}(y; \beta, \mathscr{N})$; the notations "$\mathrm{mem}(x; \alpha, \mathscr{N})$" and "$\mathrm{mem}(y; \beta, \mathscr{N})$", which take order of system (8.3.7) and modified truncation point $\mathscr{N}$ as parameters are the latest part of memories for $x$ and $y$, respectively.

We now turn to the Julia set of the real dynamical system (8.3.7). It is introduced by using the method in fractal theory for complex systems.

**Definition 8.4** Let $K$ be the set of point $(x, y)$ in $\mathbb{R}^2$ whose trajectory is limited, i.e.,

$$K = \left\{(x, y) : \{\mathbf{F}^n(x, y)\}_{n \in \mathbb{Z}_+} \text{ is bounded}\right\}.$$

Then the set $K$ is called the filled-in Julia set corresponding to the map $\mathbf{F}$. The boundary of $K$ is called the Julia set of the map $\mathbf{F}$, which is denoted by $J_{\mathbf{F}}$, i.e., $J_{\mathbf{F}} = \partial K$.

From [218] and Definition 8.4, the Julia set of system (8.3.7) has the following properties:

(1) $J_{\mathbf{F}}$ is non-empty;
(2) $J_{\mathbf{F}}$ is full-invariant, that is, $J_{\mathbf{F}} = \mathbf{F}(J_{\mathbf{F}}) = \mathbf{F}^{-1}(J_{\mathbf{F}})$;
(3) $J_{\mathbf{F}} = J_{\mathbf{F}^p}$ for any positive integer $p$;
(4) Let $s$ be an attractive fixed point of $\mathbf{F}$. Then $J_{\mathbf{F}} = \partial A(s)$, where $A(s)$ is the attractive domain of the fixed point $s$. The same is true if $s = \infty$.

Next, we discuss fixed points of system (8.3.7) and the stability of these fixed points. To evaluate fixed points of the system, let

$$\begin{cases} u^* = r_1 u^* - a_1 u^{*2} - b_1 u^* v^* - \sum\limits_{k=1}^{\mathscr{N}} \omega_k^{(\alpha)} u^*, \\ \\ v^* = -r_2 v^* + a_2 u^* v^* - b_2 v^{*2} - \sum\limits_{k=1}^{\mathscr{N}} \omega_k^{(\beta)} v^* \end{cases}$$

from which we can get the fixed points $(u^*, v^*)$ as follow:

(1) the trivial state: $P_t = (0, 0)$;
(2) the axial state: $P_a = \left(0, \frac{-r_2 - 1 - \sum_{k=1}^{\mathscr{N}} \omega_k^{(\beta)}}{b_2}\right)$ or $P_a = \left(\frac{r_1 - 1 - \sum_{k=1}^{\mathscr{N}} \omega_k^{(\alpha)}}{a_1}, 0\right)$;

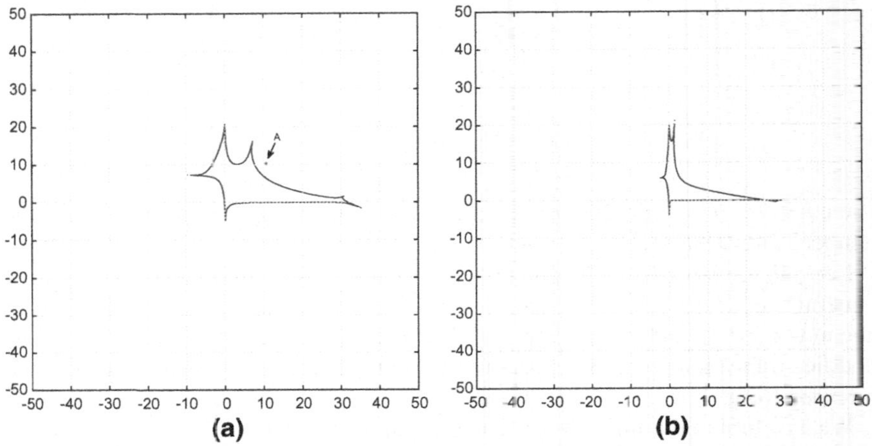

**Fig. 8.9** Original Julia sets of system (8.3.7) when $\alpha = \beta = 0.50, r_1 = 0.20, a_1 = 0.03, b_1 = 0.15$, $r_2 = 0.11$ and $b_2 = 0.04$. **a** $a_2 = 0.11$; **b** $a_2 = 0.51$

(3) the coexistent state:

$$P_c = \left( \frac{b_1 r_2 + b_2 r_1 + (b_1 - b_2) - \Omega_b}{a_1 b_2 + a_2 b_1}, \frac{a_2 r_1 - a_1 r_2 - (a_1 + a_2) - \Omega_a}{a_1 b_2 + a_2 b_1} \right)$$

where $\Omega_a = \sum_{k=1}^{\mathcal{N}} \left[ a_2 \omega_k^{(\alpha)} + a_1 \omega_k^{(\beta)} \right]$ and $\Omega_b = \sum_{k=1}^{\mathcal{N}} \left[ b_2 \omega_k^{(\alpha)} - b_1 \omega_k^{(\beta)} \right]$. In particular, it is one of the most common cases that the two species can exist for a long time and neither of them becomes extinct. Therefore only the coexistent state is considered in this work, that is, the fixed point of the model is taken as $P_c$.

For the convenience of study, set

$$\begin{aligned} \alpha = 0.50,\ r_1 = 0.20,\ a_1 = 0.03,\ b_1 = 0.15, \\ \beta = 0.50,\ r_2 = 0.11,\ a_2 = 0.10,\ b_2 = 0.04 \end{aligned} \tag{8.3.8}$$

in system (8.3.7). Then its Julia set is shown in Fig. 8.9a. Through the previous discussion and simple calculation, we know that the fixed point $P_c = (1.6905, 0.7294)$ is stable due to the spectral radius $\rho$ of the Jacobian matrix $\mathbf{J}$ of system (8.3.7) is 0.9725, i.e., $\rho(\mathbf{J}) = 0.9725 < 1$. Another Julia set with a different coefficient $a_2$ is presented in Fig. 8.9b.

Specifically, take $a_1 = b_2 = 0$. Then the fixed point $P_c = (1.3987, 1.0675)$ is stable because of $\rho(\mathbf{J}) = 0.9728 < 1$. More specifically, take $\alpha = \beta = 1$ further. Then the fixed point $P_c = (1, 1.3333)$ is unstable and the Julia set of system (8.3.7) showed in Fig. 8.10a is reduced to the case in [179]. The introduction of fractional derivatives rides the original unstable fixed point stable. This indicates that fractional operators expand the stability region of fixed points (equilibrium points). There is a corresponding result in the field of differential equation [388].

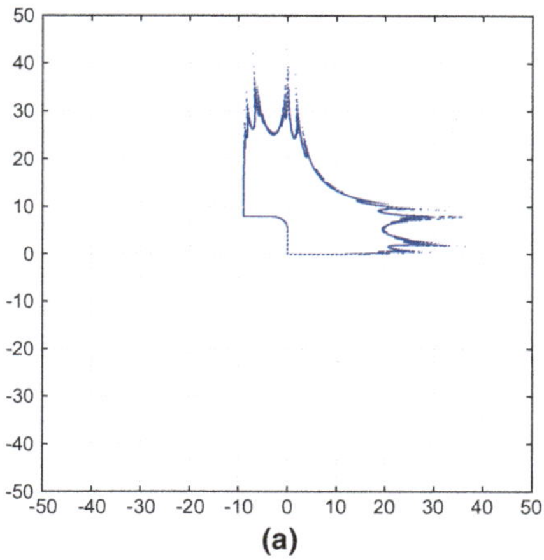


**Fig. 8.10** Original Julia sets of system (8.3.7) when $r_1 = 0.20, b_1 = 0.15, r_2 = 0.11$ and $a_2 = 0.10$. **a** $\alpha = \beta = 1.00$, $a_1 = b_2 = 0$; **b** $\alpha = \beta = 0.81$, $a_1 = 0.04$, $b_2 = 0.05$

In Fig. 8.10b, another Julia set with different orders is displayed, from which we can find that the structure and size of a Julia set are closely related to not only coefficients but also orders of a system.

### 8.3.3 *Theory Analysis*

From the discussion of discretization of fractional derivatives in Sect. 8.3.2, system (8.3.7) can be seen as a truncation of the following system:

$$\begin{cases} u_{n+1} = r_1 u_n - a_1 u_n^2 - b_1 u_n v_n - \sum_{k=1}^{n} \omega_k^{(\alpha)} u_{n-k+1}, \\ v_{n+1} = -r_2 v_n + a_2 u_n v_n - b_2 v_n^2 - \sum_{k=1}^{n} \omega_k^{(\beta)} v_{n-k+1}, \end{cases} \tag{8.3.9}$$

where $n \in \mathbb{N}_+$. It must be pointed out that system (8.3.9) is a nonlinear Volterra system in which the future state $(u_{n+1}, v_{n+1})$ depends on not only the present state $(u_n, v_n)$ but also all past states from $(u_{n-1}, v_{n-1})$ to $(u_0, v_0)$ [393]. This kind of system that has been widely used as mathematical models in population dynamics has memory effect like a fractional system. Once given the initial condition $(u(0), v(0)) = (u_0, v_0)$, one can generate the solution to systems (8.3.7) and (8.3.9). As a matter of fact, system (8.3.7) can be treated as a nonlinear system containing two 200th-order nonlinear difference equations, and the initial conditions $(u_0, v_0)$, $(u_1, v_1), \cdots, (u_{199}, v_{199})$ of this system can be totally generated by $(u_0, v_0)$.

Let

$$\begin{cases} z_1(n) = u_{n-199}, \quad z_2(n) = u_{n-198}, \cdots, z_{200}(n) = u_n, \\ z_{201}(n) = v_{n-199}, \; z_{202}(n) = v_{n-198}, \cdots, z_{400}(n) = v_n. \end{cases}$$

Then system (8.3.7) can be converted to the system

$$\mathbf{z}(n+1) = \mathbf{H}(\mathbf{z}(n)), \quad n \in \mathbb{N}, \tag{8.3.10}$$

where

$$\mathbf{z}(n) = (z_1(n), z_2(n), \cdots, z_{400}(n))^{\mathrm{T}}, \quad \mathbf{H} = (h_1, h_2, \cdots, h_{400})^{\mathrm{T}}$$

and

$$\begin{cases} h_i(z_1, z_2, \cdots, z_{400}) = z_{i+1}, \quad i = 1, 2, \cdots, 199, 201, 202, \cdots, 399, \\ h_{200}(z_1, z_2, \cdots, z_{400}) = r_1 z_{200} - a_1 z_{200}^2 - b_1 z_{200} z_{400} - \sum\limits_{k=1}^{N} \omega_k^{(\alpha)} z_{201-k}, \\ h_{400}(z_1, z_2, \cdots, z_{400}) = -r_2 z_{400} + a_2 z_{200} z_{400} - b_2 z_{400}^2 - \sum\limits_{k=1}^{N} \omega_k^{(\beta)} z_{401-k}. \end{cases}$$

Point $(u^*, v^*) \in \mathbb{R}^2$ is a fixed point of system (8.3.7) if $\mathbf{F}(u^*, v^*) = (u^*, v^*)$. This corresponds to the fixed point

$$\mathbf{z}^* := (\underbrace{u^*, \cdots, u^*}_{200}, \underbrace{v^*, \cdots, v^*}_{200})^{\mathrm{T}} \in \mathbb{R}^{400}$$

for system (8.3.10). Therefore a fixed point of system (8.3.7) is stable if and only if the corresponding fixed point of system (8.3.10) is stable.

Denote

$$\xi_1 := \left.\frac{\partial h_{200}}{\partial z_{200}}\right|_{\mathbf{z}^*} = r_1 - \omega_1^{(\alpha)} - 2a_1 u^* - b_1 v^*,$$

$$\xi_2 := \left.\frac{\partial h_{400}}{\partial z_{400}}\right|_{\mathbf{z}^*} = -r_2 - \omega_1^{(\beta)} + a_2 u^* - 2b_2 v^*.$$

Then the Jacobian matrix of system (8.3.10) at the fixed point $\mathbf{z}^*$ is obtained. Noticing the equivalence between systems (8.3.7) and (8.3.10), the matrix $\mathbf{J}$ can also be called the Jacobian matrix of system (8.3.7) at the fixed point $(u^*, v^*)$.

Let $\alpha$, $\beta$, $r_i$, $a_i$, and $b_i$ $(i = 1, 2)$ satisfy (8.3.8). Then the spectral radius of the Jacobian matrix $\mathbf{J}$ at $P_c = (1.6905, 0.7294)$ is 0.9725, denoting $\rho(\mathbf{J}) = 0.9725$. From Corollary 4.34 of [393], the fixed point $P_c$ is stable. Further, let $a_1 = b_2 = 0$. Then $P_c = (1.3987, 1.0675)$ and $\rho(\mathbf{J}) = 0.9728$.

### 8.3.4 Control of the Julia Set of System

Different features are usually required in the same biological system according to different situations. So we need to add some controllable objects in original system (8.3.7) and adjust them depending upon need. Introduce auxiliary reference feedback control into system (8.3.7). Then we have

$$\begin{cases} u_{n+1} = r_1 u_n - a_1 u_n^2 - b_1 u_n v_n - \sum_{k=1}^{\mathcal{N}} \omega_k^{(\alpha)} u_{n-k+1} + \theta_1 (u_n - u_{\text{ref}}), \\ v_{n+1} = -r_2 v_n + a_2 u_n v_n - b_2 v_n^2 - \sum_{k=1}^{\mathcal{N}} \omega_k^{(\beta)} v_{n-k+1} + \theta_2 (v_n - v_{\text{ref}}), \end{cases} \tag{8.3.11}$$

where $u_{\text{ref}}$ and $v_{\text{ref}}$ are auxiliary reference terms, $\theta_1$ and $\theta_2$ are feedback control parameters.

Suppose the control objective $(\overline{u}, \overline{v})$ has been realized by this control method. Then we have

$$\begin{cases} \overline{u}_{n+1} = r_1 \overline{u}_n - a_1 \overline{u}_n^2 - b_1 \overline{u}_n \overline{v}_n - \sum_{k=1}^{\mathcal{N}} \omega_k^{(\alpha)} \overline{u}_{n-k+1} + \theta_1 (\overline{u}_n - u_{\text{ref}}), \\ \overline{v}_{n+1} = -r_2 \overline{v}_n + a_2 \overline{u}_n \overline{v}_n - b_2 \overline{v}_n^2 - \sum_{k=1}^{\mathcal{N}} \omega_k^{(\beta)} \overline{v}_{n-k+1} + \theta_2 (\overline{v}_n - v_{\text{ref}}). \end{cases}$$

According to the discussion in Sect. 8.3.2, it needs to be assured that $(u^*, v^*)$ is still a fixed point of controlled system (8.3.11) and remains its stability. Therefore the control objective is taken to be the fixed point $(u^*, v^*)$, i.e., $(\overline{u}, \overline{v}) = (u^*, v^*)$. Then the fixed point $(u^*, v^*)$ must satisfy

$$\begin{cases} u^* = r_1 u^* - a_1 u^{*2} - b_1 u^* v^* - \sum_{k=1}^{\mathcal{N}} \omega_k^{(\alpha)} u^* + \theta_1 (u^* - u_{\text{ref}}), \\ v^* = -r_2 v^* + a_2 u^* v^* - b_2 v^{*2} - \sum_{k=1}^{\mathcal{N}} \omega_k^{(\beta)} v^* + \theta_2 (v^* - v_{\text{ref}}). \end{cases}$$

Since the fixed point $(u^*, v^*)$ also satisfies system (8.3.7), we conclude that the auxiliary reference point $(u_{\text{ref}}, v_{\text{ref}})$ must be chosen as the fixed point $(u^*, v^*)$. So controlled system (8.3.11) becomes

$$\begin{cases} u_{n+1} = r_1 u_n - a_1 u_n^2 - b_1 u_n v_n - \sum\limits_{k=1}^{\mathscr{N}} \omega_k^{(\alpha)} u_{n-k+1} + \theta_1 (u_n - u^*), \\ v_{n+1} = -r_2 v_n + a_2 u_n v_n - b_2 v_n^2 - \sum\limits_{k=1}^{\mathscr{N}} \omega_k^{(\beta)} v_{n-k+1} + \theta_2 (v_n - v^*). \end{cases} \tag{8.3.12}$$

It is still assumed that the parameters of system (8.3.7) satisfy (8.3.8). Notice that $P_c = (1.6905, 0.7294)$ is also the fixed point of controlled system (8.3.12). We will make sure that its stability will not be destroyed in system (8.3.12) when the Julia set changes with different feedback control parameters.

According to the analysis in Sect. 8.3.3, controlled system (8.3.12) is handled with the same method. In this case,

$$\overline{\xi}_1 = \theta_1 + r_1 - \omega_1^{(\alpha)} - 2a_1 u^* - b_1 v^*,$$
$$\overline{\xi}_2 = \theta_2 - r_2 - \omega_1^{(\beta)} + a_2 u^* - 2b_2 v^*.$$

Substitute $\xi_1$ and $\xi_2$ with $\overline{\xi}_1$ and $\overline{\xi}_2$, respectively. Then one gets the Jacobian matrix of controlled system (8.3.12) at the fixed point $(u^*, v^*)$, and we denote it as $\overline{\mathbf{J}}(\theta_1, \theta_2)$. Obviously, $\overline{\mathbf{J}}(\theta_1, \theta_2)$ is continuous about $\theta_1$ and $\theta_2$. According to Corollary 4.34 of [393] and discussion in Sect. 8.3.3, the fixed point $(u^*, v^*)$ is stable if $\rho\big(\overline{\mathbf{J}}(\theta_1, \theta_2)\big) < 1$. Setting $\theta_1 = \theta_2 = \theta$, we have $\theta \in [-1.5749, 0.6139]$ that satisfies $\rho\big(\overline{\mathbf{J}}(\theta, \theta)\big) < 1$ with the help of computer aided calculation.

Thus, parameters $\theta_1$ and $\theta_2$ must satisfy $\rho\big(\overline{\mathbf{J}}(\theta_1, \theta_2)\big) \leq 1$. For convenience and without loss of generality, let $\theta_1 = \theta_2 = \theta$. Then we obtain that the range of the control parameter is $E = \{\theta : -1.5749 \leq \theta \leq 0.6139\}$.

From Fig. 8.11 we can see that the Julia set of controlled system (8.3.12) changes as the value of the control parameter $\theta$ changes:

(1) The shape of the Julia set is like a piece of ink mark and it shrinks with the increasing of the control parameter $\theta$.
(2) The Julia set has a complex "broken" structure when the parameter is small, such as $\theta = -1.57$ and $\theta = -1.52$. This means that the connectedness of the Julia set is related to the control parameter $\theta$. In other words, the connectedness of the set can be controlled by changing the parameter.
(3) Overall, the structure of the Julia set becomes more and more simple with the increase of the parameter.

**Remark 8.5** From the perspective of ecological applications of feedback control, a system can keep stable for a long time by this method. Under the current initial conditions, we can make the original explosive (unbounded) solution to a system become a stable one by artificially adjusting the number of individuals in two populations. For example, put the numbers of predator and prey be $u_0 = 9.9954$ and $v_0 = 10.0083$, respectively, at the initial time. It is an initial value of an explosive solution to original system (8.3.7) because the point (9.9954, 10.0083) is outside the Julia set, which is shown as point $A$ in Fig. 8.9a. After introducing feedback

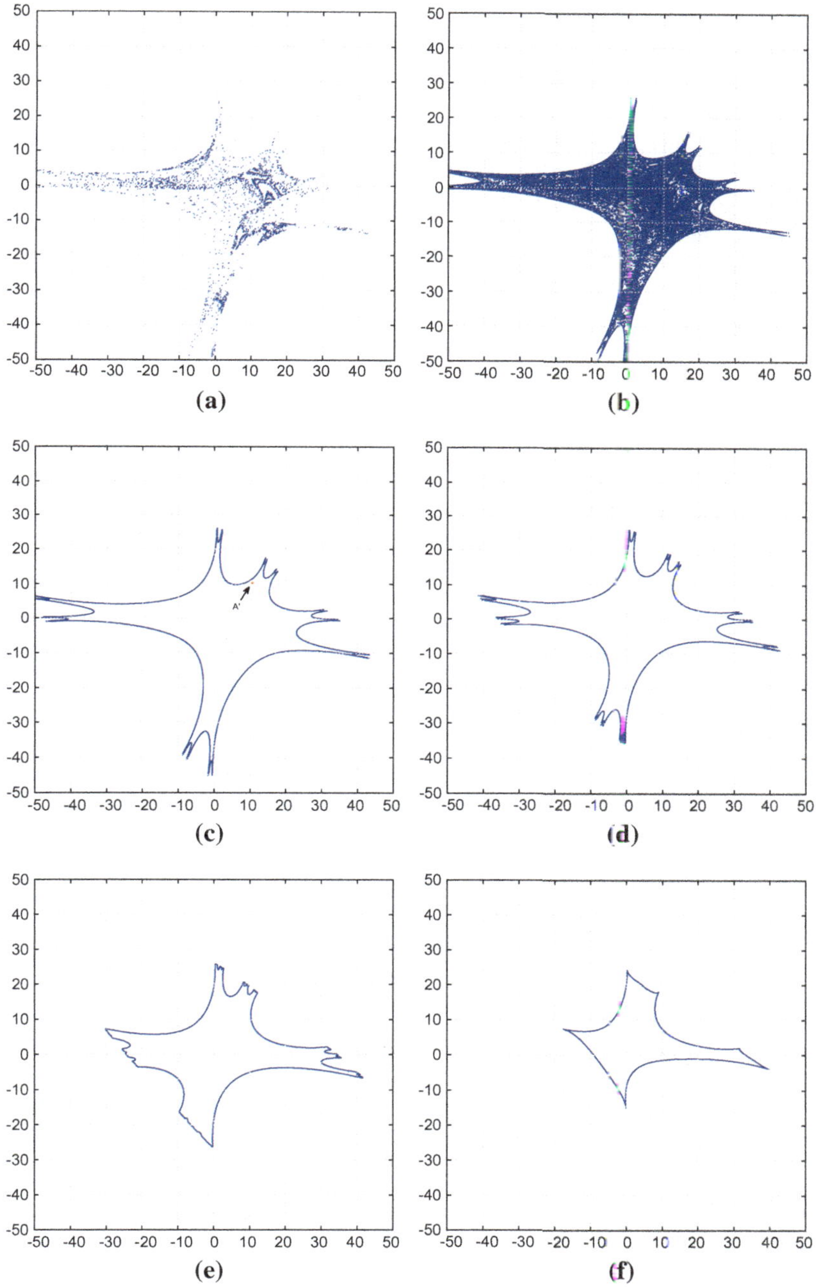


**Fig. 8.11** The change of Julia sets of controlled system (8.3.12) when $\alpha = \beta = 0.50$, $r_1 = 0.20$, $a_1 = 0.03$, $b_1 = 0.15$ and $r_2 = 0.10$, $a_2 = 0.10$, $b_2 = 0.04$. **a** $\theta = -1.58$; **b** $\theta = -1.52$; **c** $\theta = -1.21$; **d** $\theta = -0.86$; **e** $\theta = -0.51$; **f** $\theta = -0.16$

control, that is, increasing/decreasing (depending on the symbols of control items) the numbers of predators and prey by $|\theta(u_n - u^*)|$ and $|\theta(v_n - v^*)|$, respectively, the system becomes stable, seeing Fig. 8.11c. Finally, the numbers of predator and prey can remain stable with the feedback control.

Furthermore, take $a_1 = b_2 = 0$ in system (8.3.12). Then stable the fixed point $P_c = (1.3987, 1.0675)$. The range of the control parameter is $E = \{\theta : -1.6125 \leq \theta \leq 0.5636\}$. Compare with $E = (-1.9899, -0.0101)$ in [179], we can see that the positive range of the feedback control parameter is extended greatly due to the introduction fractional operator, even though the negative range is reduced to a certain extent.

### 8.3.5 *Synchronization of the Julia Set of System (8.3.7)*

Similar or even the same feature is required in different biological systems sometimes. It is necessary to consider the synchronization in different systems.

To facilitate the expression of the problem, we first provide the definition of the synchronization of different Julia sets. Consider two driving systems

$$\begin{cases} u_{n+1} = f_1\big(u_n, v_n, \mathrm{mem}(u_n; \alpha, \mathcal{N})\big), \\ v_{n+1} = f_2\big(u_n, v_n, \mathrm{mem}(v_n; \beta, \mathcal{N})\big) \end{cases} \tag{8.3.13}$$

and

$$\begin{cases} x_{n+1} = g_1\big(x_n, y_n, \mathrm{mem}(x_n; \alpha, \mathcal{N})\big), \\ y_{n+1} = g_2\big(x_n, y_n, \mathrm{mem}(y_n; \beta, \mathcal{N})\big). \end{cases} \tag{8.3.14}$$

Set the couple system (or called response system)

$$\begin{cases} x_{n+1} &= g_1\big(x_n, y_n, \mathrm{mem}(u_n; \alpha, \mathcal{N})\big) \\ &\quad + h_1\big(u_n, v_n, x_n, y_n, \mathrm{mem}(u_n; \alpha, \mathcal{N}), \mathrm{mem}(x_n; \alpha, \mathcal{N}); l_1\big), \\ y_{n+1} &= g_2\big(x_n, y_n, \mathrm{mem}(v_n; \beta, \mathcal{N})\big) \\ &\quad + h_2\big(u_n, v_n, x_n, y_n, \mathrm{mem}(v_n; \beta, \mathcal{N}), \mathrm{mem}(y_n; \beta, \mathcal{N}); l_2\big) \end{cases} \tag{8.3.15}$$

where $h_1$ and $h_2$ are coupling terms relating to the current system state and the memory terms; $l_1$ and $l_2$ are coupling parameters. Denote the Julia sets of systems (8.3.13)–(8.3.15) by $J^*$, $J$ and $J_l$, respectively.

**Definition 8.5** ([207, 345]) Julia sets synchronization of driving system (8.3.13) and response system (8.3.15) occurs if

$$\lim_{l \to l_0} (J_l \cup J^* - J_l \cap J^*) = \varnothing$$

for some $l_0$ ($l_0$ may be $\infty$).

Now consider a driving system which has the same structure to system (8.3.7) but with different system parameters

$$\begin{cases} x_{n+1} = r_1 x_n - a_1 x_n^2 - b_1 x_n y_n - \sum\limits_{k=1}^{\mathcal{N}} \omega_k^{(\alpha)} x_{n-k+1}, \\ y_{n+1} = -r_2 y_n + \widehat{a}_2 x_n y_n - b_2 y_n^2 - \sum\limits_{k=1}^{\mathcal{N}} \omega_k^{(\beta)} y_{n-k+1}, \end{cases} \tag{8.3.16}$$

where $a_2 \neq \widehat{a}_2 > 0$. In order to associate the Julia sets of systems (8.3.7) and (8.3.16), two coupled terms are added to system (8.3.16). Then we have

$$\begin{cases} x_{n+1} = r_1 x_n - a_1 x_n^2 - b_1 x_n y_n - \sum\limits_{k=1}^{\mathcal{N}} \omega_k^{(\alpha)} x_{n-k+1} + h_1, \\ y_{n+1} = -r_2 y_n + \widehat{a}_2 x_n y_n - b_2 y_n^2 - \sum\limits_{k=1}^{\mathcal{N}} \omega_k^{(\beta)} y_{n-k+1} + h_2, \end{cases} \tag{8.3.17}$$

where

$$h_1 = l_1 \left[ b_1(x_n y_n - u_n v_n) + a_1(x_n^2 - u_n^2) + \sum_{k=1}^{\mathcal{N}} \omega_k^{(\alpha)} (x_{n-k+1} - u_{n-k+1}) \right],$$

$$h_2 = l_2 \left[ a_2 u_n v_n - \widehat{a}_2 x_n y_n + b_2(y_n^2 - v_n^2) + \sum_{k=1}^{\mathcal{N}} \omega_k^{(\beta)} (y_{n-k+1} - v_{n-k+1}) \right].$$

Obviously, different Julia sets are obtained for different $l_1$ and $l_2$.

**Theorem 8.2** *If $r_i < 1\,(i = 1, 2)$, then $|x_n - u_n| \to 0$ and $|y_n - v_n| \to 0$ when $n \to +\infty$ and $l_i \to 1$ $(i = 1, 2)$, that is, the synchronization of trajectories of driving system (8.3.7) and response system (8.3.17) is achieved.*

***Proof*** From driving system (8.3.7) and response system (8.3.17), one gets

$$\begin{aligned} x_{n+1} - u_{n+1} &= r_1(x_n - u_n) - a_1(x_n^2 - u_n^2) - b_1(x_n y_n - u_n v_n) \\ &\quad - \textstyle\sum_{k=1}^{\mathcal{N}} \omega_k^{(\alpha)} (x_{n-k+1} - u_{n-k+1}) + h_1 \\ &= r_1(x_n - u_n) + b_1(l_1 - 1)(x_n y_n - u_n v_n) + a_1(l_1 - 1)(x_n^2 - u_n^2) \\ &\quad + (l_1 - 1) \textstyle\sum_{k=1}^{\mathcal{N}} \omega_k^{(\alpha)} (x_{n-k+1} - u_{n-k+1}). \end{aligned}$$

Based on the Definition 8.4, we know Julia sets are bounded. Denote a bounded region as $\Omega$. Then Julia sets can be derived by calculating the iteration trajectory of points inside of $\Omega$. It is because that, assuming $(x, y) \notin J_{\mathbf{F}}$, there is an $n_0 \in \mathbb{N}_+$ such that $\mathbf{F}^{n_0}(x, y) \notin \Omega$ from property (2) of Julia sets. Hence, for any $n \in \mathbb{N}$ and $k \in \{1, 2, \cdots, \mathcal{N}\}$, there exist three positive numbers $M_i$ $(i = 1, 2, 3)$ such that

$$|x_n y_n| + |u_n v_n| \leq M_1, \quad |x_n^2| + |u_n^2| \leq M_2, \quad |x_{n-k+1}| + |u_{n-k+1}| \leq M_3.$$

Then we have

$$\begin{aligned}
&|x_{n+1} - u_{n+1}| \\
&\leq r_1|x_n - u_n| + b_1|l_1 - 1| \cdot |x_n y_n - u_n v_n| + a_1|l_1 - 1| \cdot |x_n^2 - u_n^2| \\
&\quad + |l_1 - 1| \textstyle\sum_{k=1}^{\mathcal{N}} \left|\omega_k^{(\alpha)}(x_{n-k+1} - u_{n-k+1})\right| \\
&\leq r_1|x_n - u_n| + b_1|l_1 - 1|\big(|x_n y_n| + |u_n v_n|\big) + a_1|l_1 - 1|\big(|x_n^2| + |u_n^2|\big) \\
&\quad + |l_1 - 1| \textstyle\sum_{k=1}^{\mathcal{N}} |\omega_k^{(\alpha)}|\big(|x_{n-k+1}| + |u_{n-k+1}|\big) \\
&\leq r_1|x_n - u_n| + |l_1 - 1|M \\
&\leq r_1\big(r_1|x_{n-1} - u_{n-1}| + |l_1 - 1|M\big) + |l_1 - 1|M \\
&= r_1^2|x_{n-1} - u_{n-1}| + (r_1 + 1)|l_1 - 1|M \\
&\leq r_1^{n+1}|x_0 - u_0| + (r_1^n + r_1^{n-1} + \cdots + 1)|l_1 - 1|M \\
&= r_1^{n+1}|x_0 - u_0| + \tfrac{1-r_1^n}{1-r_1}|l_1 - 1|M,
\end{aligned}$$

where $M := b_1 M_1 + a_1 M_2 + M_3 \sum_{k=1}^{\mathcal{N}} |\omega_k^{(\alpha)}| < +\infty$. Notice that $|x_0 - u_0|$ is bounded since the initial values $x_0$ and $u_0$ are taken inside in the bounded region $\Omega$. Obviously, with $r_1 < 1$, if $|1 - l_1| \to 0$, i.e., $l_1 \to 1$, we have $|x_n - u_n| \to 0$ as $n \to +\infty$. In a similar way, one can get that $|y_n - v_n| \to 0$ if $l_2 \to 1$ and $n \to +\infty$. As a result, the synchronization of trajectories of driving system (8.3.7) and response system (8.3.17) is achieved. The proof is completed. □

According to Definition 8.4 and Theorem 8.2, the Julia set of response system (8.3.17) is changing into the Julia set of driving system (8.3.7) as $l_i \to 1$ if $r_i < 1$ $(i = 1, 2)$. Therefore the synchronization of Julia sets of systems (8.3.7) and (8.3.17) is obtained. And coupling parameters $l_i$ $(i = 1, 2)$ also reflect the degree of synchronization, which can be seen in the following simulations.

The assumption still remains that $\alpha$, $\beta$, $r_i$, $a_i$ and $b_i$ $(i = 1, 2)$ satisfy (8.3.8). Put $\widehat{a}_2 = 0.50$. Then Julia sets of systems (8.3.7) and (8.3.16) are shown in Fig. 8.9. The change of the Julia sets of response system (8.3.17) with different coupling parameters $l_i$ $(i = 1, 2)$ is showing in Fig. 8.12: the blue curve is the object set as showed in Fig. 8.9a determined by driving system (8.3.7), and red curves are the Julia

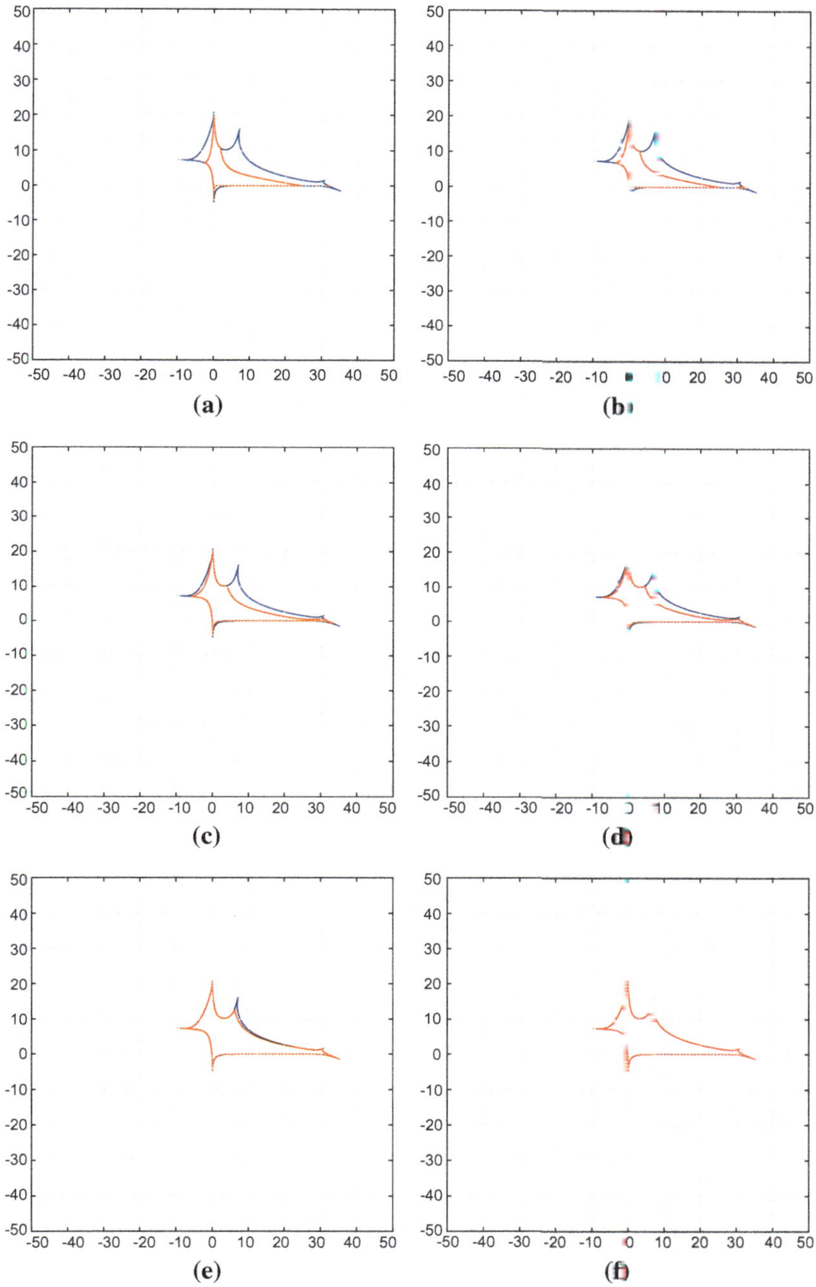


**Fig. 8.12** Julia set synchronization of systems (8.3.7) and (9.1.7). **a** $l_1 = 0.20, l_2 = 0.11$; **b** $l_1 = 0.40, l_2 = 0.31$; **c** $l_1 = 0.50, l_2 = 0.51$; **d** $l_1 = 0.70, l_2 = 0.51$; **e** $l_1 = 0.80, l_2 = 0.71$; **f** $l_1 = 0.90, l_2 = 0.81$

sets of response system (8.3.17) with different parameters. It is obvious that Julia sets of response system (8.3.17) are changing into the object set with the increasing of coupling parameters.

**Remark 8.6** From the viewpoint of ecological applications of coupling control, we may synchronize evolutions of two ecological systems containing the same or even similar species. It needs one system to participate in the evolution of another one as catalysts. This is a meaningful topic. But it is foreseeable that this is also a quite complicated problem in practical operation.

## 8.4 Fractal Dimension Analysis and Control of Julia Set Generated by Fractional Lotka–Volterra Model

### *8.4.1 Julia Set, Fixed Point and Dimension*

It is understood that an attractive domain (basin) is a set of the initial states of a system, and the trajectories from this set converge to an attractor (fixed point, limit cycle, chaotic attractor, etc.). A Julia set is the common boundary of a filled-in Julia set and the attractive domain of infinity. Therefore, the stability of a biological system can be reflected by the system's Julia set to a certain extent.

On the basis of the discussion in Sect. 8.3.3, we denote the spectral radius of a matrix $\mathbf{A}$ as $\rho(\mathbf{A})$.

**Theorem 8.3** *If $\rho(\mathbf{J}) < 1$, Then, fixed point of system (8.3.7) is (exponentially) stable.*

***Proof*** According to Corollary 4.34 of [393], it is easy to know that the fixed point of system (8.3.10) is exponentially stable if $\rho(\mathbf{J}) < 1$. Notice the equivalence between system (8.3.7) and system (8.3.10), then, we have the result immediately. The proof is completed. □

Without special instructions, the coexistence of the two species in the model is considered, the parameters $r_1, a_1, b_1, r_2, a_2, b_2$ in system (8.3.7) are taken as (8.3.8), and set $\alpha = \beta := r$. Then, fixed points $P_c = (u^*, v^*)$ and spectral radii $\rho = \rho(\mathbf{J})$ with different orders are shown in Table 8.1.

From Table 8.1, we can see that

(1) with the order $r$ increasing in [0, 1.2], the component $u^*$ shows a decreasing trend while the component $v^*$ shows an increasing trend;
(2) the spectral radius $\rho(\mathbf{J}) < 1$ when $r \in [0.2, 1.0]$, that is, fixed points of system (8.3.7) are stable when the system's order $r \in [0.2, 1.0]$.

Now we recall briefly the escape-time algorithm in [394] as it is applied for the numerical calculation of the Julia set of system (8.3.7) and its dimension. Initially,

**Table 8.1** Fixed points and spectral radii with different orders $r$ $(= \alpha = \beta)$

| $r$ | 0.00 | 0.10 | 0.20 | 0.30 | 0.40 | 0.50 | 0.60 | 0.70 |
|---|---|---|---|---|---|---|---|---|
| $u^*$ | 8.2099 | 5.1596 | 3.4403 | 2.4865 | 1.9671 | 1.6905 | 1.5471 | 1.4753 |
| $v^*$ | -6.9753 | -3.3704 | -1.3385 | -0.2113 | 0.4025 | 0.7294 | 0.8989 | 0.9837 |
| $\rho$ | 1.9797 | 1.0296 | 0.9800 | 0.9769 | 0.9745 | 0.9725 | 0.9704 | 0.9681 |
| $r$ | 0.80 | 0.90 | 1.00 | 1.10 | 1.20 | 1.30 | 1.40 | 1.50 |
| $u^*$ | 1.4411 | 1.4258 | 1.4198 | 1.4179 | 1.4177 | 1.4182 | 1.4187 | 1.4191 |
| $v^*$ | 1.0242 | 1.0422 | 1.0494 | 1.0516 | 1.0518 | 1.0513 | 1.0507 | 1.0502 |
| $\rho$ | 0.9656 | 0.9625 | 0.9693 | 1.1362 | 1.2874 | 1.4313 | 1.5713 | 1.7090 |

**Table 8.2** Box-counting dimension $D$ with different orders $r$ $(= \alpha = \beta)$

| $r$ | 0.00 | 0.10 | 0.20 | 0.30 | 0.40 | 0.50 | 0.60 | 0.70 |
|---|---|---|---|---|---|---|---|---|
| $D$ | 1.6141 | 1.6048 | 1.5669 | 1.5228 | 1.4569 | 1.4083 | 1.4026 | 1.4185 |
| $r$ | 0.80 | 0.90 | 1.00 | 1.10 | 1.20 | 1.30 | 1.40 | 1.50 |
| $D$ | 1.4325 | 1.4464 | 1.4360 | 1.1964 | 0.7821 | NaN[1] | NaN | NaN |

a lattice of the space of initial points $p_0 = (u_0, v_0)$ is fixed as $u_0 \in [\zeta_1, \zeta_2]$ and $v_0 \in [\eta_1, \eta_2]$ with $\zeta_1 < \zeta_2$, $\eta_1 < \eta_2$. Then, for any lattice resolution $M \in \mathbb{N}_+$, we have obtained $M^2$ points $p_0$ on the lattice with coordinates defined via the algorithm: $u_0 = \zeta_1 + (i_1 - 1)\frac{\zeta_2 - \zeta_1}{M-1}$ and $v_0 = \eta_1 + (i_2 - 1)\frac{\eta_2 - \eta_1}{M-1}$, where $1 \le i_1, i_2 \le M$. Take $T$ as the escape time limit i.e., the maximum number of iteration counts. Set $R$ as the escape radius. The distance $r$ of the $T$th iteration from the original point $p_0$ in the region $\Omega$ is calculated. If $r \le R$, Then, $p_0 \in K$, and we keep the point in a file. Otherwise, the point is abandoned. Eventually, the new set $K$ is obtained and we plot the resulting file, which gives us the corresponding Julia set on the given lattice of points.

Take $\zeta_1 = \eta_1 = -50$ and $\zeta_2 = \eta_2 = 50$, i.e., $\Omega = [-50, 50] \times [-50, 50]$. Fix $M = 1500$, $T = 200$ and $R = 10^5$. The box-counting dimension of system's Julia sets is shown in Table 8.2.

The relationship between[1] box-counting dimension $D$ and order $r$ can be acquired by an eighth-degree polynomial fitting as following:

$$D(r) = 4.582r^8 - 21.02r^7 + 40.35r^6 - 46.08r^5 + 31.96r^4 - 10.15r^3 + 0.2076r^2 - 0.04761r + 1.614. \tag{8.4.1}$$

[1]It is a MATLAB® value that stands for "not a number". This is the case because the Julia set is "empty" under the current situation, which is largely due to the restriction of lattice resolution. Although it is difficult to further enhance the resolution under our existing hardware conditions, it is good to ignore these, which will not affect the analysis of our main results.

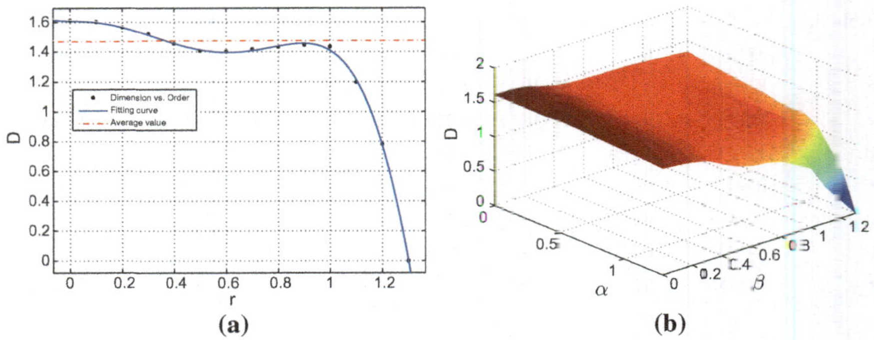


**Fig. 8.13** Relationship between box-counting dimension and order: **a** polynomial fitting when $\alpha = \beta$; **b** 3-dimensional image of box-counting dimension and orders

The fitting curve is shown in Fig. 8.13a. The sum of squares due to error (SSE) is $1.50 \times 10^{-3}$; root mean squared error (RMSE) is $1.71 \times 10^{-2}$, which indicate that fitting expression (8.4.1) is an efficient one. Figure 8.13b shows the change of box-counting dimension $D$ with order of system (8.3.7) in more detail.

From Fig. 8.13 we can see that

(1) the box-counting dimension of Julia sets of system (8.3.7) is bounded, and it is roughly decreasing as the order increases;
(2) the change of the box-counting dimension is relatively gentle when $\alpha, \beta \in [0, 1]$, and the average figure works out at 1.4784;
(3) the change of the box-counting dimension falls rapidly when the order exceeds 1, which indicates that the space-filling capacity of this set determined by system (8.3.7) will decline drastically with the increase of system's order in this case.

Some specific examples about Julia sets of system (8.3.7) with different orders are shown in Fig. 8.14, from which it is easy to see that the structure and size of Julia sets are closely related to the system's order.

### 8.4.2 Control of the Julia Set of System (8.3.7)

It must be pointed out that in theoretical researches we can constrain the size of attractive domain to make systems show different behaviors. The control of the behavior of the solution with a given initial value point can be achieved by controlling the Julia set of the system, that is, design a certain control strategy such that the point "enters" or "exits" the Julia set. As viewed from ecology, the Julia set provides a "map" for the stability of a biological system. People can control the Julia set to enable biological populations towards the desired direction, such as persistence or extinction. Therefore, reasonable control items should be designed and introduced to achieve control objectives.

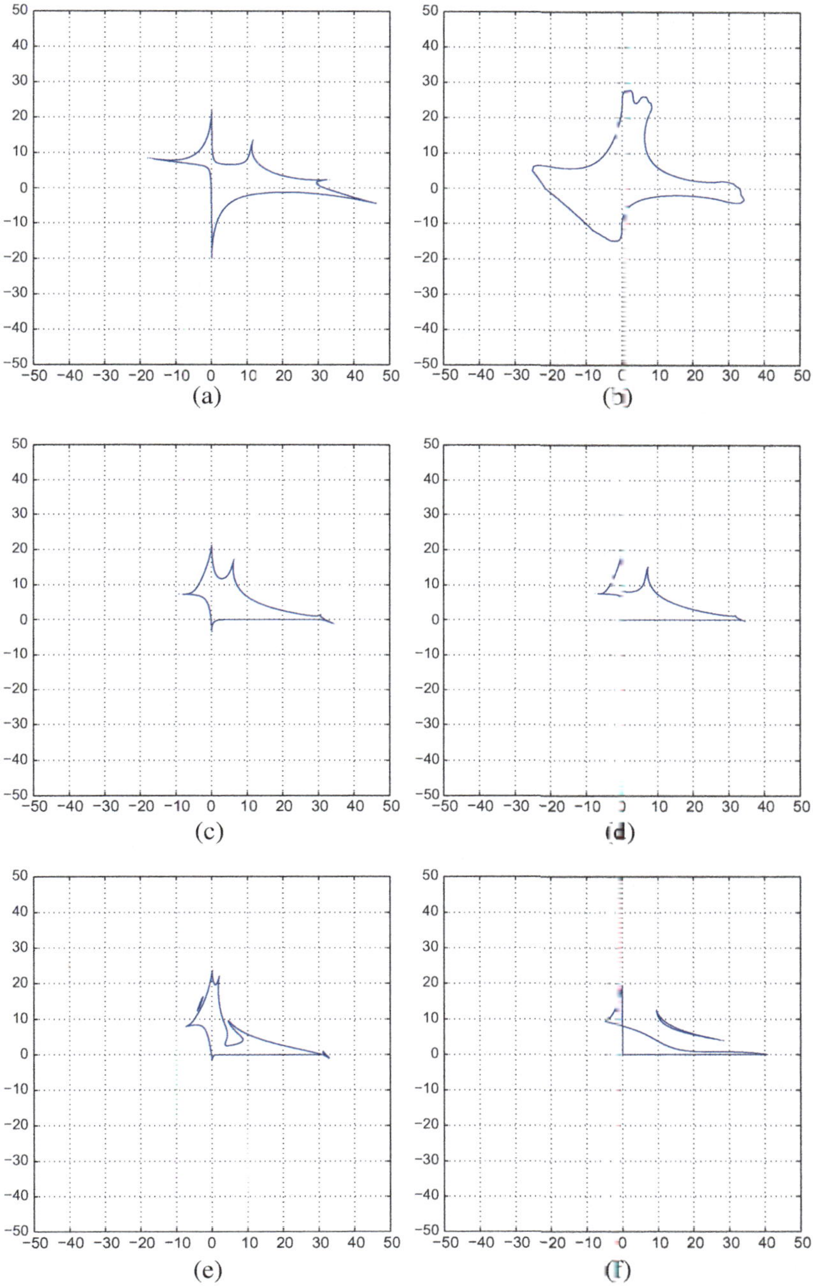


**Fig. 8.14** Original Julia sets of system (8.3.7): **a** $\alpha = 0.5,\ \beta = 0.11$; **b** $\alpha = 0.1,\ \beta = 0.51$; **c** $\alpha = 0.5,\ \beta = 0.61$; **d** $\alpha = 0.6,\ \beta = 0.51$; **e** $\alpha = 0.5,\ \beta = 1.11$; **f** $\alpha = 1.1,\ \beta = 0.51$

Our purpose is to introduce an idea of control into the Julia set of system (8.3.7) and discuss the effect of the control terms. Set controlled system as

$$\begin{cases} u_{n+1} = r_1 u_n - a_1 u_n^2 - b_1 u_n v_n - \sum_{k=1}^{\mathcal{N}} \omega_k^{(\alpha)} u_{n-k+1} + \theta_1 x_n, \\ v_{n+1} = -r_2 v_n + a_2 u_n v_n - b_2 v_n^2 - \sum_{k=1}^{\mathcal{N}} \omega_k^{(\beta)} v_{n-k+1} + \theta_2 y_n, \end{cases} \tag{8.4.2}$$

where $x_n$ and $y_n$ are control terms; $\theta_1$ and $\theta_2$ are control parameters. Then, the control terms will be designed to enable the fixed point $(u^*, v^*)$ of controlled system (8.4.2) stable. After this, the purpose can be achieved with different control strategies.

**Gradient Control Strategy**

Motivated by [207, 395–398], we take the gradient control strategy to achieve the purpose that the fixed point $(u^*, v^*)$ of original system (8.3.7) is also a fixed point of controlled system (8.4.2), and remains stable. Setting $\theta_1 = \theta_2 = 1$, we choose the objective vector function

$$\mathbf{Q}(u, v) := \begin{bmatrix} q(u) \\ q(v) \end{bmatrix} = \frac{1}{2} \begin{bmatrix} (u - u^*)^2 \\ (v - v^*)^2 \end{bmatrix}$$

and write the general form of the gradient control algorithm as

$$\mathbf{w}_n := \begin{bmatrix} x_n \\ y_n \end{bmatrix} = -\theta_g \nabla_{\mathbf{w}_n} \mathbf{Q}(u_{n+1}, v_{n+1}) = \begin{bmatrix} -\theta_g q'_{x_n}(u_{n+1}) \\ -\theta_g q'_{y_n}(v_{n+1}) \end{bmatrix},$$

where $\nabla_{\mathbf{w}_n}$ denotes the gradient operator with respect to control vector $\mathbf{w}_n$ and $\theta_g$ is the gradient control parameter. It is simple to show that

$$\begin{cases} x_n = -\frac{\theta_g}{\theta_g+1} \left( r_1 u_n - a_1 u_n^2 - b_1 u_n v_n - \sum_{k=1}^{\mathcal{N}} \omega_k^{(\alpha)} u_{n-k+1} - u^* \right), \\ y_n = -\frac{\theta_g}{\theta_g+1} \left( -r_2 v_n + a_2 u_n v_n - b_2 v_n^2 - \sum_{k=1}^{\mathcal{N}} \omega_k^{(\beta)} v_{n-k+1} - v^* \right). \end{cases} \tag{8.4.3}$$

By substituting expression (8.4.3) into controlled system (8.4.2), it becomes

$$\begin{cases} u_{n+1} = \frac{1}{\theta_g+1} \left( r_1 u_n - a_1 u_n^2 - b_1 u_n v_n - \sum_{k=1}^{\mathcal{N}} \omega_k^{(\alpha)} u_{n-k+1} \right) + \frac{\theta_g}{\theta_g+1} u^*, \\ v_{n+1} = \frac{1}{\theta_g+1} \left( -r_2 v_n + a_2 u_n v_n - b_2 v_n^2 - \sum_{k=1}^{\mathcal{N}} \omega_k^{(\beta)} v_{n-k+1} \right) + \frac{\theta_g}{\theta_g+1} v^*. \end{cases} \tag{8.4.4}$$

**Table 8.3** Box-counting dimension $D$ with different gradient control parameter $\theta_g$

| $\theta_g$ | –3.0 | –2.9 | –2.8 | –2.7 | –2.6 | –2.5 | –2.4 | –2.3 | –2.2 | –2.1 |
|---|---|---|---|---|---|---|---|---|---|---|
| $D$ | 1.8000 | 1.7873 | 1.7737 | 1.7574 | 1.7423 | 1.7289 | 1.7144 | 1.6932 | 1.6709 | 1.6516 |
| $\theta_g$ | –2.0 | –1.9 | –1.8 | –1.7 | –1.6 | –0.4 | –0.3 | –0.2 | –0.1 | 0.0 |
| $D$ | 1.6231 | 1.5891 | 1.5556 | 1.5355 | 1.4690 | NaN | NaN | NaN | 0.5575 | 1.4083 |
| $\theta_g$ | 0.1 | 0.2 | 0.3 | 0.4 | 0.5 | 0.6 | 0.7 | 0.8 | 0.9 | 1.0 |
| $D$ | 1.5209 | 1.5866 | 1.6070 | 1.6412 | 1.6764 | 1.7027 | 1.7292 | 1.7430 | 1.7626 | 1.7747 |

It is obvious that, $(u^*, v^*)$ is also a fixed point of controlled system (8.4.4) if it is a fixed point of original system (8.3.7). The comparison between systems (8.3.7) and (8.4.4) shows that the control decreases at a rate of $\theta_g + 1$ times the uncontrolled value of the next step, and adds, simultaneously, constant terms are related to gradient control parameters.

Denote

$$\widetilde{\omega}_i^{(\alpha)} := \frac{\omega_i^{(\alpha)}}{\theta_g + 1}, \quad \widetilde{\omega}_i^{(\beta)} := \frac{\omega_i^{(\beta)}}{\theta_g + 1}, \quad i = 2, 3, \cdots, 200,$$

$$\xi_g := \frac{r_1 - \omega_1^{(\alpha)} - 2a_1 u^* - b_1 v^*}{\theta_g + 1}, \quad \eta_g := \frac{-r_2 - \omega_1^{(\beta)} + a_2 u^* - 2b_2 v^*}{\theta_g + 1}.$$

Then, the Jacobian matrix of system (8.4.4) at the fixed point $P_c = (u^*, v^*)$ can be obtained.

In order to maintain the stability of the fixed point $P_c = (1.6905, 0.7294)$ unchanged in controlled system (8.4.4), parameter $\theta_g$ should satisfy $\rho\big(\mathbf{J}(\theta_g)\big) < 1$ according to Theorem 8.3. With the help of computer aided calculation, we have $\theta_g \in (-\infty, -1.5979] \cup [-0.4787, +\infty)$ which meets the requirement. The box-counting dimensions of Julia sets of system (8.4.4) are shown in Table 8.3.

As the range of parameter $\theta_g$ consists of two disjoint parts, we fit them separately. The relationship between box-counting dimension $D$ and parameter $\theta_g$ ranged in $[-3, -1.6]$ can be acquired by quadratic fitting (SSE $= 6.93 \times 10^{-4}$, RMSE $= 7.60 \times 10^{-3}$), seeing Fig. 8.15a; the relationship with $\theta_g \in [-0.1, 1]$ can be acquired by exponential fitting with two terms (SSE $= 1.66 \times 10^{-3}$, RMSE $= 1.44 \times 10^{-2}$), seeing Fig. 8.15b. Moreover, some specific examples about Julia sets of controlled system (8.4.4) with different parameters are shown in Fig. 8 16.

From Table 8.3, Figs. 8.15 and 8.16 we can see that the Julia set of controlled system (8.4.4) and its box-counting dimension change as the value of parameter $\theta_g$ changes:

(i) The Julia set has different shapes when $\theta_g$ lies in different ranges. The shape, by contrast, is more "regular" in appearance if $\theta_g \in [0\ 1]$.

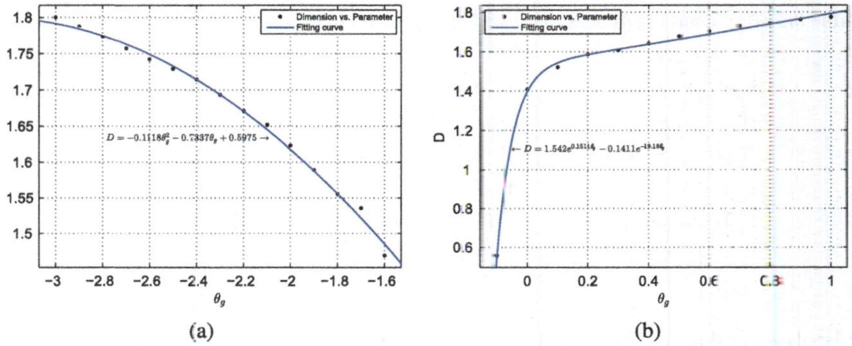


**Fig. 8.15** Relationship between the box-counting dimension and the gradient control parameter: **a** $\theta_g \in [-3, -1.6]$; **b** $\theta_g \in [-0.1, 1]$

(ii) The box-counting dimension is bounded in this control strategy. Its range is about from 0 to 1.8.

(iii) With increase of the parameter ranged in $[-3, -1.6]$, the Julia set shrinks and its box-counting dimension reduces; with increase of the parameter ranged in $[0, 1]$, the Julia set expands and its box-counting dimension increases. This shows that the size of the Julia set can be reflected by the dimension of the set.

### *8.4.3 Optimal Function Control Strategy*

According to the optimal control function method [208, 330], we take the control terms as

$$\begin{cases} x_n = r_1 u_n - a_1 u_n^2 - b_1 u_n v_n - \sum\limits_{k=1}^{\mathcal{N}} \omega_k^{(\alpha)} u_{n-k+1} - u_n, \\ y_n = -r_2 v_n + a_2 u_n v_n - b_2 v_n^2 - \sum\limits_{k=1}^{\mathcal{N}} \omega_k^{(\beta)} v_{n-k+1} - v_n. \end{cases}$$

For convenience of computation, set $\theta_1 = \theta_2 := \theta_o$. Then, controlled system (8.4.2) becomes

$$\begin{cases} u_{n+1} = (\theta_o + 1)\left( r_1 u_n - a_1 u_n^2 - b_1 u_n v_n - \sum\limits_{k=1}^{\mathcal{N}} \omega_k^{(\alpha)} u_{n-k+1} \right) - \theta_o u_n, \\ v_{n+1} = (\theta_o + 1)\left( -r_2 v_n + a_2 u_n v_n - b_2 v_n^2 - \sum\limits_{k=1}^{\mathcal{N}} \omega_k^{(\beta)} v_{n-k+1} \right) - \theta_o v_n. \end{cases} \quad (8.4.5)$$

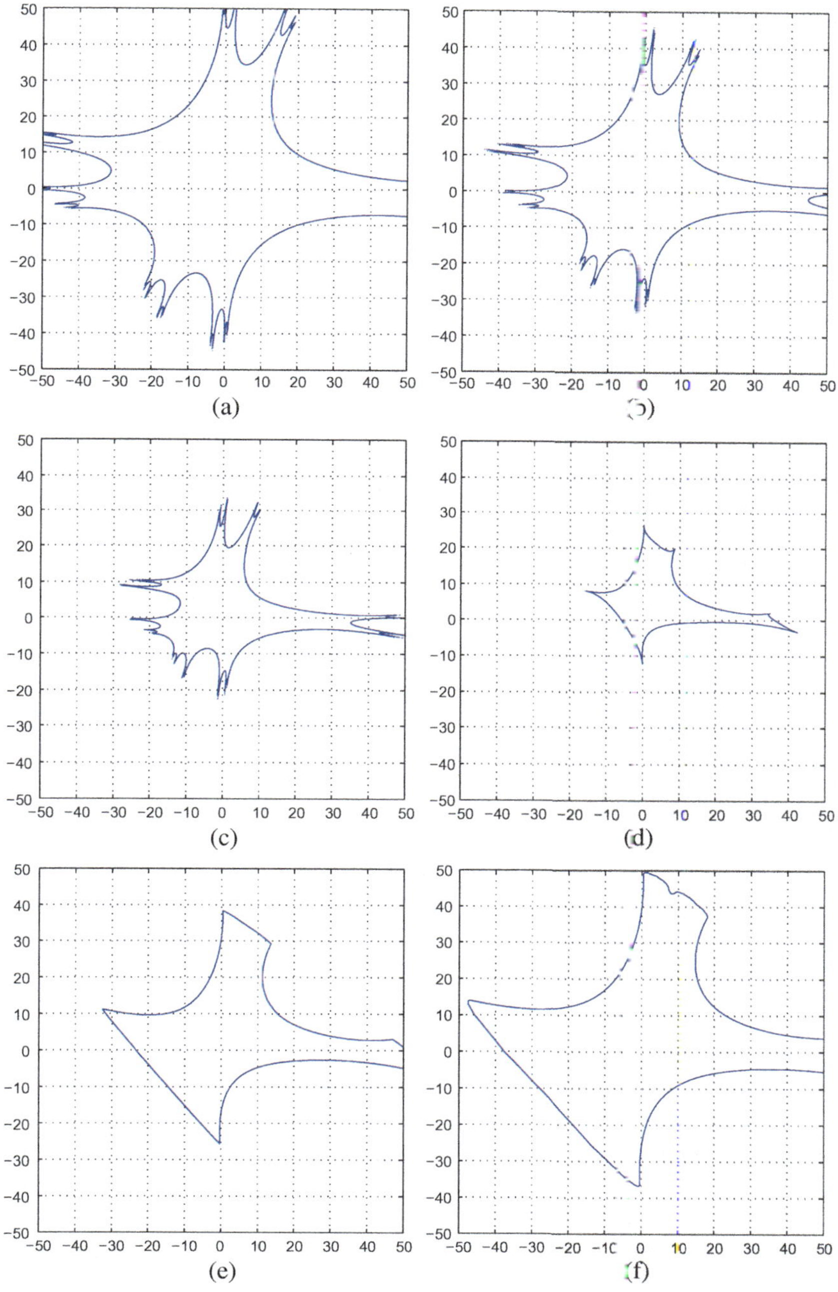


**Fig. 8.16** The changing of the Julia set of controlled system (8.5.10): **a** $\theta_g = -2.71$; **b** $\theta_g = -2.29$; **c** $\theta_g = -1.91$; **d** $\theta_g = 0.11$; **e** $\theta_g = 0.51$; **f** $\theta_g = 0.91$

**Table 8.4** Box-counting dimension $D$ with different optimal control parameter $\theta_o$

| $\theta_o$ | –0.6 | –0.5 | –0.4 | –0.3 | –0.2 | –0.1 | 0.0 | 0.1 |
|---|---|---|---|---|---|---|---|---|
| $D$ | 1.7096 | 1.6496 | 1.6051 | 1.5479 | 1.4960 | 1.4506 | 1.4083 | 1.38[illegible]3 |
| $\theta_o$ | 0.2 | 0.3 | 0.4 | 0.5 | 0.6 | 0.7 | 0.8 | 0.9 |
| $D$ | 1.3469 | 1.3356 | 1.3093 | 1.2838 | 1.2600 | 1.2379 | 1.2178 | 1.2086 |
| $\theta_o$ | 1.0 | 1.1 | 1.2 | 1.3 | 1.4 | 1.5 | 1.6 | |
| $D$ | 1.2571 | 1.2448 | 1.2300 | 1.1555 | 1.2107 | 1.1931 | 1.1990 | |

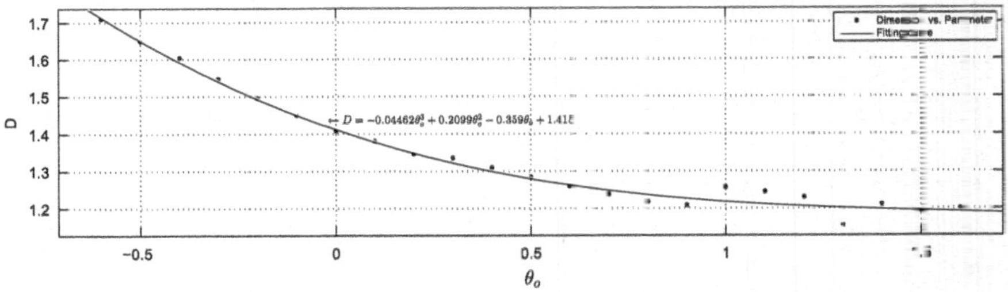


**Fig. 8.17** Relationship between the box-counting dimension and the optimal control parameter

where $\theta_o$ is the optimal control parameter. it is well known that $(u^*, v^*)$ is also a fixed point of controlled system (8.4.5) if it is a fixed point of original system (8.3.7). The comparison between systems (8.3.7) and (8.4.5) shows that the control increases at a rate of $\theta_g + 1$ times the uncontrolled value of the next step, and subtracts, simultaneously, $\theta_o$ times the current value. In addition, one advantage of this control strategy is that there is no need to know the fixed point of the system beforehand.

Denote

$$\widehat{\omega}_i^{(\alpha)} := (\theta_o + 1)\omega_i^{(\alpha)}, \quad \widehat{\omega}_i^{(\beta)} := (\theta_o + 1)\omega_i^{(\beta)}, \quad i = 2, 3, \cdots, 200,$$

$$\xi_o := (\theta_o + 1)(r_1 - \omega_1^{(\alpha)} - 2a_1u^* - b_1v^*) - \theta_o, \quad \eta_o := (\theta_o - 1)(-r_2 - \omega_1^{(\beta)} + a_2u^* - 2b_2v^*) - \theta_o.$$

In order to maintain the stability of the fixed point $P_c = (1.6905, 0.7294)$ unchanged in controlled system (8.4.5), parameter $\theta_o$ should satisfy $\rho(J(\theta_o)) < 1$ according to Theorem 8.3. Hence $\theta_o \in [-0.6046, 1.6221]$ can meet the requirement. The box-counting dimensions of Julia sets of system (8.4.5) are shown in Table 8.4.

The relationship between box-counting dimension $D$ and parameter $\theta_o$ ranged in $[-0.6, 1.6]$ can be acquired by a third-degree polynomial fitting, seeing Fig. 8.17. The SSE is $6.69 \times 10^{-3}$ and RMSE is $1.88 \times 10^{-2}$. Some specific examples about Julia sets of controlled system (8.4.5) with different parameters are shown in Fig. 8.18.

From Table 8.4, Figs. 8.17 and 8.18 we can see that the Julia set of controlled system (8.4.5) and its box-counting dimension change as the value of parameter $\theta_o$ changes:

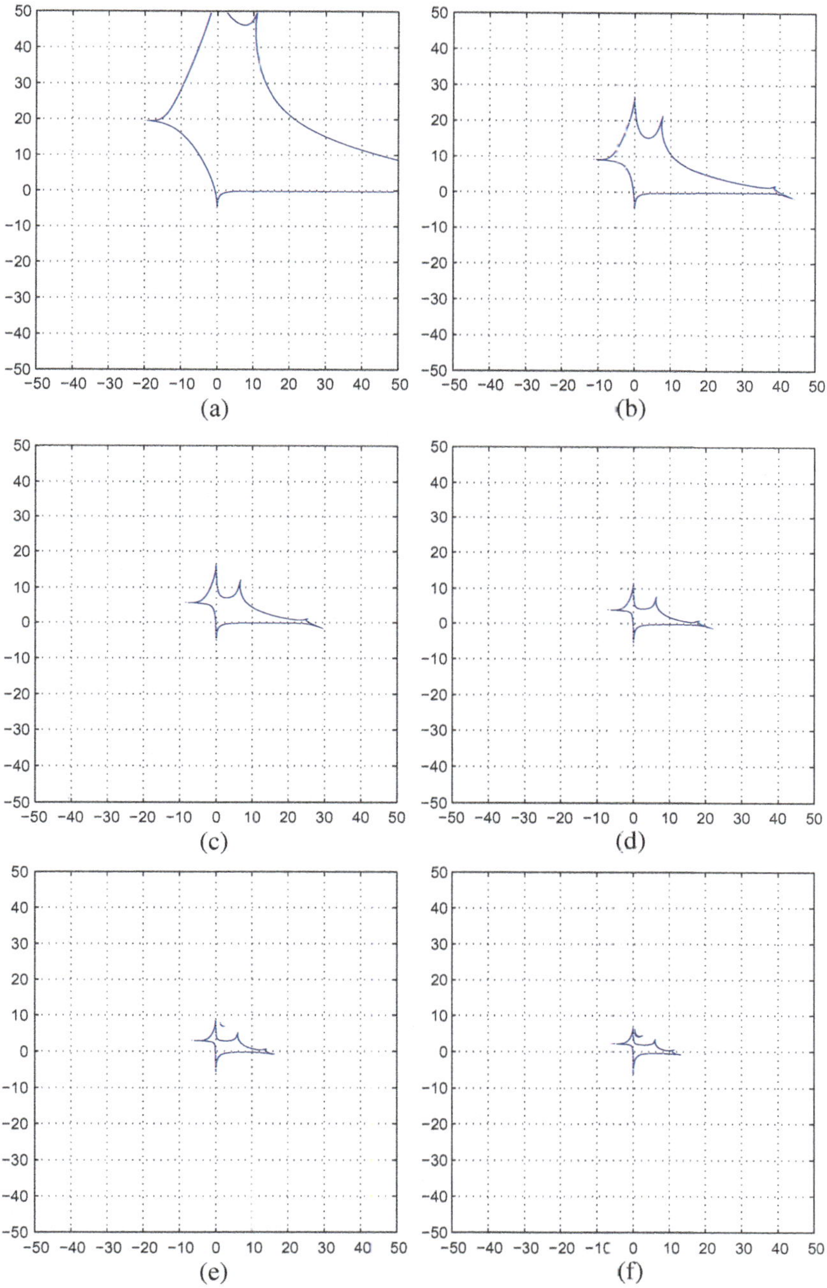


**Fig. 8.18** The changing of the Julia set of controlled system (8.5.11): **a** $\theta_o = -0.61$; **b** $\theta_o = -0.21$; **c** $\theta_o = 0.21$; **d** $\theta_o = 0.61$; **e** $\theta_o = 1.01$; **f** $\theta_o = 1.41$

(i) With increase of the parameter range in [−0.5, 0.5], the Julia set shrinks and its box-counting dimension decreases speedily, but its shape is basically unchanged.
(ii) When $\theta_o \in [0.5, 1.6]$, the change trend of the set size is slightly, and the average of its box-counting dimension is 1.2249.

## 8.5 On the Spatial Julia Set Generated by Fractional Lotka–Volterra System with Noise

It must be admitted that in mathematics, a fractal is an infinitely recursive and detailed set [242] . Julia set, proposed by Gaston Julia [189] in 1918, is considered as one of the most attractive fractal sets. The Julia set is not only an attractor in the sense that the iteration sequence generated from a point in the set is still in it, but also represents a basin of attraction where the iteration of a dynamical system converges. Hence, the Julia set can be considered as a "map" for the boundedness or stability of nonlinear systems. In terms of the population dynamics, the most obvious advantage for knowing about Julia sets is to grasp the long-term behavior of population dynamics, and then control or avoid explosive solutions [399–401] that mean the "collapse" of the system. In biology and other interdisciplinary fields, Julia set also provides eximious explanations for many complicated phenomena [165, 174, 402]. The study of noise-perturbed Julia sets has attracted interest in the early 2000s [403, 404]. Although more work has been developed on this topic, we mainly refer the readers to the following development. In 2010, Andreadis et al. built a Julia deviation distance and a Julia deviation plot to quantify and visualize the effect of noise on a Julia set [405], respectively. Then, the authors studied the structural stability of Julia sets of a complex logistic system and found a power-law behavior of the Julia deviation distance. In 2012, Andreadis et al. calculated and obtained the critical noise strength that makes the Julia set generated by a family of complex quadratic systems completely lose its original (noise-free) structure [172]. In 2017, Wang et al. studied the Julia set of complex Lorenz system [394]

$$\begin{cases} z_{n+1} = (1 + a\tau)z_n - \tau z_n w_n, \\ w_{n+1} = (1 - \tau)w_n + \tau z_n^2, \end{cases}$$

where $z_n, w_n, a, \tau \in \mathbb{C}$ and $n \in \mathbb{N}$. The authors extended the study of noise-perturbed planar Julia sets to the spatial case, while proposed an index so called "symmetry criterion" to quantitatively analyze the influence of dynamic noise on the symmetry of system's Julia sets. On the other hand, fractional operators are increasingly considered in modeling processes of biological systems due to their unique advantages in describing complex phenomena and behaviors related to memory or fractals [384] that are essentially nonlinear, highly complex and hard to be explained by integer-order models. Ahmed et al. introduced a fractional model that emphasizes the nonlocal interaction besides the local interaction in an epidemic spreading [406].

These two actions are observed in foot-and-mouth disease, SARS and avian flu outbreaks in 2007. Rivero et al. found that the order of fractional derivation is an excellent controller of the velocity how a system's trajectories approach to the critical point, which represents the anomalous reality of the competition among some species [407]. Hamdan et al. discovered that the fractional differential equation is more appealing and realistic in modeling the dengue transmission as it possesses memory [408]. Moreover, fractional models allow greater degrees of freedom in systems, which enable us to consider new intricate situations in population dynamics. In terms of the Lotka–Volterra model in ecology, it has been considered by Ahmed et al. [379] since 2007. Afterwards, qualitative properties and numerical solutions to this kind of fractional models are widely studied. Interestingly enough, the authors found that the corresponding discretized version of this model exhibits much richer dynamical behaviors. In 2018, Wang et al. investigated a discrete fractional Lotka–Volterra system [400, 401]

$$\begin{cases} u_{n+1} = r_1 h_N^{\alpha} u_n - a_1 h_N^{\alpha} u_n^2 - b_1 h_N^{\alpha} u_n v_n - \sum\limits_{k=1}^{\mathscr{N}} \omega_k^{(\alpha)} u_{n-k+1}, \\ v_{n+1} = -r_2 h_N^{\beta} v_n + a_2 h_N^{\beta} u_n v_n - b_2 h_N^{\beta} v_n^2 - \sum\limits_{k=1}^{\mathscr{N}} \omega_k^{(\beta)} v_{n-k+1} \end{cases} \tag{8.5.1}$$

from the fractal viewpoint for the first time, which extends the work of Sun et al. [179] in 2017 by introducing of fractional elements into the classical fractal sets of discrete dynamical systems. After discussing the stability of fixed points of system (8.5.1), control and synchronization of system's Julia sets are realized by several control strategies. Motivated by the work above, system (8.5.1) will be expanded to the complex field. Two kinds of systems with dynamic multiplicative noise are also generalized from system (8.5.1). Then, it is studied that features of the spatial fractal structures of these systems' Julia sets.

### *8.5.1 Preliminaries*

**Remark 8.7** In Definition 8.1, the derivative is defined by a series. This nonlocal operation embodies the memory effect of this fractional derivative. So a fundamental property of the fractional derivative is taking the "past" into account [409].

Now, the derivation process of system (8.5.1) is reviewed briefly. See publication [400, 401] for more details.

In the beginning, based on the Grünwald–Letnikov fractional derivatives of order $\alpha$ and $\beta$, satisfying $\alpha, \beta \in (0, 1]$, the fractional Lotka–Volterra system that we develop is as follows:

$$\begin{cases} {}^{GL}D_{0+}^{\alpha} u(t) = r_1 u(t) \left(1 - \frac{u(t)}{N_1} - \sigma_1 \frac{v(t)}{N_2}\right), \\ {}^{GL}D_{0+}^{\beta} v(t) = r_2 v(t) \left(-1 + \sigma_2 \frac{u(t)}{N_1} - \frac{v(t)}{N_2}\right), \end{cases} \tag{8.5.2}$$

where $u(t)$ and $v(t)$ be the densities of prey and predator populations at time $t$, respectively; $r_i$, $N_i$ and $\sigma_i$ $(i = 1, 2)$ are nonnegative numbers as explained in [400, 401]. Compared with most of the existing work, system (8.5.2) contains logistic growth of both predator and prey, which can reveal the nonlinear dynamical behavior caused by interaction or interference between individuals per unit time

**Remark 8.8** The Grünwald coefficients, also called historical correlation coefficients, play a significant role in truncation (8.3.3). It is because of these coefficients that the previous memory term (or called hereditary biologically) must be taken into account to evaluate the present value. Moreover, it is also noteworthy to regard the following two aspects:

- The Grünwald coefficients tend to be zero with growing $k$ [391]. This reasonable and natural feature is called the "fading memory" or "short memory principle" [387, 388] that guarantees the reasonableness of truncation (8.3.3). In this part, two hundred memory terms are considered, i.e., truncation point $N = 200$ as in [391].
- From the Grünwald coefficients, we can see that truncation (8.3.3) becomes a classical difference scheme when $\alpha \in \mathbb{N}$. In fact, Grünwald–Letnikov fractional derivative is reduced to the classical (integer-order) one and its memory effect vanishes in this case.

Let $u(t - kh_N) = u_{n+1-k}$ and $v(t - kh_N) = v_{n+1-k}$. Substitute reduced parameters and discretization (8.3.3) into system (8.5.2). Then, the discrete version is obtained as

$$\begin{cases} u_{n+1} = r_1 h_N^{\alpha} u_n - a_1 h_N^{\alpha} u_n^2 - b_1 h_N^{\alpha} u_n v_n - \sum\limits_{k=1}^{\mathscr{N}} \omega_k^{(\alpha)} u_{n-k+1}, \\ v_{n+1} = -r_2 h_N^{\beta} v_n + a_2 h_N^{\beta} u_n v_n - b_2 h_N^{\beta} v_n^2 - \sum\limits_{k=1}^{\mathscr{N}} \omega_k^{(\beta)} v_{n-k+1}, \end{cases} \tag{8.5.3}$$

where $n \in \mathbb{N}$, $\mathscr{N} = \min\{200, n+1\}$ is the correction of the truncation point $N$ about the number of iterations $n$. For this kind of system, the future state $(u_{n+1}, v_{n+1})$ depends on not only the present state $(u_n, v_n)$ but also all past states $(u_{n-1}, v_{n-1})$, $(u_{n-2}, v_{n-2})$, $\cdots$, $(u_{n-\mathscr{N}+1}, v_{n-\mathscr{N}+1})$. Once given the initial condition $(u(0), v(0)) = (u_0, v_0)$, one can generate the solution to system (8.5.3) immediately. Next, we will discuss system (8.5.3) in the complex field, i.e., putting $u = \overline{u} + \mathrm{i}\widetilde{u}$ and $v = \overline{v} + \mathrm{i}\widetilde{v}$, where $\overline{u}, \widetilde{u}, \overline{v}, \widetilde{v} \in \mathbb{R}$ and $\mathrm{i} = \sqrt{-1}$ means the imaginary unit, to define system's Julia set. Let

$$\mathbf{F}(x, y) = [f_1(x, y), f_2(x, y)]^{\mathrm{T}},$$

where

$$\begin{cases} f_1(x, y) = r_1 h_N^{\alpha} x - a_1 h_N^{\alpha} x^2 - b_1 h_N^{\alpha} xy + \mathrm{mem}(x; \alpha, \mathscr{N}), \\ f_2(x, y) = -r_2 h_N^{\beta} y + a_2 h_N^{\beta} xy - b_2 h_N^{\beta} y^2 + \mathrm{mem}(y; \beta, \mathscr{N}), \end{cases}$$

notations "mem$(x; \alpha, \mathscr{N})$" and "mem$(y; \beta, \mathscr{N})$" with $\alpha, \beta$ being the order of system (8.5.3) and $\mathscr{N}$ being the modified truncation point, are the latest part of memories for $x$ and $y$, respectively.

**Definition 8.6** (*Julia set*) Let $K$ be the set of point $(x, y)$ in $\mathbb{C}^2$ whose trajectory is limited, i.e.,

$$K = \left\{(x, y) : \{\mathbf{F}^n(x, y)\}_{n\in\mathbb{Z}_+} \text{ is bounded}\right\}.$$

Then, the set $K$ is called the filled-in Julia set corresponding to the map $\mathbf{F}$. The boundary of $K$ is called the Julia set of the map $\mathbf{F}$, which is denoted by $J_{\mathbf{F}}$, i.e., $J_{\mathbf{F}} = \partial K$.

In the following, the escape-time algorithm [410] is applied for the numerical calculation of the Julia set of fractional Lotka–Volterra system in this part. Initially, by dividing system (8.5.3) into real part and imaginary parts, we get

$$\begin{cases}
\overline{u}_{n+1} = r_1 h_N^\alpha \overline{u}_n - a_1 h_N^\alpha (\overline{u}_n^2 - \widetilde{u}_n^2) - b_1 h_N^\alpha (\overline{u}_n \overline{v}_n - \widetilde{u}_n \widetilde{v}_n) - \sum\limits_{k=1}^{\mathscr{N}} \omega_k^{(\alpha)} \overline{u}_{n-k+1}, \\
\widetilde{u}_{n+1} = r_1 h_N^\alpha \widetilde{u}_n + 2a_1 h_N^\alpha \overline{u}_n \widetilde{u}_n - b_1 h_N^\alpha (\overline{u}_n \widetilde{v}_n + \widetilde{u}_n \overline{v}_n) - \sum\limits_{k=1}^{\mathscr{N}} \omega_k^{(\alpha)} \widetilde{u}_{n-k+1}, \\
\overline{v}_{n+1} = r_2 h_N^\beta \overline{v}_n + a_2 h_N^\beta (\overline{u}_n \overline{v}_n - \widetilde{u}_n \widetilde{v}_n) - b_2 h_N^\beta (\overline{v}_n^2 - \widetilde{v}_n^2) - \sum\limits_{k=1}^{\mathscr{N}} \omega_k^{(\beta)} \overline{v}_{n-k+1}, \\
\widetilde{v}_{n+1} = r_2 h_N^\beta \widetilde{v}_n + a_2 h_N^\beta \overline{v}_n \widetilde{v}_n - \sum\limits_{k=1}^{\mathscr{N}} \omega_k^{(\beta)} \widetilde{v}_{n-k+1}.
\end{cases} \tag{8.5.4}$$

Now, we fix a lattice of the space of initial points $p_0 = (u_0, v_0) := (\overline{u}_0, \widetilde{u}_0, \overline{v}_0, \widetilde{v}_0)$ as follows: $\overline{u}_0 \in [\overline{\zeta}_1, \overline{\zeta}_2]$, $\widetilde{u}_0 \in [\widetilde{\zeta}_1, \widetilde{\zeta}_2]$, $\overline{v}_0 \in [\overline{\eta}_1, \overline{\eta}_2]$ and $\widetilde{v}_0 \in [\widetilde{\eta}_1, \widetilde{\eta}_2]$, where $\overline{\zeta}_i, \widetilde{\zeta}_i, \overline{\eta}_i, \widetilde{\eta}_i \in \mathbb{R}$ $(i = 1, 2)$. Then, for any lattice resolution $\rho \in \mathbb{N}_+$, there are $\rho^4$ points $p_0$ in the lattice with coordinates $(\overline{u}_0, \widetilde{u}_0, \overline{v}_0, \widetilde{v}_0)$ defined via the following algorithm:

$$\begin{cases}
\overline{u}_0 = \overline{\zeta}_1 + (i_1 - 1)\frac{\overline{\zeta}_2 - \overline{\zeta}_1}{\rho - 1}, & \widetilde{u}_0 = \widetilde{\zeta}_1 + (i_2 - 1)\frac{\widetilde{\zeta}_2 - \widetilde{\zeta}_1}{\rho - 1}, \\
\overline{v}_0 = \overline{\eta}_1 + (i_3 - 1)\frac{\overline{\eta}_2 - \overline{\eta}_1}{\rho - 1}, & \widetilde{v}_0 = \widetilde{\eta}_1 + (i_4 - 1)\frac{\widetilde{\eta}_2 - \widetilde{\eta}_1}{\rho - 1},
\end{cases} \tag{8.5.5}$$

where $i_1, i_2, i_3, i_4$ are positive integers but not exceeding $\rho$. Take $T \in \mathbb{N}_+$ as the escape time limit (the maximum number of iteration counts) and $R$ as the escape radius. Then, the distance $r$ of the $T$th iteration from the initial point $p_0$ in $\Omega = [\overline{\zeta}_1, \overline{\zeta}_2] \times [\widetilde{\zeta}_1, \widetilde{\zeta}_2] \times [\overline{\eta}_1, \overline{\eta}_2] \times [\widetilde{\eta}_1, \widetilde{\eta}_2]$ is calculated. If $r \leq R$, then $p_0 \in K$, and the point is kept in a file. Finally, the new set $K$ is obtained and the resulting file is plotted, which gives us the corresponding Julia set in the given lattice of points. Based on the above analysis, the Julia set of system (8.5.3) should be seen as a 4-D object, which increases the complexity of simulating the set and runs short of the intuition. To avoid those troubles, it is necessary to slice Julia sets to simulate them with some 3-D graphics. Furthermore, by fixing one of the four coefficients in representation (8.5.5), for instance $\widetilde{v}_0 \equiv \widetilde{v}^*$, the points with the form $(\overline{u}_0, \widetilde{u}_0, \overline{v}_0, \widetilde{v}^*)$ in $K$ are found

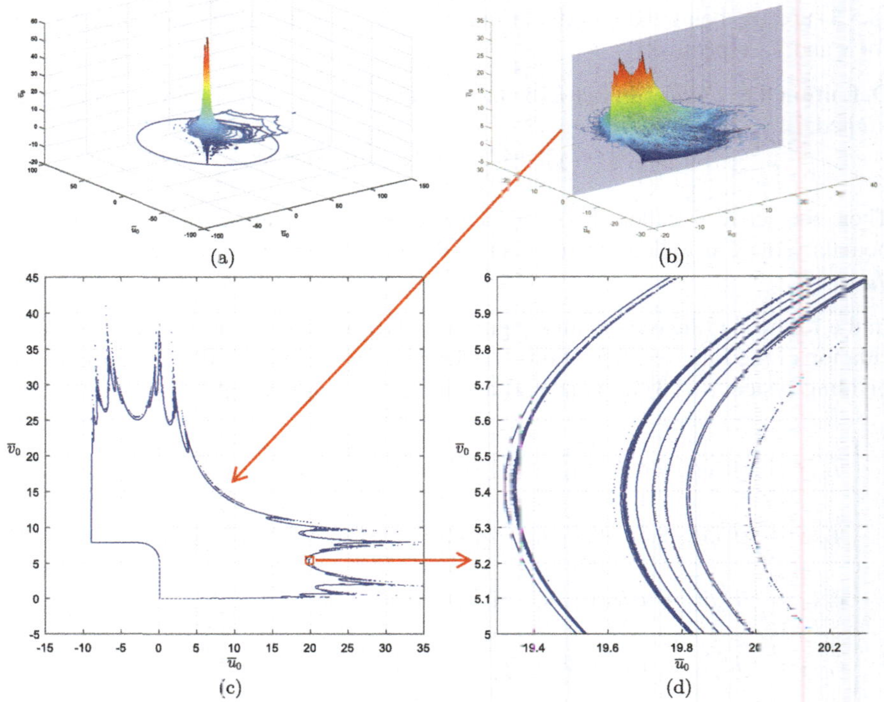


**Fig. 8.19** Slices of spatial Julia set of system (8.5.3): **a** $J_{\mathbf{F},\widetilde{v}_0\equiv 0}$ with $\alpha=\beta=0.50$; **b** $J_{\mathbf{F},\widetilde{v}_0\equiv 0}$ with $\alpha=\beta=1.00$; **c** $J_{\mathbf{F},\widetilde{u}_0\equiv 0,\widetilde{v}_0\equiv 0}$; **d** partially enlarged details of $J_{\mathbf{F},\widetilde{u}_0\equiv 0,\widetilde{v}_0\equiv 0}$

and recombined to the set $K_{\widetilde{v}^*}$. Thus, the 3-D slice of the Julia set of the map **F**, defined as $J_{\mathbf{F},\widetilde{v}^*}$, i.e., $J_{\mathbf{F},\widetilde{v}^*}=\partial K_{\widetilde{v}^*}$, can be simulated in lattice $(\overline{u}_0,\widetilde{u}_0,\overline{v}_0)$. Subsequently, take $\Omega=[-80,120]\times[-80,80]\times[-15,80]$, $\widetilde{v}_0\equiv 0$, and fix $\rho=800$, $R=10^5$. Then, $J_{\mathbf{F},\widetilde{v}_0\equiv 0}$ with $h_N^{\alpha}=h_N^{\beta}=1$, $r_1=0.20$, $r_2=0.10$, $a_1=0.00$, $a_2=0.10$, $b_1=0.15$ and $b_2=0.00$ is presented by (a) and (b) in Fig. 8.19. Furthermore, a 2-D slice of Fig. 8.19b with $\widetilde{u}_0\equiv 0$, i.e., $J_{\mathbf{F},\widetilde{u}_0\equiv 0,\widetilde{v}_0\equiv 0}$ is illustrated in Fig. 8.19c. Partially enlarged details of Fig. 8.19c are shown in Fig. 8.19d. The simulation fully displays the symmetry and complex fractal structure of $J_{\mathbf{F},\widetilde{v}_0\equiv 0}$.

From Fig. 8.19, we can see that

(i) the order of the fractional system seriously influences the structure of its Julia set;
(ii) when $\alpha=\beta=1.00$, the 2-D slice of the Julia set of system (8.5.3) degenerates into the integer-order case in Fig. 1 of [179];
(iii) the Julia set has an infinitely detailed fascinating structure and so it does have fractal property.

### 8.5.2 The Noise-Perturbed Fractional Lotka–Volterra System

From the viewpoint of population dynamics, there are two sources of noise: the intrinsic stochasticity associated with the dynamics of the individuals (intrinsic noise) and the random variability of the environment (environmental noise). The intrinsic noise can be interpreted as the fluctuation of model's parameters, while the environmental noise is a direct disturbance to system's states. Multiplicative noise source attracts more attention because of its better physical meaning in population models. Lots of meaningful work has been done in this area, for example, spatio-temporal pattern [411–413], epidemiological infection [414], noise-induced phenomenon [415], stability [416] and chaos [417]. One the other hand, noise perturbations in fractal has been examined by many scholars. However, there is no relevant work on fractional systems. In this section, we will consider fractional Lotka–Volterra systems with dynamic multiplicative noise. The Julia set $J_{\mathbf{F}}$ with dynamic state multiplicative noise (environmental noise), denoted by $J_{\mathbf{F}}^{\mathrm{ms},k}$, is generated from

$$\begin{cases} u_{n+1} = r_1 u_n - a_1 u_n^2 - b_1 u_n v_n - \sum\limits_{k=1}^{\mathscr{N}} \omega_k^{(\alpha)} u_{n-k+1} + k_1 \delta_{1,n} u_n, \\ v_{n+1} = -r_2 v_n + a_2 u_n v_n - b_2 v_n^2 - \sum\limits_{k=1}^{\mathscr{N}} \omega_k^{(\beta)} v_{n-k+1} + k_2 \delta_{2,n} v_n, \end{cases} \tag{8.5.6}$$

where $n \in \mathbb{N}$, $\delta_{i,n} = \overline{\delta}_{i,n} + \mathrm{i}\widetilde{\delta}_{i,n}$ $(i = 1, 2)$ is the state noise of the $n$th iteration. The noise $\overline{\delta}_{i,n}$ and $\widetilde{\delta}_{i,n}$ are on the real and complex axes, respectively. The parameters $k_1, k_2 \in \mathbb{R}$ represent the noise strength. The Julia set $J_{\mathbf{F}}$ with dynamic order multiplicative noise (intrinsic noise), denoted by $J_{\mathbf{F}}^{\mathrm{mo},k}$, is generated from

$$\begin{cases} u_{n+1} = r_1 u_n - a_1 u_n^2 - b_1 u_n v_n - \sum\limits_{k=1}^{\mathscr{N}} \omega_k^{(\alpha+\alpha k_1 \varrho_{1,n})} u_{n-k+1}, \\ v_{n+1} = -r_2 v_n + a_2 u_n v_n - b_2 v_n^2 - \sum\limits_{k=1}^{\mathscr{N}} \omega_k^{(\beta+\beta k_2 \varrho_{2,n})} v_{n-k+1}, \end{cases} \tag{8.5.7}$$

where the orders $\alpha, \beta \in \mathbb{R}$ are treated as parameters of the system; $\varrho_{i,n} \in \mathbb{R}$ $(i = 1, 2)$ is the order noise of the $n$th iteration. Let $k_1 = k_2 = k$ for convenience. Considering the Gaussian white noise that conforms to the standard normal distribution, i.e., $\overline{\delta}_{i,n}, \widetilde{\delta}_{i,n}, \varrho_{i,n} \sim N(0, 1)$, where $i = 1, 2$ and $n \in \mathbb{N}$. Based on the escape-time algorithm, we present $J_{\mathbf{F},\widetilde{v}_0=0}^{\mathrm{ms},k}$ and $J_{\mathbf{F},\widetilde{v}_0=0}^{\mathrm{mo},k}$ with different noise strengths in Fig. 8.20.

From Fig. 8.20, we can conclude that

(i) with the same noise strength, the state multiplicative noise has a greater effect than the order multiplicative noise in destroying the structure of the original (noise-free) Julia set;
(ii) for the former, the structural damage of system's Julia set becomes more and more apparent and the set tends to disappear as the noise strength increases;

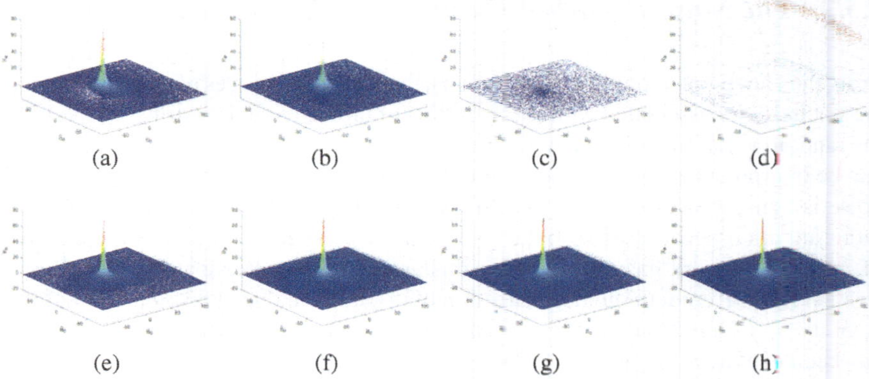

**Fig. 8.20** Julia sets of noise-perturbed fractional Lotka–Volterra systems (8.5.6) and (8.5.7): **a** $J^{\mathrm{ms},0.25}_{\mathbf{F},\widetilde{v}_0\equiv 0}$; **b** $J^{\mathrm{ms},0.50}_{\mathbf{F},\widetilde{v}_0\equiv 0}$; **c** $J^{\mathrm{ms},0.75}_{\mathbf{F},\widetilde{v}_0\equiv 0}$; **d** $J^{\mathrm{ms},1.00}_{\mathbf{F},\widetilde{v}_0\equiv 0}$; **e** $J^{\mathrm{mo},0.25}_{\mathbf{F},\widetilde{v}_0\equiv 0}$; **f** $J^{\mathrm{mo},0.50}_{\mathbf{F},\widetilde{v}_0\equiv 0}$; **g** $J^{\mathrm{mo},0.75}_{\mathbf{F},\widetilde{v}_0\equiv 0}$; **h** $J^{\mathrm{mo},1.00}_{\mathbf{F},\widetilde{v}_0\equiv 0}$

(iii) for the latter, the influence of noise under different strength is not obvious and it is mainly concentrated in plane $\overline{v}_0 = 0$, at a first glance.

To support the above observations, structural changes and symmetry damages from the two types of noise will be investigated further in the following two sections.

## 8.5.3 *The Structural Changes of Julia Sets of Noise-Perturbed Fractional Lotka–Volterra System*

This section contains our main results on quantification and visualization of noise on Julia sets. A new deviation index and an auxiliary ratio index are given to characterize and analyze these effects of noise.

**The Julia Deviation Index**

In order to quantify the effect of noise on the Julia set of a system, Andreadis et al. proposed a mathematical framework called Julia deviation distance [405] defined as

$$\mathbf{DistJ}(J_{\mathbf{F}}, J_{\mathbf{G}}) = \frac{n(J_{\mathbf{F}}) + n(J_{\mathbf{G}}) - 2n(J_{\mathbf{F}} \cap J_{\mathbf{G}})}{\rho^3}, \tag{8.5.8}$$

where $n(\cdot)$ denotes the cardinal of a set. It is clear that the two Julia sets, $J_{\mathbf{F}}$ and $J_{\mathbf{G}}$, are close to each other when their Julia deviation distance is a small number. In general, the Julia deviation distance between the two Julia sets satisfies the inequality $0 \le \mathbf{DistJ}(J_{\mathbf{F}}, J_{\mathbf{G}}) < 1$ [219, 394, 405]. In fact, the distance is usually much less than 1, as $n(J_{\mathbf{F}} \cap J_{\mathbf{G}}) \le n(J_{\mathbf{F}}) \ll \rho^3$. The distance defined by representation (9.1.11) is related to the resolution of the lattice. This is bound to cause some inconvenience when dealing with specific situations. Therefore, we advance the Julia deviation index **JDI** defined by

$$\mathbf{JDI}(J_{\mathbf{F}}, J_{\mathbf{G}}) := \frac{n(J_{\mathbf{F}}) + n(J_{\mathbf{G}}) - 2n(J_{\mathbf{F}} \cap J_{\mathbf{G}})}{n(J_{\mathbf{F}}) + n(J_{\mathbf{G}})}. \tag{8.5.9}$$

**Remark 8.9** In the Julia deviation index defined by representation (9.1.12), "$\rho^3$" in the denominator of expression (9.1.11) is replaced by "$n(J_{\mathbf{F}}) + n(J_{\mathbf{G}})$". By contrast, the new index obviously has the following advantages:

(i) Immanence. Each component of representation (9.1.12) is only related to the number of points in the two Julia sets, not to the size of the space $\Omega$ or the resolution $\rho$ of the lattice. That is to say, the Julia deviation index is entirely determined by the two Julia sets.

(ii) Normalization. For two identical Julia sets, i.e., $J_{\mathbf{F}} = J_{\mathbf{G}}$, we have $\mathbf{JDI}(J_{\mathbf{F}}, J_{\mathbf{G}}) = 0$, which inherits the nature of expression (9.1.11). And more importantly, when the two Julia sets are totally different from each other, i.e., $J_{\mathbf{F}} \cap J_{\mathbf{G}} = \varnothing$, we have $\mathbf{JDI}(J_{\mathbf{F}}, J_{\mathbf{G}}) = 1$. However, it is not available for the Julia deviation distance ($\mathbf{DistJ}(J_{\mathbf{F}}, J_{\mathbf{G}}) \ll 1$ in this case, generally). It is this normalization that embodies advantages of the new version and makes it possible to compare Julia sets generated by systems with different parameters or even different structures.

In order to reflect the influence of multiplicative noise on Julia sets by comparison, we additionally introduce a system with state additive noise

$$\begin{cases} u_{n+1} = r_1 u_n - a_1 u_n^2 - b_1 u_n v_n - \sum\limits_{k=1}^{\mathcal{N}} \omega_k^{(\alpha)} u_{n-k+1} + k_1 \delta_{1,n}, \\ v_{n+1} = -r_2 v_n + a_2 u_n v_n - b_2 v_n^2 - \sum\limits_{k=1}^{\mathcal{N}} \omega_k^{(\beta)} v_{n-k+1} + k_2 \delta_{2,n} \end{cases} \tag{8.5.10}$$

and a system with order additive noise

$$\begin{cases} u_{n+1} = r_1 u_n - a_1 u_n^2 - b_1 u_n v_n - \sum\limits_{k=1}^{\mathcal{N}} \omega_k^{(\alpha + k_1 \varrho_{1,n})} u_{n-k+1}, \\ v_{n+1} = -r_2 v_n + a_2 u_n v_n - b_2 v_n^2 - \sum\limits_{k=1}^{\mathcal{N}} \omega_k^{(\beta + k_2 \varrho_{2,n})} v_{n-k+1}. \end{cases} \tag{8.5.11}$$

Julia sets of systems (8.5.10) and (8.5.11) are denoted as $J_{\mathbf{F}}^{\mathrm{as},k}$ and $J_{\mathbf{F}}^{\mathrm{ao},k}$, respectively. Note that introducing such noise source is only for mathematical analysis and comparison. Limit the range of noise strength on [0, 10]. In the following, the Julia deviation index and distance of the Julia sets with four kinds of noise mentioned above through calculations is shown in Fig. 8.21, by which the following claims are supported.

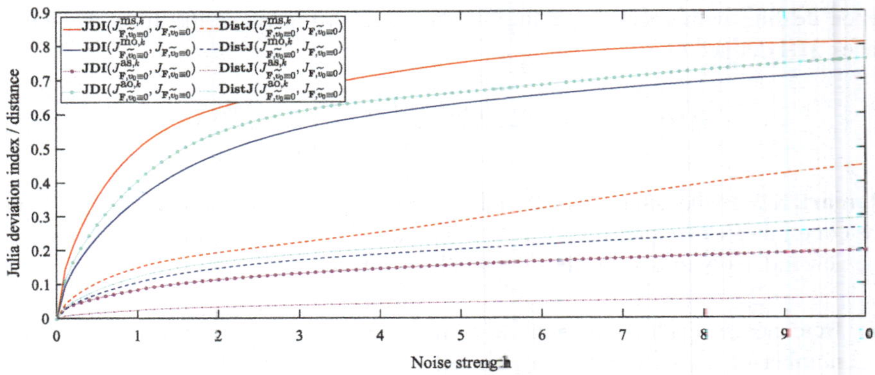


**Fig. 8.21** Julia deviation index/distance of Julia sets of noise-perturbed fractional Lotka–Volterra systems (8.5.6), (8.5.7), (8.5.10) and (8.5.11)

(1) For each perturbed system, the Julia deviation index is closer to 1 than the Julia deviation distance under the same noise strength.
(2) The Julia deviation index/distance increases rapidly with $k \in [0, 2]$ approximately and then slow down, from which it the illustrate that the influence on the Julia set of the original system gradually shows a slow variation with the noise strength increasing.
(3) From $\mathbf{JDI}(J^{ms,k}_{\mathbf{F},\tilde{v}_0=0}, J_{\mathbf{F},\tilde{v}_0=0}) > \mathbf{JDI}(J^{mo,k}_{\mathbf{F},\tilde{v}_0=0}, J_{\mathbf{F},\tilde{v}_0=0})$, it can be seen that the state multiplicative noise destroy the original system more than the order multiplicative noise. This signifies that for population dynamics, disturbances from the environment have a greater impact on the system than internal individual fluctuations in the system. But the situation of additive noise is just the opposite.
(4) From $\mathbf{JDI}(J^{ms,k}_{\mathbf{F},\tilde{v}_0=0}, J_{\mathbf{F},\tilde{v}_0=0}) > \mathbf{JDI}(J^{as,k}_{\mathbf{F},\tilde{v}_0=0}, J_{\mathbf{F},\tilde{v}_0=0})$, it can be seen that the state multiplicative noise disturb the original system more than the state additive noise. It is shown that the structure of Julia sets of systems with multiplicative noise sources is more unpredictable.

Moreover, the relationship $f$ between $\mathbf{JDI}(\cdot, J_{\mathbf{F},\tilde{v}_0=0})$ and noise strength $k$ can be acquired by the exponential fitting with two terms as following:

$$f(k) = ae^{bk} + ce^{dk}. \tag{8.5.12}$$

Coefficients and goodness of fitting (8.5.12) are collected in Table 8.5, in which the goodness includes the coefficient of determination (R-square), the sum of squares due to error (SSE) and the root mean squared error (RMSE).

From Table 8.5, we can see that R-squares are very close to 1, besides, neither SSEs nor RMSEs exceed 2.2%, which indicate that fitting expression (8.5.12) is an efficient one. Besides, it can be seen that the fitting of multiplicative noise case is not as good as that of additive order noise case. This confirms once again that

**Table 8.5** Exponential fitting for $\mathbf{JDI}(\cdot, J_{\mathbf{F},\widetilde{v}_0\equiv 0})$

| Set | $a$ | $b$ | $c$ | $d$ | R-square | SSE | RMSE |
|---|---|---|---|---|---|---|---|
| $J^{\mathrm{ms},k}_{\mathbf{F},\widetilde{v}_0\equiv 0}$ | 0.6613 | 0.0231 | –0.5899 | –1.1730 | 0.9911 | 0.0215 | 0.0149 |
| $J^{\mathrm{mo},k}_{\mathbf{F},\widetilde{v}_0\equiv 0}$ | 0.5589 | 0.0267 | –0.5084 | –0.8199 | 0.9981 | 0.0046 | 0.0069 |
| $J^{\mathrm{as},k}_{\mathbf{F},\widetilde{v}_0\equiv 0}$ | 0.1267 | 0.0456 | –0.1053 | –0.7631 | 0.9954 | 0.0008 | 0.0029 |
| $J^{\mathrm{ao},k}_{\mathbf{F},\widetilde{v}_0\equiv 0}$ | 0.5807 | 0.0279 | –0.5394 | –1.0650 | 0.9987 | 0.0031 | 0.0057 |

multiplicative noise disturbs the system more severely, which makes the Julia set more unpredictable.

**Remark 8.10** For the sake of intuition, the Julia deviation index is studied by taking the 3-D slice of Julia sets as a specific example here. Nevertheless, from the above process, it is not difficult to see that this method, in fact, is completely applicable to the 4-D case and even a higher dimensional situation.

### The Julia Deviation Plot

Although the Julia deviation index provides a quantification of the difference between a noise-perturbed Julia set and the original one, it does not give us any information or geometric intuition about the way that the deformation takes place from one Julia set to the other. To connect with this, this will be achieved with a graphical tool named Julia deviation plot [219, 394, 405]. Let us consider two Julia sets $J_{\mathbf{F}}$ and $J_{\mathbf{G}}$ defined on the same lattice of points. Then, every point $p$ in the lattice belongs to one of the following four disjoint sets:

$$S := J_{\mathbf{F}} \cap J_{\mathbf{G}}, \quad D := J_{\mathbf{F}} \cap J^c_{\mathbf{G}}, \quad B := J^c_{\mathbf{F}} \cap J_{\mathbf{G}}, \quad N := J^c_{\mathbf{F}} \cap J^c_{\mathbf{G}},$$

where $J^c_{\mathbf{F}}$ ($J^c_{\mathbf{G}}$) denotes the complement of the set $J_{\mathbf{F}}$ ($J_{\mathbf{G}}$). Correspondingly, each point $p$ is coloured according to the following rule:

$$\mathbf{JDP}(p) = \begin{cases} \text{yellow (stable)}, & p \in S, \\ \text{red (death)}, & p \in D, \\ \text{blue (birth)}, & p \in B, \\ \text{white (neutral)}, & p \in N. \end{cases}$$

Thus it can be viewed that $J_{\mathbf{F}} = S \cup D$ and $J_{\mathbf{G}} = S \cup B$. By aid of this tool, noise-perturbed Julia sets of systems (8.5.6) and (8.5.7) with $k = 0.5$ and 1.0 can be re-colored point by point, which is presented in Fig. 8.22.

From Fig. 8.22, it further validates the results that have been shown in Fig. 8.20.

(1) Compared with the original Julia set shown in Fig. 8.19**a**, noise makes the Julia set more diverse. Noise with different types and different strength has different effects on Julia sets in different ways and degrees.

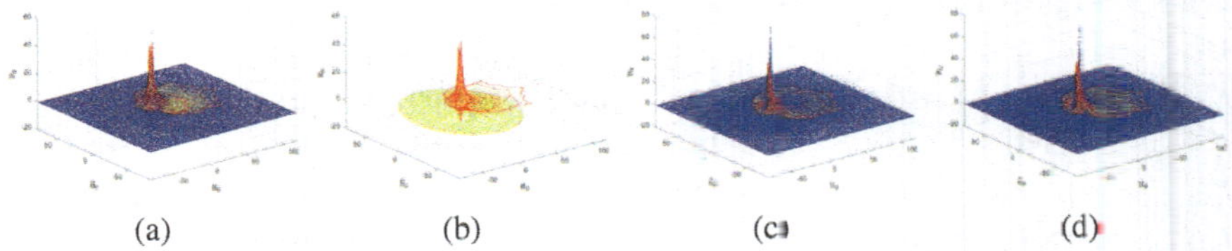


**Fig. 8.22** Julia deviation plot of Julia sets of noise-perturbed fractional Lotka–Volterra systems (8.5.6) and (8.5.7): **a** $J^{ms,0.5}_{\mathbf{F},\widetilde{v}_0\equiv 0}$; **b** $J^{ms,1.0}_{\mathbf{F},\widetilde{v}_0\equiv 0}$; **c** $J^{mo,0.5}_{\mathbf{F},\widetilde{v}_0\equiv 0}$; **d** $J^{mo,1.0}_{\mathbf{F},\widetilde{v}_0\equiv 0}$

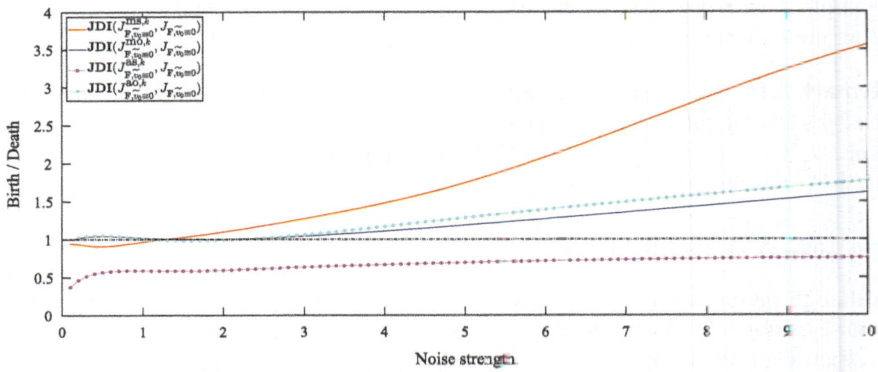


**Fig. 8.23** The ratio of births point to death points

(2) For the state multiplicative noise, perturbations from birth points and death points are distinct under different noise strength, seeing (a) and (b). In detail, birth points are still the main factor of disturbance when $k = 0.5$ while death points occupy a dominant position obviously when $k = 1.0$.
(3) For the order multiplicative noise, birth and death points are evenly distributed around the original set, and birth points are more concentrated in plane $\widetilde{v}_0 = 0$.

What is more, the distribution of death points and birth points could provide us insight into the way in which the Julia set of noise-perturbed fractional Lotka-Volterra system is deformed due to noise changes. In order to get a better look at the Julia deviation plot, we introduce an index $B/D$, i.e., the ratio of birth points to death points, seeing Fig. 8.23.

The following conclusions can be drawn combining Figs. 8.22 and 8.23.

(i) For state multiplicative noise, at a lower noise strength, $B/D < 1$, while after about $k = 1.2$, $B/D > 1$ and then it rapidly rises. The former can explain the degeneration of the Julia set as previously showed in Figs. 8.20 and 8.22; the latter indicates that the system will have new complex dynamics. However, for state additive noise, death points are always more numerous than birth points ($B/D < 1$).

(ii) For the two types of order noise, there are fewer points of death than birth ($B/D > 1$). The number of these two kinds of points is approximately the same ($B/D \approx 1$) initially. The joint action from them results in a great change in the structure of Julia sets during this period. Then, index $B/D$ grows steadily and slowly, which shows birth points gradually occupy the dominant position. It is a large number of birth points that can proffer an interpretation for the high values of the Julia deviation index/distance in Fig. 8.21.

Moreover, it is noted that formula (9.1.12) for the Julia deviation index is equivalent to

$$\mathbf{JDI}(J_{\mathrm{F}}, J_{\mathrm{G}}) = \frac{n(D) + n(B)}{n(D) + n(B) + 2n(S)} \tag{8.5.13}$$

by using the above notations in this subsection. From expression (8.5.13), it can also be seen that the Julia deviation index eliminates neutral points in the denominator of expression (9.1.11), which intuitively reflects the immanence of the index, because a neutral point is neither the element of the original Julia set nor the element of the noise-perturbed Julia set.

### *8.5.4 The Symmetry Damages of Julia Sets of Noise-Perturbed Fractional Lotka–Volterra System*

The investigation of the symmetry of Julia sets can be found in [190, 394, 418]. In this part, Fig. 8.19 has already preliminarily shown that the Julia set of system (8.5.3) has a symmetrical structure, which can be treated as a visual evidence for the following theorem.

**Theorem 8.4** *$J_{\mathrm{F},\widetilde{v}_0 \equiv 0}$ is symmetric about the $\overline{u}_0\overline{v}_0$-plane.*

***Proof*** For two initial points

$$\mu_0 = (\mu_{01}, \mu_{02}, \mu_{03}, 0) \quad \text{and} \quad \nu_0 = (\nu_{01}, \nu_{02}, \nu_{03}, 0),$$

set them be symmetric about the $\overline{u}_0\overline{v}_0$-plane, i.e., $\mu_{01} = \nu_{01}$, $\mu_{02} = -\nu_{02}$ and $\mu_{03} = \nu_{03}$. Then, denote the $n$th iterations of $\mu_0$ and $\nu_0$ as

$$\mu_n = (\mu_{n1}, \mu_{n2}, \mu_{n3}, \mu_{n4}) \quad \text{and} \quad \nu_n = (\nu_{n1}, \nu_{n2}, \nu_{n3}, \nu_{n4}), \quad n \in \mathbb{N}_+.$$

For $n = 1$, from Eq. (8.5.4) with $\mu_0$ and $\nu_0$, we directly have

$$\mu_1 = (\mu_{11}, \mu_{12}, \mu_{13}, \mu_{14}) \quad \text{and} \quad \nu_1 = (\mu_{11}, -\mu_{12}, \mu_{13}, -\mu_{14}),$$

which means that after the first iteration, the first and the third coordinates of the two points stay the same, while second and the fourth coordinates turn the contrary.

For $n = m > 1$, assume that $\mu_{m1} = \nu_{m1}$, $\mu_{m2} = -\nu_{m2}$, $\mu_{m3} = \nu_{m3}$ and $\mu_{m4} = -\nu_{m4}$, namely,

$$\mu_m = (\mu_{m1}, \mu_{m2}, \mu_{m3}, \mu_{m4}) \quad \text{and} \quad \nu_m = (\mu_{m1}, -\mu_{m2}, \mu_{m3}, -\mu_{m4}). \tag{8.5.14}$$

Then, for $n = m + 1$, by substituting (8.5.14) into (8.5.4), an elementary calculation gives us

$$\mu_{m+1} = (\mu_{m+1,1}, \mu_{m+1,2}, \mu_{m+1,3}, \mu_{m+1,4}) \quad \text{and}$$
$$\nu_{m+1} = (\mu_{m+1,1}, -\mu_{m+1,2}, \mu_{m+1,3}, -\mu_{m+1,4}).$$

According to mathematical induction, we reach the following conclusion:

$$\mu_n = (\mu_{n1}, \mu_{n2}, \mu_{n3}, \mu_{n4}) \quad \text{and} \quad \nu_n = (\mu_{n1}, -\mu_{n2}, \mu_{n3}, -\mu_{n4}), \quad n \in \mathbb{N}.$$

The above fact implies that $\lim_{n\to\infty} |\mu_n| = \lim_{n\to\infty} |\nu_n|$, which ensures that $\mu_n$ and $\nu_n$ (do not) belong to $J_{\mathbf{F},\widetilde{v}_0\equiv 0}$ simultaneously. Therefore, the two parts of $J_{\mathbf{F},\widetilde{v}_0\equiv 0}$ cut by the $\overline{u}_0\overline{v}_0$-plane are symmetric. The proof is completed. □

It must be pointed that Theorem 8.4 qualitatively shows the symmetry of the Julia set generated by system (8.5.3) about the $\overline{u}_0\overline{v}_0$-plane. Next, we will further analyze quantitatively the effect of noise on the symmetry of $J_{\mathbf{F},\widetilde{v}_0\equiv 0}$ by utilizing a symmetry criterion [394]. The two parts of the 3-D slice of $\Omega$ that are divided by the $\overline{u}_0\overline{v}_0$-plane are denoted as $D_+ := \{p \in \Omega : \widetilde{u}_0 > 0, \widetilde{v}_0 \equiv 0\}$ and $D_- := \{p \in \Omega : \widetilde{u}_0 < 0, \widetilde{v}_0 \equiv 0\}$, where $p = (\overline{u}_0, \widetilde{u}_0, \overline{v}_0, \widetilde{v}_0)$. For any two initial points $\mu_0 \in D_+$ and $\nu_0 \in D_-$ that are symmetric about the $\overline{u}_0\overline{v}_0$-plane, a symmetry index in terms of the Julia set is defined by:

$$\text{sym}(\mu_0, \nu_0) := \begin{cases} 2, & \mu_0 \in J_{\mathbf{F},\widetilde{v}_0\equiv 0} \text{ and } \nu_0 \in J_{\mathbf{F},\widetilde{v}_0\equiv 0}, \\ 0, & \text{other cases.} \end{cases}$$

Then, the symmetry criterion of $J_{\mathbf{F},\widetilde{v}_0\equiv 0}$ about the $\overline{u}_0\overline{v}_0$-plane, denoted as $\mathbf{SC}_{\overline{u}_0\overline{v}_0}(J_{\mathbf{F}})$, is calculated as:

$$\mathbf{SC}_{\overline{u}_0\overline{v}_0}(J_{\mathbf{F},\widetilde{v}_0\equiv 0}) := \frac{1}{n(J_{\mathbf{F},\widetilde{v}_0\equiv 0})} \sum_{\mu_0 \in D_+} \text{sym}(\mu_0, \nu_0) = \frac{1}{n(J_{\mathbf{F},\widetilde{v}_0\equiv 0})} \sum_{\nu_0 \in D_-} \text{sym}(\mu_0, \nu_0).$$

The symmetry criterion of the Julia sets with four kinds of noise through calculations is shown in Fig. 8.24. It indicates the evolution of the symmetry criterion with the change of noise strength.

From Fig. 8.24, some conclusions can be summarized as follows.

(i) For the state multiplicative noise, the symmetry of the noise-perturbed Julia set rebounds with noise strength increasing, that is, it decreases fast at first and then increases slowly. The inflection point is near $k = 4.0$. Connected with Fig. 8.23, we speculate that this phenomenon is due to the large number of birth points.

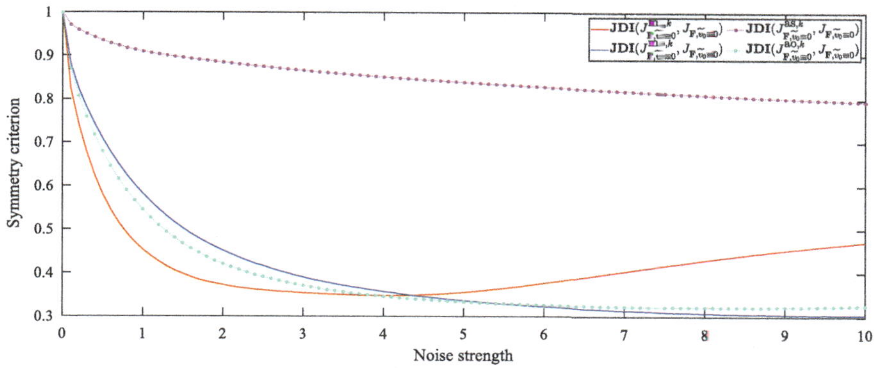


**Fig. 8.24** Symmetry criterion of Julia sets of noise-perturbed fractional Lotka–Volterra systems (8.5.6), (8.5.7), (8.5.10) and (8.5.11)

**Table 8.6** Exponential fitting for $\mathbf{SC}_{\overline{u}_0\overline{v}_0}(\cdot)$

| set | $a$ | $b$ | $c$ | $d$ | R-square | SSE | RMSE |
|---|---|---|---|---|---|---|---|
| $J^{\mathrm{ms},k}_{\mathbf{F},\widetilde{v}_0\equiv 0}$ | 0.6340 | −1.5000 | 0.2950 | 0.0459 | 0.9818 | 0.0177 | 0.0135 |
| $J^{\mathrm{mo},k}_{\mathbf{F},\widetilde{v}_0\equiv 0}$ | 0.5557 | −0.9860 | 0.3885 | −0.0283 | 0.9965 | 0.0067 | 0.0083 |
| $J^{\mathrm{as},k}_{\mathbf{F},\widetilde{v}_0\equiv 0}$ | 0.1296 | −1.2650 | 0.8438 | −0.0113 | 0.9941 | 0.0013 | 0.0036 |
| $J^{\mathrm{ao},k}_{\mathbf{F},\widetilde{v}_0\equiv 0}$ | 0.6392 | −1.0670 | 0.3151 | 0.0458 | 0.9968 | 0.0035 | 0.0060 |

Besides, the multiplicative noise evidently has a greater impact on the Julia set compared with the additive one, which could lead to more complex behavior of the set.

(ii) As for the order noise, the effects of multiplicative noise and additive noise are not much different. The symmetry of the noise-perturbed Julia set decreases with the noise strength increasing, and eventually tends to fixed values on the whole. It is in accordance with the conclusion of an integer-order model such as Wang and Liu [394].

Similarly, the relationship $f$ between $\mathbf{SC}_{\overline{u}_0\overline{v}_0}(\cdot)$ and noise strength $k$ can be acquired by the exponential fitting with two terms as expression (8.5.12). The values of fitting coefficients and goodness are collected in Table 8.6.

From Table 8.6, we can see that R-squares are very close to 1, meanwhile SSEs and RMSEs are basically no more than 1.8%, which indicate that fitting expression (8.5.12) is also efficient for the symmetry criterion.

## 8.6 Adaptive Synchronization of Julia Sets Generated by Mittag-Leffler Function

### 8.6.1 Background

The term "fractal" was first proposed in 1975 by Benoit Mandelbrot [419]. In fact, as early as the late 1910s, the Julia set, as an important class of fractal objects, had been acquired by Gaston Julia [189] and Pierre Fatou [420] when they developed the basic iteration theory of a complex quadratic polynomial function/map

$$\begin{aligned} f: \mathbb{C} &\to \mathbb{C} \\ z &\mapsto z^2 + c. \end{aligned} \tag{8.6.1}$$

Thereafter, with the development of computer technology, a lot of promotional work and profound studies have been made in this area for various systems, such as generalized map [190], alternating map [185, 421], Lorenz system [221, 394], Lotka–Volterra system [179], fractional system [400, 401, 422], etc. However, these are mainly concentrated on iterations of polynomial functions.

Although the research on Julia sets of non-polynomial functions is relatively few, there is still some excellent work has been carried out. Early as 1981, the Julia set $J$ of complex system

$$z_{n+1} = e^{z_n} \tag{8.6.2}$$

was investigated by Misiurewicz [423] who proved that this set is the whole complex plane, i.e., $J = \mathbb{C}$, which shows that periodic points of map $f: z \to e^z$ are dense in $\mathbb{C}$.

In 1984, Devaney et al. [424] the dynamics of complex system

$$z_{n+1} = \lambda e^{z_n}, \tag{8.6.3}$$

where $\lambda \in \mathbb{R}$. They found that a bifurcation occurs at $\lambda = e^{-1}$ and the Julia set is a Cantor set of curves in $\mathbb{C}$, i.e., a so-called Cantor bouquet, when $0 < \lambda < e^{-1}$. In the following, the case of $\lambda \in \mathbb{C}$ was also considered in [425–427], etc. Besides, Zhang et al. [428] realized the control of geometric characteristics of Julia sets for system (8.6.3) by using an auxiliary reference feedback control method.

In 1987, McMullen [429] studied the Julia set of the following system with sine function:

$$z_{n+1} = \lambda \sin z_n. \tag{8.6.4}$$

This Julia set has positive area and the action of map $f: z \to \lambda \sin(z)$ on its Julia set is not ergodic. In 2015, Sun et al. [430] used a zero-asymptotical sliding variable discrete control method to control and synchronize Julia sets generated by two kinds of sine functions with unknown parameters.

In 2005, Sajid et al. [431] studied the Julia set and bifurcation phenomena of the following system involving hyperbolic sine function:

$$z_{n+1} = \lambda \frac{\sinh z_n}{z_n^2}, \tag{8.6.5}$$

where $\lambda > 0$. Dynamic and fractal properties of system (8.6.5) are also compared with those of a system based on map $f : z \to \lambda(e^z - 1)/z$.

On the other hand, generally speaking, as advantages of fractional calculus have been increasingly found in modeling for various practical problems in the past two decades, more and more attention has been paid to "fractional functions" among which one of the most famous certainly is the Mittag-Leffler function. The function, named after Gösta Magnus Mittag-Leffler, sometimes is also called "fractional exponential function", because it provides a generalization of the exponential function [432]. Another reason for its wide reputation is that it successfully constructs or refines mathematical models of various physical, chemical, economical, management and bioengineering phenomena. However, as an analytic (holomorphic) function, as far as we know, Mittag-Leffler function has not been investigated with respect to the fractal aspect till now.

### 8.6.2 Preliminaries

Throughout this section, some necessary knowledge about fractional calculus theory, Mittag-Leffler function and Julia set is reviewed briefly. More relevant contents can be found in monographs [218, 387, 402, 419, 432, 433] and references contained therein.

**Fractional Calculus**

Recently, Riemann–Liouville fractional integral (Definition 8.1), Caputo (Definition 8.2), Riemann–Liouville and modified Riemann–Liouville fractional differential operators are frequently used. The standard definitions of the latter two are listed here and some basic relations are introduced briefly.

**Definition 8.7** The Riemann–Liouville fractional integral of order $\alpha > 0$ of a function $y : (0, +\infty) \to \mathbb{R}$ is given by

$$I_{0+}^{\alpha} y(t) = \frac{1}{\Gamma(\alpha)} \int_0^t (t - s)^{\alpha-1} y(s) ds$$

provided that the right-hand side is pointwise defined on $(0, -\infty)$, where $\Gamma(z)$ is the Gamma function defined as $\Gamma(z) = \int_0^{\infty} t^{z-1} e^{-t} dt$.

**Definition 8.8** ([434]) The modified Riemann–Liouville fractional derivative of order $\alpha > 0$ of a function $y$ is given by

$$ {}^{mRL}D_{0+}^{\alpha}y(t) = \frac{1}{\Gamma(n-\alpha)}\left(\frac{d}{dt}\right)^n \int_0^t (t-s)^{n-\alpha-1}\big(y(s)-y(0)\big)ds $$

provided that the right-hand side is pointwise defined on $(0, +\infty)$.

The function $(t-s)^{n-\alpha-1}$ determines a time interval that reflects the memory of the derivatives. what's more, Definitions 8.7 and 8.8 are most popular among various definitions of fractional derivatives, although the revised definition makes up for some shortcomings of the two. Definition 8.7 is usually used in theoretical models especially in pure mathematics, while Definition 8.8 appears more in the field of engineering modeling and numerical computation. Later, Caputo fractional derivative will be considered to investigate some properties of Mittag-Leffler functions.

**Mittag-Leffler Function, Fractional Trigonometric Function and Fractional Hyperbolic Function**

As generalizations of the classical exponential function, Mittag-Leffler functions are important in fractional differential equations and control systems. Their single parameter form and dual parameter form are defined as [432]

$$ E_\alpha(t) = \sum_{k=0}^{\infty} \frac{t^k}{\Gamma(\alpha k+1)} \quad \text{and} \quad E_{\alpha,\beta}(t) = \sum_{k=0}^{\infty} \frac{t^k}{\Gamma(\alpha k+\beta)}, $$

respectively, where $\alpha, \beta \in \mathbb{C}$ and $\alpha$'s real part $\Re(\alpha) > 0$. In particular, (i) $E_{\alpha,1}(t) = E_\alpha(t)$; (ii) $E_1(t) = \sum_{k=0}^{\infty} \frac{t^k}{k!} = e^t$; (iii) $E_0(t) = \sum_{k=0}^{\infty} t^k = \frac{1}{1-t}$, $|t| < 1$.

Mittag-Leffler functions not only contain some classical functions, but also have stronger applicability due to their adjustable parameters. When the parameter $\alpha$ is non-integer, such functions exhibit a power law that can describe phenomena related to certain scaling laws or heavy-tailed distributions [435].

It is noticed that similar to the exponential function, function $E_\alpha(\cdot)$ has the following interesting property ([433] Theorem 5.12):

$$ {}^{C}D_{0+}^{\alpha} E_\alpha(\lambda t^\alpha) = \lambda E_\alpha(\lambda t^\alpha). \tag{8.6.6} $$

This illustrated that the Mittag-Leffler function to the fractional derivative is a counterpart of the exponential function to the classical derivative.

As a generalization to trigonometry functions, fractional sine and cosine functions ([436] Definition 5.2 and [437] ) can be defined by the Mittag-Leffler function as following:

$$ \sin_\alpha(t) := \frac{E_\alpha(\mathrm{i}t^\alpha) - E_\alpha(-\mathrm{i}t^\alpha)}{2\mathrm{i}} = t^\alpha \sum_{k=0}^{\infty} \frac{(-t^{2\alpha})^k}{\Gamma(2\alpha k+\alpha+1)} = t^\alpha E_{2\alpha,\alpha+1}(-t^{2\alpha}) \tag{8.6.7} $$

and

$$\cos_\alpha(t) := \frac{E_\alpha(\mathrm{i}t^\alpha) + E_\alpha(-\mathrm{i}t^\alpha)}{2} = \sum_{k=0}^{\infty} \frac{(-t^{2\alpha})^k}{\Gamma(2\alpha k + 1)} = E_{2\alpha}(-t^{2\alpha}), \tag{8.6.8}$$

where $\alpha > 0$ and $\mathrm{i} = \sqrt{-1}$ is the imaginary unit.

Analogously, fractional hyperbolic sine and cosine functions [438, 439] are defined as

$$\sinh_\alpha(t) := \frac{E_\alpha(t^\alpha) - E_\alpha(-t^\alpha)}{2} = t^\alpha \sum_{k=0}^{\infty} \frac{t^{2\alpha k}}{\Gamma(2\alpha k + \alpha + 1)} = t^\alpha E_{2\alpha,\alpha+1}(t^{2\alpha}) \tag{8.6.9}$$

and

$$\cosh_\alpha(t) := \frac{E_\alpha(t^\alpha) + E_\alpha(-t^\alpha)}{2} = \sum_{k=0}^{\infty} \frac{t^{2\alpha k}}{\Gamma(2\alpha k + 1)} = E_{2\alpha,\beta}(t^{2\alpha}), \tag{8.6.10}$$

respectively.

**Remark 8.11** When $\alpha = 1$, functions defined by expressions (8.6.7)–(8.3.7) degenerate into the classical situations, i.e., sine function, cosine function, hyperbolic sine function and hyperbolic cosine function. In addition, from formula (8.6.6), the following relationships can be seen immediately:

$$\begin{cases} {}^C D_{0+}^\alpha \sin_\alpha(t) = \frac{\mathrm{i}E_\alpha(\mathrm{i}t^\alpha)+\mathrm{i}E_\alpha(-\mathrm{i}t^\alpha)}{2\mathrm{i}} = \cos_\alpha(t), \\ {}^C D_{0+}^\alpha \cos_\alpha(t) = \frac{\mathrm{i}E_\alpha(\mathrm{i}t^\alpha)-\mathrm{i}E_\alpha(-\mathrm{i}t^\alpha)}{2} = -\sin_\alpha(t), \\ {}^C D_{0+}^\alpha \sinh_\alpha(t) = \frac{E_\alpha(t^\alpha)+E_\alpha(-t^\alpha)}{2} = \cosh_\alpha(t), \\ {}^C D_{0+}^\alpha \cosh_\alpha(t) = \frac{E_\alpha(t^\alpha)-E_\alpha(-t^\alpha)}{2} = \sinh_\alpha(t), \end{cases}$$

which is another inheritance from properties of corresponding classical functions.

### 8.6.3 *Main Results*

In the beginning, three instances of Julia sets are presented and compared with the existing classical results to reflect the significant impact from fractional functions.

Consider the following three discrete complex dynamical systems:

$$z_{n+1} = \lambda E_\alpha(z_n) + c, \tag{8.6.11}$$

$$z_{n+1} = \lambda \sin_\alpha(z_n^2) + c, \tag{8.6.12}$$

$$z_{n+1} = \lambda \sinh_\alpha(z_n) + c, \tag{8.6.13}$$

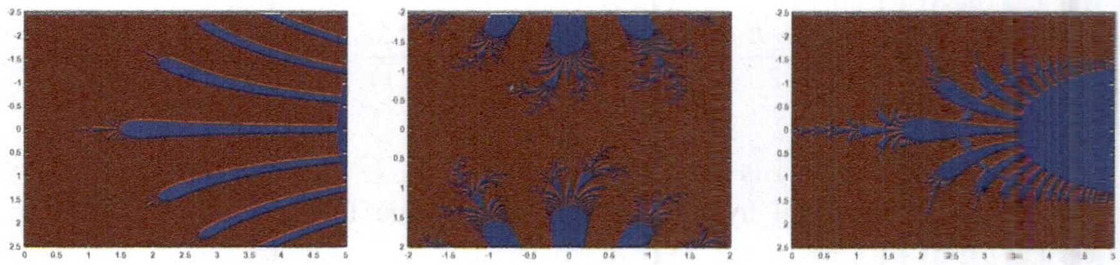

(a) $J^{0.5}_{(8.7.1)}$; $\lambda = \lambda^{*1}, c = 0$ (b) $J^{0.5}_{(8.7.2)}$; $\lambda = 1, c = -0.7 + 0.3i$ (c) $J^{0.5}_{(8.7.3)}$; $\lambda = 0.7, c = 0$

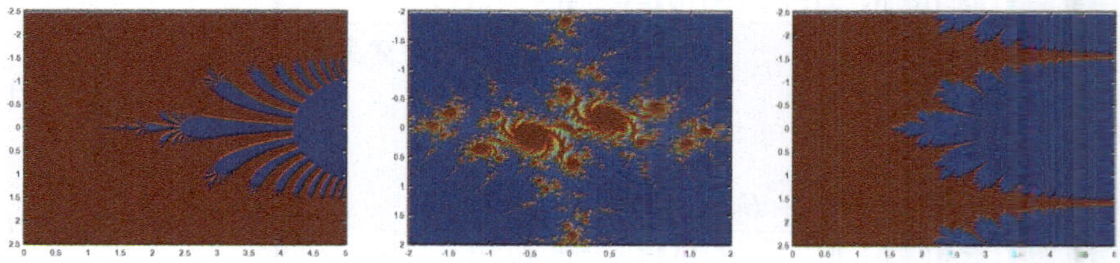

(d) $J^{1.0}_{(8.7.1)}$; $\lambda = e^{-1}, c = 0$ (e) $J^{1.0}_{(8.7.2)}$; $\lambda = 1, c = -0.7 + 0.3i$ (f) $J^{1.0}_{(8.7.3)}$; $\lambda = 0.7, c = 0$

**Fig. 8.25** Julia sets of systems (9.1.1)–(9.1.3)

where $\alpha, \lambda \in \mathbb{R}$, $c, z_0 \in \mathbb{C}$ and $n \in \mathbb{N}$. For clarity and convenience, Julia sets of systems (9.1.1), (9.1.2) and (9.1.3) are recorded as $J^{\alpha}_{(8.7.1)}$, $J^{\alpha}_{(8.7.2)}$ and $J^{\alpha}_{(8.7.3)}$, respectively.

Given parameters $\alpha$, $\lambda$ and $c$, the Julia sets of systems (9.1.1)–(9.1.3) can be calculated by the escape-time algorithm [410] and then shown in Fig. 8.25.

In Fig. 8.25, it shows that "fractional elements" introduced can increase the freedom and complexity of systems:

(1) Julia sets generated by systems (9.1.1)–(9.1.3) have extraordinarily intricate shape and greatly attractive fractal characteristics. At the same time, the selection of parameter $\alpha$ has a tremendous impact on these sets. Generally speaking, it mainly affects the "hair" of Julia sets: the smaller the parameter, the thinner the hair.

(2) The existing results are only some special cases contained in systems (9.1.1)–(9.1.3). Specifically, $J^{\alpha}_{(8.7.1)} \to J^{1.0}_{(8.7.1)}$, $J^{\alpha}_{(8.7.2)} \to J^{1.0}_{(8.7.2)}$ and $J^{\alpha}_{(8.7.3)} \to J^{1.0}_{(8.7.3)}$ as $\alpha \to 1$, where $J^{1.0}_{(8.7.1)}$, $J^{1.0}_{(8.7.2)}$ and $J^{1.0}_{(8.7.3)}$ are the situations that have been shown in [428] Fig. 2, [430] Fig. 5a and [429] Fig. 1.2, respectively.

From another point of view, Mittag-Leffler function gives a unified and generalized representation including the exponential function ($\alpha = 1$) and a rational[2] function ($\alpha = 0$). Therefore, parameter $\alpha$ also makes the Julia set of the system have flexible and changeable "intermediate" features.

[2]Here number $\lambda^* \approx 0.2610$ is the critical parameter of system (9.1.1) with $\alpha = 0.5$. See Appendix for details and specific algorithm.

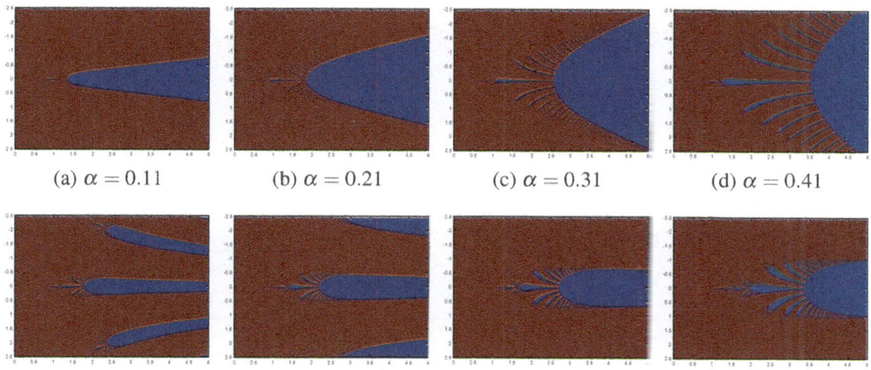

(a) $\alpha = 0.11$ (b) $\alpha = 0.21$ (c) $\alpha = 0.31$ (d) $\alpha = 0.41$

(e) $\alpha = 0.61$ (f) $\alpha = 0.71$ (g) $\alpha = 0.81$ (h) $\alpha = 0.91$

**Fig. 8.26** Julia sets of system (9.1.1) with different parameter $\alpha$

Figure 8.26 further shows how the shape and structure of Julia set $J^{\alpha}_{(8.7.1)}$ vary with parameter $\alpha$ of the Mittag-Leffler function. (The values of critical parameter $\lambda^*$ corresponding to parameter $\alpha$ are shown in Table 8.7 in Appendix.) It must be pointed out that the effect of the parameter on the pattern of Julia sets approximately satisfies a certain scaling law: the larger the parameter value is, the greater the magnification of the pattern is, and thus the more exquisite structure presented is.

By using numerical calculation,[3] box-counting dimension of Julia set $J^{\alpha}_{(8.7.1)}$ with respect to parameter $\alpha$ is shown in Fig. 8.27. With the increase of the parameter, in general, the dimension of the Julia set decreases rapidly and evenly at first and then increases gradually with some fluctuates in the meantime; the inflection point (also the minimum point) appears at $\alpha = 0.31$. The mean value of dimension is 1.8813, against to which the fluctuation is less than 1.6820%. These indicate that the set is indeed a fractal ($1 < 1.8560 \leq D \leq 1.9129 < 2$). And as the parameter changing, it exhibits extremely complex dynamics. Therefore, introducing fractional elements furnishes a new and feasible way for fractal control that has been extensively studied in [402] and references contained therein.

Occasionally, the same behavior of performance needs to be represented by different systems and the relation between behaviors of two Julia sets needs to be considered. Synchronization of Julia sets is another research content for nonlinear dynamical systems since synchronization of chaos. This synchronization enables Julia sets from different systems consistent eventually in their geometric structures and fractal characteristics. Therefore, it is necessary and important to consider such synchronization. There are some achievements on this subject, such as for complex Lorenz system [221], fractional Lotka–Volterra model [400], coupled map lattice [222], rational map [440] and Brusselator model [345]. Hence, synchronization of Julia sets generated by Mittag-Leffler function is also worth studying.

[3]Such numerical calculation processes in this work take advantage of Frederic Moisy's MATLAB® function shared on https://ww2.mathworks.cn/matlabcentral/fileexchange/13063-boxcount.

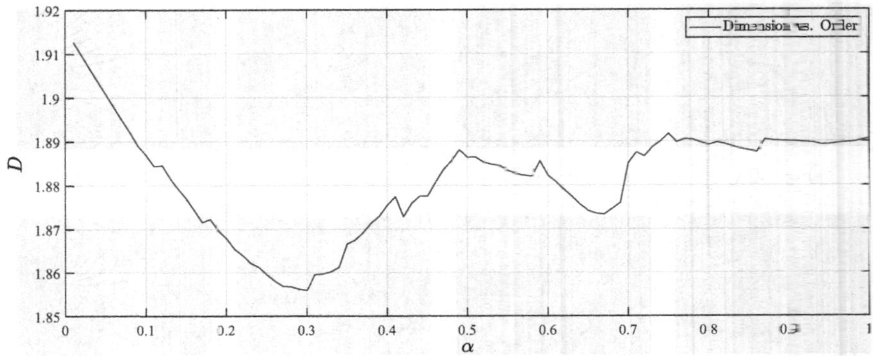


**Fig. 8.27** Box-counting dimension of Julia set $J^{\alpha}_{(8.7.1)}$

We now specialise to the control and synchronization of Julia sets generated by above-mentioned systems. More generally, take the driving system as

$$z_{n+1} = \lambda f_{\alpha}(z_n) + c, \quad z_0 \in \mathbb{C} \tag{8.6.14}$$

with a unknown parameter $\lambda \in \mathbb{R}$, where $f_\alpha$ is a function that may contain $\exp_\alpha$, $\sin_\alpha$, $\sinh_\alpha$ or other ones satisfying certain conditions. The response system is taken as

$$y_{n+1} = \gamma f_{\alpha}(y_n) + u_n + c, \quad y_0 \in \mathbb{C} \tag{8.6.15}$$

with a unknown parameter $\gamma \in \mathbb{R}$, where $u_n$ is a controller to be designed. Then, the system error at the $n$th step is

$$e_n = y_n - z_n. \tag{8.6.16}$$

**Definition 8.9** ([441, 442]) For the driving system (9.1.4) and response system (9.1.5) with any initial conditions $(z_0, y_0) \in \mathbb{C} \times \mathbb{C}$, it is said to achieve synchronization between $z_n$ and $y_n$ if there exists a controller $u_n(z_n, y_n)$ such that all trajectories $(z_n, y_n)$ of systems (9.1.4) and (9.1.5) approach $y_n \to z_n$ as $n \to \infty$, that is,

$$\lim_{n\to\infty} |e_n| = \lim_{n\to\infty} |y_n - z_n| = 0.$$

Denote Julia sets of systems (9.1.4) and (9.1.5) by $J_{(8.7.4)}$ and $J_{(8.7.5)}$, respectively. Then, synchronization of two Julia sets can be defined as follows.

**Definition 8.10** ([221]) Julia sets asymptotic synchronization of driving system (9.1.4) and response system (9.1.5) occurs if

$$\lim_{n\to\infty} \left(J_{(8.7.4)} \cup J_{(8.7.5)} - J_{(8.7.4)} \cap J_{(8.7.5)}\right)_n = \varnothing.$$

**Remark 8.12** According to Definition of Julia set, from the same initial value, if trajectories of the two systems are synchronized, then their convergence and divergence are the same. Hence, Julia sets of the two systems are synchronized as well. This means that Definition 8.10 can be derived directly from Definition 8.9.

**Theorem 8.5** *Suppose both $\gamma$ and $\lambda$ in systems (9.1.4) and (9.1.5) are full known a priori. If $|k| < 1$, then the designed adaptive controller*

$$u_n = -\gamma f_\alpha(y_n) + \lambda f_\alpha(z_n) - ke_n \tag{8.6.17}$$

*can guarantee that for any initial conditions $(z_0, y_0)$, the synchronization between the Julia set of response system (9.1.5) and that of driving system (9.1.4) is realized.*

***Proof*** Substitute controller $u_n$ in system (9.1.5) for the one designed in (9.1.7). Then, from Eqs. (9.1.4)–(9.1.6), we obtain

$$\begin{aligned} e_{n+1} &= y_{n+1} - z_{n+1} \\ &= \gamma f_\alpha(y_n) - \gamma f_\alpha(y_n) + \lambda f_\alpha(z_n) - ke_n - \lambda f_\alpha(z_n) \\ &= -ke_n. \end{aligned}$$

It then follows from $|k| < 1$ that $|e_n| \to 0$ as $n \to \infty$. According to Definition 8.9, all trajectories $(z_n, y_n)$ of systems (9.1.4) and (9.1.5) achieve synchronization. What is more, from Remark 8.12, the Julia set of response system (9.1.5) changes into that of driving system (9.1.4). Therefore, the synchronization of Julia sets of systems (9.1.4) and (9.1.5) is obtained. The proof is completed. □

**Theorem 8.6** *Suppose the parameter $\gamma$ in system (9.1.5) is known a priori. For some given $a, k \in \mathbb{R}$ satisfying $|a + k| < 1$, the designed adaptive controller*

$$u_n = -\gamma f_\alpha(y_n) + \widehat{\lambda}_n f_\alpha(z_n) - ke_n \tag{8.6.18}$$

*and the complex-valued update law of $\widehat{\lambda}_n$*

$$\widehat{\lambda}_{n+1} = \widehat{\lambda}_n - \frac{a[e_{n+1} f_\alpha(z_n) - e_n f_\alpha(z_{n+1})]}{f_\alpha(z_n) f_\alpha(z_{n+1})} \tag{8.6.19}$$

*can guarantee that for any initial conditions $(z_0, y_0)$ and $\widehat{\lambda}_0$, the synchronization between the Julia set of response system (9.1.5) and that of driving system (9.1.4) is realized, where $\widehat{\lambda}_n$ is the estimate of the unknown parameter $\lambda$.*

***Proof*** Substitute controller $u_n$ in system (9.1.5) for the one designed in (9.1.8). Then, from Eqs. (9.1.4)–(9.1.6), we obtain

$$
\begin{aligned}
e_{n+1} &= y_{n+1} - z_{n+1} \\
&= \gamma f_\alpha(y_n) - \gamma f_\alpha(y_n) + \widehat{\lambda}_n f_\alpha(z_n) - ke_n - \lambda f_\alpha(z_n) \\
&= (\widehat{\lambda}_n - \lambda) f_\alpha(z_n) - ke_n \\
&= \widetilde{\lambda}_n f_\alpha(z_n) - ke_n,
\end{aligned} \tag{8.6.20}
$$

where $\widetilde{\lambda}_n := \widehat{\lambda}_n - \lambda$. From update law (9.1.9), we have

$$
\widehat{\lambda}_{n+1} - \widehat{\lambda}_n = -a\left[\frac{e_{n+1}}{f_\alpha(z_{n+1})} - \frac{e_n}{f_\alpha(z_n)}\right]. \tag{8.6.21}
$$

Notice $\widehat{\lambda}_n - \lambda = \widetilde{\lambda}_n$. Then, there exists

$$
\widetilde{\lambda}_n = -\frac{ae_n}{f_\alpha(z_n)} \tag{8.6.22}
$$

satisfying Eq. (9.1.11). By substituting Eq. (9.1.12) into Eq. (9.1.10), we get

$$
e_{n+1} = \widetilde{\lambda}_n f_\alpha(z_n) - ke_n = -(a+k)e_n.
$$

It then follows from $|a+k| < 1$ that $|e_n| \to 0$ as $n \to \infty$. According to Definition 8.9, all trajectories $(z_n, y_n)$ of systems (9.1.4) and (9.1.5) achieve synchronization. In addition, it follows from Eq. (8.6.22) that $\widehat{\lambda}_n \to \lambda$ as $n \to \infty$. What is more, from Remark 8.12, the synchronization of Julia sets of systems (9.1.4) and (9.1.5) is obtained. The proof is completed. □

**Theorem 8.7** *For some given $a, b, k \in \mathbb{R}$ satisfying $|a+k| < 1$, the designed adaptive controller*

$$
u_n = -\widehat{\gamma}_n f_\alpha(y_n) + \widehat{\lambda}_n f_\alpha(z_n) - ke_n \tag{8.6.23}
$$

*and the complex-valued update laws of $\widehat{\lambda}_n$ and $\widehat{\gamma}_n$*

$$
\begin{cases}
\widehat{\lambda}_{n+1} = \widehat{\lambda}_n - \dfrac{(a-b)[e_{n+1} f_\alpha(z_n) - e_n f_\alpha(z_{n+1})]}{2 f_\alpha(z_n) f_\alpha(z_{n+1})}, \\
\widehat{\gamma}_{n+1} = \widehat{\gamma}_n + \dfrac{(a+b)[e_{n+1} f_\alpha(y_n) - e_n f_\alpha(y_{n+1})]}{2 f_\alpha(y_n) f_\alpha(y_{n+1})}
\end{cases} \tag{8.6.24}
$$

*can guarantee that for any initial conditions $(z_0, y_0)$ and $(\widehat{\lambda}_0, \widehat{\gamma}_0)$, the synchronization between the Julia set of response system (9.1.5) and that of driving system (9.1.4) is realized, where $\widehat{\lambda}_n$ and $\widehat{\gamma}_n$ are estimates of unknown parameters $\lambda$ and $\gamma$, respectively.*

***Proof*** Substitute controller $u_n$ in system (9.1.5) for the one designed in (8.6.23). Then, from Eqs. (9.1.4)–(9.1.6), we obtain

$$\begin{aligned} e_{n+1} &= y_{n+1} - z_{n+1} \\ &= \gamma f_\alpha(y_n) - \widehat{\gamma}_n f_\alpha(y_n) + \widehat{\lambda}_n f_\alpha(z_n) - ke_n - \lambda f_\alpha(z_n) \\ &= -(\widehat{\gamma}_n - \gamma) f_\alpha(y_n) + (\widehat{\lambda}_n - \lambda) f_\alpha(z_n) - ke_n \\ &= -\widetilde{\gamma}_n f_\alpha(y_n) + \widetilde{\lambda}_n f_\alpha(z_n) - ke_n, \end{aligned} \tag{8.6.25}$$

where $\widetilde{\lambda}_n := \widehat{\lambda}_n - \lambda$ and $\widetilde{\gamma}_n := \widehat{\gamma}_n - \gamma$. From the first formula of update laws (8.6.24), we have

$$\widehat{\lambda}_{n+1} - \widehat{\lambda}_n = -\frac{(a-b)}{2}\left[\frac{e_{n+1}}{f_\alpha(z_{n+1})} - \frac{e_n}{f_\alpha(z_n)}\right]. \tag{8.6.26}$$

Notice $\widehat{\lambda}_n - \lambda = \widetilde{\lambda}_n$. Then, there exists

$$\widetilde{\lambda}_n = -\frac{(a-b)e_n}{2f_\alpha(z_n)} \tag{8.6.27}$$

satisfying Eq. (8.6.26). Similarly, we get

$$\widetilde{\gamma}_n = \frac{(a+b)e_n}{2f_\alpha(y_n)}. \tag{8.6.28}$$

By substituting Eqs. (8.6.27) and (8.6.28) into Eq. (8.6.25), one has

$$\begin{aligned} e_{n+1} &= -\widetilde{\gamma}_n f_\alpha(y_n) + \widetilde{\lambda}_n f_\alpha(z_n) - ke_n \\ &= -\tfrac{a+b}{2}e_n - \tfrac{a-b}{2}e_n - ke_n \\ &= -(a+k)e_n. \end{aligned}$$

It then follows from $|a+k| < 1$ that $|e_n| \to 0$ as $n \to \infty$. Hence, from Definition 8.9 and Remark 8.12, the synchronization of Julia sets of systems (9.1.4) and (9.1.5) is obtained. Meanwhile, it follows from Eqs. (8.6.27) and (8.6.28) that $\widehat{\lambda}_n \to \lambda$ and $\widehat{\gamma}_n \to \gamma$ as $n \to \infty$. The proof is completed. □

**Remark 8.13** From Theorems 8.5–8.7, it can be seen that as far as system (9.1.5) itself is concerned, its Julia set can be evolved to any desired one whose system has the same structure as system (9.1.5)'s by designing appropriate controllers, regardless of whether parameter $\gamma$ is known beforehand or not.

### 8.6.4 An Example

To illustrate the effectiveness of the above theoretic results, let us consider a specific case.

In systems (9.1.4) and (9.1.5), the common half-order scenario (i.e., $\alpha = 0.5$) in fractional circuits is considered, $f_\alpha$ is taken as $E_\alpha$, and unknown parameters are selected as $\lambda = \lambda^*$ (the critical value corresponding to $\alpha$) and $\gamma = 1$ (the case in

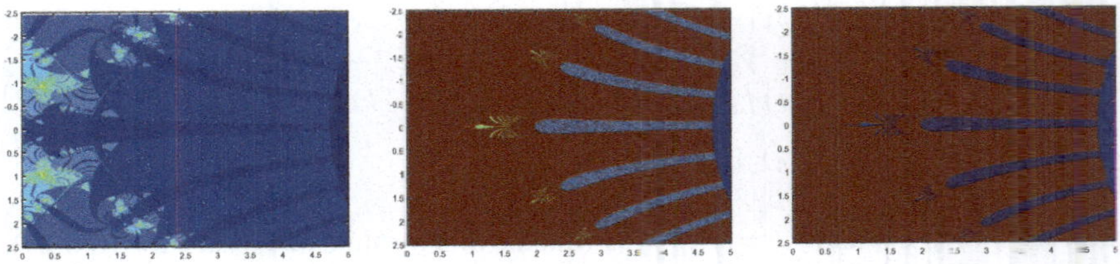

(a) without controller (b) after 5 steps (c) after 25 steps

**Fig. 8.28** The evolution of Julia set $J_{(8.7.5)}$ under the adaptive control strategy

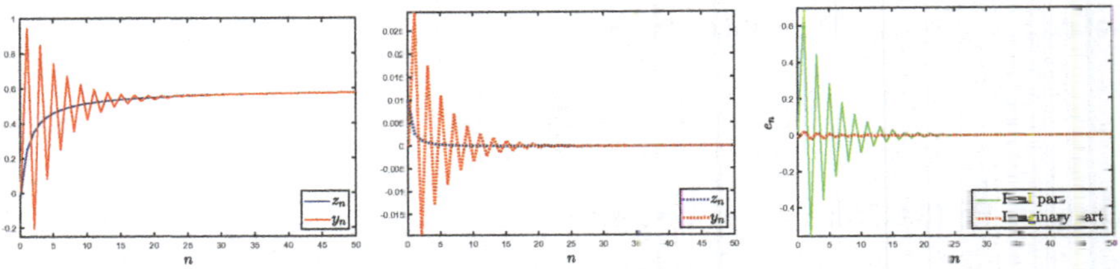

(a) Real part of states (b) Imaginary part of states (c) Error

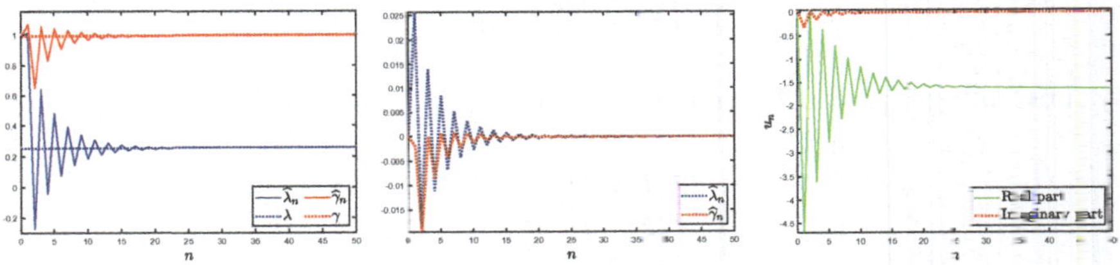

(d) Real part of parameters (e) Imaginary part of parameters (f) Controller

**Fig. 8.29** The system evolution under the adaptive control strategy

[423]). In the following, appropriate control parameters and initial values of update laws are taken, such as $a = -1$, $b = 2$, $k = 1.8$ and $\widehat{\lambda}_0 = \widehat{\gamma}_0 = 1$. The Julia set of system (9.1.5) without controller (i.e., $u_n \equiv 0$) is shown in Fig. 8.28a, from which it can be observed the enormous difference between $J_{(8.7.4)}$ shown in Fig. 8.25a and $J_{(8.7.5)}$ with $u_n \equiv 0$.

The controller and update laws are taken as (8.6.23) and (8.6.24), respectively. Then, it is plain to verify that all the conditions of the Theorem 8.7 are satisfied. The simulation is shown in Fig. 8.28: the shape of $J_{(8.7.5)}$ is very close to $J_{(8.7.4)}$'s after 5 steps; $J_{(8.7.5)}$ is almost the same as $J_{(8.7.4)}$ after 25 steps.

From the viewpoint of state evolution, Fig. 8.29 is obtained to sketch the synchronization behavior between system (9.1.4) and system (9.1.5), where $(z_0, y_0) = (-0.01 + 0.01\mathrm{i}, 0.01)$. Particularly, we see that $e_n \to 0$, $\widehat{\lambda}_n \to \lambda$ and $\widehat{\gamma}_n \to \gamma$ as $n \to \infty$, which indeed illustrates the effectiveness of the designed controller, that is, the response system (9.1.5) synchronizes with the driving system (9.1.4) eventually, and the unknown parameters can be identified gradually.

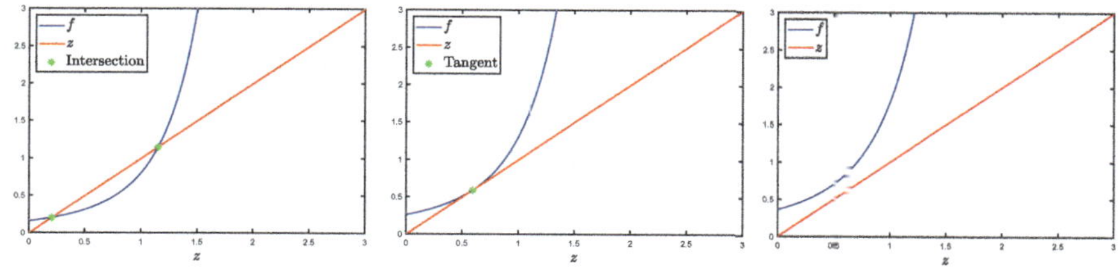


(a) $\lambda < \lambda^*$ (b) $\lambda = \lambda^*$ (c) $\lambda > \lambda^*$

**Fig. 8.30** Three cases of fixed point of system (9.1.1) with $\alpha = 0.5$, $c = 0$

## 8.6.5 Note

Consider system (9.1.1). Fix $c = 0$ and let

$$f(z) = \lambda E_\alpha(z) - z. \tag{8.6.29}$$

For given parameters $\alpha$ and $\lambda$, suppose $z^*$ is the fixed point of system (9.1.1). Then, $z^*$ satisfies $f(z^*) = 0$. What we are most interested in is the case that system (9.1.1) has and only has one fixed point. This can be achieved by adjusting parameter $\lambda$.

Differentiating both sides of Eq. (8.6.29) with respect to $z$ gives

$$f'(z) = \frac{\lambda}{\alpha} E_{\alpha,\alpha}(z) - 1. \tag{8.6.30}$$

Set $f'(z) = 0$. Then, we have

$$\lambda E_{\alpha,\alpha}(z) = \alpha. \tag{8.6.31}$$

Multiplying both sides of Eq. (8.6.31) by $E_\alpha(z)$ and then substituting system (9.1.1) with $c = 0$ into it, we get

$$z E_{\alpha,\alpha}(z) = \alpha E_\alpha(z). \tag{8.6.32}$$

Take $\alpha = 0.5$. Utilizing the bisection method to solve equation (8.6.32), one has

$$z^* = 0.593915726761207.$$

By substituting $z^*$ back into Eq. (8.6.31), one gets

$$\lambda^* = 0.261018837993562.$$

Accordingly, there are two fixed points when $\lambda < \lambda^*$; there is no fixed point when $\lambda > \lambda^*$, seeing Fig. 8.30. That is the reason that $\lambda^*$ is called critical parameter.

Table 8.7 Critical parameter $\lambda^*$

| $\alpha$ | $\lambda^*$ | $\alpha$ | $\lambda^*$ |
|---|---|---|---|
| 0.1 | 0.240888 | 0.6 | 0.275242 |
| 0.2 | 0.239277 | 0.7 | 0.292816 |
| 0.3 | 0.242713 | 0.8 | 0.313909 |
| 0.4 | 0.250120 | 0.9 | 0.338802 |
| 0.5 | 0.261019 | 1.0 | 0.367879 |

Additional critical parameter $\lambda^*$ corresponding to different $\alpha$ are presented in Table 8.7. In particular, when $\alpha = 1$, we have $\lambda^* = e^{-1}$ immediately from formula (8.6.31), which is the critical value described in literature such as [424–42[illegible], 428].

# Chapter 9
# Control of the Thermal Fractal Diffusion Systems

Fractal theory has attracted considerable research interest [163, 171, 178, 181, 183, 241, 453, 454], with the diffusion-limited aggregation (DLA) [455] model proposed by Witten and Sander receiving particular attention. The DLA model successfully simulates fractal growth processes under far-from-equilibrium conditions. It has been widely used to explain various growth and aggregation phenomena relating to fractal morphologies and has met with widespread enthusiasm from researchers [456, 457]. Besides, many external factors in real and complex environments affect the aggregation growth process of fractal diffusion, resulting in a fractal growth morphology that is complicated, variable, and difficult to predict. It is thus important to identify effective methods of predicting and controlling fractal growth theoretically. Our simulations show that the proposed methods of thermal fractal diffusion in this chapter are effective and have reference value for actual fractal growth systems.

## 9.1 On Fractional Diffusion-Limited Aggregation: Anisotropy Originating from Memory

### *9.1.1 Fractional DLA Model*

In the classical DLA model, a cluster can be created according to the following rule. In the beginning, fix a single particle (the seed) at the center of coordinates in a lattice. Then release random walkers at a long distance (e.g., the boundary) from the center and allow them to move around until they hit any particle belonging to the cluster. This process will last until the cluster of expected size is formed.


S. T. Liu et al., *Fractal Control and Its Applications*,
https://doi.org/10.1007/978-981-15-5459-9_9

**Diffusion Mechanism**

In such models, the growth of the cluster depends on the external probability field that describes the concentration, heat, pressure, electric field and so on. Consequently, at a micro level, the probability field plays an important role in determining particle migration. Take heat conduction [443] for example. In mathematics, it is usually determined by the following two-dimensional diffusion equation:

$$\frac{\partial \rho c u}{\partial t} = \nabla \cdot (k \nabla u) + F, \tag{9.1.1}$$

where $u = u(x, y, t)$ and $F = F(x, y, t)$ are the heat density (temperature) of a medium and heat sources, respectively, at given location $(x, y)$ and time $t$; $k = k(u, x, y)$ is the medium's thermal conductivity; $c$ is the specific heat; $\rho$ is the medium density. The heat equation (9.1.1) is derived from Fourier's law and conservation of energy. What is more, a similar equation is also applicable to describe Brownian motion, diffusion model and many other phenomena relevant to the Markov process.

Let us examine some special cases. If the medium is homogeneous isotropic and the change range of temperature is relatively small, then $c$, $\rho$, $k$ are constant, and Eq. (9.1.1) can be written in the linear form

$$\frac{\partial u}{\partial t} = a^2 \Delta u + f, \tag{9.1.2}$$

where $a^2 = \frac{k}{\rho c}$ is the thermal diffusivity, $\Delta := \nabla^2$ is the Laplace operator (the second order spatial derivative) and $f = \frac{F}{\rho c}$. Moreover, if the sources are absent ($f = 0$) under steady-state condition ($\frac{\partial u}{\partial t} = 0$), then Eq. (9.1.2) takes the more simple form

$$\Delta u = 0. \tag{9.1.3}$$

It is widely used for the classical DLA model and its variations because of, to a large extent, formal conciseness and computational convenience [444, 445].

From the perspective of physical modeling, the rate of change in Eqs. (9.1.1)–(9.1.3) is characterized by an integer-order derivative, which is not suitable to describe processes of historical dependence. While a fractional differential operator, by contrast, is a powerful mathematical tool for modeling the memory processes. At the same time, a diffusion equation with fractional derivatives can better characterize the nonlocal diffusion or conduction with memory that has long been regarded as "abnormal" problem. The following is a typical fractional diffusion equation:

$$\frac{\partial^\alpha u}{\partial t^\alpha} = -\gamma (-\nabla^2)^\beta u + f, \quad 0 < \alpha, \beta \leq 1, \tag{9.1.4}$$

where $\frac{\partial^\alpha}{\partial t^\alpha}$ is a certain time fractional derivative; $(-\nabla^2)^\beta$ represents the symmetric fractional Laplace operator [446] that is a spatial fractional derivative; $u$ is the physical quantity of the problem being studied (e.g., concentration or temperature); constant $\gamma$ is the corresponding normalized physical coefficient such as diffusivity, thermal conductivity and so on. Equation (9.1.4) also implies a fractional Brownian motion with statistical memory. Besides, $\alpha$ is the memory strength index of the system [447, 448]. Under the condition of $\beta = 1$, the smaller $\alpha$ is, the stronger the memory of the process is [447–449].

As diffusion mechanism of our new DLA model, the following time fractional differential equation will be considered based on the Grünwald–Letnikov fractional derivative:

$$ {}^{GL}D_{0+}^{\alpha} u = \gamma \Delta u, \quad 0 < \alpha \leq 1, \tag{9.1.5} $$

which comes from Eq. (9.1.4) by setting $\beta = 1$ and $f \equiv 0$. Since it is not easy to obtain exact analytical solutions to Eq. (9.1.5), we generally consider quantitative properties of its discrete form by using the numerical method. As a result, it is necessary to discretize the Grünwald–Letnikov fractional derivative and the Laplace operator. As in [387, 388], the fractional derivative can be discretized by truncation, which signifies,

$$ {}^{GL}D_{0+}^{\alpha} u(x, y, t) \to \frac{1}{h_N^{\alpha}} \sum_{k=0}^{N} \omega_k^{(\alpha)} u(x, y, t - kh_N). \tag{9.1.6} $$

By using the five-point stencil central difference method, the Laplace operator in two-dimensional space can be approximated as

$$ \begin{aligned} \Delta u \to\ & \frac{f(x-h_x,y)-2f(x,y)+f(x+h_x,y)}{h_x^2} \\ & + \frac{f(x,y-h_y)-2f(x,y)+f(x,y+h_y)}{h_y^2}, \end{aligned} \tag{9.1.7} $$

where the grid size is $h_x \times h_y$, so that the stencil of a point $(x, y)$ on the grid is $\{(x, y), (x - h_x, y), (x + h_x, y), (x, y - h_y), (x, y + h_y)\}$. Sometimes, expression (9.1.7) is also called the discrete Laplace operator. In particular, the result is the square lattice grid if $h_x = h_y$, which indicates that the medium is spatially isotropic.

Let $h_N = h_x = h_y = 1$ and $u(x + h_x, y - h_y, t - kh_N) = u_{m+1,n-1,\tau-k}$. Substitute discretizations (9.1.6) and (9.1.7) into system (9.1.5). Then, the discrete version of equation (9.1.5) is obtained as

$$ u_{m,n,\tau} = \frac{\gamma}{1+4\gamma}(u_{m-1,n,\tau} + u_{m+1,n,\tau} + u_{m,n-1,\tau} + u_{m,n+1,\tau}) - \frac{\gamma}{1+4\gamma} \sum_{k=1}^{\mathcal{N}} \omega_k^{(\alpha)} u_{m,n,\tau-k}, \tag{9.1.8} $$

where $m, n, \tau \in \mathbb{N}$, $\mathcal{N} = \min\{200, \tau\}$ is the correction of the truncation point $N$ about the number of iterations $\tau$. For system (9.1.8), it needs to be noticed that the present state $u_{m,n,\tau}$ depends on not only surroundings states $u_{m-1,n,\tau}$, $u_{m+1,n,\tau}$, $u_{m,n-1,\tau}$ and $u_{m,n+1,\tau}$, but also past states $u_{m,n,\tau-1}$, $u_{m,n,\tau-2}$, $\cdots$, $u_{m,n,\tau-\mathcal{N}}$. The current state is affected by past ones, which indicates that the fractional diffusion process has memory property. Therefore, the density field $u$ calculated from Eq. (9.1.8) will naturally reflect this memory effect — the density distribution can have a certain directional preference. It will further have an inevitable impact on the growth of clusters.

**Visualization Analysis**

Let's start with the aggregation mechanism proposed in [450]. As shown in Fig. 9.1, the pattern grows stepwise. At the end of each diffusion process, points $(m, n)$ in the circular domain with radius $R_l$ need to be classified according to the value of the function $u$ at this point. The specific rule is as follows:

$$(m,n) \in \begin{cases} B \text{ (boundary)}, & u_{m,n} \equiv 1, \\ C \text{ (cluster)}, & u_{m,n} \equiv 0, \\ D \text{ (diffusion region)}, & u_{m,n} \in (0,1), \end{cases}$$

where $u_{m,n}$ is the solution to Eq. (9.1.8) at point $(m, n)$ under current conditions. Specially, denote $(m, n) \in I$ (interface, i.e., the set of candidate points) if $(m, n) \in D$ and it is contiguous to $C$. The probability $P_{m,n,\tau}$ of growth at interface $I$ is given as

$$P_{m,n,\tau}(I \to C) = \frac{(u_{m,n,\tau})^{\eta}}{\sum\limits_{(m',n')\in I} (u_{m',n',\tau})^{\eta}}, \tag{9.1.9}$$

where the summation, as the normalization coefficient, represents the total probability of all the direction in which the next particle may be aggregate.

Then, the next cluster point can be chosen from all candidate points by satisfying the condition

$$\sum_{i=1}^{j} P_i < \xi \le \sum_{i=1}^{j+1} P_i, \tag{9.1.10}$$

where $P_i$ is determined by formula (9.1.9), $i = 1, 2, \cdots, n(I)$, $n(I)$ is the cardinal of set $I$; a random number $\xi \sim U(0, 1)$, which signifies that it conforms to uniform distribution on [0, 1]. In other words, the $j$th point in the interface will become the new cluster point if it meets the range of probabilities limited by inequality (9.1.10). The newborn cluster point also changes the boundary condition on the interface which in turn necessitates that the diffusion region be recomputed. The process continues until the cluster has grown to the required size or reached the annihilation boundary.

Consider a diffusion region restricted by a unit circle ($R_l = 1$) defined on a 201 × 201 grid. The aggregated total particle number is no more than 2000. Figure 9.2 visualizes the classical DLA model based on Eq. (9.1.3) and a fractional DLA model based on Eq. (9.1.8) with $\gamma = 1$.

Due to the fractional diffusion process, particles have a tendency to aggregate in the candidate points with higher density. So, memory of the fractional DLA model originates from the diffusion process and can finally be embodied by the anisotropy of the pattern. Comparing (b) to (a) in Fig. 9.2, it can be seen that the order $\alpha$ can affect the isotropy of the cluster's pattern significantly, which will be further explored

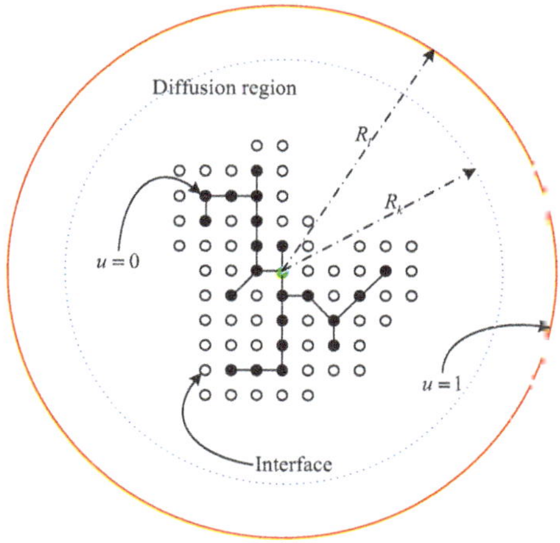


**Fig. 9.1** An example of a aggregation after a few steps: the red solid circle with radius $R_l$ represents the boundary where $u \equiv 1$, and its interior is the diffusion region; the blue break circle with radius $R_k$ represents the annihilation boundary that defines the maximum growth range; the cluster is shown with the solid dots (the seed is green) where $u \equiv 0$, and bonds showing the pattern of cluster so far; the interface is shown with the open dots where the probabilities of selection will be computed by formula (9.1.9)

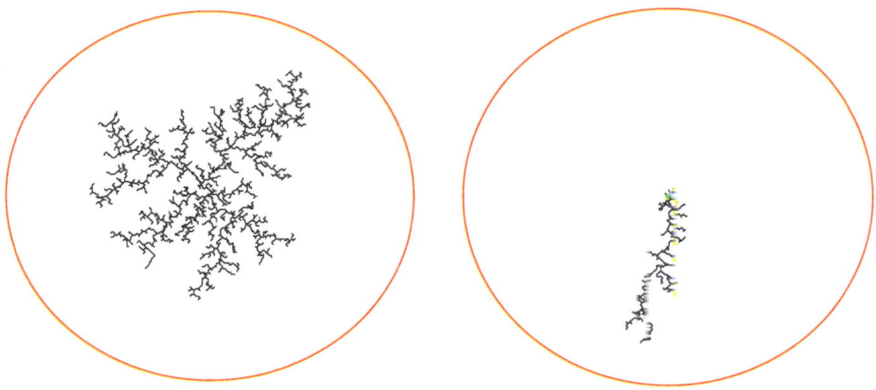

(a) (b)

**Fig. 9.2** Patterns of the DLA model. **a** classical DLA model; **b** fractional DLA model with $\alpha = 0.50$

detailedly in the next section. Because of such effect, the cluster grows almost along a certain direction, and soon reaches the annihilation boundary ($R_k = 0.8R_l$).

Furthermore, it must be noted that the major difficulty to formulate the fractional model relates to its computing time consumption. The diffusion stage, specially, takes hundreds of times as long to achieve the same accuracy as the classical model. Consequently, on the premise of explaining the problem clearly, the scale of our simulation here is not large (less than 2000 particles).

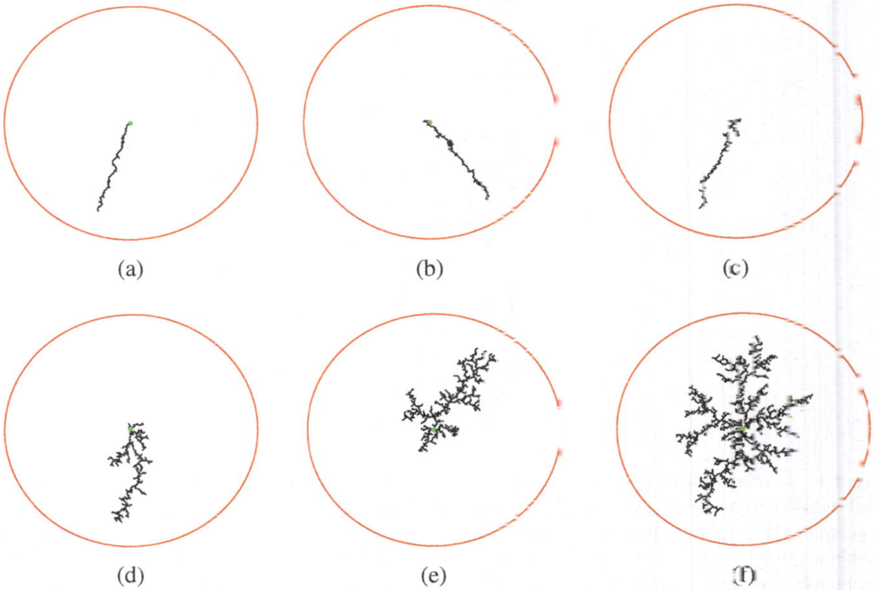


**Fig. 9.3** Patterns of the fractional DLA model on a 201 × 201 grid with $R_k = 0.8R_l$. **a** $\alpha = 0$; **b** $\alpha = 0.21$; **c** $\alpha = 0.41$; **d** $\alpha = 0.61$; **e** $\alpha = 0.81$; **f** $\alpha = 1$

## *9.1.2 Directive Property and Fractal Dimension Analysis*

Figure 9.2 has preliminarily revealed a remarkable new feature of the fractional DLA model: the order $\alpha$ has a significant impact on the pattern of clusters. This section will present some quantitative results and analysis in terms of anisotropy (or called directivity) and dimension of aggregations.

**Anisotropy Index**

In the classical DLA model, clusters can grow in the preferred directions due to lattice effects or to some anisotropy in the sticking probabilities [451]. These effects can be controlled by reduction-of-noise techniques. In the modified model, it is considered that the medium is homogeneous and isotropic, and the aggregation mechanism is random and uniform. However, anisotropy appears in aggregation, which gives us reason to believe that it is the result of introducing fractional elements.

Figure 9.3 intuitively shows the changes of the anisotropy of aggregate clusters with different model's order: within a certain range, the higher the order, the weaker the anisotropy. This coincides with the conclusion in [452]: the higher the order, the weaker the memory. Therefore, the anisotropy is an intuitive reflection of the memory of the aggregation mechanism. In other words, it is the memory effect contained in fractional diffusion mechanism that rides the aggregate clusters exhibit macroscopic anisotropy. Under the fractional mechanism, a particle remembers "paths"

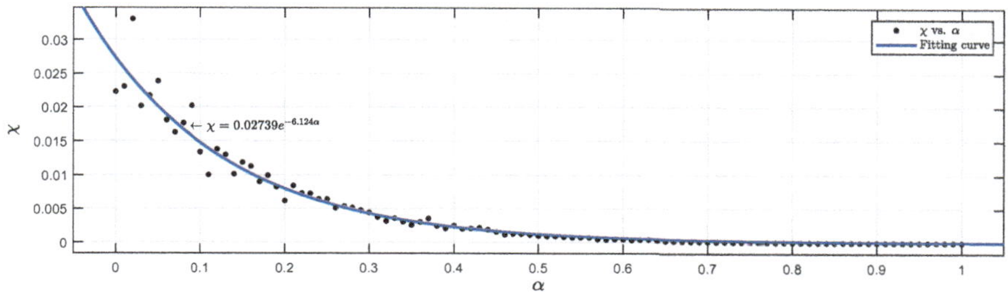


**Fig. 9.4** Relationship between the anisotropy index and the model's order

that previous particles have taken, even in an isotropic medium. In addition, patterns in Fig. 9.3 have different orientation, which is determined by the random aggregation of the first few particles at the beginning each time but not the essential difference.

In an effort to understand and grasp this relationship more deeply, an index need be introduced to quantify this directive property. First, let $n$ be the total number of particles in a pattern and $p_i(m_i, n_i)$ be the $i$th particle $p_i$ with coordinate $(m_i, n_i)$. For any two sequenced particles $p_i$ and $p_{i+1}$ $(i = 1, 2, \cdots, n-1)$, a measurement $m$ based on the Euclidean distance $d$ is defined as follows:

$$m(p_{i+1}, p_i) := \begin{cases} 1, & p_{i+1} \text{ is adjacent to } p_i, \\ d(p_{i+1}, p_i), & \text{otherwise.} \end{cases} \tag{9.1.11}$$

This ensures that the distance is always considered to be 1 (the smallest distance between two points on the grid) if two points $p_{i+1}$ and $p_i$ adjoin. So the "net deviation" of the two sequenced particles can be expressed by $m(p_{i+1}, p_i) - 1$, $i \in \{1, 2, \cdots, n-1\}$. In the following, a anisotropy index $\chi$ for patterns of aggregate clusters can be defined as

$$\chi := \frac{1}{\sum_{i=1}^{n-1} m(p_{i+1}, p_i) - n + 2}. \tag{9.1.12}$$

Evidently, the index is a real number between 0 and 1. If the particles grow one close to another sequentially, the index is 1; otherwise, the index is closer to 0. The relationship between anisotropy index $\chi$ of patterns and model's order $\alpha$ can be acquired by a exponential fitting with one term, seeing Fig. 9.4. The data in the fitting are the average of ten groups (101 data points in each group) of simulation. The SSE (sum of squares due to error) is $1.882 \times 10^{-4}$ and R-square (the coefficient of determination) is 0.9598.

The following conclusions can be drawn by combining Figs. 9.3 and 9.4.

(1) The index drops rapidly at the beginning and then slows down, which shows an exponential type distribution on the whole.

**Table 9.1** Box-counting dimension $D$ of the pattern with order $\alpha$

| $\alpha$ | 0.00 | 0.10 | 0.20 | 0.30 | 0.40 | 0.50 |
|---|---|---|---|---|---|---|
| $D$ | 1.2176 | 1.2424 | 1.2629 | 1.2641 | 1.3172 | 1.3487 |
| $\alpha$ | 0.60 | 0.70 | 0.80 | 0.90 | 1.00 | Classic |
| $D$ | 1.4223 | 1.4722 | 1.5615 | 1.5862 | 1.6979 | 1.6968 |

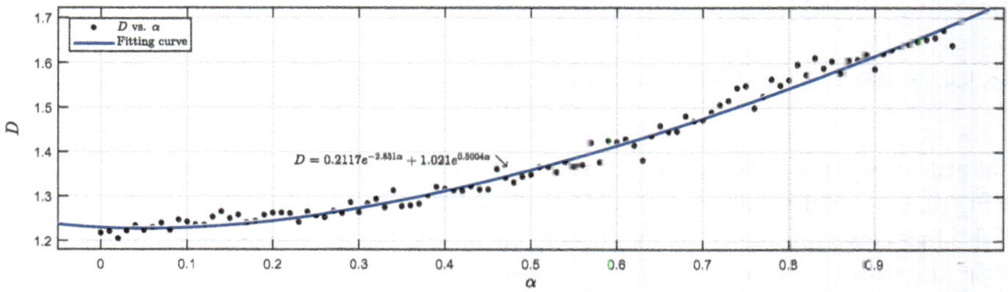


**Fig. 9.5** Relationship between the box-counting dimension and the model's order

(2) When the order exceeds 0.6, the index equals approximately zero. At this time, the anisotropy of the cluster has mostly vanished, while isotropy begins to emerge clearly enough that one can directly observe it in Fig. 9.3.

Incidentally, the fitting effect is not as perfect as we expect because of the randomness of aggregation, but the basic inherent law has been manifested.

### Fractal Dimension

By means of numerical calculation,[1] box-counting dimensions of patterns with respect to model's order are shown in Table 9.1 and Fig. 9.5. The data in the fitting are the average of five groups of simulation.

In an effort to facilitate comparison and analysis, the dimension of the classical model is also given in the table, which is marked as "classic". In the exponential fitting with two terms, SSE $= 3.005 \times 10^{-2}$ and R-square $= 0.9859$.

Table 9.1 and Fig. 9.5 as well as the analysis on the anisotropy index as above-mentioned support the following claim.

(1) The dimension increases exponentially with the increase of order, which is, viewing from its tendency, contrary to the anisotropy index.
(2) Compared with the first-order DLA model (i.e., the case of $\alpha = 1$) in Fig. 9.2a and the classical one in Fig. 9.3f, the patterns of the two models are very close, and their dimensions are not very different (about 1% only), which illustrates that it is reasonable to ignore the time derivative in the classical model. However, in

[1] Such process is based on the recorded data of particles in simulation and takes advantage of Frederic Moisy's MATLAB® function shared on https://ww2.mathworks.cn/matlabcentral/fileexchange/13063-boxcount.

the fractional model, the influence of the derivative on the model is so essential and remarkable that it cannot be ignored.

In a word, fractional model provides a higher degree of freedom for describing DLA. The fractal dimension can be converted by changing the model's order, which cannot be achieved in the classical DLA model. Therefore, fractional model furnishes a new and feasible way for fractal control of aggregate clusters that has been studied in [402].

## 9.2 Control of the Thermal Fractal Diffusion of Tightly Compressed Heterogeneous Layers of Thin Plates

As the thermal conductivity of thin plates composed of tightly compressed heterogeneous layers varies continuously in the form of an exponential function, we present a nonlinear dynamical model of the fractal growth of thermal diffusion. We also analyze the quantitative relationship between the probability of growth and the disturbance term, predict the control action of the environmental disturbance term on fractal growth, and use Matlab simulation to verify the control effectiveness of thermal fractal diffusion. The results facilitate the selection of appropriate control areas and control parameters for the thermal diffusion variable coefficients. In addition, variation in the fractal dimension reflects the influence of environmental disturbance on the complex process of thermal fractal diffusion.

### 9.2.1 *Thermal Diffusion System of the Nonhomogeneous Layers of Thin Plates*

In $W$ (see Fig. 9.6), a thin plate comprised of tightly compressed heterogeneous layers, each layer contains various substances, all with different levels of thermal conductivity. The layers are tightly compressed to remove transition resistance. The plate's thermal-conduction equation is as follows: $\rho c \frac{\partial V}{\partial t} = \frac{\partial V}{\partial x}(a(x, y)\frac{\partial V}{\partial x}) + \frac{\partial V}{\partial y}(b(x, y)\frac{\partial V}{\partial y})$, where $\rho$ refers to the density of the substance, $c$ refers to the specific heat, and $a(x, y)$ and $b(x, y)$ refer to thermal conductivity [458].

We arrange the tightly compressed heterogeneous layers of the thin plate in spatial order to enable the use of a step approximation for the plate's varying thermal conductivity. As shown in Fig. 9.7, the solid line represents the thermal conductivity, $k_i$, and the dashed line represents the approximately continuous function $K(x)$. We assume that the thermal conductivity, $x$, varies and meets the following conditions: $a(x, y) = b(x, y) = k_0 e^{\alpha x}$, $k_0 > 0$, where $\alpha$ is a constant to be identified. When $\alpha > 0$, $x$ steps up; when $\alpha < 0$, $x$ steps down; when $\alpha = 0$, the thin plate of nonhomogeneous layers is composed of the same material [459].

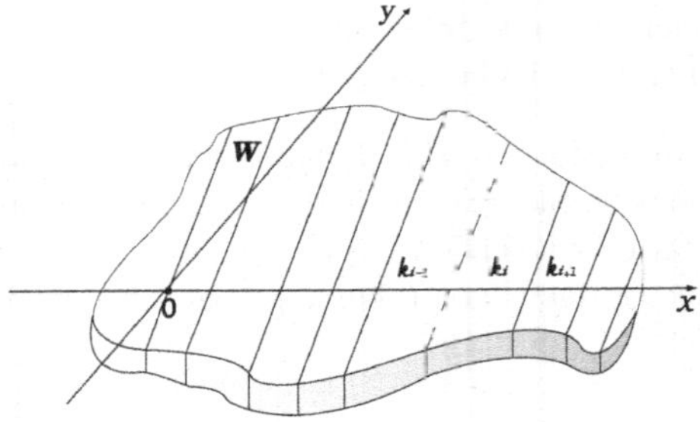


**Fig. 9.6** Thermal conductivity $k_i$ of a thin plate of tightly compressed heterogeneous layers

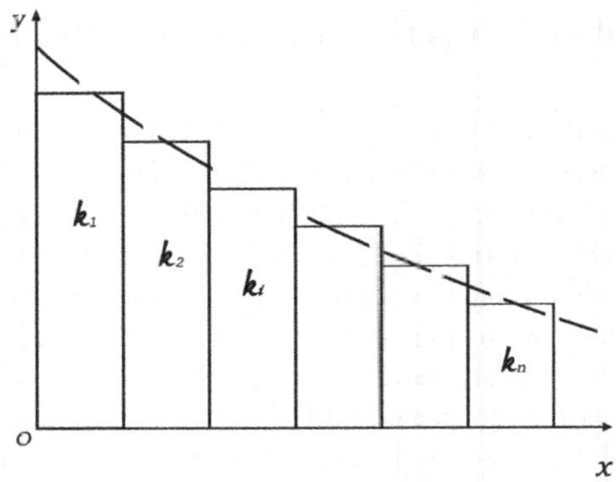


**Fig. 9.7** Spatial arrangement of the varying thermal conductivity of a thin plate comprising tightly compressed heterogeneous layers

## 9.2.2 Mathematical Model

If the tightly compressed heterogeneous layers of thin plates contain thermal sources, the dynamics of fractal growth allow us to represent the probability of growth in the thermal diffusion process as a distributed-parameter system conforming to the following boundary conditions:

$$\frac{\partial V}{\partial x}\left(a(x, y)\frac{\partial V}{\partial x}\right) + \frac{\partial V}{\partial y}\left(b(x, y)\frac{\partial V}{\partial y}\right) = u(x, y) + f(V(x, y)), \qquad (9.21)$$

where $V(x, y)$ represents the probability of thermal diffusion (temperature) [450] at a certain time, when the spatial body $W$ is located at $(x, y)$. The function $f(\cdot)$ is usually a nonlinear function known as the "forced" term, and $u$ is a quantitative initial condition called the "source" term. Both the nonlinear disturbance term $f(\cdot)$ and the source term $u$ can be considered equivalent to the environmental disturbance term $G$.

Given the complexity of Eq. (9.2.1), which depicts thermal diffusion with variable coefficients, we cannot analytically determine the solution. However, we can use the finite difference scheme to calculate a numerical solution for the thermal diffusion equation. In accordance with the finite volume method [461] the discrete dynamic structure of System (9.2.1), a nonlinear fractal growth system, can be represented as follows:

$$\begin{aligned} & a_{i+\frac{1}{2},j} V_{i+1,j} + a_{i-\frac{1}{2},j} V_{i-1,j} + b_{i,j-\frac{1}{2}} V_{i,j-1} + b_{i,j+\frac{1}{2}} V_{i,j+1} \\ & - \left( a_{i+\frac{1}{2},j} + a_{i-\frac{1}{2},j} + b_{i,j-\frac{1}{2}} + b_{i,j+\frac{1}{2}} \right) V_{i,j} = u_{i,j} + f\left(V_{i,j}\right). \end{aligned} \tag{9.2.2}$$

Without loss of generality, we assume that $u(x, y)$ and $f(x, y)$ are smooth functions, and ensure that the values of $u(x, y)$ and $f(x, y)$ within the spatial area of thermal diffusion are sufficiently small. To ensure that adding the source term and the forced term does not negate the compliance of the difference equation with the extremum principle, $V(x, y)$ can only obtain an extremum at the boundary, which requires it to meet the following boundary conditions. First, the far-from-internal $V(x, y)$ must be capable of providing the surface with a sustained total particle flow, $V(x, y) \equiv 1$. Second, the arrival probability of the internal $V(x, y)$ at the boundary point where the particles terminate their growth must be $V(x, y) = 0$. As $u(x, y)$ and $f(x, y)$ are sufficiently small, $0 \leq a_{i+r-\frac{1}{2},j} V_{i+r,j} + a_{i-r+\frac{1}{2},j} V_{i-r,j} + b_{i,j-r+\frac{1}{2}} V_{i,j-r} + b_{i,j+r-\frac{1}{2}} V_{i,j+r} < 1$ holds in the aggregation region.

### 9.2.3 *Thermal Diffusion Simulation*

The thermal diffusion of a thin plate consisting of tightly compressed heterogeneous layers is a random process controlled by a temperature gradient. The specific algorithm for thermal diffusion fractal growth is obtained as follows.

(1) Set the initial conditions. Establish the aggregation region $L \times L$ for the simulated thermal particles, the particle-movement step size, and the number of aggregation particles.
(2) Calculate the temperature gradient. In accordance with System (9.2.2), a discrete system, and combining the central particles subject to the initial boundary conditions ($V = 0$) with the surrounding points at infinity ($V = 1$), we can calculate the distribution of the temperature gradient $\nabla V$ across the whole thin plate after multiple iterations [462].
(3) Calculate the thermal-growth probability of the points awaiting growth. The growth-probability function of every possible aggregation point has a more nonlinear than linear relationship with the temperature gradient. Niemeyer et al. proposed the following model of growth probability [463]:

$$p_i = \frac{|\nabla\ V_i(x, y)|}{\sum_{i=1}^{n} |\nabla\ V_i(x, y)|}, \tag{9.2.3}$$

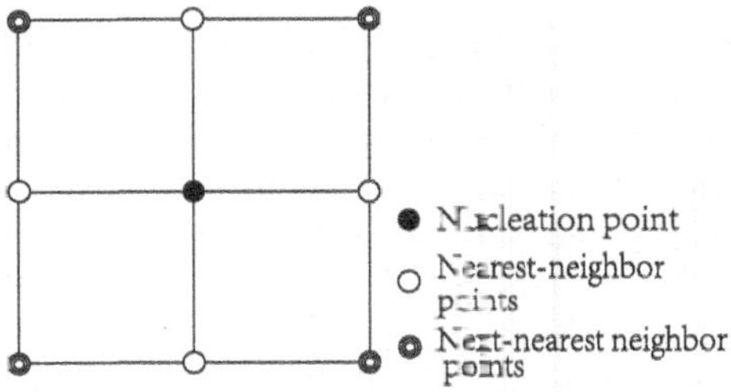


**Fig. 9.8** Nearest neighbor and next-nearest neighbor points awaiting growth

where $n$ refers to the number of points awaiting growth, and the denominator refers to the total probability and normalization coefficient of the particles growing in all directions.

(4) Choose the particle aggregation location. Around the nucleation particles, there are both nearest neighbor and next-nearest neighbor points awaiting growth (as shown in Fig. 9.8). During the practical process of conduction, the next-nearest neighbor is also of great importance.
The final aggregation position of the thermal particles occurs at random, and its probability is proportional to the probability $p_i$ of the nodes awaiting growth. We use the roulette wheel selection algorithm [464] to choose new thermal diffusion points from all of the possible points awaiting growth. $\alpha$ is a randomly produced number with a value interval of (0, 1). The comparison is as follows:
$\sum_{i=1}^{k-1} p_i \le \alpha < \sum_{i=1}^{k} p_i$.

(5) Modify the boundary conditions. When new particles aggregate, the thermal-diffusion probability falls to zero. The thermal-diffusion probability of these newly developed points is also included among the boundary conditions.

By modifying the movement-boundary conditions of the mathematical model and using the roulette wheel selection algorithm, the model described in this part more accurately simulates the thermal diffusion of thin plates consisting of tightly compressed heterogeneous layers.

### *9.2.4 Fractal Diffusion Control of the Thermal Sources of Thin Plates Consisting of Homogeneous Layers*

The polynomial function approximation is widely used in the fields of fractal graphics, voice-processing systems, geotechnical engineering, and electrical engineering, and in real physical systems such as ion-exchange glass, ion chemistry, and thin-film preparation [465].

According to the Weierstrass theorem, any continuous function defined in a closed interval can be uniformly approximated by a polynomial function. Therefore, we assume that system (9.2.2) takes the nonlinear function $f(V_{i,j}) = \xi(\omega V_{i,j})^p + \mu$, $p = 1, 2, \ldots, n$. We can thus represent the system as follows:

$$\begin{aligned} & a_{i+\frac{1}{2},j} V_{i+1,j} + a_{i-\frac{1}{2},j} V_{i-1,j} \\ & \quad + b_{i,j-\frac{1}{2}} V_{i,j-1} + b_{i,j+\frac{1}{2}} V_{i,j+1} - \omega V_{i,j} = u_{i,j} + \mu + \xi(\omega V_{i,j})^p + \omega V_{i,j}. \end{aligned} \tag{9.2.4}$$

For simplicity of presentation, the constant $\mu = 0$ is assumed, and the following notation is used:

$$E(r) = E_{i,j}(r) = a_{i+r-\frac{1}{2},j} V_{i+r,j} + a_{i-r+\frac{1}{2},j} V_{i-r,j} + b_{i,j-r+\frac{1}{2}} V_{i,j-r} + b_{i,j+r-\frac{1}{2}} V_{i,j+r}. \tag{9.2.5}$$

Consequently, System (9.2.2) takes the following form: $E( ) = u_{i,j} + \xi E^p(0) + E(0)$, $p = 1, 2, \ldots, n$. $p \in Z^+$.

Without loss of generality, we represent the system as follows:

$$E(r) = u(x, y) + \xi E^p(r-1) + E(r-1), \ \ r = 1, 2, \ldots, n. \tag{9.2.6}$$

After a simple iteration of System (9.2.6), we obtain the following:

$$\begin{aligned} E(r) &= r u_{i,j} + \xi E^p(r-1) + \xi E^p(r-2) + \cdots + \xi E^p(0) + E(0) \\ &= \xi(E^p(r-1) + E^p(r-2) + \cdots + E^p(0)) + E(0) + r u_{i,j}. \end{aligned} \tag{9.2.7}$$

With regard to System (9.2.7), we discuss the control action of environmental disturbance on thermal diffusion fractal growth in thin plates composed of tightly compressed heterogeneous layers. With regard to System (9.2.2), we expand $f(t) = \xi t^p$ into a Fourier series within the interval $[-1, 1]$, as follows: $f(t) = \frac{a_0}{2} + \sum_{n=1}^{n=\infty} (a_n \cos n\pi t + b_n \sin n\pi t), n = 1, 2, \ldots$.

In the following discussion, we compare the effects of an odd $p$ value with those of an even $p$ value.

**Control Action of Environmental Disturbance on Thermal Fractal Diffusion When $p$ Is Even**

$a_0 = \frac{1}{p+1}$, $a_n = \int_{-1}^{1} f(t) \cos n\pi t dt = \sum_{n=1}^{\infty} [-2 \cos n\pi (\frac{p}{n^2\pi^2} + \cdots + \frac{p!}{n^p \pi^p})]$, $b_n = 0$.

$$f(t) = \frac{\xi}{p+1} + \xi \sum_{n=1}^{\infty} \left[ -2 \cos n\pi \left( \frac{p}{n^2\pi^2} + \cdots + \frac{p!}{n^p \pi^p} \right) \cos n\pi t \right], n = 1, 2, \ldots \tag{9.2.8}$$

Therefore, system (9.2.6) takes the following form:

$$E(r) = \frac{\xi}{p+1} + \xi \sum_{n=1}^{\infty} \left[ -2\cos n\pi \left( \frac{p}{n^2\pi^2} + \cdots + \frac{p!}{n^p \pi^p} \right) \cos(n\pi E(r-1)) \right] + E(r-1) + u,$$

and

$$\begin{aligned} E(r) &= \frac{\xi}{p+1} + \xi \sum_{n=1}^{\infty} \left[ -2\cos n\pi \left( \frac{p}{n^2\pi^2} + \cdots + \frac{p!}{n^p \pi^p} \right) \cos n\pi E(r-1) \right] + E(r-1) - u, \\ &\leq \frac{r\xi}{p+1} + r\xi \sum_{n=1}^{\infty} \left[ -2\cos n\pi \left( \frac{p}{n^2\pi^2} + \cdots + \frac{p!}{n^p \pi^p} \right) \right] + E(0) + ru. \end{aligned}$$

$\sum_{n=1}^{\infty} [-2\cos n\pi (\frac{p}{n^2\pi^2} + \cdots + \frac{p!}{n^p\pi^p})]$ appears to be an alternating series. According to Leibnitz's theorem [466], this type of series is convergent and meets the following conditions: $\sum_{n=1}^{\infty} [-2\cos n\pi (\frac{p}{n^2\pi^2} + \cdots + \frac{p!}{n^p\pi^p})] = g, (\frac{p}{\pi^2} > g > 0)$.

Therefore,

$$\begin{aligned} E(r) &\leq \frac{r\xi}{p+1} + r\xi g + E(0) + ru, \\ &= r\xi \left( \frac{1}{p+1} + g \right) + E(0) + ru. \end{aligned}$$

Let $G(\xi, u, r) = r\xi(\frac{1}{p+1} + g) + E(0) + ru$, where $u(x, y)$, $\xi$ is the linear function of $G(\xi, u, r)$, $G(\xi, u, r) = r\xi(\frac{1}{p+1} + g) + E(0) + ru$, and $\frac{\partial G}{\partial \xi} = r(\frac{1}{p+1} - g) > 0$, $\frac{\partial G}{\partial u} = r > 0$.

According to the "judgment" method of increasing and decreasing functions in mathematical analysis, $E(r)$ increases monotonically with respect to $V_{i,j}$, $\xi$ and $u(x, y)$, respectively.

**Conclusion 9.1** *$V_{i,j}$ increases with source item $u_{i,j}$, which means that the introduction of source item $u(x, y)$ significantly increases the particle concentration in the control directions. Thus, the particles have obvious growth advantages in the control directions.*

**Conclusion 9.2** *$V_{i,j}$ increases with source item $\xi$, which means that the introduction of the nonlinear forcing term $\xi(\omega V_{i,j})^p$ intensifies particle aggregation in the whole region.*

As $V_{i,j}$ can only reach its extreme at the boundaries, extremum conditions can only be met by modifying control parameter $\xi$ and source item $u(x, y)$ to make $V(x, y) \ll 1$. This also ensures that the system's controlled quantity is much less than its input quantity.

$$E(r) < r\xi\left(\frac{1}{p+1}+g\right)+E(0)+ru<1. \tag{9.2.9}$$

From Conclusions 9.1 and 9.2 and Formula (9.2.9), the quantitative relations between source item $u(x, y)$ and the nonlinear forced term can be identified. When the action region of the source item is circular, Deduction 9.1 holds.

**Deduction 9.1** *When the action region of a source item is circular, the center of the control region is the center of the circle, the radius is R, and the control function is as follows:* $u(x, y)|_{x^2+y^2\le R^2} = \lambda(x^2+y^2)$. *When* $\lambda > \frac{1-E(0)}{r(\frac{1}{p+1}+g+1)R^2}$, $\xi < u_{\max}$, *that is, the particles gain an obvious control advantage in the control region.*

***Proof*** When the number of iterations is $r$, the source item is $u(x, y) = \lambda(x^2+y^2)$, where $x^2+y^2 \le R^2$, then $u_{\max} = \lambda R^2 > \frac{1-E(0)}{r(\frac{1}{p+1}+g+1)R^2}R^2 = \frac{1-E(0)}{r(\frac{1}{p+1}+g+1)}$.

It follows from Formula (9.2.9) that $E(r) \le r\xi(\frac{1}{p+1}+g)+E(0)+ru < 1$, i.e., $r\xi(\frac{1}{p+1}+g)+E(0)+ru \le r\xi(\frac{1}{p+1}+g)+E(0)+ru_{\max} < 1$.

Then $\xi < \frac{1-E(0)-ru_{\max}}{r(\frac{1}{p+1}+g)}$, because $u_{\max} > \frac{1-E(0)}{r(\frac{1}{p+1}+g+1)}$, $-ru_{\max} < -\frac{1-E(0)}{\frac{1}{p+1}+g+1}$.

It follows that

$$\begin{aligned}\xi &< \frac{1-E(0)-ru_{\max}}{r\left(\frac{1}{p+1}+g\right)},\\ &< \frac{1-E(0)}{r\left(\frac{1}{p+1}+g\right)} - \frac{1-E(0)}{r\left(\frac{1}{p+1}+g+1\right)\left(\frac{1}{p+1}+g\right)},\\ &= \frac{1-E(0)}{r\left(\frac{1}{p+1}+g+1\right)} = u_{\max}\,.\end{aligned}$$

Therefore, when $u_{\max}(x, y) > \xi$, the increase in particle concentration in the control directions clearly exceeds that of the nonlinear item in the whole region. The particles thus have obvious growth advantages in the control directions.

**Control Action of Environmental Disturbance on Thermal Fractal Diffusion When *p* Is Odd**

$a_n = 0$, $b_n = \int_{-1}^{1} f(t)\sin n\pi t dt = \sum\limits_{n=1}^{\infty}[-2\cos n\pi(\frac{p}{n\pi}+\frac{p(p-1)}{n^2\pi^2}+\cdots+\frac{p!}{n^p\pi^p})]$.

$b_n = \int_{-1}^{1} f(t)\sin n\pi t dt = \sum\limits_{n=1}^{\infty}[-2\cos n\pi(\frac{p}{n\pi}+\frac{p(p-1)}{n^2\pi^2}+\cdots+\frac{p!}{n^p\pi^p})]$.

System (9.2.6) takes the following form:

$$E(r) = \xi\sum_{n=1}^{\infty}\left[-2\cos n\pi\left(\frac{p}{n^2\pi^2}+\cdots+\frac{p!}{n^p\pi^p}\right)\sin n\pi E(r-1)\right]+E(r-1)+u,$$

and as a result,

$$E(r) = \xi \sum_{n=1}^{\infty} \left[ -2\cos n\pi \left( \frac{p}{n^2\pi^2} + \cdots + \frac{p!}{n^p\pi^p} \right) \sin n\pi E(r-1) \right] + E(r-1) + u,$$

$$\le r\xi \sum_{n=1}^{\infty} \left[ -2\cos n\pi \left( \frac{p}{n^2\pi^2} + \cdots + \frac{p!}{n^p\pi^p} \right) \right] + E(0) + ru.$$

According to Dirichlet's theorem [466], $\sum_{n=1}^{\infty} [-2\cos n\pi (\frac{p}{n^2\pi^2} + \cdots + \frac{p!}{n^p\pi^p})] = g$ belongs to a converging series. Then, $E(r) \le r\xi g + E(0) + ru$.

We can then state the following: $G'(\xi, u, r) = r\xi g + E(0) + ru$. $G'(\xi, u, \cdot)$ is clearly a monotonically increasing function of the variables $\xi$ and $u$.

It is easy to show that Conclusions 9.1 and 9.2 hold when the nonlinear forced term is an odd function. Proceeding according to the proof for Deduction 9.1 gives Deduction 9.2.

**Deduction 9.2** *When the action region of the source item is circular, the center of the control region is the center of the circle, the radius is, and the control function is as follows:* $u(x, y)|_{x^2+y^2 \le R^2} = \lambda(x^2 + y^2)$. *When* $\lambda > \frac{1-E(0)}{r(1+g)R^2}$, $\xi < u_{max}$, *that is, the particles gain an obvious control advantage in the control region.*

In short, a similar conclusion is reached whether $p$ is even or odd, but the condition parameters differ.

## 9.2.5 *Numerical Analysis of the Thermal Fractal Diffusion of Thin Plates Consisting of Homogeneous Layers*

We simulate the fractal diffusion and aggregation of 10000 thermal particles by constructing a 101 × 101 grid on the thin plate $(W)$ of heterogeneous layers, with the thermal source in the center of the grid $(51, 51)$. In addition, we set various values for $u(x, y)$ and the radius of the action region of the source item to simulate the control action on thermal diffusion fractal growth under conditions of environmental disturbance. Figure 9.9 shows the morphology of thermal diffusion aggregates on the thin plate when $f = 0, u = 0$.

**Control of Thermal Diffusion Under Conditions of Environmental Disturbance**

According to Conclusion 9.1 and Deduction 9.1, the particles have an obvious growth advantage in the control region when $p = 2$. When the control region is enlarged, the corresponding $V_{i,j}$ also grows, and the particles gain a distinct growth advantage in the circular region. Figure 9.10 shows that when $p = 2$, $a(x, y) = b(x, y) = e^{0.2x}$, $u_{max}(x, y) > \xi$, the thermal fractal form increases with the action region of the source item. The figure clearly illustrates that when $u_{max}(x, y) > \xi$, the growing particles become more closely aggregated as the radius of the domain of the source item

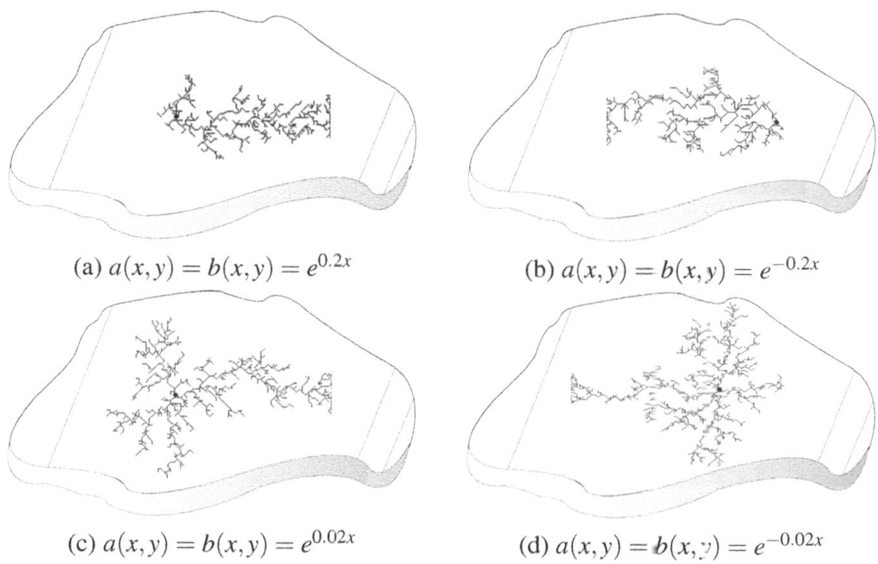


**Fig. 9.9** Thermal fractal diffusion on plate $W$ when $f = 0, u = 0$

increases. When the control region of the source item is a fixed constant, we can construct fractal-aggregation graphs for various control radiuses, as shown in Fig. 9.11. According to the graph for the control region in Fig. 9.11, particles aggregate more closely than the particles in the control region in Fig. 9.10. This result indicates that constant control strength has a greater control action than uneven control strength.

In Figs. 9.10 and 9.11, thermal conductivity to the right of the central point increases with $x$ more obviously than to the left of the central point. The thermal particles on the right grow rapidly in the direction of $x$, such that thermal conductivity in the right control region exceeds that in the left. Compared with Fig. 9.9a, which depicts particle aggregation in the absence of environmental disturbance, the disturbance term in Figs. 9.10 and 9.11 aggregates thermal particles much more obviously in the action region.

According to Conclusion 9.1 and Deduction 9.2, particles have an obvious growth advantage in the control region when $p = 3$. When the control region is enlarged, the corresponding $V_{i,j}$ also grows, and the particles gain a distinct growth advantage in the circular region. Figure 9.12 shows that when $p = 3$, $a(x, y) = b(x, y) = e^{0.02x}$, $u_{\max}(x, y) > \xi$, and the thermal fractal form increases with the action region of the source item. Figure 9.13 presents the particle aggregation in various control regions when the control strength is a constant. The graph for the control region in Fig. 9.13 shows closer aggregation in this region than that in Fig. 9.12. Compared with Fig. 9.9c, which presents the thermal particle aggregation in the absence of environmental disturbance, the thermal particles in the control regions in Figs. 9.12 and 9.13 are more clearly aggregated. As a result, our conclusions are similar to those

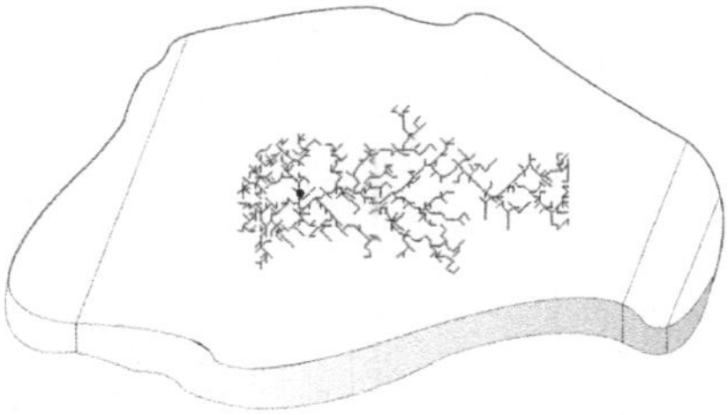

(e) $R = 10$

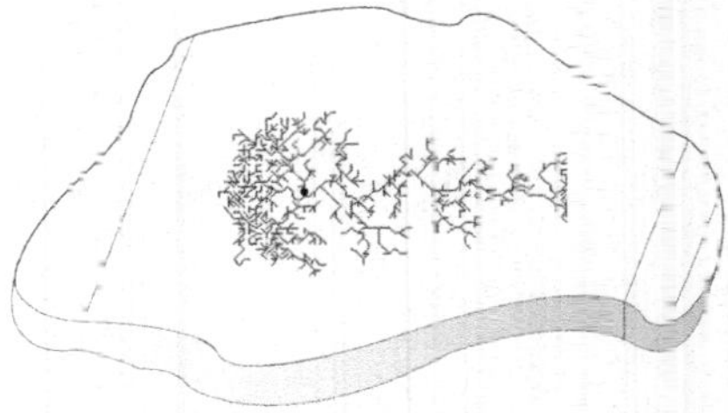

(f) $R = 14$

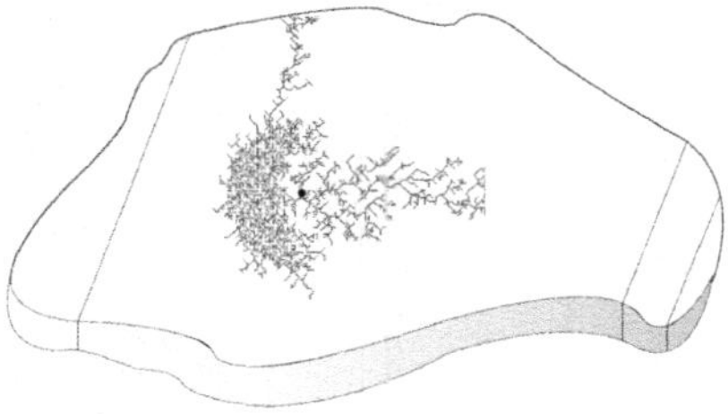

(g) $R = 20$

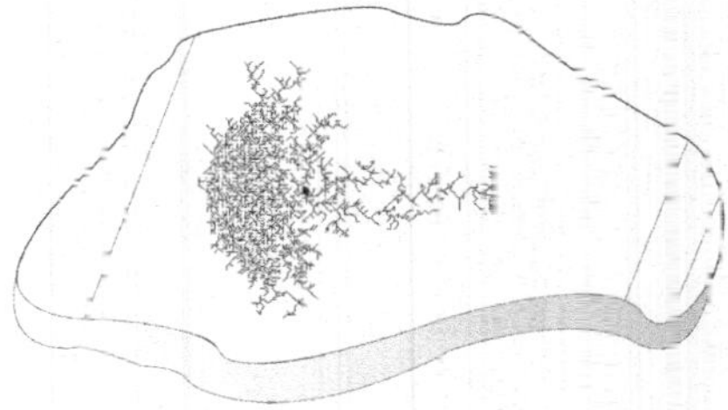

(h) $R = 26$

**Fig. 9.10** Fractal aggregation when $u = 5 \times 10^{-5}(x^2 + y^2)$, $\xi = 1 \times 10^{-7}$, $p = 2$

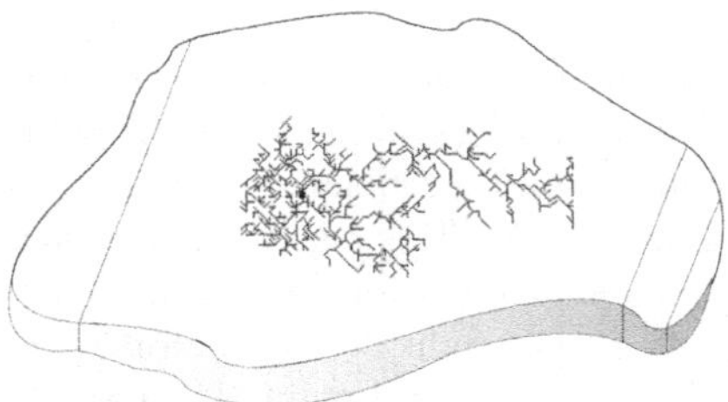

(i) $u = 5 \times 10^{-3}, R = 10, \xi = 1 \times 10^{-7}$

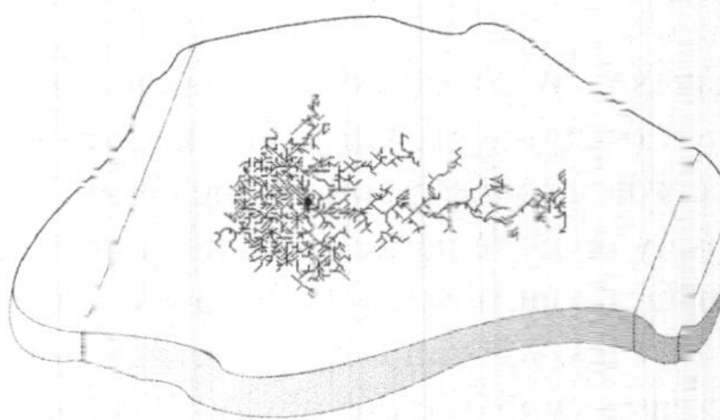

(j) $u = 10^{-2}, R = 14, \xi = 1 \times 10^{-6}$

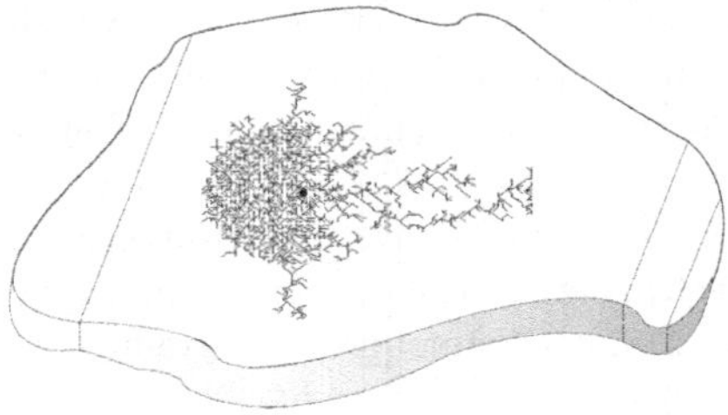

(k) $u = 2 \times 10^{-2}, R = 20, \xi = 1 \times 10^{-6}$

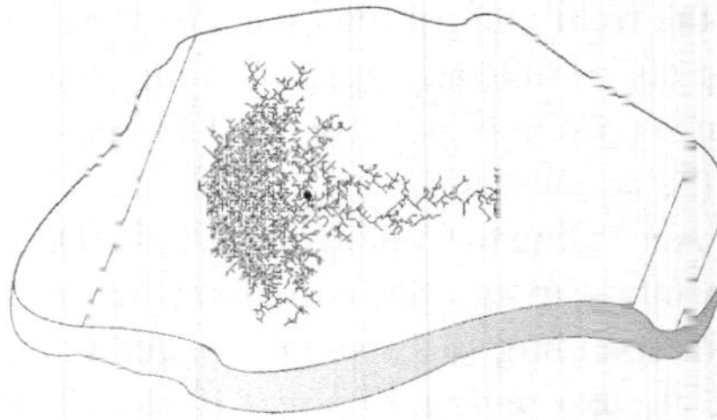

(l) $u = 5 \times 10^{-3}, R = 26, \xi = 1 \times 10^{-7}$

**Fig. 9.11** Fractal aggregation when $u$ is a constant and $p = 2$

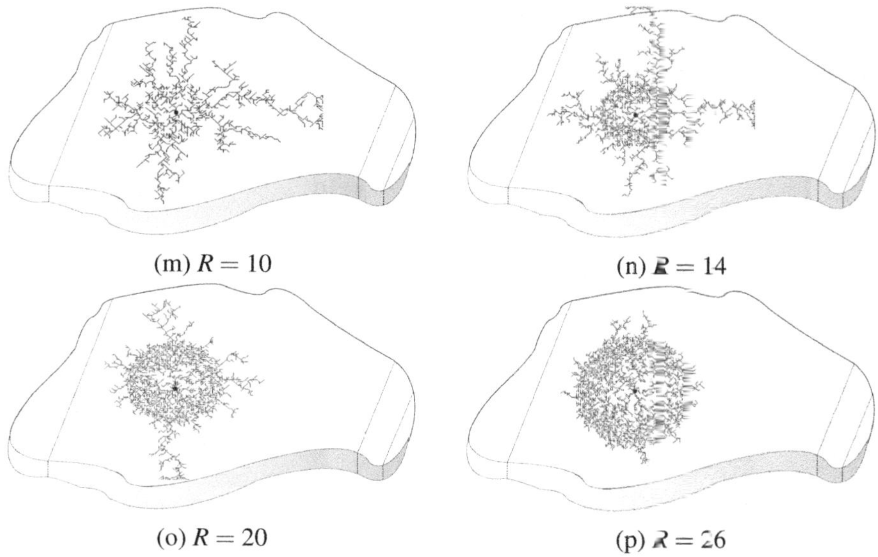


**Fig. 9.12** Fractal aggregation when $u = 5 \times 10^{-5}(x^2 + y^2)$, $\xi = 1 \times 10^{-7}$, $p = 3$

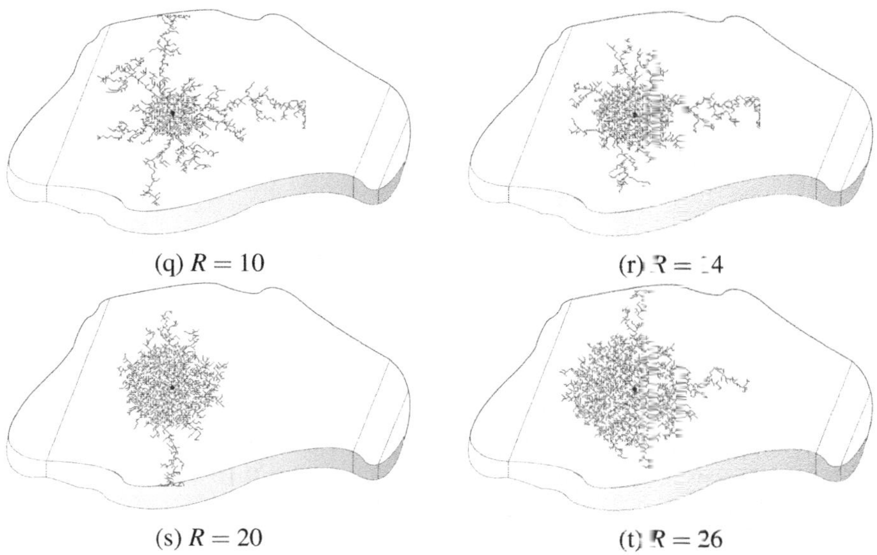


**Fig. 9.13** Fractal aggregation when $u = 0.01$, $\xi = 1 \times 10^{-7}$, $p = 3$

**Table 9.2** $a(x, y) = b(x, y) = e^{0.2x}$ fractal dimension of thermal diffusion under different environmental disturbances

| Figure number | Parameter | | | Fractal dimension |
|---|---|---|---|---|
| | $\xi$ | u | R | |
| 4a | 0 | 0 | 0 | 1.2102 |
| 5a | $10^{-7}$ | $5 * 10^{-5}(x^2 + y^2)$ | 10 | 1.3557 |
| 5b | $10^{-7}$ | $5 * 10^{-5}(x^2 + y^2)$ | 14 | 1.4763 |
| 5c | $10^{-7}$ | $5 * 10^{-5}(x^2 + y^2)$ | 20 | 1.7419 |
| 5d | $10^{-7}$ | $45 * 10^{-5}(x^2 + y^2)$ | 26 | 1.8522 |
| 6a | $10^{-7}$ | $5 * 10^{-3}$ | 10 | 1.2746 |
| 6b | $10^{-6}$ | 0.01 | 14 | 1.3347 |
| 6c | $10^{-6}$ | $5 * 10^{-4}$ | 20 | 1.4334 |
| 6d | $10^{-7}$ | $5 * 10^{-3}$ | 26 | 1.7437 |

**Table 9.3** $a(x, y) = b(x, y) = e^{0.02x}$ fractal dimension of thermal diffusion under different environmental disturbances

| Figure number | Parameter | | | Fractal dimension |
|---|---|---|---|---|
| | $\xi$ | u | R | |
| 4c | 0 | 0 | 0 | 1.4456 |
| 7a | $10^{-7}$ | $5 * 10^{-5}(x^2 + y^2)$ | 10 | 1.1569 |
| 7b | $10^{-7}$ | $5 * 10^{-5}(x^2 + y^2)$ | 14 | 1.5921 |
| 7c | $10^{-7}$ | $5 * 10^{-5}(x^2 + y^2)$ | 20 | 1.795 |
| 7d | $10^{-7}$ | $5 * 10^{-5}(x^2 + y^2)$ | 26 | 1.9432 |
| 8a | $10^{-7}$ | $10^{-2}$ | 10 | 1.4995 |
| 8b | $10^{-7}$ | $10^{-2}$ | 14 | 1.5276 |
| 8c | $10^{-7}$ | $10^{-2}$ | 20 | 1.553 |
| 8d | $10^{-7}$ | $10^{-2}$ | 26 | 1.7866 |

for the even term. However, as thermal conductivity does not increase very clearly with $x$, the aggregation of the particles to the right of the central point in the direction of $x$ is less obvious than the equivalent aggregation in Figs. 9.10 and 9.11.

**Fractal Dimension of Thermal Diffusion Under Conditions of Environmental Disturbance**

A fractal dimension is an index for characterizing fractal patterns or sets by quantifying their complexity as a ratio of the change in detail to the change in scale. It has also been characterized as a measure of the space-filling capacity of a pattern that tells how a fractal scales differently from the space it is embedded in [206]. As the thermal diffusion of a spatial point source has an obvious central growth point, we use Sandbox to measure the fractal dimension with different variable coefficients and circular regions, as shown in Tables 9.2 and 9.3. With different variable coefficient functions, and as the environmental disturbance term increases in scope and intensity, the corresponding fractal dimensions expand to different degrees. This suggests that the environmental disturbance term not only controls the surface morphological changes that occur during thermal fractal diffusion in flat objects, but also increases the complexity of fractal diffusion growth.

## 9.3 Directional Region Control of the Thermal Fractal Diffusion of a Space Body

In this section, we present a directional region control (DRC) model of thermal diffusion fractal growth with active heat diffusion in three-dimensional space. It must be pointed out that this model can be applied to predict the space body heat fractal growth and study its directional region control. When the nonlinear interference term and the inner heat source term are generalized functions, the relationship between the particle aggregation probability and the interference terms can be obtained using the norm theory. We can then predict the aggregation form of particles in different regions. The nonlinear interference terms in the model are expressed as a trigonometric function and its composite function.

### 9.3.1 *Space Body Thermal Diffusion System*

Considering the thermal diffusion of a space body $W$ (as shown in Fig. 9.14) in an actual complex environment, we take the thermal diffusion process as a distributed parameter system meeting the boundary conditions, [462]

$$\nabla^2 V(x) = u(x) + f\left(V(x), \frac{\partial V(x)}{\partial x}\right), \quad x \in \mathbf{R}^3, \tag{9.3.1}$$

where $V(x)$ refers to the space body $W$'s temperature (the probability of thermal diffusion) at certain point $x$, and $u(x)$ refers to the density of the inner heat resource. The $f$ is a nonlinear function, referred to as the forced term in physics. $u(x)$ is a quantitative initial condition, referred to as the source term. Suppose that $G = u(x) + f(V(x), \partial V(x)/\partial x)$, both the nonlinear interference term $f(V(x), \partial V(x)/\partial x)$ and the source term $u(x)$ can be regarded as a part of the environmental interference term $G$, which is assumed to be an arbitrary function. Figure 9.15 shows the thermal fractal diffusion aggregation cluster without environmental interference.

Generalized functions can be regarded as continuous linear functionals of function space and are applied extensively in mathematics, physics, mechanics, and engineering [467]. Supposing both $f(V(x), \partial V(x)/\partial x)$ and $u(x)$ are generalized functions, and that these generalized functions are all elements in a basic function space $K$.

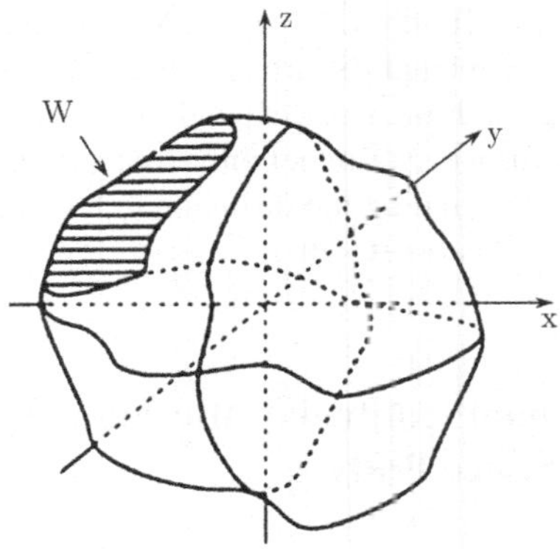


**Fig. 9.14** A space body $W$

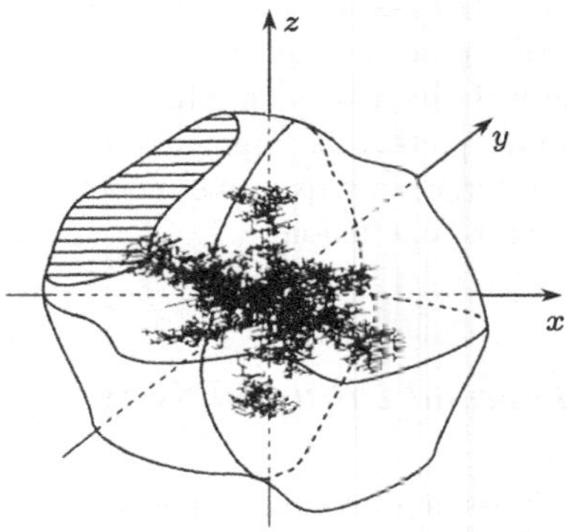


**Fig. 9.15** The thermal diffusion of the space body $W(f = 0, u = 0)$

In general, it is very difficult to find an analytical solution to the nonlinear partial differential equation (9.3.1). By studying the weak solution to system (9.3.1), we discuss the quantitative relationship between the particle aggregation probability and the environmental interference term $G$.

### 9.3.2 *Directional Region Control of a Space Body Thermal Fractal Diffusion System*

**Definition 9.1** ($L^2$ *space*) Assume that $\Omega$ is a bounded domain of $\mathbf{R}^3$. We use $L^2(\Omega)$ to express the function class consisting of all the measurable functions $V(x)$ which are defined in $\Omega$ and meet $\int_\Omega |V(x)|^2 x < \infty$. The $L^2(\Omega)$ is a vector space with norms defined as $\|V(x)\|_{L^2(\Omega)} = (\int_\Omega V^2(x)x)^{1/2}$.

**Definition 9.2** (*Sobolev space $H^k$*) Assume that $\Omega$ is a bounded domain of $\mathbf{R}^3$. We use $H^k(\Omega)$ to express the function class consisting of all the measurable functions which are defined in $\Omega$ and meet $\int_\Omega \sum\limits_{0\le i\le k} |\nabla^i V(x)|^2 x < \infty$. The $H^k(\Omega)$ is a vector space with norms defined as $\|V(x)\|_{H^k(\Omega)} = (\sum\limits_{0\le i\le k} \|\nabla^i V(x)\|^2_{L^2(\Omega)})^{1/2}$.

**Definition 9.3** Assume that $a(V, v)$ is a bilinear functional in Hilbert space $H$, i.e., $V$ and $v$ are both linear.

(i) $a(V, v)$ is bounded, if here is a positive constant $M \ge 0$, such that

$$|a(V, v)| \le M \|V\|_{L^2(\Omega)} \|v\|_{L^2(\Omega)}, \forall V, v \in H.$$

(ii) $|a(V, v)|$ is compulsory, if there is a positive constant $\delta > 0$, such that

$$a(V, V) \ge \delta \|V\|^2_{L^2(\Omega)}, \forall V \in H.$$

Now we can consider the existence and uniqueness of the weak solution of the following system:

$$\begin{cases} \Delta V = f(V, DV) + u(x), \ x \in \mathbf{R}^3, \\ V|_{\partial\Omega} = 0. \end{cases} \tag{9.3.2}$$

**Theorem 9.1** *Assume that $u(x) \in L^2(\Omega)$, $f(V, DV) \in L^2(\Omega)$. There exists a unique weak solution $V \in H_0^1(\Omega)$ to system (9.3.2), and the following estimation exists:*

$$\|V\|_{H_0^1(\Omega)} \le \frac{1}{\lambda} \|u\|_{L^2(\Omega)},$$

*where $\lambda$ is a positive constant.*

***Proof*** Denote $a(V, \psi) = \int_\Omega D_i V \cdot D_i \psi x, \forall V, \psi \in H_0^1(\Omega)$.

Obviously, $a(V, \psi)$ is bilinear. First, we prove the boundedness of $a(V, \psi)$,

$$\begin{aligned}
&|a(V, \psi)| \\
&= \left| \int_\Omega f(V, DV)\psi + D_i V \cdot D_j \psi x \right| \\
&\le C \int_\Omega |f(V, DV)\psi| x + C \int_\Omega |D_i V| |D_j \psi| x \\
&\le C \|f(V, DV)\|_{L^2(\Omega)} \|\psi\|_{L^2(\Omega)} + C \|\nabla V\|_{L^2(\Omega)} \|\nabla \psi\|_{L^2(\Omega)} \\
&\le C \|V\|_{H_0^1(\Omega)} \|\psi\|_{H_0^1(\Omega)}.
\end{aligned}$$

Next, due to the ellipticity condition, we have $a(V, V) \ge \lambda \|\nabla V\|^2_{L^2}$ for $V \in H_0^1(\Omega)$. So $a(V, \psi)$ is compulsory. Let us prove that $F(\psi) = \int_\Omega u\psi x$ is a bounded functional in $H_0^1(\Omega)$. The $F(\psi)$ is linear. According to the Hölder inequality, we have

$$|F(\psi)| = \left| \int_{\Omega} u\psi x \right| \leq C\|u\|_{L^2(\Omega)} \|\psi\|_{L^2(\Omega)} \leq C\|u\|_{L^2(\Omega)} \|\psi\|_{H_0^1(\Omega)}.$$

Therefore, according to the Lax–Milgram Theorem [468], there exists a unique $V \in H_0^1(\Omega)$, such that $\|V\|_{H_0^1(\Omega)} \leq \|u\|_{L^2(\Omega)}/\lambda$. Thus, $V \in H_0^1(\Omega)$ is the only weak solution to system (9.3.2).

In order to improve the regularity of the weak solution for system (9.3.2) we introduce the difference operator. If $V(x)$ is a function defined on $\mathbf{R}^3$, then $\Delta_h^i V = [V(x+he_i) - u(x)]/h$, where $e_i$ is the unit vector in the direction of $x_i$ and $\Delta_h^i$ is the difference operator along the direction of $x_i$. In the following, we introduce some of the basic properties of the difference operator.

**Proposition 9.1** *We have the following basic properties of the difference operator.*

*(1) The adjoint operator of $\Delta_h^i$ is $\left(\Delta_h^i\right)^* = -\Delta_{-h}^i$; for any integral functions $f(x)$, $g(x) \in L^2(\Omega)$, there is an integral equation*

$$\int_{R^3} f(x)\Delta_h^i g(x)x = -\int_{R^3} g(x)\Delta_{-h}^i f(x)x.$$

*(2) The differential operator and $\Delta_h^i$ are exchangeable, for any weak differentiable function $\phi$,*

$$D_j \Delta_h^i V = \Delta_h^i D_j V, \quad (j = 1, 2, 3).$$

*(3) The difference of products can be expressed as*

$$\Delta_h^i(f(x)g(x)) = \Delta_h^i f(x) T_h^i g(x) + f(x)\Delta_h^i g(x),$$

*where $T_i$ is the shift operator in the direction of $x_i$, $T_h^i V(x) = V(x + he_i)$.*

**Proposition 9.2** *Assume that $\Omega \subset \mathbf{R}^3$ is a region.*

*(1) Assume that $V \subset H^1(\Omega), \Omega_1 \subset\subset \Omega$. For $|h| > 0$, when sufficiently small, $\Delta_h^i \Phi \in L^2(\Omega_1)$, giving the following estimation formula $\left\|\Delta_h^i V\right\|_{L^2(\Omega_1)} \leq \|D_i V\|_{L^2(\Omega)}$.*

*(2) Assume that $V \subset L^2(\Omega)$ and there exists a constant $K$ such that $\Omega_1 \subset\subset \Omega$. When $|h| > 0$ is small enough, $\left\|\Delta_h^i V\right\|_{L^2(\Omega)} \leq K$, where $K$ is unrelated to $h$. Then $D_h^i V \in L^2(\Omega_1)$, giving the following estimation formula $\|D_i V\|_{L^2(\Omega')} \leq K$.*

*We can obtain Theorem 9.2 using the basic properties of the difference operator.*

**Theorem 9.2** *If $V(x) \in H_0^1(\Omega)$ is the weak solution of system (9.3.2) and $\Omega_1 \subset\subset \Omega$, then $V \in H_0^1(\Omega_1)$ and we have*

$$\|V\|_{H_0^1(\Omega)} \leq C\left(\|V\|_{H^1(\Omega)} + \|u\|_{L^2(\Omega)}\right), \tag{9.3.3}$$

*where C is a positive constant that only depends on the distance* $dist\{\Omega_1, \partial\Omega\}$ *from* $\Omega_1$ *to* $\partial\Omega$.

***Proof*** Supposing $\Omega_1 \subset \Omega$, we note that $d = \text{dist}\{\Omega_1, \partial\Omega\}/4$. The cut-off function $\eta(x) \in C_0^\infty(\Omega), 0 \leq \eta(x) \leq 1, \eta(x) \equiv 1, \text{dist}\{\text{supp}\eta, \partial\Omega\} \geq 2d$.

Based on the definition of the weak solution and the density of $C_0^\infty$ in $H_0^1(\Omega)$, we conclude that

$$-\int_\Omega \nabla V \nabla \psi x = \int_\Omega (f+u)\cdot \psi x, \quad \forall \psi \in H_0^1(\Omega).$$

If $h$ meets the condition $0 < |h| < d$ and $\psi = (\Delta_h^i)^*(\eta^2 \Delta_h^i \Phi)$, then dist $\{\text{supp}\psi, \partial\Omega\} \geq d$, and $\phi \in H_0^1(\Omega)$, therefore, $\psi \in H_0^1(\Omega)$. Substituting $\psi$ into the above integral equation, we can obtain the following formula:

$$-\int_\Omega \nabla V \nabla (\Delta_h^i)^*(\eta^2 \Delta_h^i V)x = \int_\Omega (f+u)(\Delta_h^i)^*(\eta^2 \Delta_h^i V)x.$$

By applying the exchangeability of the difference operator and the differential operator, and the relationship between the difference operator and its adjoint operator, the above integral relationship can be rewritten as

$$-\int_\Omega \nabla \Delta_h^i V \cdot \nabla(\eta^2 \Delta_h^i V)x = \int_\Omega (f+u)(\Delta_h^i)^*(\eta^2 \Delta_h^i V)x. \tag{9.3.4}$$

The following formula is obtained from formula (9.3.5) by calculation:

$$\begin{aligned}
&\int_\Omega \eta^2 \left|\nabla \Delta_h^i V\right|^2 x \\
&= -2\int_\Omega \nabla \Delta_h^i V \eta \cdot \nabla \eta \Delta_h^i V x - \int_\Omega (f+u)(\Delta_h^i)^*(\eta^2 \Delta_h^i V)x \\
&\leq \frac{1}{4}\int_\Omega \eta^2 \left|\nabla \Delta_h^i V\right|^2 x + 4\int_\Omega |\nabla \eta|^2 \left|\Delta_h^i V\right|^2 x \\
&\quad + \frac{1}{8}\int_\Omega \left|(\Delta_h^i)^*(\eta^2 \Delta_h^i V)\right|^2 x + 2\int_\Omega (f+u)^2 x
\end{aligned} \tag{9.3.5}$$

For further calculations, we need to estimate the third term on the right side of formula (9.3.6). From Proposition 9.1, we conclude that

$$
\begin{aligned}
&\int_{\Omega} \left|(\Delta_h^i)^*(\eta^2 \Delta_h^i V)\right|^2 x \\
&= \int_{\text{supp}\psi} \left|\Delta_{-h}^i(\eta^2 \Delta_h^i V)\right|^2 x \\
&\le \int_{\Omega} \left|\eta^2 \nabla \Delta_h^i V + 2\eta \Delta_h^i V \nabla \eta\right|^2 x \\
&\le 2\int_{\Omega} \eta^2 \left|\nabla \Delta_h^i V\right|^2 x + 8\int_{\Omega} |\nabla \eta^2| \left|\Delta_h^i V\right|^2 x.
\end{aligned} \tag{9.3.5}
$$

Substituting formula (9.3.7) into formula (9.3.6), we obtain

$$
\begin{aligned}
&\int_{\Omega} \eta^2 \left|\nabla \Delta_h^i V\right|^2 \\
&\le \frac{1}{4}\int_{\eta^2} \left|\nabla \Delta_h^i V\right|^2 x + 4\int_{\Omega} |\nabla \eta|^2 \left|\Delta_h^i V\right|^2 x \\
&\quad + \frac{1}{4}\int_{\Omega} \eta^2 \left|\nabla \Delta_h^i V\right|^2 x + \int_{\Omega} |\nabla \eta|^2 \left|\Delta_h^i V\right|^2 x \\
&\quad + 2\int_{\Omega} (f+u)^2 x,
\end{aligned}
$$

and then

$$
\begin{aligned}
&\int_{\Omega} \eta^2 \left|\nabla \Delta_h^i V\right|^2 x \\
&\le 10\int_{\Omega} |\nabla \eta|^2 \left|\Delta_h^i V\right|^2 x + 4\int_{\Omega} (f+u)^2 x.
\end{aligned} \tag{9.3.7}
$$

Next, if we estimate the first term in formula (9.3.8) by using the cut-off function and Proposition 9.1, then we have

$$
\begin{aligned}
\int_{\Omega} \eta^2 \left|\nabla \Delta_h^i V\right|^2 &\le C\int_{\text{supp}\psi} \left|\Delta_h^i V\right|^2 x \\
&\le C\int_{\Omega} |D_i V|^2 x \le C\int_{\Omega} |\nabla u|^2 x.
\end{aligned} \tag{9.3.8}
$$

Substituting formula (9.2.8) into formula (9.3.8), we conclude that

$$
\int_{\Omega} \eta^2 \left|\nabla \Delta_h^i V\right|^2 x \le C\int_{\Omega} |\nabla V|^2 x - C\int_{\Omega} (f^2+u^2) x.
$$

Based on the above formula and using the cut-off function, we have

$$
\int_{\Omega_1} \left|\nabla \Delta_h^i V\right|^2 x \le C\int_{\Omega} |\nabla V|^2 x + C\int_{\Omega} (f^2+u^2) x.
$$

Based on Proposition 9.2, we can conclude that formula (9.33) is correct. The proof is completed.

### 9.3.3 Simulation of the Thermal Fractal Diffusion in Directional Regions

Based on Theorem 9.1, it can be concluded that when $f(V(x, y, z), DV(x, y, z) \in L^2(\Omega)$, $u(x, y, z) \in L^2(\Omega)$, and $\|V(x, y, z)\|_{H_0^1(\Omega)} \leq \|u(x, y, z)\|_{L^2(\Omega)} / \lambda$. The source term $u(x)$ can restrain the nonlinear influence of the generalized function term $f(V(x), \partial V(x)/\partial x)$ on the thermal fractal diffusion. This allows the thermal particle to be aggregated in different directions or regions, and achieves directed control.

Assume that $u(x)$ is a constant $k$, $f(V(x), \partial V(x)/\partial x) = a\sin^2(V(x))$, and $f(V(x), \partial V(x)/\partial x = br^2(x) + c\sin(V(x))$, the control region $P$ of the source item $u(x)$ can be taken as a one-eighth space, one half space, cuboid region, spherical domain, or sectional region.

The thermal diffusion of a space body is a random process induced by a temperature gradient. In our recent work, [462] the thermal diffusion fractal growth was simulated. The growth-probability function of every possible aggregation point has a nonlinear relationship with the temperature gradient. With the environmental disturbance, the boundary condition $V(x)$ is set to be zero at any points on the aggregation surface, and for the boundary at infinity, $V(x)$ is set to be one, which provides a stable particle flow. By modifying the movement-boundary conditions of the mathematical model, the temperature gradient can be calculated from the Poisson equation. What is more, by using the roulette wheel selection algorithm, the particle aggregation location can be determined.

Considering environmental disturbances, the thermal diffusions in different regions are simulated by using the simulation method in the literature [462]. As shown in Figs. 9.15, 9.16, 9.17 and 9.18, the interference source item controls the aggregation of the thermal particles in different directions. This result shows that our control method is valid.

What is more, a fractal dimension is an index for characterizing fractal patterns or sets by quantifying their complexity as a ratio of the change in detail to the change in scale. It has also been considered as a measure of the space-filling capacity of a pattern, which tells how a fractal scales differently from the space it is embedded in [206]. As the thermal diffusion system of a space point source has an obvious central point, the Sandbox method is used to calculate the fractal dimension ($D_f$) of the thermal aggregation cluster with different interference actions [469]. Figure 9.19 shows the comparison between the thermal diffusion fractal dimensions of different sectional control regions. Besides, Fig. 9.18 shows the results predicted by our DRC model for spherical domain, where the spherical interference control region is

$$H_1 = \{(r, \theta, \varphi)|r \text{ change},\ 0 \leq \theta \leq \pi,\ 0 \leq \varphi \leq 2\pi\},$$

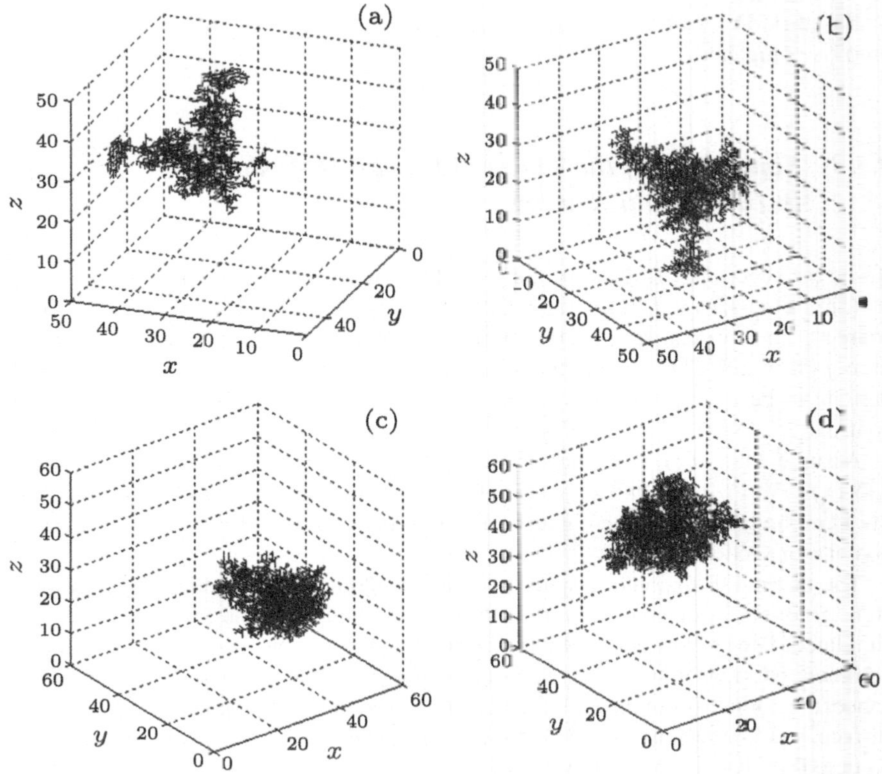


**Fig. 9.16** Thermal fractal diffusion in different interference space regions: **a** $0 \le \theta \le \pi/2, 0 \le \varphi \le \pi/2$, **b** $\pi/2 \le \theta \le \pi, \pi \le \varphi \le 3\pi/2$, **c** $\pi/2 \le \theta \le \pi, 3\pi/2 \le \varphi \le 2\pi$, **d** $0 \le \theta \le \pi/2, 0 \le \varphi \le \pi$

and the sectional control region is

$$
\begin{aligned}
H_1 &= \{(r, \theta, \varphi)|r \text{ change},\ 0 \le \theta \le \pi,\ 0 \le \varphi \le 2\pi\}, \\
H_2 &= \left\{(r, \theta, \varphi)|r \in \mathbb{R},\ \theta = i\frac{\pi}{2},\ \varphi = j\frac{\pi}{2},\ i = 0, 1, 2,\ j = 0, 1, 2, 3\right\}.
\end{aligned}
$$

According to Fig. 9.19, the aggregations of thermal particles based on the spherical control region are more complex than that controlled in different directions in addition to the spherical domain. Although they share the same control spherical domain, the directed region control beyond spherical domain determines the growth direction and also reduces the growth complexity. Apart from that, Fig. 9.19 shows that the fractal dimensions of clusters aggregated in the multidirectional region are larger than those of clusters aggregated in the unidirectional region for the same spherical

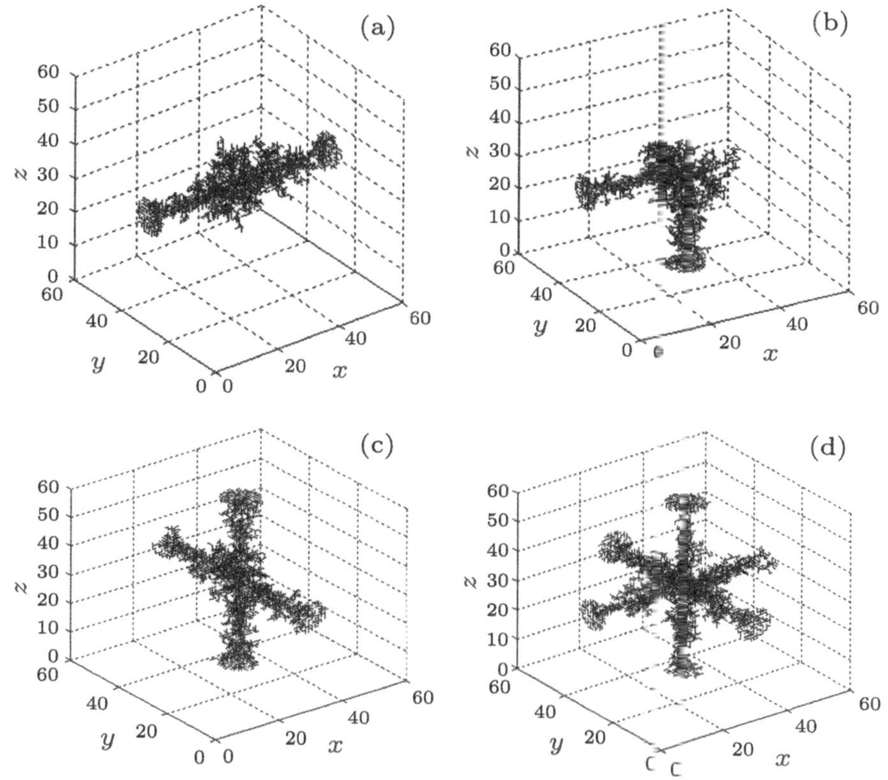


**Fig. 9.17** Thermal fractal diffusion under control conditions in differently directed regions: **a** $\theta = \pi/2, \varphi = j\pi/2, j = 0, 2$, **b** $\theta = \pi; \theta = \pi/2, \varphi = \pi$, **c** $\theta = i\pi/2, i = 0, 2; \theta = \pi/2, \varphi = j\pi/2, j = 1, 3$, **d** $\theta = i\pi/2, i = 0, 2; \theta = \pi/2, \varphi = j\pi/2, j = 0, 1, 2, 3$

domain. This indicates that the complexity of thermal aggregation clusters increases along with the increase in multidirectional control.

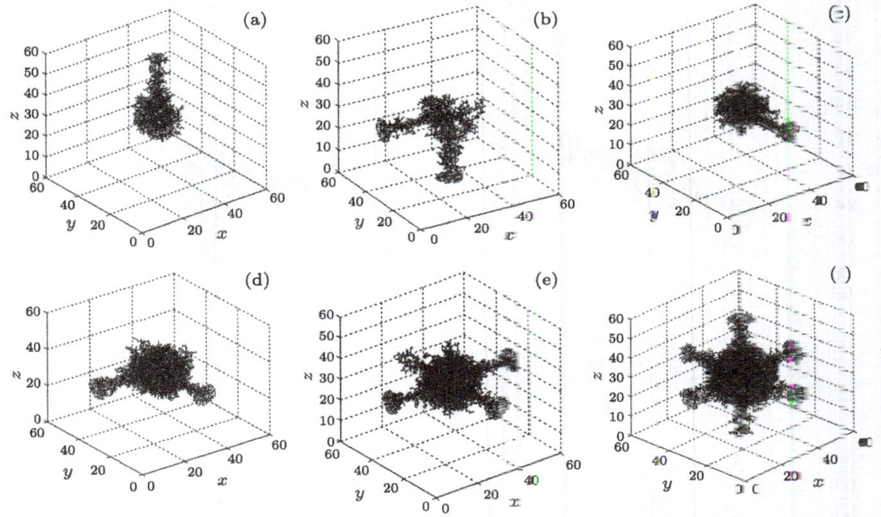


**Fig. 9.18** Thermal fractal diffusion with variation of the spherical domain and sectional region: **a** $r \leq 8; \theta = 0$, **b** $r \leq 8; \theta = \pi; \theta = \pi/2, \varphi = j\pi/2, j = 2, 3$, **c** $r \leq 9; \theta = \pi/2, \varphi = 3\pi/2$, **d** $r \leq 10; \theta = \pi/2, \varphi = j\pi/2, j = 2, 3$, **e** $r \leq 11; \theta = \pi/2, \varphi = j\pi/2, j = 0, 1, 2, 3$, **f** $r \leq 14; \theta = i\pi/2, i = 0, 2; \theta = \pi/2, \varphi = j\pi/2, j = 0, 1, 2, 3$

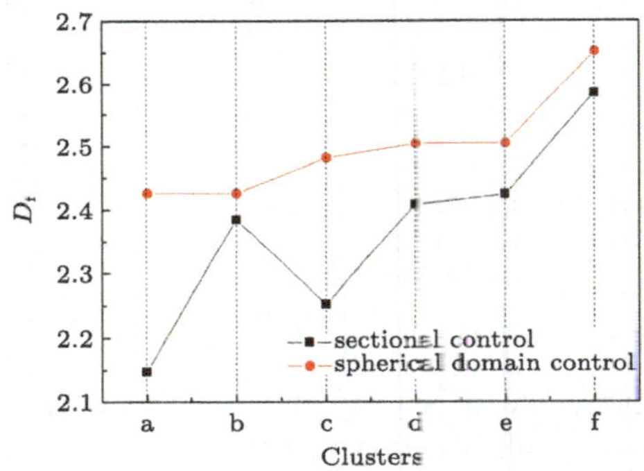


**Fig. 9.19** The fractal dimensions for the thermal diffusion clusters in Fig. 9.18 under variant spherical domain controls, and those under variant sectional controls

# Chapter 10
# Fractal Analysis and Control of the SIRS Models

As for disease dynamics, SIRS model is one of classical models, which divides animal states into three categories: susceptible state (S), infectious state (I), and removed state (R). The SIRS model describes the temporary immunity after cure, when contacting with infected people; infectious diseases may be transmitted from infected people to susceptible ones with a certain probability. People in an infectious state that has recovered over time are no longer infected and pose no threat to others, which means they are in a state being removed from the whole. After a period of time, the removed individual restores a susceptible state.

## 10.1 Fractal Dimension and Synchronization of the Controlled Julia Sets of the SIRS Model

### *10.1.1 Fractal Dimension of the Controlled Julia Sets of the SIRS Model*

The following SIRS models with constant inoculation rate and nonlinear infection rate [470] are considered.

$$\begin{cases} \dfrac{d\tau_1}{dt} = \beta\tau_1^2(1-\tau_1-\tau_2) - (q+v)\tau_1 \\ \dfrac{d\tau_2}{dt} = p + v\tau_1 - (q+\varepsilon)\tau_2, \end{cases} \tag{10.1.1}$$

where $\tau_1$ is the representative of infected persons and their number, $\tau_2$ represents the persons who are rehabilitated from infection and their number. Supposing during


S. T. Liu et al., *Fractal Control and Its Applications*,
https://doi.org/10.1007/978-981-15-5459-9_10

the epidemic period, the birth rate coefficient and the natural mortality coefficient of the population are equal to the constant $q$, $\ell$ denotes the effective contact rate coefficient, $p$ means the inoculation rate, $v$ represents the removal rate coefficient and $\varepsilon$ denotes the immune loss rate coefficient. Furthermore, regardless of the death due to illness, we suppose all the newborns are susceptible.

Practically, the seasonal outbreak of disease is a common phenomenon, such as influenza, a frequently-used form of the contact rate $\beta$ is expressed as $\beta = \beta_0(1 + \varphi\phi(t))$, the common form of seasonal factors can be approximately recognized as a single sine function, see [471]. As a result, we consider the situation that the effective contact rate coefficient is affected by seasonal factors, which means $\phi(t) = \sin(\omega t)$.

$$\begin{cases} \dfrac{d\tau_1}{dt} = \beta_0(1 + \varphi \sin(\omega t))\tau_1^2(1 - \tau_1 - \tau_2) - (q + v)\tau_1, \\ \dfrac{d\tau_2}{dt} = p + v\tau_1 - (q + \varepsilon)\tau_2, \end{cases} \tag{10.1.2}$$

where $\beta_0$ denotes the effective contact rate coefficient, $\varphi$ is the seasonal influence factor. Take $\tau_3 = \sin(\omega t)$, $\tau_4 = \cos(\omega t)$, then $\frac{d\tau_3}{dt} = \omega\tau_4$ and $\frac{d\tau_4}{dt} = -\omega\tau_3$. Thus, system (10.1.2) is as follows.

$$\begin{cases} \dfrac{d\tau_1}{dt} = \beta_0(1 + \varphi\tau_3)\tau_1^2(1 - \tau_1 - \tau_2) - (q + v)\tau_1, \\ \dfrac{d\tau_2}{dt} = p + v\tau_1 - (q + \varepsilon)\tau_2, \\ \dfrac{d\tau_3}{dt} = \omega\tau_4, \\ \dfrac{d\tau_4}{dt} = -\omega\tau_3, \end{cases} \tag{10.1.3}$$

and the initial values of $\tau_3$ and $\tau_4$ are $\tau_3(0) = 0$ and $\tau_4(0) = 1$. Then, substitute these discretizations in the system [472], replace the form of the $x_n, y_n, s_n, t_n$ into $\tau_1, \tau_2, \tau_3, \tau_4$ in this part and $\Delta t$ is expressed by $r$, so the equation is as follows,

$$\begin{cases} \dfrac{x_{n+1} - x_n}{r} = \beta_0(1 + \varphi s_n){x_n}^2(1 - x_n - y_n) - (q + v)x_n, \\ \dfrac{y_{n+1} - y_n}{r} = p + vx_n - (q + \varepsilon)y_n. \\ \dfrac{s_{n+1} - s_n}{r} = \omega t_n, \\ \dfrac{t_{n+1} - t_n}{r} = -\omega s_n. \end{cases}$$

So, the equation presents as

$$\begin{cases} x_{n+1} = (1 - rq - rv)x_n + r\beta_0(1 + \varphi s_n)(1 - x_n - y_n){x_n}^2, \\ y_{n+1} = rvx_n + (1 - rq - r\varepsilon)y_n + p, \\ s_{n+1} = s_n + rwt_n, \\ t_{n+1} = t_n - rws_n. \end{cases}$$

Change the representation of the parameters, then the discrete version is

$$\begin{cases} x_{n+1} = (1 - c - d)x_n + a(1 + bs_n)(1 - x_n - y_n)x_n^2, \\ y_{n+1} = dx_n + (1 - c - e)y_n + \upsilon, \\ s_{n+1} = s_n + ut_n, \\ t_{n+1} = t_n - us_n, \end{cases} \tag{10.1.4}$$

and $s_0 = 0$, $t_0 = 1$, $a, b, c, d, e, \upsilon, u$ are system parameters. Now, fractal theory is used to introduce the Julia set of the real dynamic system (10.1.4).

**Definition 10.1** Let

$$F(x, y) = (1 - c - d)x + a(1 + bs)(1 - x - y)x^2, G(x, y) = dx + (1 - c - e)y + v,$$

$$H(x, y) = (F(x, y), G(x, y), 0, 1),$$

$$D = \{(x, y, 0, 1) | \{H^n(x, y, 0, 1)\}_{n=1}^{\infty} \text{ remains bounded}\}.$$

The set $D$ is called the filled Julia set corresponding to the map $H(x, y, 0, 1)$. And the boundary of $D$ is called the Julia set of the map $H(x, y, 0, 1)$, which is denoted by $J_H$, i.e, $J_H = \partial D$.

Obviously, it is inconvenient to simulate the system (10.1.4) because it has 4 dimensions. However, the initial values of the third and the forth variables are fixed, so the equation can be seen as a map on the $x - y$ plane.

The parameters of model (10.1.4) are taken to be $a = 0.08$, $b = 0.02$, $c = 0.005$, $d = 0.005$, $e = 0.05$, $\upsilon = 0.005$, $u = 0.001$, then the corresponding Julia set is shown in Fig. 10.1.

The irregular curve in Fig. 10.1 represents the initial numbers of infected and rehabilitated people. If these values are inside the curve, the numbers of the two clusters will be stable and bounded, and if the values of the initial numbers of two clusters are outside the curve, then at least one of the number of infected or rehabilitated people will approach to infinity.

In the following, the fixed point is used to consider the stability of system (10.1.4). Let the fixed point of the equation be $(x^*, y^*, s^*, t^*)$. Here, if $u = 0$, $s^*$ and $t^*$ can be arbitrary number, and if $u \neq 0$, thus $s^* = 0$, $t^* = 0$. There is no significance of studying in the case of $u = 0$, so just consider the fixed point that both $s^*$ and $t^*$ are zero. The solutions are as follows.

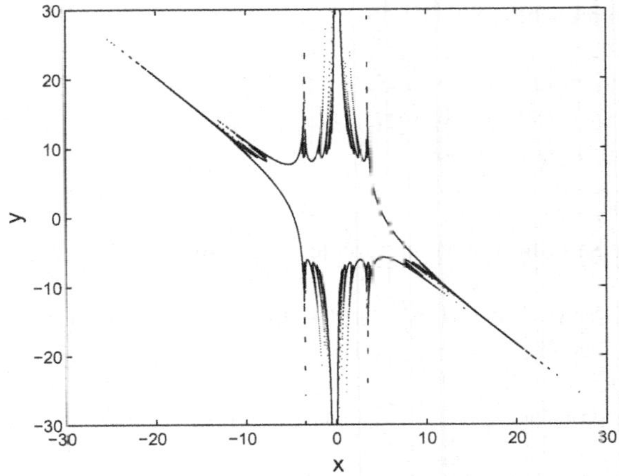


**Fig. 10.1** Original Julia sets of system (10.1.4) when $a = 0.08, b = 0.02, c = 0.005, d = 0.005, e = 0.05, v = 0.005, u = 0.001$

**Proposition 10.1** *The values of $x^*$ and $y^*$ in the fixed point $(x^*, y^*, s^*, t^*)$ is as follows:*

*(1) If $x^* = 0$ and $y^* \neq 0$, so $(c+e)y^* = v$ and $(x^*, y^*) = (0, \dfrac{v}{c+e})$.*

*(2) If $x^* \neq 0$ and $y^* = 0$, so $dx^* + v = 0$, $(c+d)x^* + a(x^* - 1)x^{*2} = 0$, $vd^3 + cvd^2 + av^2d + av^3 = 0$ and $(x^*, y^*) = (\dfrac{-v}{d}, 0)$.*

*(3) If $x^* \neq 0$ and $y^* \neq 0$, so*

$$x^* = \frac{a(c+e-v) - \sqrt{a^2(v-c-e)^2 - 4a(c+e-d)(c+d)(c+e)}}{2a(c+e+d)},\ y^* = \frac{dx^* + v}{c+e}.$$

In particular, when disease breaks out, the coexistence of infected persons and persons who are rehabilitated from infection is the most common example, they can co-exist over a long period of time and neither of them will suddenly go to zero. There is no practical significance for other situation, thus, we only consider the condition that $x^* \neq 0$, $y^* \neq 0$ and $s^* = 0$, $t^* = 0$. Introduce the control items to control Julia set of the model (10.1.4), so the following controlled system (10.1.5) is obtained.

$$\begin{cases} x_{n+1} = (1-c-d)x_n + a(1+bs_n)(1-x_n-y_n)x_n^2 + k(x_n^3 - x^{*3}), \\ y_{n+1} = dx_n + (1-c-e)y_n + v + k(y_n - y^*), \\ s_{n+1} = s_n + ut_n + k(s_n - s^*), \\ t_{n+1} = t_n - us_n + k(t_n - t^*), \end{cases} \tag{10.1.5}$$

where $k$ is the control parameter. Let

$$\begin{cases} F(x^*, y^*, s^*, t^*) = (1-c-d)x_n + a(1+bs_n)(1-x_n-y_n)x_n^2 + k(x_n^3 - x^{*3}), \\ G(x^*, y^*, s^*, t^*) = dx_n + (1-c-e)y_n + v + k(y_n - y^*), \\ M(x^*, y^*, s^*, t^*) = s_n + ut_n + k(s_n - s^*), \\ N(x^*, y^*, s^*, t^*) = t_n - us_n + k(t_n - t^*). \end{cases} \tag{10.1.6}$$

Obviously, the fixed point $(x^*, y^*, s^*, t^*)$ of system (10.1.4) is also a fixed point of system (10.1.5) according to the definition of the fixed point, which means $F(x^*, y^*, s^*, t^*) = x^*, G(x^*, y^*, s^*, t^*) = y^*, M(x^*, y^*, s^*, t^*) = s^*, N(x^*, y^*, s^*, t^*) = t^*$. Therefore the Jacobi matrix of the controlled model (10.1.6) at this fixed point is

$$J = \frac{\partial(F, G, M, N)}{\partial(x, y, s, t)}|_{(x^*,y^*,s^*,t^*)} = \begin{bmatrix} a_1 & a_2 & a_3 & 0 \\ d & a_4 & 0 & 0 \\ 0 & 0 & a_5 & u \\ 0 & 0 & -u & a_5 \end{bmatrix},$$

where $a_1 = (1-c-d) + a(1+bs^*)(-3x^{*2} + 2x^* - 2x^*y^*) + 3kx^{*2}$, $a_2 = -a(1+bs^*)x^{*2}$, $a_3 = ab(1-x^*-y^*)x^{*2}$, $a_4 = 1-c-e+k$, $a_5 = 1+k$, and the corresponding characteristic equation is

$$|\lambda E - J| = \begin{vmatrix} \lambda - a_1 & -a_2 & -a_3 & 0 \\ -d & \lambda - a_4 & 0 & 0 \\ 0 & 0 & \lambda - a_5 & -u \\ 0 & 0 & u & \lambda - a_5 \end{vmatrix} = 0.$$

That is,

$$\lambda^4 + \alpha\lambda^3 + \gamma\lambda^2 + \eta\lambda + \varepsilon = 0,$$

where $\alpha = -(2a_5 + a_1 + a_4)$, $\gamma = 2a_5(a_1 + a_4) + a_1a_4 - a_2d + a_5{}^2 + u^2$, $\eta = 2a_5(a_2d - a_1a_4) - (a_5{}^2 + u^2)(a_1 + a_4)$, $\varepsilon = (a_1a_4 - a_2d)(a_5{}^2 + u^2)$. Then, the same parameters are taken to be $a = 0.08$, $b = 0.02$, $c = 0.005$, $d = 0.005$, $e = 0.05$, $v = 0.005$, $u = 0.001$, the fixed point of the system is $(x^*, y^*, s^*, t^*) = (0.6596, 0.1509, 0, 0)$, so according to the previous discussion, the Jacobi matrix of the control model (10.1.6) at the fixed point is

$$J = \begin{bmatrix} 0.9752 + 1.3052k & -0.0348 & 0.0001 & 0 \\ 0.005 & 0.945 + k & 0 & 0 \\ 0 & 0 & 1+k & 0.001 \\ 0 & 0 & -0.001 & 1+k \end{bmatrix}.$$

The corresponding characteristic equation is

$$|\lambda E - J| = \begin{vmatrix} \lambda - 0.9752 - 1.3052k & 0.0348 & -0.0001 & 0 \\ -0.005 & \lambda - 0.945 - k & 0 & 0 \\ 0 & 0 & \lambda - 1 - k & -0.001 \\ 0 & 0 & 0.001 & \lambda - 1 - k \end{vmatrix} = 0.$$

Then the above equation becomes,

$$m(\lambda) = |\lambda E - J| = \lambda^4 + \alpha^*\lambda^3 + \gamma^*\lambda^2 + \eta^*\lambda + \varepsilon^* = 0,$$

where $\alpha^* = -4.3052k - 3.9202$, $\gamma^* = 6.9156k^2 + 12.6594k - 5.7621$, $\eta^* = -4.9156k^3 - 13.5582k^2 - 12.4062k - 3.7636$, $\varepsilon^* = 1.3052k^4 + 4.819k^3 + 6.6441k^2 + 4.0524k + 0.9217$.

Based on the Jury criterion, when $m(1) > 0$, $m(-1) > 0$ and $\varepsilon^* < 1$, the fixed point $(x^*, y^*, s^*, t^*)$ is attractive. The condition that the modules of the eigenvalues of the Jacobian matrix at the fixed point are entirely less than 1 guarantees the attractivity of the fixed point. Thus, the range of the system parameters $k$ can be obtained.

$$\begin{aligned}
m(1) &= 1 + \alpha^* + \gamma^* + \eta^* + \varepsilon^* = 1.3052k^4 - 0.0966k^3 + 0.0015k^2 + 0.0004k > 0,\\
m(-1) &= 1 - \alpha^* + \gamma^* - \eta^* + \varepsilon^*\\
&= 1.3052k^4 + 9.7346k^3 + 27.1179k^2 + 33.4232k + 15.3676 > 0,\\
\varepsilon^* &= 1.3052k^4 + 4.819k^3 + 6.6441k^2 + 4.0524k + 0.9217 < 1.
\end{aligned}$$

The fixed point $(x^*, y^*, s^*, t^*)$ is attractive as long as the value of $k$ can satisfy the inequality $m(1) > 0$, $m(-1) > 0$ and $\varepsilon^* < 1$. Then denote that the control of the system is accomplished. The corresponding Julia sets are shown in Fig. 10.2.

Irregular curves in Fig. 10.2 are the changing process diagrams of Julia sets with the change of parameters, it is evident that these diagrams mirror Julia sets are closely related to parameters, and as the parameters slowly increase, the structure of the Julia set becomes flatter and flatter, which seems like there is a pair of invisible hand has stretched the figure like dough. The control of Julia set enables some initial values outside the set be inside the set, which is equal to control some people with infectious diseases, in other words, the disease to some extent has controlled.

According to the Definition 6.1 of box-counting dimension, we calculate the box-counting dimension of the controlled Julia sets to describe the complexity and diversity of Julia set. The changes of box-counting dimension of the Julia set with the changes of the values of the controlled parameter $k$ are shown in Table 10.1 or Fig. 10.3.

The box-counting dimension decreases with the increase of the control coefficient $k$. When $k$ is in the range of 0-0.06, the box dimension tends to be stable.

### 10.1.2 Synchronization of Julia Sets of the SIRS Model

It is noticed that the method of chaos synchronization is also applied to the study of Julia sets. In this section, a nonlinear coupled item is designed to make one Julia set of the SIRS model change to be another. Consider a system with the same form about Eq. (10.1.4) but with different parameters.

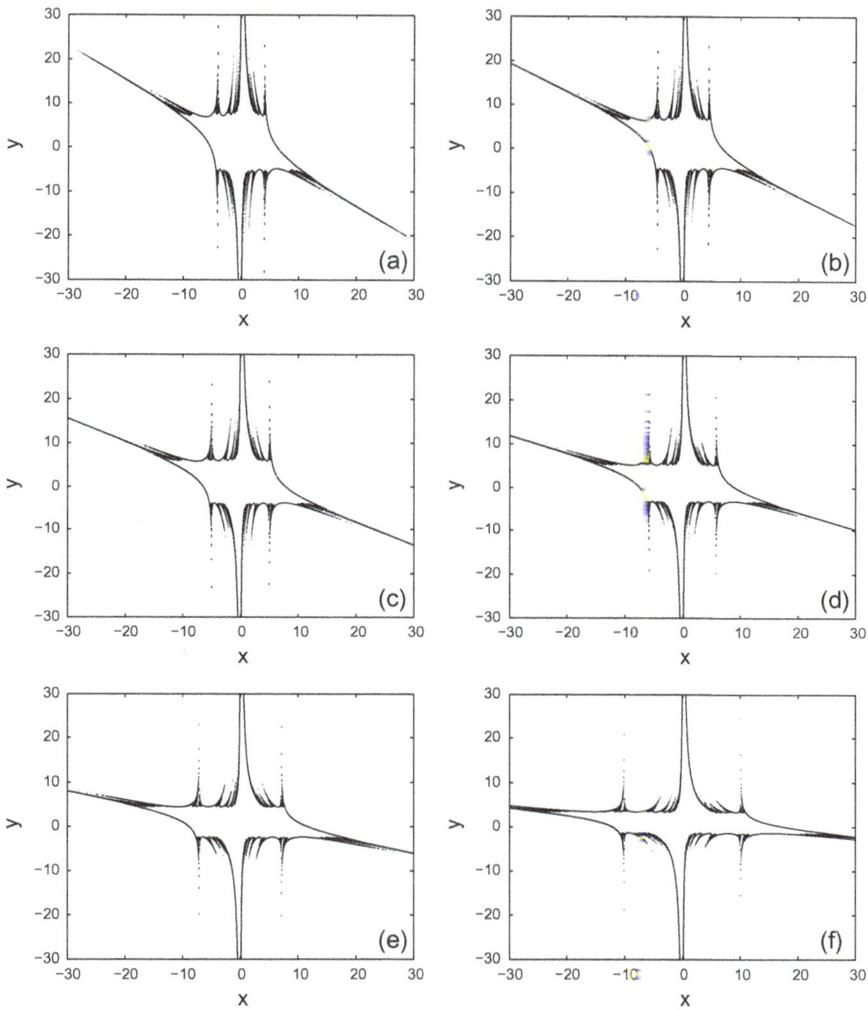


**Fig. 10.2** The changing of Julia sets of the controlled system (10.1.5) when **a** $k = 0.021$; **b** $k = 0.031$; **c** $k = 0.041$; **d** $k = 0.051$; **e** $k = 0.061$; **f** $k = 0.071$

$$\begin{cases} C_{n+1} = (1-c-d)C_n + a(1+bA_n)(1-C_n-D_n)C_n^2, \\ D_{n+1} = dC_n + (1-c-e)D_n + \upsilon, \\ A_{n+1} = A_n + uB_n, \\ B_{n+1} = B_n - uA_n. \end{cases} \tag{10.1.7}$$

In order to connect these two Julia sets together, for example, to make one Julia set change to be another, introduce four coupled items into Eq. (10.1.7). Take $E = (1-c-d)C_n + a(1+bA_n)(1-C_n-D_n)C_n^2$. Then, system (10.1.8) is as follows.

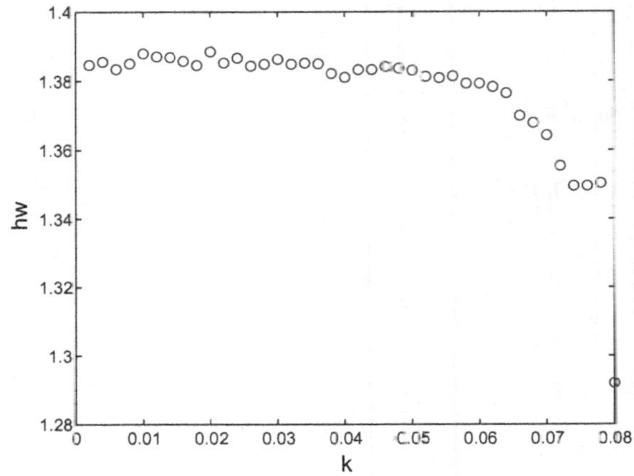


**Fig. 10.3** Box-counting dimensions of Julia sets with different control parameters $k$

**Table 10.1** The changes of box-counting dimension with changes of $k$

| $k$ | 0.002 | 0.004 | 0.006 | 0.008 | 0.01 | 0.012 | 0.014 | 0.016 | 0.018 | 0.02 |
|---|---|---|---|---|---|---|---|---|---|---|
| $hw$ | 1.3848 | 1.3857 | 1.3836 | 1.3852 | 1.3881 | 1.3372 | 1.387 | 1.3859 | 1.3847 | 1.3[illegible]86 |
| $k$ | 0.022 | 0.024 | 0.026 | 0.028 | 0.03 | 0.032 | 0.034 | 0.036 | 0.038 | 0.04 |
| $hw$ | 1.3854 | 1.3868 | 1.3845 | 1.385 | 1.3864 | 1.335 | 1.3853 | 1.3851 | 1.3823 | 1.3[illegible]12 |
| $k$ | 0.042 | 0.044 | 0.046 | 0.048 | 0.05 | 0.052 | 0.054 | 0.056 | 0.058 | 0.06 |
| $hw$ | 1.3834 | 1.3834 | 1.3843 | 1.3838 | 1.3832 | 1.3814 | 1.381 | 1.3815 | 1.3794 | 1.3[illegible]94 |
| $k$ | 0.062 | 0.064 | 0.066 | 0.068 | 0.07 | 0.072 | 0.074 | 0.076 | 0.078 | 0.08 |
| $hw$ | 1.3784 | 1.3766 | 1.3701 | 1.368 | 1.3644 | 1.3555 | 1.3497 | 1.3497 | 1.3505 | 1.2[illegible]21 |

$$
\begin{cases}
C_{n+1} = E + 0.6 + i + l(((1-c-d)x_n + a(1+bs_n)(1-x_n-y_n)x_n^2) \\
-(E+0.6+i)), \\
D_{n+1} = dC_n + (1-c-e)D_n + v + l((dx_n + (1-c-e)y_n + v) \\
-(dC_n + (1-c-e)D_n + v)), \\
A_{n+1} = A_n + uB_n + l((s_n + ut_n) - (A_n + uB_n)), \\
B_{n+1} = B_n - uA_n + l((t_n - us_n) - (B_n - uA_n)).
\end{cases}
\tag{10.1.8}
$$

**Theorem 10.1** *$|C_{n+1} - x_{n+1}| \to 0$, $|D_{n+1} - y_{n+1}| \to 0$, $|A_{n+1} - s_{n+1}| \to 0$ and $|B_{n+1} - t_{n+1}| \to 0$ when $n \to \infty$ and $l \to 1$, which signifies the synchronization of system (10.1.4) and system (10.1.8) is achieved.*

***Proof*** From Eqs. (10.1.4) and (10.1.8), we have

$$
\begin{aligned}
C_{n+1} - x_{n+1} = {} & (1-l)(1-c-d)(C_n - x_n) + a(1-l)((1+bA_n)(1-C_n-D_n)C_n^2 \\
& - (1+bs_n)(1-x_n-y_n)x_n^2) + (1-l)(0.6+i).
\end{aligned}
$$

Thus,

$$
\begin{aligned}
|C_{n+1}-x_{n+1}| &= |1-l||1-c-d||C_n-x_n| + |a||1-l|((1+|bA_n|)|1-C_n-D_n|C_n^2 \\
&\quad + (1+|bs_n|)|1-x_n-y_n|x_n^2) + |1-l||0.6+i|.
\end{aligned}
$$

Because of the bounded Julia set, to get the Julia set, just consider the iteration of points in the bounded area $D$. If there is a $m_0$ satisfying $h^{m_0}(z) \notin D$ so that $z \notin J$, the Julia set is unaltered. Moreover, there always exists a $T > 0$ to satisfy $|z| < T$ for arbitrary $z \in D$ due to the bounded region $D$.

Since the value of the system parameters $b$, $d$ and $u$ are bounded, there are $P > 0$, $Q > 0$, $W > 0$ and $R > 0$, so that $(1+|bA_n|)|1-C_n-D_n|C_n^2 + (1+|bs_n|)|1-x_n-y_n|x_n^2 < P$, $|d|(|C_n|+|x_n|) < Q$, $|u|(|B_n|+|t_n|) < W$, $|u|(|s_n|+|A_n|) < R$. Here, take the same iterative original values in system (10.1.4) and (10.1.8). Then

$$
\begin{aligned}
|C_{n+1}-x_{n+1}| &\le |1-l||1-c-d||C_n-x_n| + |1-l||0.6+i+|a|P| \\
&\le |1-l||1-c-d|(|1-l||1-c-d||C_{n-1}-x_{n-1}| + |1-l||0.6+i \\
&\quad + |a|P|) + |1-l||0.6+i+|a|P| = |1-l|^2|1-c-d|^2|C_{n-1}-x_{n-1}| \\
&\quad + |1-l||0.6+i+|a|P|(1+|1-l||1-c-d|) \le \cdots \\
&\le |1-l|^n|1-c-d|^n|C_1-x_1| + |1-l||0.6+i \\
&\quad + |a|P|(1+|1-l||1-c-d| + \cdots + |1-l|^{n-1}|1-c-d|^{n-1}) \\
&= |1-l|^n|1-c-d|^n|C_1-x_1| + |1-l||0.6+i \\
&\quad + |a|P|\frac{1-|1-l|^n|1-c-d|^n}{1-|1-l||1-c-d|}.
\end{aligned}
$$

Take $c+d<1$, then the limitation of the right side of the above equation is $|1-l||0.6+i+|a|P|\frac{1}{1-|1-l||1-c-d|}$ as $n \to \infty$. It is certain that $|C_{n+1}-x_{n+1}| \to 0, n \to \infty$ with $|1-l| \to 0$, that is $l \to 1$. Since

$$
D_{n+1}-y_{n+1} = (1-l)(1-c-e)(D_n-y_n) + d(1-l)(C_n-x_n),
$$

therefore

$$
\begin{aligned}
|D_{n+1}-y_{n+1}| &\le |1-l||1-c-e||D_n-y_n| + |d||1-l||C_n-x_n| \\
&\le |1-l||1-c-e||D_n-y_n| + |1-l||d|(|C_n|+|x_n|) \\
&\le |1-l||1-c-e||D_n-y_n| + |1-l|Q \\
&\le |1-l||1-c-e|(|1-l||1-c-e||D_{n-1}-y_{n-1}| \\
&\quad + |1-l|Q) + |1-l|Q \\
&\le \cdots \le |1-l|^n|1-c-e|^n|D_1-y_1| + |1-l|Q(1+|1-l||1-c-e| \\
&\quad + \cdots + |1-l|^{n-1}|1-c-e|^{n-1}) \\
&= |1-l|^n|1-c-e|^n|D_1-y_1| + |1-l|Q\frac{1-|1-l|^n|1-c-e|^n}{1-|1-l||1-c-e|}.
\end{aligned}
$$

Take $c + e < 1$, then the limitation of the right side of the above equation is $1 - l|Q\frac{1}{1-|1-l||1-c-e|}$ as $n \to \infty$. It is certain that $|D_{n+1} - y_{n+1}| \to 0, n \to \infty$ with $1 - l| \to 0$, that is $l \to 1$.

Since

$$A_{n+1} - s_{n+1} = (1-l)(A_n - s_n) + u(1-l)(B_n - t_n),$$

therefore

$$\begin{aligned}
|A_{n+1} - s_{n+1}| &\le |1-l||A_n - s_n| + |u||1-l||B_n - t_n| \\
&\le |1-l||A_n - s_n| + |1-l||u|(|B_n| + |t_n|) \\
&\le |1-l||A_n - s_n| + |1-l|W \\
&\le |1-l|(|1-l||A_{n-1} - s_{n-1}| + |1-l|W) + |1-l|W \le \cdots \\
&\le |1-l|^n|A_1 - s_1| + |1-l|W(1 + |1-l| + \cdots + |1-l|^{n-1}) \\
&= |1-l|^n|A_1 - s_1| + |1-l|W\frac{1-|1-l|^n}{1-|1-l|}.
\end{aligned}$$

Take $l < 1$, the limitation of the right side of the above equation is $|1-l|W\frac{1}{1-|1-l|}$ as $n \to \infty$. It is certain that $|A_{n+1} - s_{n+1}| \to 0, n \to \infty$ with $|1-l| \to 0$, that is $l \to 1$.

Since

$$B_{n+1} - t_{n+1} = (1-l)(B_n - t_n) - u(1-l)(s_n - A_n),$$

therefore

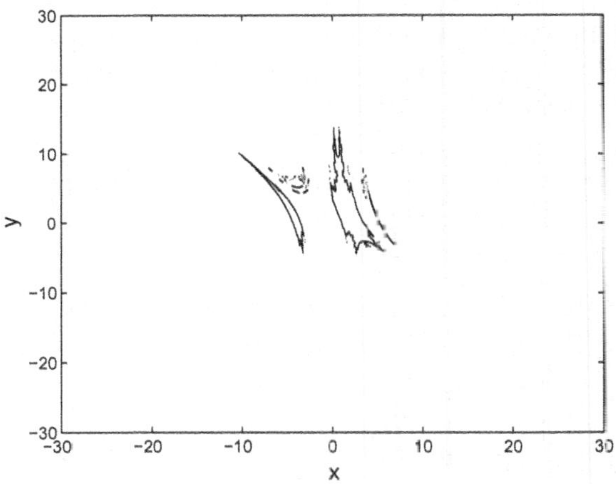


**Fig. 10.4** Original Julia sets of system (10.1.8) when $a = 0.08$, $b = 0.02$, $c = 0.005$, $d = 0.005$, $e = 0.05$, $v = 0.005$, $u = 0.001$ and $l = 0$

$$\begin{aligned}
|B_{n+1} - t_{n+1}| &\leq |1-l||B_n - t_n| + |u||1-l||s_n - A_n| \\
&\leq |1-l||B_n - t_n| + |1-l||u|(|s_n| + |A_n|) \\
&\leq |1-l||B_n - t_n| + |1-l|R \\
&\leq |1-l|(|1-l||B_{n-1} - t_{n-1}| + |1-l|R) + |1-l|R \leq \cdots \\
&\leq |1-l|^n|B_1 - t_1| + |1-l|R(1 + |1-l| + \cdots + |1-l|^{n-1}) \\
&= |1-l|^n|B_1 - t_1| + |1-l|R\frac{1-|1-l|^n}{1-|1-l|}.
\end{aligned}$$

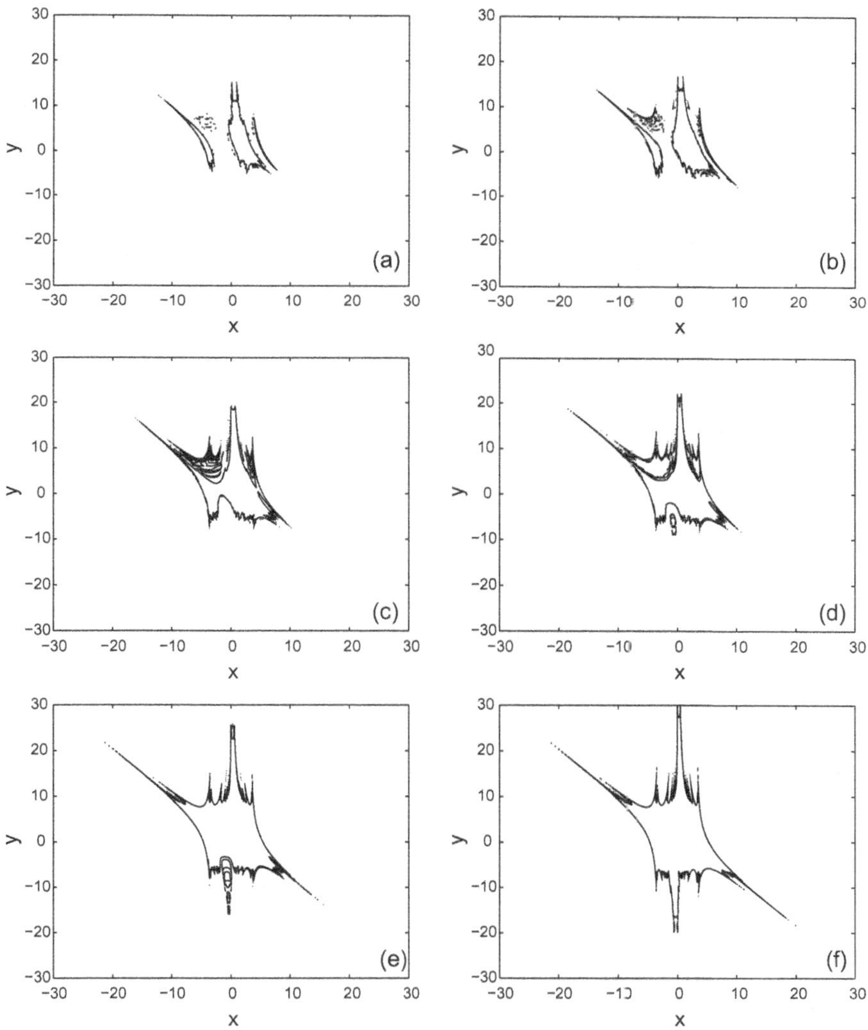

**Fig. 10.5** The changing process of Julia sets of the controlled system when **a** $l = 0.31$; **b** $l = 0.51$; **c** $l = 0.66$; **d** $l = 0.71$; **e** $l = 0.76$; **f** $l = 0.81$

Take $l < 1$, the limitation of the right side of the above equation is $|1 - l| R \frac{1}{1-|1-l|}$ as $n \to \infty$. It is certain that $|B_{n+1} - t_{n+1}| \to 0$, $n \to \infty$ with $|1 - l| \to 0$, that is $l \to 1$. Therefore, the synchronization of the trajectories of system (10.1.4) and (10.1.8) is achieved.

It obeys the definition of Julia set and the above theorems so that Julia sets of controlled system (10.1.8) are changing to be the Julia set of system (10.1.4). Therefore, the synchronization of Julia sets of systems (10.1.4) and (10.1.8) is acquired.

Regardless of the Julia sets are determined when the parameters are given, the changing process of Julia sets is exhibited by taking different coupled parameters $l$, which are approved by the following simulations. Figure 10.4 is the Julia set of system (10.1.8) when $a = 0.08$, $b = 0.02$, $c = 0.005$, $d = 0.005$, $e = 0.05$, $v = 0.005$, $\iota = 0.001$ and $l = 0$. Figure 10.5 is the changes of the Julia sets of system (10.1.8) with the changes of the different controlled parameters $l$. In Fig. 10.5, when $l = 0.8$, the $(f)$ curve is almost the same as Fig. 10.1. Obviously, Julia sets of system (10.1.8) are changing to be the Julia set of system (10.1.4) as indicated in Fig. 10.1.

## 10.2 Julia Sets and the Control of Discrete Fractional SIRS Models

### *10.2.1 Julia Set of Discrete Fractional Order SIRS Model*

In this section, the following SIRS model with constant inoculation rate and nonlinear infection rate [470] is considered.

$$\begin{cases} \dfrac{d\tau_1}{dt} = \beta\tau_1^2(1 - \tau_1 - \tau_2) - (d + \gamma)\tau_1, \\ \dfrac{d\tau_2}{dt} = p + \gamma\tau_1 - (d + \varepsilon)\tau_2, \end{cases} \tag{10.2.1}$$

where $\tau_1$ is the representative of infected person and its number, $\tau_2$ represents the persons who are rehabilitated from infection and its number, $i = 1, 2, \ldots$ Assume that during the epidemic period, the birth rate coefficient and the natural mortality coefficient of the population are equal to the constant $d$, $\beta$ denotes the effective contact rate coefficient, $p$ means the inoculation rate, $\gamma$ represents the removal rate coefficient, $\varepsilon$ denotes the immune loss rate coefficient. Moreover, regardless of the death due to illness, we suppose all the newborns are susceptible.

Practically, the seasonal outbreak of disease is a common phenomenon, such as influenza. The specific form of seasonal factors can be approximately expressed as a single sine or cosine function. Here, we consider the situation that the effective contact rate coefficient is affected by seasonal factors.

$$\begin{cases} \dfrac{d\tau_1}{dt} = \beta_0(1+\varphi\sin(\omega t))\tau_1^2(1-\tau_1-\tau_2)-(d+\gamma)\tau_1, \\ \dfrac{d\tau_2}{dt} = p+\gamma\tau_1-(d+\varepsilon)\tau_2, \end{cases} \tag{10.2.2}$$

where $\beta_0$ denotes the effective contact rate coefficient and $\varphi$ is the seasonal influence factor. Take $\tau_3 = \sin(\omega t)$, $\tau_4 = \cos(\omega t)$, then $\frac{d\tau_3}{dt} = \omega\tau_4$ and $\frac{d\tau_4}{dt} = -\omega\tau_3$. Thus system (10.2.3) is as follows.

$$\begin{cases} \dfrac{d\tau_1}{dt} = \beta_0(1+\varphi\tau_3)\tau_1^2(1-\tau_1-\tau_2)-(d+\gamma)\tau_1, \\ \dfrac{d\tau_2}{dt} = p+\gamma\tau_1-(d+\varepsilon)\tau_2, \\ \dfrac{d\tau_3}{dt} = \omega\tau_4, \\ \dfrac{d\tau_4}{dt} = -\omega\tau_3, \end{cases} \tag{10.2.3}$$

and the initial values of $\tau_3$ and $\tau_4$ are $\tau_3(0) = 0$ and $\tau_4(0) = 1$.

Julia set is a notion in fractal theory. The trajectory will be bounded if the initial values are taken in the interior of the Julia set. And the trajectory will be unbounded if the initial values are taken in the exterior of the Julia set. We hope we could obtain a preliminary judgment from initial values of the model. Since Julia set is obtained by iteration in fractal, we will give the discrete version of system (10.2.4). For the derivative in time we use

$$\dot{\tau_i} \to \frac{\tau_i(t+\Delta t)-\tau_i(t)}{\Delta t}, i = 1, 2, 3, 4.$$

Substitute these discretizations in the system, then the discrete version is

$$\begin{cases} x_{n+1} = (1-c-q)x_n + \eta(1+bs_n)(1-x_n-y_n)x_n^2, \\ y_{n+1} = qx_n + (1-c-e)y_n + z, \\ s_{n+1} = s_n + \delta t_n, \\ t_{n+1} = t_n - \delta s_n, \end{cases} \tag{10.2.4}$$

and $s_0 = 0, t_0 = 1, \eta, b, c, q, e, z, \delta$ are system parameters. Now, the ideas and methods of Julia set in fractal theory is introduced to the real dynamic system (10.2.4).

**Definition 10.2** Let $G(x, y, s, t) = ((1-c-d)x + \eta(1+bs)(1-x-y)x^2,$ $qx + (1-c-e)y + z,\ s + \delta t,\ t - \delta s)$. The set

$$K = \{(x, y) | \{G^n(x, y, 0, 1)\}_{n \in Z_+} \text{ is bounded}\}$$

is called the filled Julia set of the map $G$. The boundary of $K$ is called the Julia set of the map $G$, which is denoted by $J_G$, that is $J_G = \partial K$.

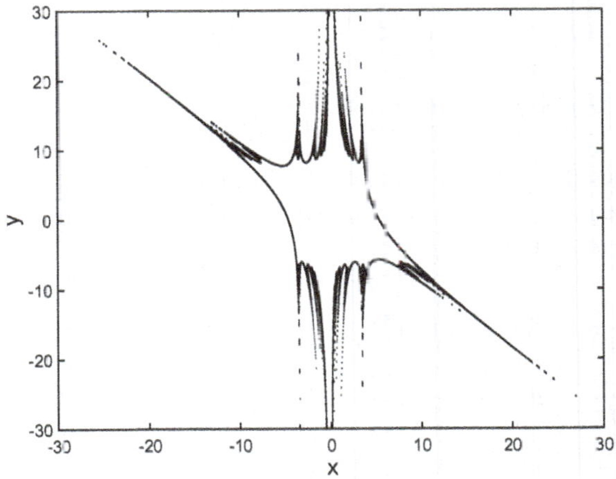


**Fig. 10.6** Original Julia sets of system (10.2.4) when $\eta = 0.08,\ b = 0.02,\ c = 0.005,\ q = 0.005,\ e = 0.05,\ z = 0.005,\ \delta = 0.001$

Consider the discrete version of the SIRS models with constant inoculation rate and nonlinear infection rate. The following four difference equations show the relationship between the number of the persons who are vulnerable to infection and the number of the infected persons. The parameters of model are taken to be $\eta = 0.08, b = 0.02, c = 0.005, q = 0.005, e = 0.05, z = 0.005, \delta = 0.001$. so the corresponding model is as follow and the homologous Julia set is shown in Fig. 10.6.

The irregular curve in Fig. 10.6 is the Julia set of the system (10.2.5) and represents the initial numbers $x$ of infected and the number $y$ of rehabilitated people. If these values are inside the Julia set, the numbers of the two clusters will be stable and bounded, and if the values of the initial numbers of two clusters are outside the Julia set, then at least one of the number of infected and rehabilitated people will approach to infinity.

$$\begin{cases} x_{n+1} = 0.99x_n + 0.08(1 + 0.02s_n)(1 - x_n - y_n)x_n^2, \\ y_{n+1} = 0.005x_n + 0.945y_n + 0.005, \\ s_{n+1} = s_n + 0.001r_n, \\ r_{n+1} = r_n - 0.001s_n. \end{cases} \tag{10.2.5}$$

Now we will give the fractional difference to the first and the second equations of system (10.2.4).

Subtract $x_n$ and $y_n$ from both sides of the first and second equations, then the system can be represented to the following one.

$$\begin{cases} \nabla x_{n+1} = -(c+q)x_n + \eta(1+bs_n)(1-x_n-y_n)x_n^2, \\ \nabla y_{n+1} = qx_n - (c+e)y_n + z, \\ s_{n+1} = s_n + \delta r_n, \\ r_{n+1} = r_n - \delta s_n. \end{cases} \tag{10.2.6}$$

Rewrite the equation as a fractional one from the discrete fractional calculus, and the momentums $x(n)$ and $y(n)$ are taken into account.

$$\begin{cases} {}_c\Delta_a^{\mu} x(t) = -(c+q)x(t+\mu-1) \\ \qquad +\eta(1+bs(t+\mu-1))(1-x(t+\mu-1)-y(r+\mu-1))x(r+\mu-1)^2, \\ {}_c\Delta_a^{\mu} y(t) = qx(t+\mu-1)-(c+e)y(t+\mu-1)+z, \\ s(n) = s(n-1)+\delta r(n-1), \\ r(n) = r(n-1)-\delta s(n-1). \end{cases} \tag{10.2.7}$$

From Theorem 10.2 in the appendix, the discrete integral form for $0 < \mu \le 1$ is obtained.

$$\begin{cases} x(\mathrm{t}) = x(a) + \frac{1}{\Gamma(\mu)} \sum\limits_{p=a+1-\mu}^{t-\mu} \frac{\Gamma(t-p)}{\Gamma(t-p+1-\mu)}(-(c+q)x(p+\mu-1) \\ \qquad +\eta(1+bs(p+\mu-1))(1-x(p+\mu-1)-y(p+\mu-1))x(p+\mu-1)^2), \\ y(t) = y(a) + \frac{1}{\Gamma(\mu)} \sum\limits_{p=a+1-\mu}^{t-\mu} \frac{\Gamma(t-p)}{\Gamma(t-p+1-\mu)}(qx(p+\mu-1)-(c+e)y(p+\mu-1)+z), \\ s(n) = s(n-1)+\delta r(n-1), \\ r(n) = r(n-1)-\delta s(n-1). \end{cases} \tag{10.2.8}$$

Consequently, the numerical equations can be accurately proposed.

$$\begin{cases} x(n) = x(a) + \frac{1}{\Gamma(\mu)} \sum\limits_{j=a+1}^{n} \frac{\Gamma(n-j+\mu)}{\Gamma(n-j+1)}(-(c+q)x(j-1) \\ \qquad +\eta(1+bs(j-1))(1-x(j-1)-y(j-1))x(j-1)^2), \\ y(n) = y(a) + \frac{1}{\Gamma(\mu)} \sum\limits_{j=a+1}^{n} \frac{\Gamma(n-j+\mu)}{\Gamma(n-j+1)}(qx(j-1)-(c+e)y(j-1)+z), \\ s(n) = s(n-1)+\delta r(n-1), \\ r(n) = r(n-1)-\delta s(n-1). \end{cases} \tag{10.2.9}$$

In particular, if $a = 0$ and the summation starts with $j = 1$, the model can be written as

$$\begin{cases} x(n) = x(0) + \frac{1}{\Gamma(\mu)} \sum\limits_{j=1}^{n} \frac{\Gamma(n-j+\mu)}{\Gamma(n-j+1)} (-(c+q)x(j-1) \\ \quad +\eta(1+bs(j-1))(1-x(j-1)-y(j-1))x(j-1)^2), \\ y(n) = y(0) + \frac{1}{\Gamma(\mu)} \sum\limits_{j=1}^{n} \frac{\Gamma(n-j+\mu)}{\Gamma(n-j+1)} (qx(j-1) - (c+e)y(j-1) + z), \\ s(n) = s(n-1) + \delta r(n-1), \\ r(n) = r(n-1) - \delta s(n-1). \end{cases} \tag{10.2.10}$$

Obviously, it is inconvenient to simulate the system (10.2.10) because it has 4 dimensions. However, the initial values of the third and the forth variables are fixed, so the Julia set of system (10.2.10) can be depicted on the $x - y$ plane.

**Definition 10.3** Let

$$F(x, y, s, r) = x + \frac{1}{\Gamma(\mu)} \sum_{j=1}^{n} \frac{\Gamma(n-j+\mu)}{\Gamma(n-j+1)} (-(c+q)x + \eta(1+bs)(1-x-y)x^2,$$

$$G(x, y, s, r) = y + \frac{1}{\Gamma(\mu)} \sum_{j=1}^{n} \frac{\Gamma(n-j+\mu)}{\Gamma(n-j+1)} (qx - (c+e)y + z),$$

$$S(x, y, s, r) = s + \delta r,$$

$$R(x, y, s, r) = r - \delta s,$$

and

$$H(x, y, s, r) = (F(x, y, s, r), G(x, y, s, r), S(x, y, s, r), R(x, y, s, r)).$$

The set

$$D = \{(x, y, 0, 1) | \{H^n(x, y, 0, 1)\}_{n=1}^{\infty} \text{ remains bounded}\}.$$

is called the filled Julia set corresponding to the map $H(x, y, s, r)$. And the boundary of $D$ is called the Julia set of the map $H(x, y, s, r)$, which is denoted by $J_H$, i.e, $J_H = \partial D$.

The fractional orders of the model are taken to be $(a)$ $\mu = 0.001$; $(b)$ $\mu = 0.5$; $(c)$ $\mu = 0.8$; $(d)$ $\mu = 1$, then the corresponding Julia set is shown in Fig. 10.7.

As can be seen from the Fig. 10.7, Julia sets are closely related to the parameter $\mu$, and the image of Julia sets change with the selection of different fractional orders in the system. As $\mu$ decreases, the area inside the image gradually diminishes.

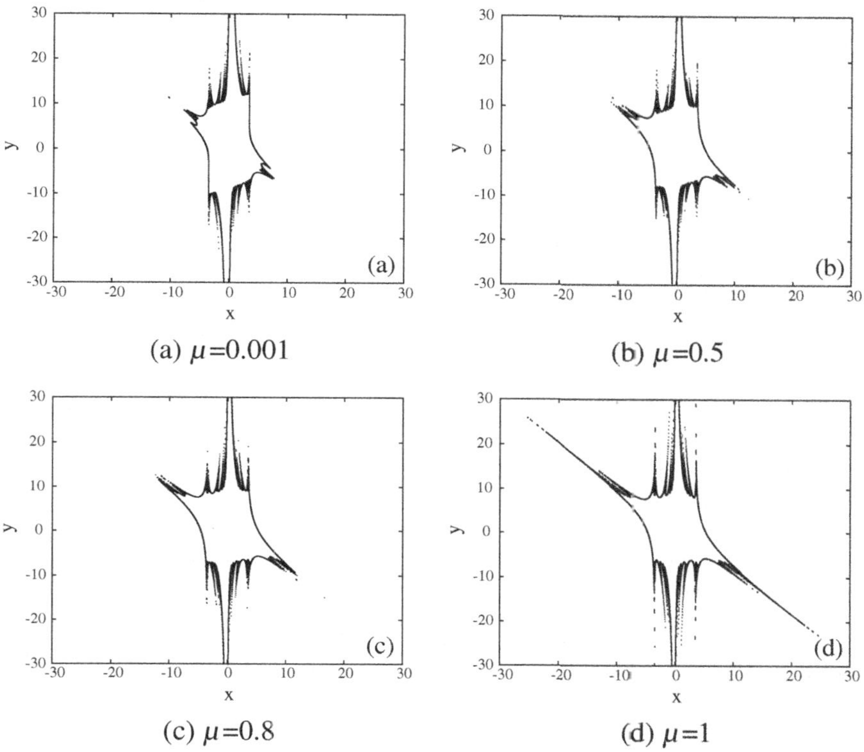


**Fig. 10.7** Julia sets of the controlled system (10.2.10) when **a** $\mu = 0.001$; **b** $\mu = 0.5$; **c** $\mu = 0.8$; **d** $\mu = 1$

### *10.2.2 Control of Julia Sets of the Discrete Fractional Order SIRS Model*

Considering relevant literatures [473, 474] about different ways of adding control items, the following three ways of adding control items under fractional order iteration are proposed, respectively, to achieve the control of Julia set. Controllers are added into different positions of the system as a whole, a part, and a product factor. The design methods are not uniform.

Since infinite terms are involved in the sum, we apply continuous iteration to get the latter item $x(n+1)$ through the previous one $x(n)$. Julia sets vary with fractional orders, and assume $\mu = 0.001, 0.8, 0.9, 0.99$ to carry out the following study. And a function $g(n) = g(n-1) + \alpha(y(n) - x(n))$ that ties the iteration of $x(n)$ and $y(n)$ closely together was set up. Besides, over the iteration, $x(n)$ and $y(n)$ always change. Keep other parameters in the system unaltered, and the initial value of $g(n)$ is taken as $g(0) = 1$. Furthermore, the evolution of Julia set is exhibited in the diagram when different parameters $\alpha$ and disparate fractional orders $\mu$ are taken.

### (I) Controller Is Added into the System as a Product Factor

In this section, we will design the controller for discrete fractional order SIRS model as below, and controller is added into the system as a product factor.

$$\begin{cases} x(n) = x(0) + \frac{g(n-1)}{\Gamma(\mu)} \sum\limits_{j=1}^{n} \frac{\Gamma(n-j+\mu)}{\Gamma(n-j+1)}(-(c+q)x(j-1) \\ \quad +\eta(1+bs(j-1))(1-x(j-1)-y(j-1))x(j-1)^2), \\ y(n) = y(0) + \frac{1}{\Gamma(\mu)} \sum\limits_{j=1}^{n} \frac{\Gamma(n-j+\mu)}{\Gamma(n-j+1)}(qx(j-1)-(c+e)y(j-1)+z), \\ s(n) = s(n-1) + \delta r(n-1), \\ r(n) = r(n-1) - \delta s(n-1), \\ g(n) = g(n-1) + \alpha(y(n)-x(n)). \end{cases} \tag{10.211}$$

The control parameters are taken as $(a)\ \alpha = 0.01$; $(b)\ \alpha = 0.02$; $(c)\ \alpha = 0.03$; $(d)\ \alpha = 0.05$; $(e)\ \alpha = 0.1$; $(f)\ \alpha = 0.5$ and the corresponding Julia sets for different values of $\mu$ are shown in Fig. 10.8.

As demonstrated in the Fig. 10.8, the discrete fractional order SIRS system is very sensitive to the control parameter $\alpha$. Only slight adjustment of control parameters is needed to control the system with this method. As the control coefficient $\alpha$ gradually increases, the image keeps shrinking.

### (II) Controller Is Added into the System as a Whole

In this part, the controller is added to the outside of the summation sign. This method is utilized to construct the following system and control of Julia set is implemented.

$$\begin{cases} x(n) = x(0) + \frac{1}{\Gamma(\mu)} \sum\limits_{j=1}^{n} \frac{\Gamma(n-j+\mu)}{\Gamma(n-j+1)}(-(c+q)x(j-1) \\ \quad +\eta(1+bs(j-1))(1-x(j-1)-y(j-1))x(j-1)^2) + g(n-1), \\ y(n) = y(0) + \frac{1}{\Gamma(\mu)} \sum\limits_{j=1}^{n} \frac{\Gamma(n-j+\mu)}{\Gamma(n-j+1)}(qx(j-1)-(c+e)y(j-1)+z) + g(n-1), \\ s(n) = s(n-1) + \delta r(n-1), \\ r(n) = r(n-1) - \delta s(n-1), \\ g(n) = g(n-1) + \alpha(y(n)-x(n)). \end{cases} \tag{10.212}$$

The control parameters are taken as $(a)\ \alpha = 0.1$; $(b)\ \alpha = 0.2$; $(c)\ \alpha = 0.3$; $(d)\ \alpha = 0.5$; $(e)\ \alpha = 0.7$; $(f)\ \alpha = 1$ and the corresponding Julia sets for different values of $\mu$ are shown in Fig. 10.9.

Although the Julia sets are established when the parameters are given, the changing process of Julia sets can be exhibited by choosing different parameters $\alpha$. Compared with the span of $\alpha$ selected in Sect. (I), which is from 0.01 to 0.5, the span of $\alpha$ in Sect. (II) is from 0.1 to 1. When $\alpha$ changes, the image changes obviously.

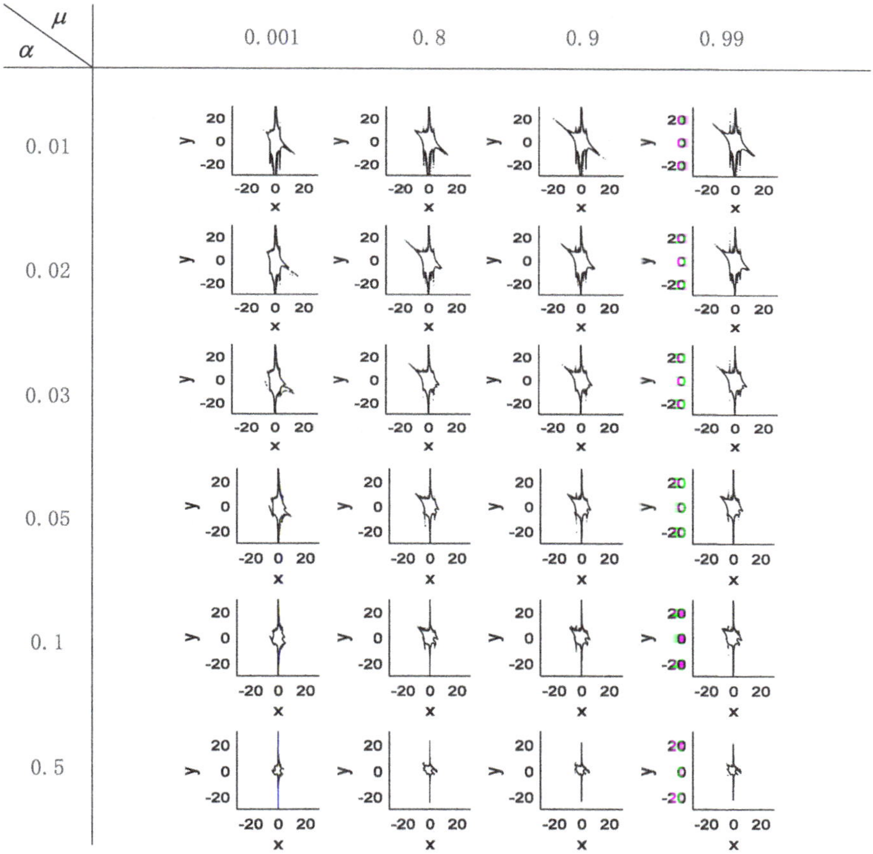


**Fig. 10.8** The changing of Julia sets of the controlled system (10.2.11) when **a** $\alpha = 0.01$; **b** $\alpha = 0.02$; **c** $\alpha = 0.03$; **d** $\alpha = 0.05$; **e** $\alpha = 0.1$; **f** $\alpha = 0.5$ and $\mu = 0.001, 0.8, 0.9, 0.99$

## (III) Controller Is Added into the System as a Part

In the perspective of this approach, the controller is added to the interior of the summation sign to implement control over Julia set.

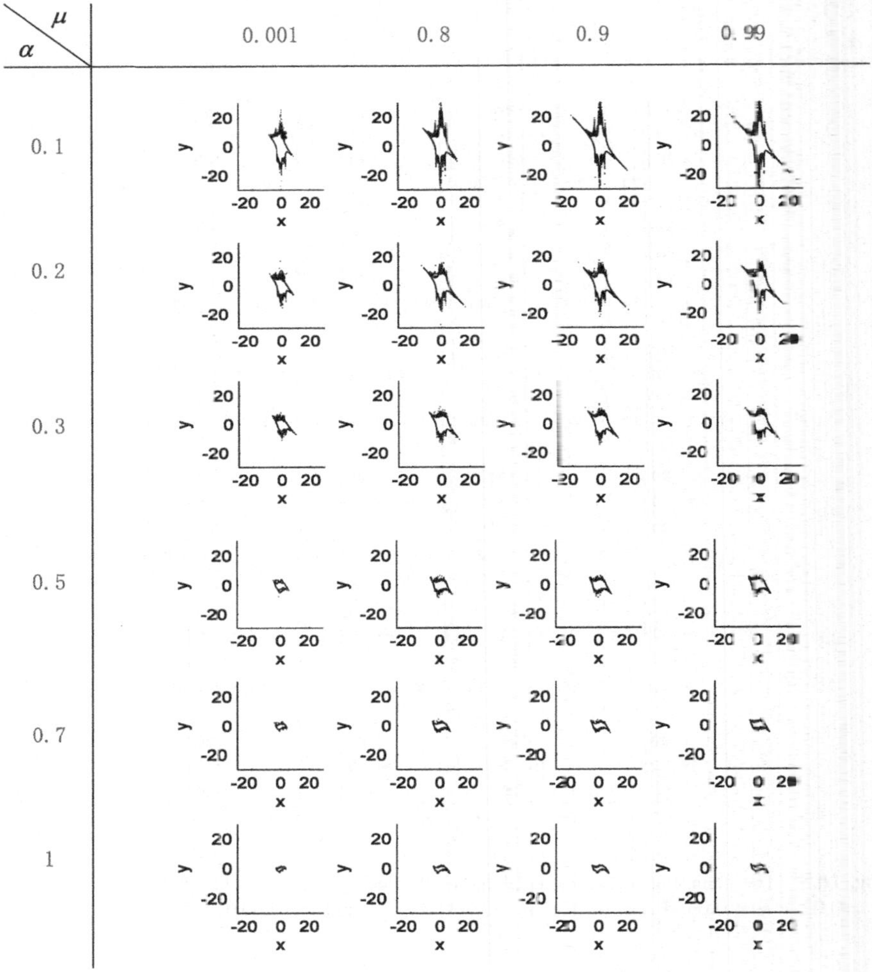


**Fig. 10.9** The changing of Julia sets of the controlled system (10.2.12) when **a** $\alpha = 0.1$; **b** $\alpha = 0.2$; **c** $\alpha = 0.3$; **d** $\alpha = 0.5$; **e** $\alpha = 0.7$; **f** $\alpha = 1$ and $\mu = 0.001,\ 0.8,\ 0.9,\ 0.99$

$$\begin{cases} x(n) = x(0) + \frac{1}{\Gamma(\mu)} \sum\limits_{j=1}^{n} \frac{\Gamma(n-j+\mu)}{\Gamma(n-j+1)} (-(c+q)x(j-1) \\ \qquad +\eta(1+bs(j-1))(1-x(j-1)-y(j-1))x(j-1)^2 + g(j-1)), \\ y(n) = y(0) + \frac{1}{\Gamma(\mu)} \sum\limits_{j=1}^{n} \frac{\Gamma(n-j+\mu)}{\Gamma(n-j+1)} (qx(j-1) - (c+e)y(j-1) + z + g(j-1)) \\ s(n) = s(n-1) + \delta r(n-1), \\ r(n) = r(n-1) - \delta s(n-1), \\ g(j) = g(j-1) + \alpha(y(j-1) - x(j-1)). \end{cases} \tag{10.2.13}$$

The control parameters are taken as $(a)\,\alpha = 0.01$; $(b)\,\alpha = 0.05$; $(c)\,\alpha = 0.1$; $(d)\,\alpha = 0.2$; $(e)\,\alpha = 0.3$; $(f)\,\alpha = 0.5$ and the corresponding Julia sets for different values of $\mu$ are shown in Fig. 10.10.

Small changes of coefficient $\alpha$ can have a great impact on control results. Compared with the figures above, Julia set variation diagrams are obtained by increasing the control parameter $\alpha$ and fractional orders $\mu$. The transformation of the position of control items will affect the Julia set control. Primarily, comparison of methods of Sects. (III) and (II), controller is severally added into the outside of the system as a whole in Sect. (II) and another is added into the interior of the summation sign as a part in Sect. (III). We can see that graphically if the adding way is as a whole, image change is particularly evident when the control parameter changes dramatically. Nevertheless in Sect. (III), the images have significant change when the values of the control parameter are in a small fluctuations. So adding control items in the interior of the summation sign can greatly change numerical values, thus Julia set is observably controlled. In addition, the second method (II) can be selected if we want to slowly change the system. As for method (I), the control is added to the system as a product factor, which the control parameter $\alpha$ has the greatest impact on the system. Therefore, the image of Julia sets will change as the control parameter $\alpha$ changes slightly.

### 10.2.3 Basis for Discrete Fractional Calculus

The following are some necessary definitions of the discrete fractional calculus.

**Definition 10.4** ([475]) Let $u : Na \to R$ and $0 < v$ be given. Then the fractional sum of $v$ order is defined by

$$\Delta_a^{-v} u(t) = \frac{1}{\Gamma(v)} \sum_{s=a}^{t-v} (t - \sigma(s))^{v-1} u(s),\, t \in N_{a+v}, \tag{10.2.14}$$

where $a$ is the starting point, $\sigma(s) = s + 1$ and $t^{(v)}$ is the falling function defined as

$$t^{(v)} = \frac{\Gamma(t+1)}{\Gamma(t+1-v)}. \tag{10.2.15}$$

**Definition 10.5** ([476]) For $0 < v, v \notin N$ and $u(t)$ defined on $N_a$, the Caputo-like delta difference is defined by

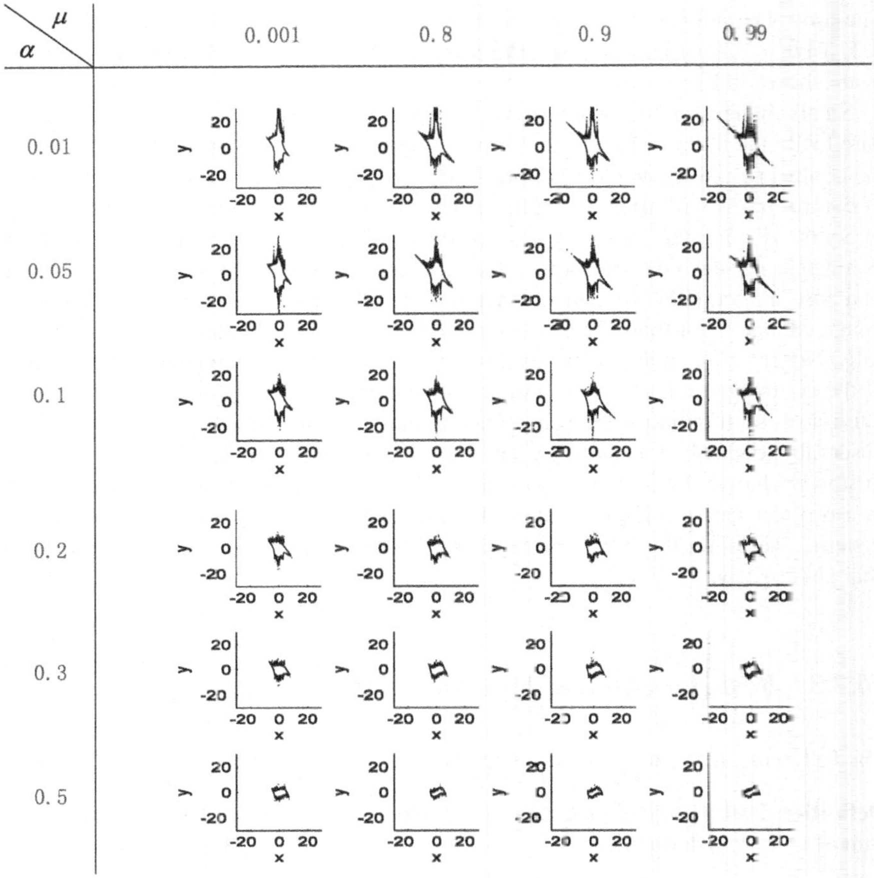


**Fig. 10.10** The changing of Julia sets of the controlled system (10.2.13) when **a** $\alpha = 0.01$; **b** $\alpha = 0.05$; **c** $\alpha = 0.1$; **d** $\alpha = 0.2$; **e** $\alpha = 0.3$; **f** $\alpha = 0.5$ and $\mu = 0.001,\ 0.8,\ 0.9,\ 0.99$

$$
{}^{C}\Delta_{a}^{\nu}u(t) = \Delta_{a}^{-(m-\nu)}\Delta^{m}u(t) = \frac{1}{\Gamma(m-\nu)}\sum_{s=a}^{t-(m-\nu)}(t-\sigma(s))^{(m-\nu-1)}\Delta_{s}^{m}u(s), \tag{10.2.16}
$$

where $t \in N_{a+m-\nu}$, $m = [\nu] + 1$.

**Theorem 10.2** ([477]) *For the delta fractional difference equation*

$$
{}^{C}\Delta_{a}^{\nu}u(t) = f(t+\nu-1, u(t+\nu-1)),\ \Delta^{k}u(a) = u_{k},\ m = [\nu]+1,\ k = 0, \ldots, m-1, \tag{10.2.17}
$$

*the equivalent discrete integral equation can be obtained as*

$$u(t) = u_0(t) + \frac{1}{\Gamma(\nu)} \sum_{s=a+m-\nu}^{t-\nu} (t-\sigma(s))^{\nu-1} \times f(s+\nu-1, u(s+\nu-1)), t \in N_{a+m}, \tag{10.2.18}$$

*where the initial iteration $u_0(t)$ reads*

$$u_0(t) = \sum_{k=0}^{m-1} \frac{(t-a)^{(k)}}{k!} \Delta^k u(a). \tag{10.2.19}$$

In [477], the authors have already debated existence results for the nonlinear fractional difference equation.

The domains of (10.2.16) and (10.2.18) are disparate. The former is $N_{a+m-p}$ and the latter is $N_{a+m}$. The function $u(t)$ is defined on the isolated time scale $N_a$. From this viewpoint, it is commendable to use the discrete fractional calculus to initialize fractional difference equations.

## 10.3 Fractal Control and Synchronization of the Discrete Fractional SIRS Model

### *10.3.1 Julia Set of Discrete Fractional Order SIRS Model and Fixed Point Control*

Consider the system (10.2.2) of SIRS model in [478] which the effective contact rate coefficient is affected by seasonal factors.

$$\begin{cases} \dfrac{d\tau_1}{dt} = \beta_0(1+\varphi\sin(\omega t))\tau_1^2(1-\tau_1-\tau_2)-(d+v)\tau_1, \\ \dfrac{d\tau_2}{dt} = p+v\tau_1-(d+\varepsilon)\tau_2, \end{cases} \tag{10.3.1}$$

where $\tau_1$ represents the number of persons susceptible to infection, $\tau_2$ expresses the number of infected person. Suppose the birth and natural death coefficients of the population are equivalent to the constant $d$ during epidemics, $p$ signifies the inoculation rate, the removal rate coefficient is shown by $v$, the immune loss rate coefficient is denoted by $\varepsilon$, $\beta_0$ means the effective contact rate coefficient, the seasonal influence factor is $\varphi$.

Take $\tau_3 = \sin(\omega t)$, $\tau_4 = \cos(\omega t)$, thus $\frac{d\tau_3}{dt} = \omega\tau_4$ and $\frac{d\tau_4}{dt} = -\omega\tau_3$. The following system (10.3.2) is obtained.

$$\begin{cases} \dfrac{d\tau_1}{dt} = \beta_0(1+\varphi\tau_3)\tau_1^2(1-\tau_1-\tau_2)-(d+v)\tau_1, \\ \dfrac{d\tau_2}{dt} = p+v\tau_1-(d+\varepsilon)\tau_2, \\ \dfrac{d\tau_3}{dt} = \omega\tau_4, \\ \dfrac{d\tau_4}{dt} = -\omega\tau_3, \end{cases} \tag{10.3.2}$$

besides, the initial values of $\tau_3$ and $\tau_4$ are $\tau_3(0) = 0$ and $\tau_4(0) = 1$. According to the method and discretization process of literature [478], the discrete version of system (10.3.2) is acquired.

$$\begin{cases} x_{n+1} = (1-c-d)x_n + a(1+bs_n)(1-x_n-y_n)x_n^2, \\ y_{n+1} = dx_n + (1-c-e)y_n + \upsilon, \\ s_{n+1} = s_n + ut_n, \\ t_{n+1} = t_n - us_n, \end{cases} \tag{10.3.3}$$

where $s_0 = 0$, $t_0 = 1$, and $a, b, c, d, e, \upsilon, u$ are system parameters. Now, refer to the method of converting integer order equation into fractional order equation in [478], subtract $x_n$, $y_n$, $s_n$ and $r_n$ from both sides of the 4 equations of system (10.3.3), rewrite the equation as a fractional one from the perspective of discrete fractional calculus, take the momentums $x(n)$, $y(n)$, $s(n)$ and $r(n)$ into account. Therefore, the numerical equations can be accurately proposed to the following one, the fractional difference about 4 equations is introduced.

$$\begin{cases} x(n) = x(0) + \dfrac{1}{\Gamma(\mu)}\displaystyle\sum_{j=1}^{n}\dfrac{\Gamma(n-j+\mu)}{\Gamma(n-j+1)}(-(c+q)x(j-1) \\ \quad + \eta(1+bs(j-1))(1-x(j-1)-y(j-1))x(j-1)^2), \\ y(n) = y(0) + \dfrac{1}{\Gamma(\mu)}\displaystyle\sum_{j=1}^{n}\dfrac{\Gamma(n-j+\mu)}{\Gamma(n-j+1)}(qx(j-1)-(c+e)y(j-1)+z), \\ s(n) = s(0) + \dfrac{1}{\Gamma(\mu)}\displaystyle\sum_{j=1}^{n}\dfrac{\Gamma(n-j+\mu)}{\Gamma(n-j+1)}(\delta r(j-1)), \\ r(n) = r(0) + \dfrac{1}{\Gamma(\mu)}\displaystyle\sum_{j=1}^{n}\dfrac{\Gamma(n-j+\mu)}{\Gamma(n-j+1)}(-\delta s(j-1)). \end{cases} \tag{10.3.4}$$

Next, the thought of Julia set in fractal theory is used to study fractional order dynamic systems (10.3.4).

*Definition* Let

$$
\begin{cases}
F(x,y,s,r) = x + \dfrac{1}{\Gamma(\mu)} \displaystyle\sum_{j=1}^{n} \dfrac{\Gamma(n-j+\mu)}{\Gamma(n-j+1)} (-(c+q)x + \eta(1+bs)(1-x-y)x^2), \\
G(x,y,s,r) = y + \dfrac{1}{\Gamma(\mu)} \displaystyle\sum_{j=1}^{n} \dfrac{\Gamma(n-j+\mu)}{\Gamma(n-j+1)} (qx - (c+e)y + z), \\
S(x,y,s,r) = s + \dfrac{1}{\Gamma(\mu)} \displaystyle\sum_{j=1}^{n} \dfrac{\Gamma(n-j+\mu)}{\Gamma(n-j+1)} (\delta r), \\
R(x,y,s,r) = r + \dfrac{1}{\Gamma(\mu)} \displaystyle\sum_{j=1}^{n} \dfrac{\Gamma(n-j+\mu)}{\Gamma(n-j+1)} (-\delta s),
\end{cases}
$$

and

$$H(x,y,s,r) = (F(x,y,s,r), G(x,y,s,r), S(x,y,s,r), R(x,y,s,r)).$$

The set

$$D = \{(x,y,0,1) | \{H^n(x,y,0,1)\}_{n=1}^{\infty} \text{ remains bounded}\}$$

is referred to the filled Julia set which is correspondent with the map $H(x, y, 0, 1)$. Besides, the boundary of $D$ is the Julia set of the map $H(x, y, 0, 1)$, and it is indicated by $J_H$, which means, $J_H = \partial D$.

It is obvious that it is difficult to simulate the system (10.3.4) since it is four dimensions. Fortunately, the equation can be viewed as a mapping on the $x - y$ plane because the initial values of the third and fourth variables are already settled.

The fractional order of the model are taken to be $\mu = 0.8$ and the system parameters are taken to be $a = 0.08, b = 0.02, c = 0.005, d = 0.005, e = 0.05$, $\upsilon = 0.005, u = 0.001$, Fig. 10.11 shows the original correlative Julia set of system (10.3.4).

Now, the fixed point is utilized to consider the stability of system (10.3.4). Let the fixed point of the equation be $(x^*, y^*, s^*, t^*)$, so the following equation is presented.

$$
\begin{cases}
x^* = x^* + \frac{1}{\Gamma(\mu)} \sum_{j=1}^{n} \frac{\Gamma(n-j+\mu)}{\Gamma(n-j+1)} (-(c+q)x^* + \eta(1+bs^*)(1-x^*-y^*)x^{*2}), \\
y^* = y^* + \frac{1}{\Gamma(\mu)} \sum_{j=1}^{n} \frac{\Gamma(n-j+\mu)}{\Gamma(n-j+1)} (qx^* - (c+e)y^* + z), \\
s^* = s^* + \frac{\delta}{\Gamma(\mu)} \sum_{j=1}^{n} \frac{\Gamma(n-j+\mu)}{\Gamma(n-j+1)} r^*, \\
r^* = r^* - \frac{\delta}{\Gamma(\mu)} \sum_{j=1}^{n} \frac{\Gamma(n-j+\mu)}{\Gamma(n-j+1)} s^*.
\end{cases}
\tag{10.3.5}
$$

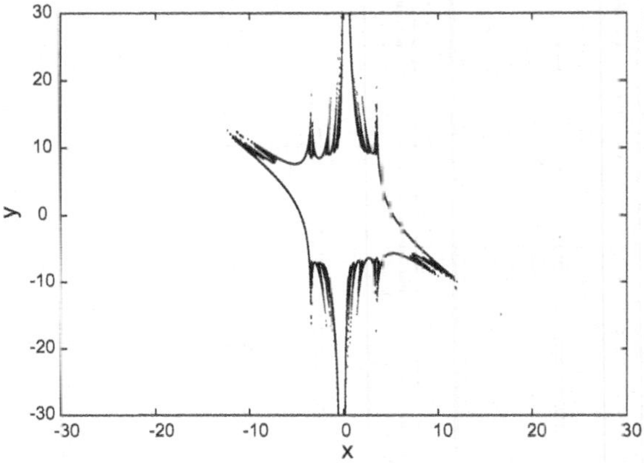


**Fig. 10.11** Original Julia sets of system (10.3.4) when $\mu = 0.8$, $c = 0.08$, $? = ?.02$, $c = 0.005$, $d = 0.005$, $e = 0.05$, $v = 0.005$, $u = 0.001$

From [479], for $\alpha \in R$ and $n \in N_0$ one has $\sum_{j=0}^{n} \binom{j-\alpha-1}{j} = \binom{n-\alpha}{n}$. Besides, this equation is equivalent to $\frac{1}{\Gamma(\mu)} \sum_{j=0}^{n} \frac{\Gamma(n-j+\mu)}{\Gamma(n-j+1)} = \binom{n+\mu}{n} = \frac{\Gamma(n+\mu+1)}{\Gamma(n-1)\Gamma(\mu+1)}$ on the condition of $-\alpha = \mu$, because $\binom{a}{b} := \frac{\Gamma(a+1)}{\Gamma(b+1)\Gamma(a-b+1)}$. Also, refer to [430], this most striking conclusion can be reached. So, the equation below is derived.

$$\begin{cases} x^* = x^* + \frac{\Gamma(n+\mu)}{\Gamma(n)\Gamma(\mu+1)}(-(c+q)x^* + \eta(1+bs^*)(1-x^*-y^*)x^{*2}), \\ y^* = y^* + \frac{\Gamma(n+\mu)}{\Gamma(n)\Gamma(\mu+1)}(qx^* - (c+e)y^* + z), \\ s^* = s^* + \delta\frac{\Gamma(n+\mu)}{\Gamma(n)\Gamma(\mu+1)}r^*, \\ r^* = r^* - \delta\frac{\Gamma(n+\mu)}{\Gamma(n)\Gamma(\mu+1)}s^*. \end{cases}$$

For the above equation, if $\delta = 0$, $s^*$ and $r^*$ can be arbitrary number, and if $\delta \neq 0$, thus $s^* = 0, r^* = 0$. There is no significance of studying in the case of $\delta = 0$, so just consider the fixed point that both $s^*$ and $t^*$ are zero. The fixed point $(x^*, y^*)$ is as follows:

(1) If $x^* = 0$ and $y^* \neq 0$, so $(c+e)y^* = \upsilon$ and $(x^*, y^*) = (0, \frac{\upsilon}{c+e})$.
(2) If $x^* \neq 0$ and $y^* = 0$, so $dx^* + \upsilon = 0$, $(c+d)x^* + a(x^*-1)x^{*2} = 0$, $\upsilon d^3 + c\upsilon d^2 + a\upsilon^2 d + a\upsilon^3 = 0$ and $(x^*, y^*) = (\frac{-v}{d}, 0)$.
(3) If $x^* \neq 0$ and $y^* \neq 0$, so

$$x^* = \frac{(c+e-v)}{2(c+e+d)} - \frac{\sqrt{a^2(v-c-e)^2-4a(c+e+d)(c+d)(c+e)}}{2a(c+e+d)},\ y^* = \frac{dx^*+v}{c+e}.$$

In general, cases (1) and (2) have no actual significance. And the common phenomenon is that people who are infected co-exist with people have rehabilitated from the infection when an outbreak occurs. Furthermore, they can co-exist for a long time and neither of them will suddenly go to zero. Therefore, we only consider case (3). Take the same system parameters, thus the fixed point of the system is $(x^*, y^*, s^*, t^*) = (0.6596, 0.1509, 0, 0)$. There are three ways to lead in control items, see [478]. Here, in order to control Julia set of the model (10.3.4), the control items by use of the fixed points are introduced into different parts of the model as a whole and a part. So the following controlled system (10.3.6) and system (10.3.7) are obtained.

Firstly, control items with the fixed point $l(x^2(j) - x^{*2})(y(j) - y^*)$, $l(y(j) - y^*)$, $l(s(j) - s^*)$ and $l(r(j) - r^*)$ are introduced into the model as a whole, that is to say that the controllers are added to the inside of the summation sign to control the Julia set.

$$\begin{cases} x(n) = x(0) + \dfrac{1}{\Gamma(\mu)} \displaystyle\sum_{j=1}^{n} \dfrac{\Gamma(n-j+\mu)}{\Gamma(n-j+1)}(-(c+q)x(j-1) \\ \quad + \eta(1+bs(j-1))(1-x(j-1)-y(j-1))x(j-1)^2 + l(x^2(j)-x^{*2})(y(j)-y^*)), \\ y(n) = y(0) + \dfrac{1}{\Gamma(\mu)} \displaystyle\sum_{j=1}^{n} \dfrac{\Gamma(n-j+\mu)}{\Gamma(n-j+1)}(qx(j-1)-(c+e)y(j-1)+z+l(y(j)-y^*)), \\ s(n) = s(0) + \dfrac{1}{\Gamma(\mu)} \displaystyle\sum_{j=1}^{n} \dfrac{\Gamma(n-j+\mu)}{\Gamma(n-j+1)}(\delta r(j-1)+l(s(j)-s^*)), \\ r(n) = r(0) + \dfrac{1}{\Gamma(\mu)} \displaystyle\sum_{j=1}^{n} \dfrac{\Gamma(n-j+\mu)}{\Gamma(n-j+1)}(-\delta s(j-1)+l(r(j)-r^*)). \end{cases} \tag{10.3.6}$$

Take the fractional order $\mu$ as $\mu = 0.8$ and the control parameter $l$ as $(a)\, l = 0.02$; $(b)\, l = 0.04$; $(c)\, l = 0.05$; $(d)\, l = 0.06$; $(e)\, l = 0.07$; $(f)\, l = 0.08$, Fig. 10.12 shows the relevant Julia set with different control parameters $l$.

With the constant tiny increase of parameter $l$, the structure of Julia set becomes more and more delicate. Compared with fig(e), fig(f) shrinks and the control of Julia set has been achieved.

Secondly, the fixed point control items $l(x^2(n) - x^{*2})(y(n) - y^*)$, $l(y(n) - y^*)$, $l(s(n) - s^*)$ and $l(r(n) - r^*)$ are added into the model as a part, more precisely, the controller is imported to the outside of the summation sign to implement control over Julia set.

$$
\begin{cases}
x(n) = x(0) + \dfrac{1}{\Gamma(\mu)} \displaystyle\sum_{j=1}^{n} \dfrac{\Gamma(n-j+\mu)}{\Gamma(n-j+1)} (-(c+q)x(j-1) \\
\quad + \eta(1+bs(j-1))(1-x(j-1)-y(j-1))x(j-1)^2) + l(x^2(n) - x^{*2})(y(n) - y^*), \\
y(n) = y(0) + \dfrac{1}{\Gamma(\mu)} \displaystyle\sum_{j=1}^{n} \dfrac{\Gamma(n-j+\mu)}{\Gamma(n-j+1)} (qx(j-1) - (c+e)y(j-1) + z) + l(y(n) - y^*), \\
s(n) = s(0) + \dfrac{1}{\Gamma(\mu)} \displaystyle\sum_{j=1}^{n} \dfrac{\Gamma(n-j+\mu)}{\Gamma(n-j+1)} (\delta r(j-1)) + l(s(n) - s^*), \\
r(n) = r(0) + \dfrac{1}{\Gamma(\mu)} \displaystyle\sum_{j=1}^{n} \dfrac{\Gamma(n-j+\mu)}{\Gamma(n-j+1)} (-\delta s(j-1)) + l(r(n) - r^*).
\end{cases}
\tag{10.3.7}
$$

Also, the fractional order $\mu$ is taken as $\mu = 0.8$ and the control parameters $l$ are fetched as $(a)\, l = 0.001$; $(b)\, l = 0.002$; $(c)\, l = 0.005$ $(d)\, l = 0.008$; $(e)\, l = 0.015$; $(f)\, l = 0.017$, the images about correlative Julia set with disparate control parameters $l$ are exhibited in Fig. 10.13.

In Fig. 10.13, as the parameter $l$ increases, the area of the filled-in Julia set becomes smaller and smaller. In fig (e) and fig (f), some points are not in the interior of the Julia set, which means that the connectivity of the Julia set is closely interrelated with the control parameter $l$.

If the values are inside the Julia set, the trajectory is bounded, and the number of infected people is limited. Relatively, if the values are outside the Julia set, the trajectory is unbounded, at this moment, the number of infected people is unlimited and the epidemic is out of control. In this section, the fixed point control methods are designed for the discrete fractional order SIRS model, from these figures, some initial values are iterated from points outside the set into the set, which means the infected person is controlled. The control of infectious diseases is realized by controlling Julia sets of models.

### *10.3.2 Synchronization of Julia Sets of the Discrete Fractional Order SIRS Model*

Different biological systems sometimes require similar or even identical characteristics. Consider synchronization in different systems is necessary. In this section, the synchronization methods are applied to the investigation of Julia set.

**Theorem 10.3** *If $|C(n) - x(n)|$, $|D(n) - y(n)|$, $|A(n) - s(n)|$ and $|B(n) - r(n)|$ are bounded when $l \to 1$ and $n \to \infty$, then the synchronization of system (10.3.4) and the system (10.3.8)/(10.3.13) is realized.*

***Proof*** Take $C(n)$ and $x(n)$ as an example. $|C(n) - x(n)|$ is bounded, which means $C(n)$ is bounded equals to $x(n)$ is bounded, $C(n)$ and $x(n)$ have the same bounded-

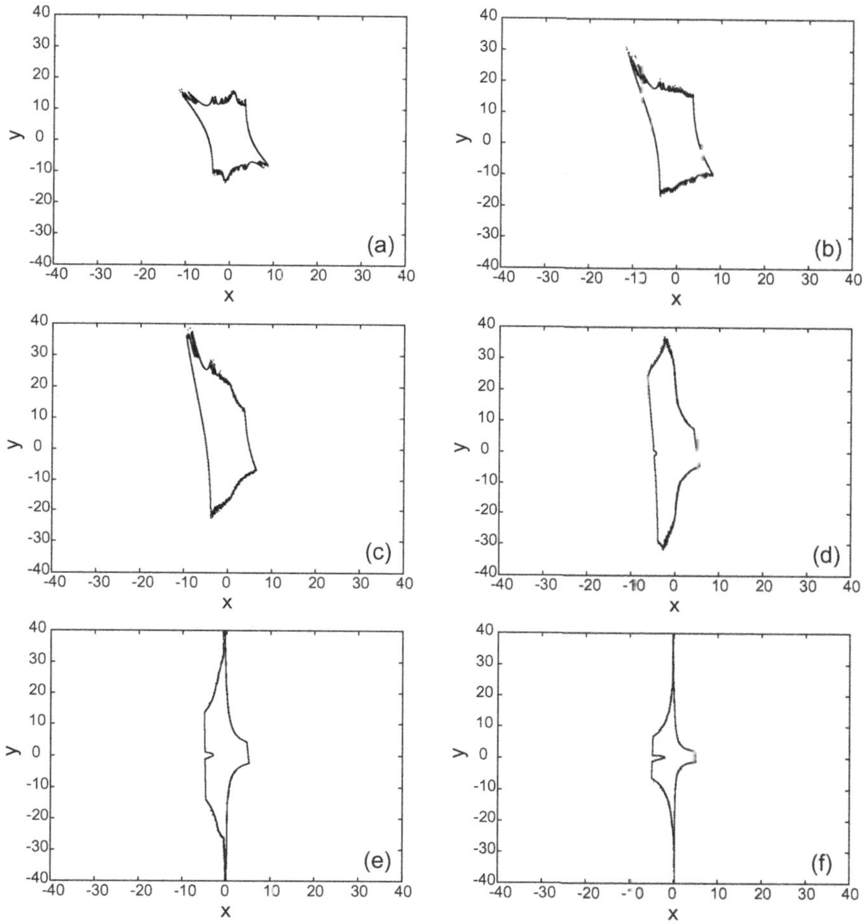

**Fig. 10.12** The changing process of Julia sets of the controlled system (10.3.6) when **a** $l = 0.02$; **b** $l = 0.04$; **c** $l = 0.05$; **d** $l = 0.06$; **e** $l = 0.07$; **f** $l = 0.08$

ness. And according to the definition of Julia set, $C(n)$ is synchronized with $x(n)$. Similarly, the same cases as $D(n)$ and $y(n)$, $A(n)$ and $s(n)$, $B(n)$ and $r(n)$. Therefore, the synchronization of Julia sets of systems (10.3.4) and system (10.3.8)/(10.3.13) is acquired.

In this part, different coupling terms are designed to achieve synchronization between the initial system and the target system. By using the idea of the Julia set, the synchronization process is shown in the graphs of our simulation results.

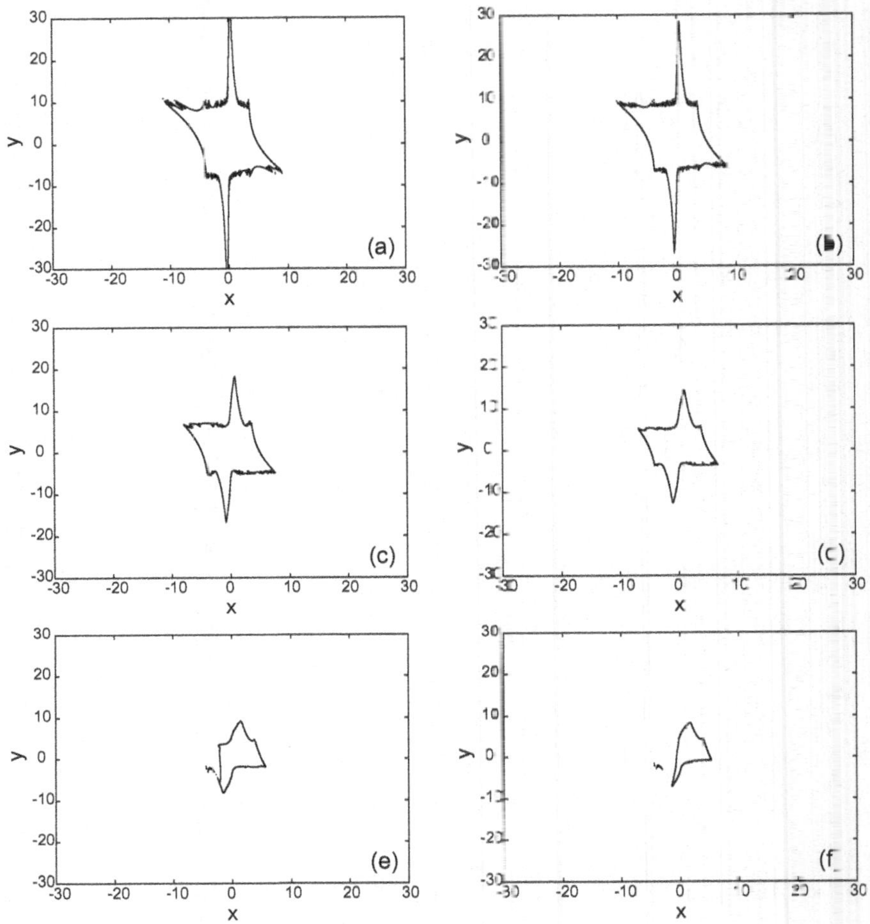


**Fig. 10.13** The changing process of Julia sets of the controlled system (10.3.7) when **a** $l = 0.001$; **b** $l = 0.002$; **c** $l = 0.005$; **d** $l = 0.008$; **e** $l = 0.015$; **f** $l = 0.017$

### The First Synchronization of Julia Sets

Consider a system (10.3.8) with the same structure of equation form (10.3.4) but with different parameter $\hat{e}$.

$$\begin{cases} C(n) = C(0) + \dfrac{1}{\Gamma(\mu)} \displaystyle\sum_{j=1}^{n} \dfrac{\Gamma(n-j+\mu)}{\Gamma(n-j+1)} (-(c+q)C(j-1) \\ \quad + \eta(1+bA(j-1))(1-C(j-1)-D(j-1))C(j-1)^2), \\ D(n) = D(0) + \dfrac{1}{\Gamma(\mu)} \displaystyle\sum_{j=1}^{n} \dfrac{\Gamma(n-j+\mu)}{\Gamma(n-j+1)} (qC(j-1) - (c+\hat{e})D(j-1) + z), \\ A(n) = A(0) + \dfrac{1}{\Gamma(\mu)} \displaystyle\sum_{j=1}^{n} \dfrac{\Gamma(n-j+\mu)}{\Gamma(n-j+1)} (\delta B(j-1)), \\ B(n) = B(0) + \dfrac{1}{\Gamma(\mu)} \displaystyle\sum_{j=1}^{n} \dfrac{\Gamma(n-j+\mu)}{\Gamma(n-j+1)} (-\delta A(j-1)), \end{cases} \tag{10.3.8}$$

where $\hat{e} \neq e > 0$. In an effort to tie the system (10.3.4) and (10.3.8) together, nonlinear coupling terms are devised to enable one Julia set of SIRS model change to be another. We added coupling terms $h_1$ and $h_2$ into the Eq. (10.3.8), so we get

$$\begin{cases} C(n) = C(0) + \dfrac{1}{\Gamma(\mu)} \displaystyle\sum_{j=1}^{n} \dfrac{\Gamma(n-j+\mu)}{\Gamma(n-j+1)} (-(c+q)C(j-1) \\ \quad + \eta(1+bA(j-1))(1-C(j-1)-D(j-1))C(j-1)^2) + h_1, \\ D(n) = D(0) + \dfrac{1}{\Gamma(\mu)} \displaystyle\sum_{j=1}^{n} \dfrac{\Gamma(n-j+\mu)}{\Gamma(n-j+1)} (qC(j-1)) - (c+\hat{e})D(j-1) + z) + h_2, \\ A(n) = A(0) + \dfrac{1}{\Gamma(\mu)} \displaystyle\sum_{j=1}^{n} \dfrac{\Gamma(n-j+\mu)}{\Gamma(n-j+1)} (\delta B(j-1)), \\ B(n) = B(0) - \dfrac{1}{\Gamma(\mu)} \displaystyle\sum_{j=1}^{n} \dfrac{\Gamma(n-j+\mu)}{\Gamma(n-j+1)} (\delta A(j-1)), \end{cases} \tag{10.3.9}$$

where

$$\begin{aligned} h_1 &= l\left(\frac{1}{\Gamma(\mu)} \sum_{j=1}^{n} \frac{\Gamma(n-j+\mu)}{\Gamma(n-j+1)} (-\eta(1+bA(j-1))(1-C(j-1)-D(j-1))C(j-1)^2 \right. \\ &\quad \left. + \eta\,(1+bs(j-1))(1-x(j-1)-y(j-1))x(j-1)^2)\right), \\ h_2 &= l\left(\frac{1}{\Gamma(\mu)} \sum_{j=1}^{n} \frac{\Gamma(n-j+\mu)}{\Gamma(n-j+1)} ((c+\hat{e})D(j-1) - (c+e)y(j-1)\right). \end{aligned}$$

It is evident that different Julia sets are obtained by selecting different coupling parameters $l$, take the control parameter as $(a) l = 0.09$; $(b) l = 0.095$; $(c) l = 0.1$; $(d) l = 0.2$; $(e) l = 0.5$; $(f) l = 0.7$, other system parameters remain unchanged, the following Fig. 10.14 is obtained.

In Figs. 10.14, 10.11 is shown in blue, which is the background, and the synchronization process diagrams of Julia set of this method are shown in red. As can be seen from Fig. 10.14, Julia set in figure (a) looks like a star with four horns. In addition, with the increase of the control parameter $l$, the graph lines become mellow and full, and the area of the Julia sets becomes larger and larger, eventually it overlaps with original Fig. 10.11.

Next, we give a proof about this synchronization method to achieve the synchronization of trajectories of system (10.3.4) and (10.3.8). In order to promote the expression of this demonstration, take

$$\begin{aligned} E_{1,j-1} &= \eta(1+bs(j-1))(1-x(j-1)-y(j-1))x(j-1)^2, \\ E_{2,j-1} &= \eta(1+bA(j-1))(1-C(j-1)-D(j-1))C(j-1)^2. \end{aligned}$$

**Theorem 10.4** *The synchronization of system (10.3.4) and system (10.3.8) can be achieved when satisfies* $|C(n)-x(n)|, |D(n)-y(n)|, |A(n)-s(n)|, |E(n)-r(n)|$ *are bounded, on the condition of* $n \to \infty$, $l \to 1$ *and* $0 < c+q < 1$.

***Proof*** Here, we take the synchronization of $C(n)$ and $x(n)$ for an example. From system (10.3.4) and system (10.3.8), we have

$$x(n) = x(0) + \frac{1}{\Gamma(\mu)} \sum_{j=1}^{n} \frac{\Gamma(n-j+\mu)}{\Gamma(n-j+1)} (-(c+q)x(j-1) + E_{1,j-1}).$$

$$\begin{aligned} C(n) = C(0) &+ \frac{1}{\Gamma(\mu)} \sum_{j=1}^{n} \frac{\Gamma(n-j+\mu)}{\Gamma(n-j+1)} (-(c+q)C(j-1) + E_{2,j-1}) \\ &+ \frac{l}{\Gamma(\mu)} \sum_{j=1}^{n} \frac{\Gamma(n-j+\mu)}{\Gamma(n-j+1)} (E_{1,j-1} - E_{2,j-1}), \end{aligned}$$

Observe that if we utilize the general binomial coefficient $\binom{a}{b} := \frac{\Gamma(a+1)}{\Gamma(b+1)\Gamma(a-b-1)}$, see [479], then $\frac{\Gamma(n-j+\mu)}{\Gamma(n-j+1)\Gamma(\mu)}$ can be rewritten as $\binom{n-1-j+\mu}{n-j}$. Thus above equations can be rewritten as

$$\begin{cases} C(n) = C(0) + \sum_{j=1}^{n} \binom{n-j-1+\mu}{n-j} (-(c+q)C(j-1) + E_{2,j-1}) \\ \qquad + l \sum_{j=1}^{n} \binom{n-j-1+\mu}{n-j} (E_{1,j-1} - E_{2,j-1}), \\ x(n) = x(0) + \sum_{j=1}^{n} \binom{n-j-1+\mu}{n-j} (-(c+q)x(j-1) - E_{1,j-1}). \end{cases} \tag{10.3.10}$$

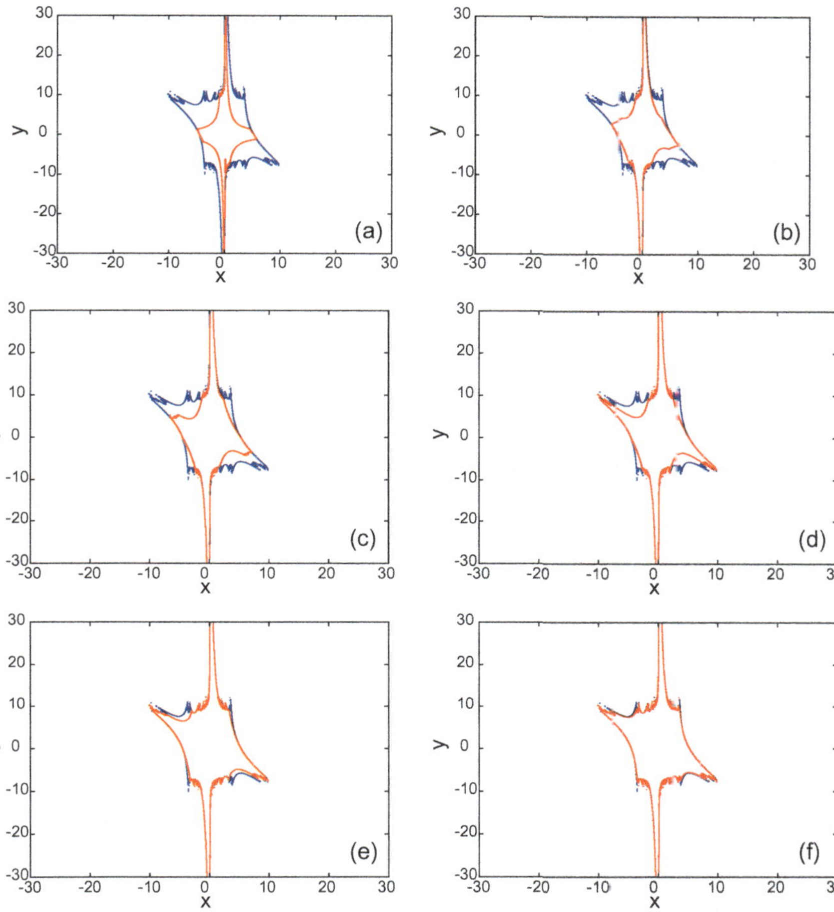


**Fig. 10.14** The synchronization process of Julia sets between the system (10.3.4) and the system (10.3.9) when **a** $l = 0.09$; **b** $l = 0.095$; **c** $l = 0.1$; **d** $l = 0.2$; **e** $l = 0.5$; **f** $l = 0.7$

Therefore, take the item as $n-1$, the $C(n-1)$ and $x(n-1)$ can be expressed as

$$\left\{\begin{aligned} C(n-1) &= C(0) + \sum_{j=1}^{n-1} \binom{n-j-2+\mu}{n-1-j} (-(c+q)C(j-1) + E_{2,j-1}) \\ &\quad + l \sum_{j=1}^{n-1} \binom{n-j-2+\mu}{n-1-j} (E_{1,j-1} - E_{2,j-1}), \\ x(n-1) &= x(0) + \sum_{j=1}^{n-1} \binom{n-j-2+\mu}{n-1-j} (-(c+q)x(j-1) + E_{1,j-1}). \end{aligned}\right. \tag{10.3.11}$$

Next, $C(n)$ in system (10.3.10) minus $C(n-1)$ in system (10.3.11) and $x(n-1)$ is subtracted by $x(n)$. And see [479], for $\alpha \in R$ and $k \in N_1$ one has $\binom{\alpha}{k} - \binom{\alpha-1}{k-1} = \binom{\alpha-1}{k}$, set $\alpha = n-1-j+\mu$ and $k = n-j$, distinctly, $\binom{n-1-j+\mu}{n-j} - \binom{n-2-j+\mu}{n-1-j} = \binom{n-2-j+\mu}{n-j}$. Then, we obtain

$$\begin{cases} C(n) - C(n-1) = -(c+q)\sum_{j=1}^{n}\binom{n-j-2+\mu}{n-j}C(j-1) \\ \qquad + \sum_{j=1}^{n}\binom{n-j-2+\mu}{n-j}(lE_{1,j-1} + (1-l)E_{2,j-1}), \\ x(n) - x(n-1) = -(c+q)\sum_{j=1}^{n}\binom{n-j-2+\mu}{n-j}x(j-1) + \sum_{j=1}^{n}\binom{n-j-2+\mu}{n-j}E_{1,j-1}, \end{cases} \tag{10.3.12}$$

where $j = 1, 2, \cdots n-1$, $\binom{n-2-j+\mu}{n-j} = \frac{(n-2-j+\mu)(n-3-j+\mu)\cdots\mu(\mu-1)}{(n-j)!} < 0$ and $j = n$, $\binom{\mu-2}{0} = 1$.

During system (10.3.12), $C(n) - C(n-1)$ minus $x(n) - x(n-1)$, so the following equation is presented.

$$C(n) - C(n-1) - (x(n) - x(n-1)) = -(c+q)\sum_{j=1}^{n-1}\binom{n-j-2+\mu}{n-j}(C(j-1) - x(j-1))$$
$$-(c+q)(C(n-1) - x(n-1)) + \sum_{j=1}^{n}\binom{n-j-2+\mu}{n-j}(1-l)(E_{2,j-1} - E_{1,j-1})$$

Take the absolute value of both sides of this equation, make $F_j = |C(j) - x(j)|$, $j = 1, 2, \cdots n$ and $0 < c+q < 1$. Furthermore, from the definition of Julia set [478], apparently, the Julia sets are bounded. Assume the bounded region is $\Omega$, and there is a $m_0$ meets $h^{m_0}(z) \notin \omega$, in that way, the Julia set is fixed. In addition, there is always a $T > 0$ which is content with $|z| < T$ for any $z \in \omega$ because of the bounded region $\omega$. Therefore, there exists positive number $M$ and $N$ which enable $F_j < M$, $j = 1, 2, \cdots, n-1$ and $\left|E_{2,j-1} - E_{1,j-1}\right| < N$, $j = 1, 2, \cdots, n$. So we get the following inequality,

$$F_n - (1-c-q)F_{n-1} \leq -(c+q)\sum_{j=1}^{n-1}\binom{n-j-2+\mu}{n-j}M$$
$$-|1-l|\sum_{j=1}^{n-1}\binom{n-j-2+\mu}{n-j}N + |1-l|N$$

Due to $\binom{n-2-j+\mu}{n-j} = \binom{n-1-j+\mu}{n-j} - \binom{n-2-j+\mu}{n-1-j}$, sum from $j=1$ to $j=n-1$, so $\sum_{j=1}^{n-1}\binom{n-2-j+\mu}{n-j} = \binom{n-2+u}{n-1} - 1$. Thus,

$$F_n - (1-c-q)F_{n-1} \leq -(c+q)\left[\binom{n-2+\mu}{n-1} - 1\right]M - |1-l|\left[\binom{n-2+\mu}{n-1} - 1\right]N + |1-l|\,N.$$

In this equation, $\binom{n-2+\mu}{n-1} = \frac{(n-2+\mu)(n-3+\mu)\cdots(1+\mu)\mu}{(n-1)!} > 0$, thus $-(c+q)\binom{n-2+\mu}{n-1}M < 0$ and $-|1-l|\binom{n-2+\mu}{n-1}N < 0$. Iterate the inequality, we gain

$$\begin{aligned}
F_n &\leq (1-c-q)F_{n-1} + (c+q)M + 2\,|1-l|\,N \\
&\leq (1-c-q)[(1-c-q)F_{n-2} + (c+q)M + 2\,|1-l|\,N] + (c+q)M + 2\,|1-l|\,N \\
&= (1-c-q)^2 F_{n-2} + [(c+q)M + 2\,|1-l|\,N]\,(1 + (1-c-q)) \\
&\leq \cdots \leq (1-c-q)^n F_0 + [(c+q)M + 2\,|1-l|\,N]\,\frac{1-(1-c-q)^{n-1}}{1-(1-c-q)} \\
&\leq (1-c-q)^n F_0 + M + 2\,|1-l|\,N\,\frac{1-(1-c-q)^{n-1}}{c+q},
\end{aligned}$$

where $0 < c+q < 1$. Notice that $|C_0 - x_0|$ is bouned since the initial values $C_0$ and $x_0$ are fetched inside in the bounded region $\omega$, thus $F_0$ is limited. Take $n \to \infty$, so $(1-c-q)^n F_0 \to 0$ and $2\,|1-l|\,N\frac{1-(1-c-q)^{n-1}}{c+q} \to \frac{2|1-l\,N}{c+q}$. When $l \to 1$, $\frac{2|1-l|N}{c+q} \to 0$.

It is obvious that $F_n = |C(n) - x(n)| \leq M$. Similarly, we can gain $|D(n) - y(n)|$, $|A(n) - s(n)|$ and $|B(n) - r(n)|$ are bounded if $l \to 1$ and $n \to \infty$. Due to the proof is analogous and simple, we do not expand the proof here. According to Theorem 10.3, the synchronization of Julia sets of systems (10.3.4) and (10.3.8) is acquired finally and the proof of theorem is finished.

### The Second Synchronization of Julia Sets

Similarly, consider a system (10.3.13) with the same structure of equation form (10.3.4) but with different parameters $\hat{\eta}$ and $\hat{q}$.

$$\begin{cases} C(n) = C(0) + \dfrac{1}{\Gamma(\mu)} \displaystyle\sum_{j=1}^{n} \dfrac{\Gamma(n-j+\mu)}{\Gamma(n-j+1)} (-(c+\hat{q})C(j-1) \\ \quad + \hat{\eta}(1+bA(j-1))(1-C(j-1)-D(j-1))C(j-1)^2), \\ D(n) = D(0) + \dfrac{1}{\Gamma(\mu)} \displaystyle\sum_{j=1}^{n} \dfrac{\Gamma(n-j+\mu)}{\Gamma(n-j+1)} (\hat{q}C(j-1) - (c+e)D(j-1) + z), \\ A(n) = A(0) + \dfrac{1}{\Gamma(\mu)} \displaystyle\sum_{j=1}^{n} \dfrac{\Gamma(n-j+\mu)}{\Gamma(n-j+1)} (\delta B(j-1)), \\ B(n) = B(0) + \dfrac{1}{\Gamma(\mu)} \displaystyle\sum_{j=1}^{n} \dfrac{\Gamma(n-j+\mu)}{\Gamma(n-j+1)} (-\delta A(j-1)), \end{cases} \tag{10.3.13}$$

where $\hat{\eta} \neq \eta > 0$ and $\hat{q} \neq q > 0$. To facilitate the description of this synchronization method, denote

$$C^{**} = C(0) + \frac{1}{\Gamma(\mu)} \sum_{j=1}^{n} \frac{\Gamma(n-j+\mu)}{\Gamma(n-j+1)} (-(c+\hat{q})C(j-1) + \hat{\eta}(1+bA(j-1))(1-C(j-1)-D(j-1))C(j-1)^2),$$

$$C^{*} = C(0) + \frac{1}{\Gamma(\mu)} \sum_{j=1}^{n-1} \frac{\Gamma(n-j+\mu)}{\Gamma(n-j+1)} (-(c+\hat{q})C(j-1) + \hat{\eta}(1+bA(j-1))(1-C(j-1)-D(j-1))C(j-1)^2),$$

$$x^{**} = x(0) + \frac{1}{\Gamma(\mu)} \sum_{j=1}^{n-1} \frac{\Gamma(n-j+\mu)}{\Gamma(n-j+1)} (-(c+q)x(j-1) + \eta(1+bs(j-1))(1-x(j-1)-y(j-1))x(j-1)^2),$$

$$D^{**} = D(0) + \frac{1}{\Gamma(\mu)} \sum_{j=1}^{n} \frac{\Gamma(n-j+\mu)}{\Gamma(n-j+1)} (\hat{q}C(j-1) - (c+e)D(j-1) + z),$$

$$D^{*} = D(0) + \frac{1}{\Gamma(\mu)} \sum_{j=1}^{n-1} \frac{\Gamma(n-j+\mu)}{\Gamma(n-j+1)} (\hat{q}C(j-1) - (c+e)D(j-1) + z),$$

$$y^{**} = y(0) + \frac{1}{\Gamma(\mu)} \sum_{j=1}^{n-1} \frac{\Gamma(n-j+\mu)}{\Gamma(n-j+1)} (qx(j-1) - (c+e)y(j-1) + z).$$

In order to connect these Julia sets of system (10.3.4) and (10.3.13) together, add two coupling terms $-C^* + x^{**} - l((C^{**} - C^*) - (x(n) - x^{**}))$ and $-D^* - y^{**} - l((D^{**} - D^*) - (y(n) - y^{**}))$ into system (10.3.13), so we obtain

$$\begin{cases} C(n) = C^{**} - C^{*} + x^{**} - l((C^{**} - C^{*}) - (x(n) - x^{**})), \\ D(n) = D^{**} - D^{*} + y^{**} - l((D^{**} - D^{*}) - (y(n) - y^{**})), \\ A(n) = A(0) + \dfrac{1}{\Gamma(\mu)} \displaystyle\sum_{j=1}^{n} \dfrac{\Gamma(n-j+\mu)}{\Gamma(n-j+1)} (\delta B(j-1)), \\ B(n) = B(0) + \dfrac{1}{\Gamma(\mu)} \displaystyle\sum_{j=1}^{n} \dfrac{\Gamma(n-j+\mu)}{\Gamma(n-j+1)} (-\delta A(j-1)). \end{cases} \tag{10.3.14}$$

Apparently, by selecting different coupling parameters $l$, disparate Julia sets of system (10.3.14) are demonstrated.

With Fig. 10.11 as the background frame, shown in blue, the synchronization process diagrams of the Julia set of this method are exhibited, shown in red. From Fig. 10.15, we can find that Julia set of figure (a) is shaped somewhat like the number 8. Besides, with the increase of control parameters $l$, the graphs expand horizontally and finally synchronize with the initial Fig. 10.11.

**Theorem 10.5** *If* $|C(n) - x(n)| \to 0$, $|D(n) - y(n)| \to 0$, *and* $|A(n) - s(n)|$, $|B(n) - r(n)|$ *are bounded when* $l \to 1$ *and* $n \to \infty$, *then the synchronization of system (10.3.4) and the system (10.3.13) is gained.*

***Proof*** From Eqs. (10.3.4) and (10.3.14), $C(n)$ minus $x(n)$, so we have

$$C(n) - x(n) = (1-l)((C^{**} - C^{*}) - (x(n) - x^{**})).$$

Take the absolute value of both sides of this equation. As we described in Theorem 10.3 because of the bounded Julia set, $|(C^{**} - C^{*}) - (x(n) - x^{**})|$ is less than $|C^{**} - C^{*}| + |x(n) - x^{**}|$, and less than a positive number, thus

$$\begin{aligned} |C(n) - x(n)| &= |1-l| \left|(C^{**} - C^{*}) - (x(n) - x^{**})\right| \\ &\le |1-l| \left(\left|C^{**} - C^{*}\right| + \left|x(n) - x^{**}\right|\right). \end{aligned}$$

The limit of the right side of this equation tends to 0 as $l \to 1$. In a semblable way, we can obtain $|D(n) - y(n)| \to 0$, $|A(n) - s(n)|$ and $|B(n) - r(n)|$ are bounded if $l \to 1$ and $n \to \infty$. Thus, the synchronization of Julia sets of systems (10.3.4) and (10.3.13) is implemented and the proof is completed.

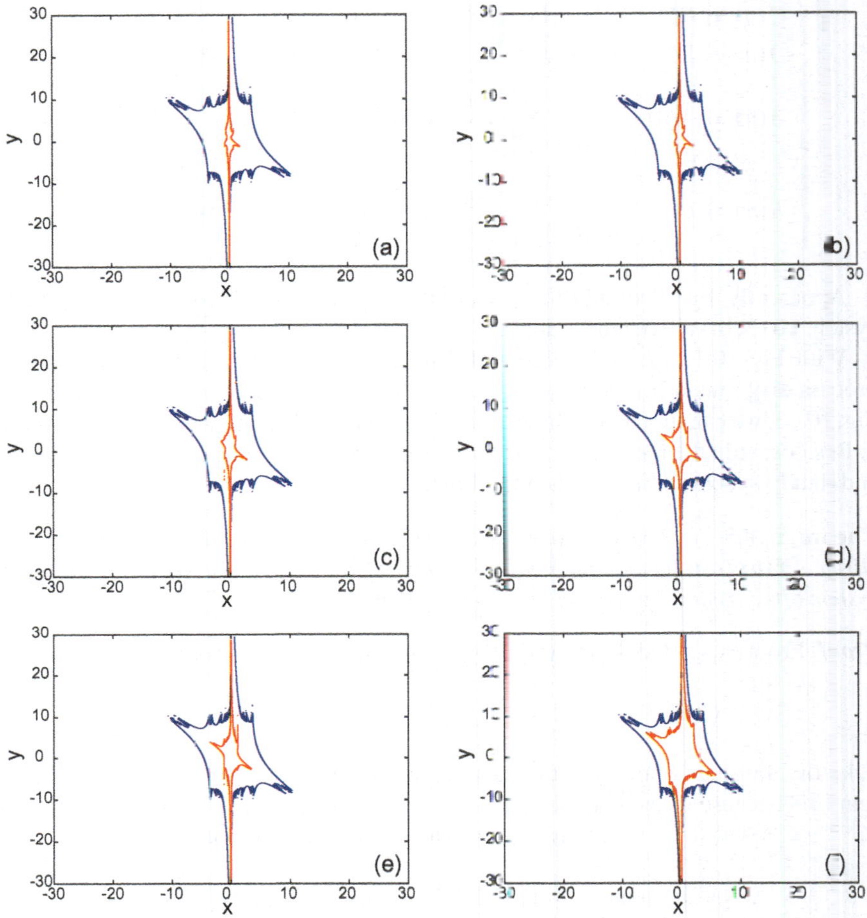


**Fig. 10.15** The synchronization process of Julia sets between the system (10.3.4) and the system (10.3.14) when **a** $l = 0$; **b** $l = 0.2$; **c** $l = 0.4$; **d** $l = 0.6$; **e** $l = 0.8$; **f** $l = 0.9$

# Chapter 11
# Application of Fractal Control in Other Fields

## 11.1 Fractal Characteristics of Heterocapsa Circularisquama and Prorocentrum Dentatum Cells Growth

### 11.1.1 Background

It is noticed that in 2011, Yamasaki et al. studied the effects on the growth interaction between H. circularisquama and P. dentatum cells because of competition among them [481]. They found that the growth of H. circularisquama was slightly inhibited when inoculated at $10^2$cells $mL^{-1}$ along with P. dentatum at $10^4$cells $mL^{-1}$. In other combinations of inoculation densities, P. dentatum density rapidly decreased to extremely low levels in the presence of H. circularisquama. For purpose of simulate growth and interactions of H. circularisquama and P. dentatum, the mathematical system was studied in [481]:

$$\begin{cases} dx/dt = r_1 x[1 - (x + ay)K_1^{-1}], \\ dy/dt = r_2 x[1 - (bx + y)K_2^{-1}], \end{cases} \tag{11.1.1}$$

where $x$ and $y$ are the cell densities of H. circularisquama and P. dentatum, respectively. The parameters r1, r2 are the growth rate of H. circularisquama and P. dentatum, respectively, and K1, K2 are the corresponding maximum cell densities of H. circularisquama and P. dentatum, when each species is in monoalgal culture. What is more, parameters $a, b$ are competition coefficients and are a measure of the degree of inhibition by the other species compared with self-interference. When each species is in a mono-algal culture, we can set $a = b = 0$.

Generally speaking, for ecological systems, persistence and extinction of populations are affected by the initial number of populations. On the one hand, the demand for resources will be too much if the initial number of one population is too large, and it will lead to a shortage of resources and inadequate supply of resources so that



S. T. Liu et al., *Fractal Control and Its Applications*,
https://doi.org/10.1007/978-981-15-5459-9_11

more intense competition between two populations will happen, and thus the two populations will extinct finally. On the other hand, if the initial number of one population is very small, the number of the other will be more and more in the process of competition so that the former will be extinct as a result of lack of competitiveness relative to the other [399]. In addition, environmental factors, for example temperature, salinity, light, pH values and so on, can dramatically affect the growth of marine algae [482–484].

In this section, based on the fact that initial cell densities of H. circularisquama and P. dentatum and culture environment in the experiment determine the final cell growth state, we study the level of initial cell densities and experimental culture environment which can make H. circularisquama and P. dentatum die on account of competition between them and extinct ultimately. For the convenience of study, the actual mathematical system is discretized for numerical simulations [485], that is

$$dx/dt \to \frac{x(t+\Delta t)-x(t)}{\Delta t}, \quad dy/dt \to \frac{y(t+\Delta t)-y(t)}{\Delta t}.$$

By plugging these discretizations into system (11.1.1), the discrete version can be obtained

$$\begin{cases} x_{n+1} = (1+r_1\Delta t)x_n - K_1^{-1}r_1 a\Delta t x_n y_n - K_1^{-1}r_1\Delta t x_n^2, \\ y_{n+1} = (1+r_2\Delta t)y_n - K_2^{-1}r_2 b\Delta t x_n y_n - K_2^{-1}r_2\Delta t y_n^2, \end{cases}$$

where $\Delta t \to 0$. For the convenience of discussion, we suppose

$$\alpha_1 = r_1\Delta t, \quad \beta_1 = K_1^{-1}r_1 a\Delta t, \quad \gamma_1 = K_1^{-1}r_1\Delta t,$$
$$\alpha_2 = r_2\Delta t, \quad \beta_2 = K_2^{-1}r_2 b\Delta t, \quad \gamma_2 = K_2^{-1}r_2\Delta t,$$

then we have the another discrete version of (11.1.1)

$$\begin{cases} x_{n+1} = (1+\alpha_1)x_n - \beta_1 x_n y_n - \gamma_1 x_n^2, \\ y_{n+1} = (1+\alpha_2)y_n - \beta_2 x_n y_n - \gamma_2 y_n^2. \end{cases} \tag{11.1.2}$$

In the fractal theory, the attractive domain is a set of the initial states, whose size can reveal the effect of the initial states on the final evolution of the biological populations with the change of system parameters [372]. For ecological systems, the set of initial cell densities of populations is the boundary of the attractive domain of the fixed point, as a result, the attractive domain can be controlled through the size control of the set of initial densities of populations. For system (11.1.2), controlling the points in the set of initial cell densities of H. circularisquama and P. dentatum outside of the set is equivalent to control H. circularisquama and P. dentatum cells with persistence to do not have persistence. Therefore, we can make H. circularisquama and P. dentatum cells do not have persistence through the size control of the set of initial cell densities and develop towards the expected direction.

For biological populations, the actual growth and multiplication are often affected by the complicated external environment, consequently, the studies on prediction and control of the growth of populations are important and have practical significance under the interference of many complex factors. In this section, we discuss the set of environmental factors disturbing the growth of H. circularisquama and P. dentatum cells of system (11.1.2). Controlling the points in the set of environmental factors disturbing the growth of H. circularisquama and P. dentatum cells outside of the set is equivalent to control H. circularisquama and P. dentatum cells with persistence in a certain initial culture environment to have no persistence through appropriate change of culture environment.

### *11.1.2 Set of Initial Cell Densities of Heterocapsa Circularisquama and Prorocentrum Dentatum*

**Definition 11.1** ([179]) Suppose $H$: $\mathbb{R}^2 \rightarrow \mathbb{R}^2$ is the real map of system (11.1.2), that is

$$H(x, y) = ((1+\alpha_1)x - \beta_1 xy - \gamma_1 x^2, (1+\alpha_2)y - \beta_2 xy - \gamma_2 y^2).$$

Let

$$K = \{(x, y) \in \mathbb{R}^2 \mid \{H^n(x, y)\}_{n=1}^{\infty} \text{ has boundary}\}.$$

where the iterative initial values are the initial cell densities of Heterocapsa circularisquama and Prorocentrum dentatum. $\partial K$ is the boundary of the set of $K$, then $\partial K$ is the set of initial cell densities of Heterocapsa circularisquama and Prorocentrum dentatum of system (11.1.2).

#### Fractal Dimension of the Set of Initial Cell Densities of H. Circularisquama and P. Dentatum

In the geometry, the dimension is the number of parameters to determine position of the points in graphics. The set of initial cell densities of H. circularisquama and P. dentatum is naturally a geometric form, and because of the irregularity and fragmentation of the geometry, the set is called fractal geometry whose fractal characteristics are described by the fractal dimension. Here, we utilize box-counting dimension in Definition 6.1. The calculation depends on the limit when the diameter $r$ tends to zero, but it is impossible to realize in the actual calculation. Thus, for the practical problem, we need to discretize the continuous variable $r$ replaced by discrete variable $\delta_r$ $(r = 1, 2, 3, \ldots)$. Theoretically, it can be based on the following lemma.

**Lemma 11.1** ([486]) *Suppose $(X, d)$ is a distance space, $G \in \Theta(X)$, $\Theta(X)$ is a space composed of all non-empty compact subsets, $(r_n, n = 1, 2, 3, \ldots)$ is a decreasing sequence with the limit zero, and $r_{n+1} > cr_n$ $(0 < c < 1)$, if there is*

$$\lim_{n\to 0} \frac{\ln N_r(G)}{-\ln r_n},$$

*then the fractal dimension of the set $G$ is*

$$\dim_B G = \lim_{r\to 0} \frac{\ln N_r(G)}{-\ln r} = \lim_{r\to 0} \frac{\ln N_{r_n}(G)}{-\ln r_n}.$$

Based on Lemma 11.1, we take a set of discrete values of the set of initial cell densities of H. circularisquama and P. dentatum $G$: $r_1, r_2, \ldots, r_n$, and calculate the number of boxes covering the set of initial cell densities of H. circularisquama and P. dentatum, denoted by $N_{r_n}(G)$. Using a line to fit the data points $(\ln r_n, \ln N_{r_n}(G))$ in the coordinate system $\ln r_n - \ln N_{r_n}(G)$, then the absolute value of the slope of the line is the approximation of fractal dimension of the set of initial cell densities of H. circularisquama and P. dentatum.

When $\alpha_1 = 0.2, \beta_1 = 0.15, \gamma_1 = 0.1, \alpha_2 = 0.3, \beta_2 = 0.25, \gamma_2 = 0.16$, the set of initial cell densities of Heterocapsa circularisquama and Prorocentrum dentatum of system (11.1.2) as shown in Fig. 11.1. The plane of the set of initial cell densities of Heterocapsa circularisquama and Prorocentrum dentatum of system (11.1.2) is

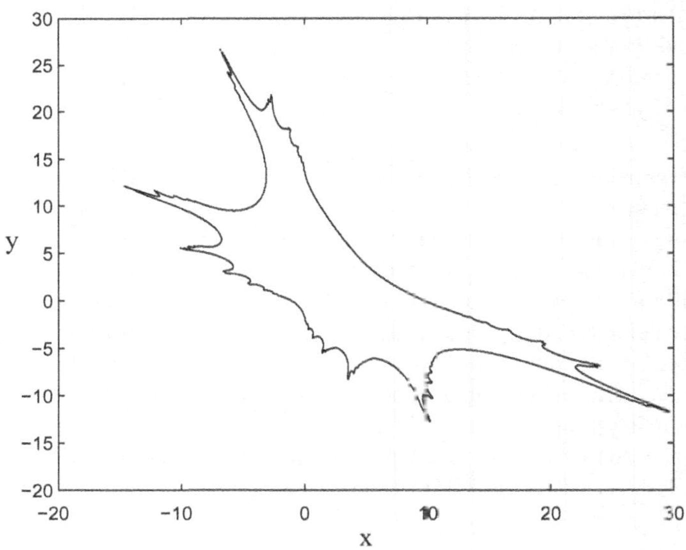


**Fig. 11.1** The set of initial cell densities of H. circularisquama and P. dentatum of system (11.1.2), when $\alpha_1 = 0.2, \beta_1 = 0.15, \gamma_1 = 0.1, \alpha_2 = 0.3, \beta_2 = 0.25, \gamma_2 = 0.16$

**Table 11.1** Data results

| r | N(r) | lnr | lnN(r) | r | N(r) | lnr | lnN(r) |
|---|---|---|---|---|---|---|---|
| 0.50 | 1236 | −0.69 | 7.12 | 2.75 | 74 | 1.01 | 4.30 |
| 0.75 | 593 | −0.28 | 6.39 | 3.00 | 65 | 1.10 | 4.17 |
| 0.80 | 520 | −0.22 | 6.25 | 3.25 | 54 | 1.13 | 3.99 |
| 1.00 | 356 | 0.00 | 5.87 | 3.50 | 49 | 1.25 | 3.89 |
| 1.25 | 246 | 0.22 | 5.51 | 3.75 | 47 | 1.32 | 3.85 |
| 1.50 | 186 | 0.41 | 5.23 | 4.00 | 41 | 1.39 | 3.71 |
| 1.75 | 138 | 0.56 | 4.93 | 4.25 | 40 | 1.45 | 3.69 |
| 2.00 | 113 | 0.69 | 4.73 | 4.50 | 36 | 1.50 | 3.58 |
| 2.25 | 95 | 0.81 | 4.55 | 4.75 | 32 | 1.55 | 3.46 |
| 2.50 | 79 | 0.92 | 4.37 | 5.00 | 30 | 1.61 | 3.40 |

divided into a set of small squares using fine grids, then record the number of small squares that even only cover a little of the set of initial cell densities of H. circularisquama and P. dentatum of system (11.1.2), denoted by $N(r)$. Through changing the side length of the small squares, 20 groups of data can be acquired as shown in Table 11.1.

We will analysis these 20 groups of data using the linear regression method to obtain the equation of linear regression.

Suppose the equation of linear regression:

$$y = px + q + \varepsilon,$$

where $p, q$ are regression coefficients, and $\varepsilon$ is an error term which reflects the influences of random factors on $y$ except the linear relationship between $x$ and $y$.

Estimate the regression coefficients using the least square method [487]. We regard the minimum of the square sum of differences between measured values $y_i$ $(i = 1, \ldots, n)$ and calculated values $\widehat{y}_i$ obtained by the linear equation $\widehat{y} = px + q$ as the optimal criterion. Let

$$\Phi = \sum_{i=1}^{n} (y_i - \widehat{y}_i)^2. \tag{11.1.3}$$

By substituting $\widehat{y}_i = \widehat{p}x_i + \widehat{q}$ $(i = 1, \ldots, n)$ into (11.1.3), we get

$$\Phi = \sum_{i=1}^{n} (y_i - \widehat{y}_i)^2 = \sum_{i=1}^{n} (y_i - \widehat{p}x_i - \widehat{q})^2. \tag{11.1.4}$$

In order to get the minimum of $\Phi$, we demand $\Phi$ on the $\widehat{p}$ and $\widehat{q}$ partial derivatives, and let the partial derivatives zero, that is

$$\frac{\partial \Phi}{\partial \widehat{p}} = 0, \quad \frac{\partial \Phi}{\partial \widehat{q}} = 0.$$

Then we have

$$\sum_{i=1}^{n} x_i y_i = \widehat{p} \sum_{i=1}^{n} x_i^2 + \widehat{q} \sum_{i=1}^{n} x_i, \quad \sum_{i=1}^{n} y_i = \widehat{p} \sum_{i=1}^{n} x_i + n\widehat{q}. \tag{11.1.5}$$

We can obtain the estimates of regression coefficients from (11.1.5), they are

$$\widehat{p} = \frac{\sum_{i=1}^{n} x_i y_i - n\overline{x}\,\overline{y}}{\sum_{i=1}^{n} x_i^2 - n\overline{x}^2}, \quad \widehat{q} = \overline{y} - \widehat{p}\,\overline{x}. \tag{11.1.6}$$

Then we can obtain the equation of linear regression as shown in Fig. 11.2 by substituting the 20 groups of data into (11.1.7), that is

$$y = -1.5844x + 5.9004.$$

From Lemma 11.1, the approximation of fractal dimension of the set of initial cell densities of H. circularisquama and P. dentatum of system (11.1.2) is 1.5844. Here the correlation coefficient $R = 0.9986$ is very close to 1, which signifies that the fitting effect and reliability are good, and the set of initial cell densities of H. circularisquama and P. dentatum of system (11.1.2) has obvious fractal characteristics.

**Relationship Between Fractal Dimensions and Size of the Sets of Initial Cell Densities of H. Circularisquama and P. Dentatum**

According to the previous discussion, we only change the value of the parameter $\beta_1$ $(0 < \beta_1 < 1)$, then 10 groups of data of fractal dimensions of the set of initial cell densities of H. circularisquama and P. dentatum of system (11.1.2) can be obtained as shown in Table 11.2, where $D$ is the fractal dimension, and $N$ is the number of small squares covering the set of initial cell densities of H. circularisquama and P. dentatum of system (11.1.2), when the side length of small squares is 0.5. It can be seen form Table 11.2 that the fractal dimensions and the size of the sets of initial cell densities

**Table 11.2** Data results

| $\beta_1$ | D | N | $\beta_1$ | D | N |
|---|---|---|---|---|---|
| 0.15 | 1.5844 | 1236 | 0.60 | 1.4216 | 493 |
| 0.20 | 1.5471 | 954 | 0.65 | 1.4103 | 469 |
| 0.30 | 1.4961 | 746 | 0.70 | 1.3926 | 455 |
| 0.40 | 1.4389 | 601 | 0.80 | 1.3878 | 433 |
| 0.50 | 1.4374 | 545 | 0.90 | 1.3589 | 396 |

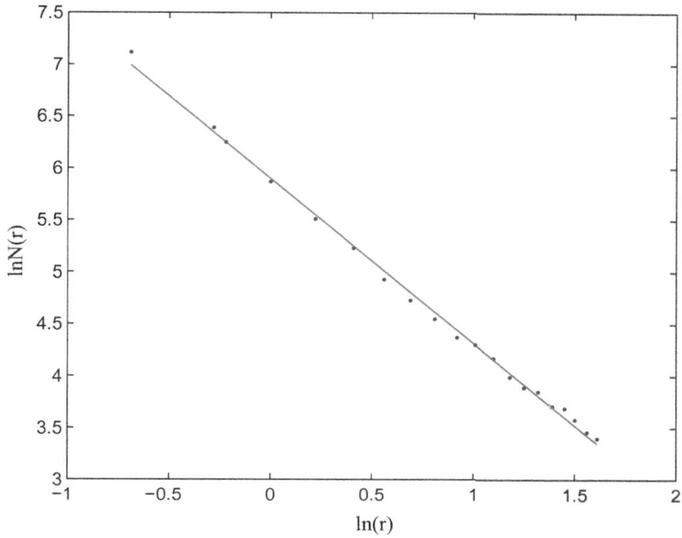


**Fig. 11.2** Fractal dimension of the set of initial cell densities of H. circularisquama and P. dentatum of system (11.1.2)

of H. circularisquama and P. dentatum of system (11.1.2) decrease gradually with the decrease of the parameter $\beta_1$. Figure 11.2 is the fitting curve of the parameter $\beta_1$ and the fractal dimension $D$, that is

$$D = 0.3595e^{-8.515\beta_1} + 1.512e^{-0.1135\beta_1}.$$

Figure 11.3 is the fitting curve of the fractal dimension $D$ and the number of small squares $N$, that is

$$N = 0.381e^{589D}.$$

**Control of the Set of Initial Cell Densities of H. Circularisquama and P. Dentatum**

In an effort to prevent the occurrence of harmful algal blooms and protect marine environment, studies on algae control are a very important topic. In this part, we will control the set of initial cell densities of H. circularisquama and P. dentatum to realize the goal of controlling the growth of H. circularisquama and P. dentatum

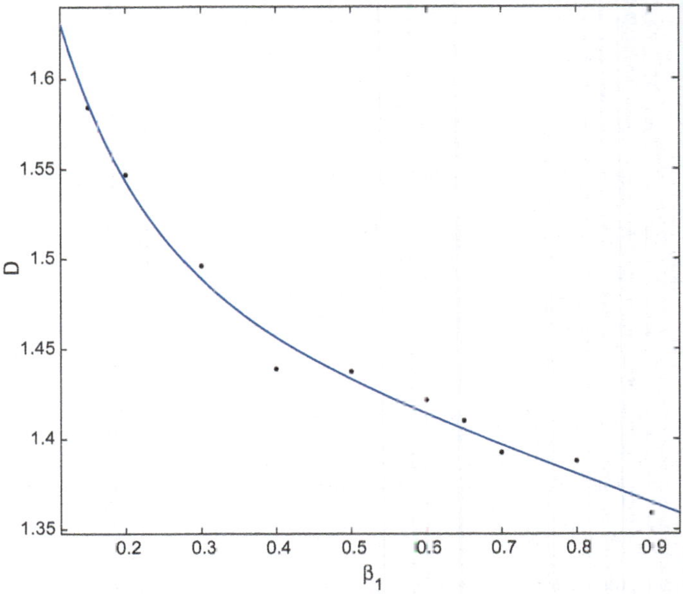


**Fig. 11.3** The changing of the fractal dimension $D$ of the set of initial cell densities of H. circularisquama and P. dentatum of system (11.1.2) with the parameter $\beta_1$

cells. Control terms are introduced to system (11.1.2) based on the optimal function method [330], then we have (Fig. 11.4)

$$\begin{cases} x_{n+1} = (1+\alpha_1)x_n - \beta_1 x_n y_n - \gamma_1 x_n^2 + p_1, \\ y_{n+1} = (1+\alpha_2)y_n - \beta_2 x_n y_n - \gamma_1 y_n^2 + p_2, \end{cases} \tag{11.1.7}$$

here the control terms are

$$p_1 = -l_1\big[(1+\alpha_1)x_n - \beta_1 x_n y_n - \gamma_1 x_n^2 - x_n\big],$$
$$p_2 = -l_2\big[(1+\alpha_2)y_n - \beta_2 x_n y_n - \gamma_2 y_n^2 - y_n\big],$$

where $l_1$, $l_2$ are the control parameters.

We hope that the fixed points of system (11.1.2) becomes stable in the controlled system (11.1.7) when the values of $l_i$ $(i = 1, 2)$ are in a certain range. The Jacobi matrix of the controlled system (11.1.7) is

$$J = \begin{bmatrix} A & C \\ B & D \end{bmatrix},$$

where

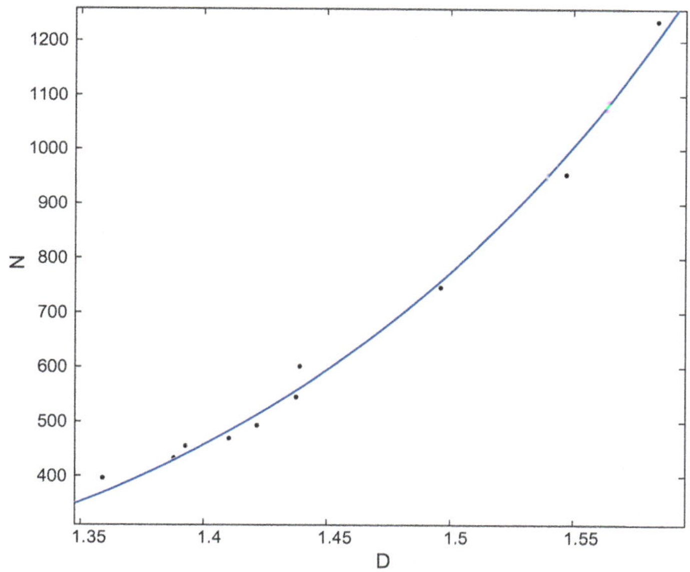


**Fig. 11.4** The changing of the number of small squares $N$ with the fractal dimension $D$ of the set of initial cell densities of H. circularisquama and P. dentatum of system (11.1.2)

$$
\begin{aligned}
A &= (1-l_1)(1+\alpha_1-\beta_1 y-2\gamma_1 x)+l_1,\\
D &= (1-l_2)(1+\alpha_2-\beta_2 x-2\gamma_2 y)+l_2,\\
B &= -\beta_2(1-l_2)y, \quad C=-\beta_1(1-l_1)x.
\end{aligned}
$$

By $|\lambda I - J| = 0$, the characteristic equation of the Jacobi matrix of $J$ is

$$\lambda^2-(A+D)\lambda+AD-BC=0.$$

Then the eigenvalues are

$$\lambda_{1,2}=\frac{A+D\pm\sqrt{(A-D)^2+4BC}}{2}. \tag{11.1.8}$$

Due to systems (11.1.2) and (11.1.7) have the same fixed points, we can get the range of the control parameters $l_i$ $(i=1,2)$ through a stable condition for the unstable fixed point:

$$|\lambda_i|<1 \ \ (i=1,2). \tag{11.1.9}$$

By calculation, the fixed points of system (11.1.2) are

$$P_1 = (0, 0), \quad P_2 = \left(0, \frac{\alpha_2}{\gamma_2}\right) \quad P_3 = \left(\frac{\alpha_1}{\gamma_1}, 0\right),$$
$$P_4 = \left(\frac{\alpha_1\gamma_2 - \alpha_2\beta_1}{\gamma_1\gamma_2 - \beta_1\beta_2}, \frac{\alpha_1}{\beta_1} - \frac{\alpha_1\gamma_1\gamma_2 - \alpha_2\beta_1\gamma_1}{\beta_1\gamma_1\gamma_2 - \beta_1^2\beta_2}\right).$$

For example, we consider the fixed point $P_4$. When $\alpha_1 = 0.2, \beta_1 = 0.15, \gamma_1 = 0.1, \alpha_2 = 0.3, \ \beta_2 = 0.25, \gamma_2 = 0.15, \ P_4 = (2/3, 8/9)$. In order to make it stable, from (11.1.8) and (11.1.9), the control parameter $l_1, l_2$ should satisfy

$$\left|\frac{\frac{27+l_1+2l_2}{15} \pm \sqrt{(\frac{1+l_1-2l_2}{15})^2 + \frac{4(1-l_1)(1-l_2)}{45}}}{2}\right| < 1. \tag{11.1.10}$$

Then we can obtain the range of the control parameter $l_1, l_2$.

For simplification, let $l_1 = l_2 = l$, from (11.1.10), we get the range of the control parameter $l$: $-6.843 < l < 1$. From Fig. 11.5, in the range of the control parameter $l$, with the gradual increase of $l$, sets of initial cell densities of H. circularisquama and P. dentatum of the controlled system (11.1.7) also gradually increase so that we can achieve the goal of controlling the size of the sets of initial cell densities of H. circularisquama and P. dentatum of system (11.1.2).

**Approximate or Same Behaviors of Sets of Initial Cell Densities of H. Circularisquama and P. Dentatum**

According to the above discussion, we only study the size control of one set of initial cell densities of H. circularisquama and P. dentatum, and now the size control of two different sets of initial cell densities of H. circularisquama and P. dentatum is investigated. One set of initial cell densities of H. circularisquama and P. dentatum is approximate or same to the another set of initial cell densities of H. circularisquama and P. dentatum, which is actually synchronization in the nonlinear science. Synchronization has always been a hot research topic, and many methods have been designed to achieve synchronization, such as coupling synchronization, generalized synchronization, gradient synchronization, optimal function synchronization [222, 230, 488, 489] and so on. In this part, we discuss approximate or same behaviors of the sets of initial cell densities of H. circularisquama and P. dentatum of system (11.1.2) using the gradient synchronization.

**Method**

Given two systems that describe the growth of H. circularisquama and P. dentatum cells on the complex plane:

$$x_{n+1} = f(x_n), \tag{11.1.11}$$
$$y_{n+1} = g(y_n). \tag{11.1.12}$$

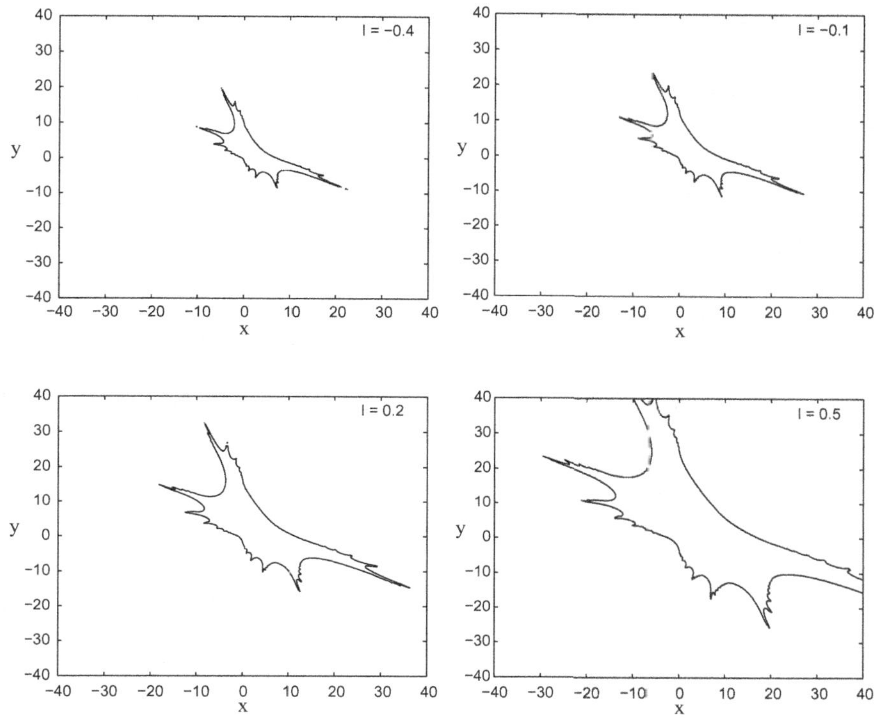


**Fig. 11.5** The changing of sets of initial cell densities of H. circularisquama and P. dentatum of the controlled system (11.1.7), when $\alpha_1 = 0.2, \beta_1 = 0.15, \gamma_1 = 0.1, \alpha_2 = 0.3, \beta_2 = 0.25, \gamma_2 = 0.16$

Once the system parameters are given, the corresponding sets of initial cell densities of H. circularisquama and P. dentatum are determinate, and there is no correlation between them. In an effort to make the sets of initial cell densities of H. circularisquama and P. dentatum of systems (11.1.11) and (11.1.12) associate, the coupling items are introduced into systems (11.1.11) and (11.1.12), then we have

$$x_{n+1} = f(x_n) + p(x_n, y_n; k), \tag{11.1.13}$$

$$y_{n+1} = g(y_n) + q(x_n, y_n; k), \tag{11.1.14}$$

where $p(x_n, y_n; k), q(x_n, y_n; k)$ are the coupling items, $k$ is the coupling strength. Let $J_1, J_2$ be the sets of initial cell densities of H. circularisquama and P. dentatum of the coupling systems (11.1.13) and (11.1.14) respectively. If

$$\lim_{k \to k_0} (J_1 \cup J_2 - J_1 \cap J_2) = \varnothing, \tag{11.1.15}$$

for some $k_0$, approximate or same behaviors of the set of initial cell densities of H. circularisquama and P. dentatum of systems (11.1.11) and (11.1.12) are achieved.

**Implementation of Approximate or Same Behaviors of Sets of Initial Cell Densities of H. Circularisquama and P. Dentatum**

For a system with the same form to (11.1.2) but with different parameters, whose the set of initial cell densities of H. circularisquama and P. dentatum is different from that of system (11.1.2). In the experiment, to begin with, we set the final growth state of H. circularisquama and P. dentatum cells of one system as the target and consider making H. circularisquama and P. dentatum cells growth of another system grow according to the given growth target in same culture environment but with different initial cell densities so that we need to achieve the goal that one set of initial cell densities of H. circularisquama and P. dentatum is approximate or same to the another set of initial cell densities of H. circularisquama and P. dentatum. For system:

$$\begin{cases} u_{n+1} = (1+\alpha_1)u_n - \beta_1^{'} u_n v_n - \gamma_1 u_n^2, \\ v_{n+1} = (1+\alpha_2)v_n - \beta_2^{'} u_n v_n - \gamma_2 v_n^2, \end{cases} \tag{11.1.16}$$

where $\beta_1^{'} \neq \beta_1, \beta_2^{'} \neq \beta_2$, for the convenience of discussion, let

$$\begin{aligned} G_1(x_n, y_n) &= (1+\alpha_1)x_n - \beta_1 x_n y_n - \gamma_1 x_n^2, \\ H_1(u_n, v_n) &= (1+\alpha_1)u_n - \beta_1^{'} u_n v_n - \gamma_1 u_n^2, \\ G_2(x_n, y_n) &= (1+\alpha_2)y_n - \beta_2 x_n y_n - \gamma_2 y_n^2, \\ H_2(u_n, v_n) &= (1+\alpha_2)v_n - \beta_2^{'} u_n v_n - \gamma_2 v_n^2. \end{aligned}$$

To associate the sets of initial cell densities of H. circularisquama and P. dentatum of systems (11.1.2) and (11.1.16), coupling items are introduced into system (11.1.16), then we have

$$\begin{cases} u_{n+1} = (1+\alpha_1)u_n - \beta_1^{'} u_n v_n - \gamma_1 u_n^2 + q_1, \\ v_{n+1} = (1+\alpha_2)v_n - \beta_2^{'} u_n v_n - \gamma_2 v_n^2 + q_2, \end{cases} \tag{11.1.17}$$

where the coupling items are

$$\begin{cases} q_1 = \frac{k}{1+k}(G_1(x_n, y_n) - H_1(u_n, v_n)), \\ q_2 = \frac{k}{1+k}(G_2(x_n, y_n) - H_2(u_n, v_n)). \end{cases}$$

**Theorem 11.1** *Approximate or same behaviors of the sets of initial cell densities of H. circularisquama and P. dentatum of systems (11.1.2) and (11.1.17) can be achieved, when the control parameter k tends to $\infty$.*

***Proof*** From (11.1.2) and (11.1.17), we get

$$\begin{aligned} & x_{n+1} - u_{n+1} \\ & = G_1(x_n, y_n) - \left[ H_1(u_n, v_n) + \frac{k}{1+k}(G_1(x_n, y_n) - H_1(u_n, v_n)) \right] \\ & = \frac{1}{1+k}\left[G_1(x_n, y_n) - H_1(u_n, v_n)\right]. \end{aligned}$$

Let $\alpha = |1 + \alpha_1|$, $\beta = \max(|\beta_1|, |\beta_1^{'}|)$, then

$$\begin{aligned} & |x_{n+1} - u_{n+1}| \\ & = \left| \frac{1}{1+k} \right| |G_1(x_n, y_n) - H_1(u_n, v_n)| \\ & \leq \left|\frac{1}{1+k}\right| \left[\alpha|x_n - u_n| + \beta(|u_n||v_n| + |x_n||y_n|) + |\gamma_1|(|u_n|^2 + |x_n|^2)\right]. \end{aligned} \tag{11.1.18}$$

Due to the set of initial cell densities of H. circularisquama and P. dentatum of system (11.1.2) is bounded, we can gain the set of initial cell densities of H. circularisquama and P. dentatum of system (11.1.2) only considering the iteration of points in a bounded region $D$, in addition, in order to get the set of initial cell densities of H. circularisquama and P. dentatum of system (11.1.2), we only need to think about the iterations of points in $D$ since $z \notin J$ if there is an $n_0$ such that $f_0^n(z) \notin D$, thus there exists a $T > 0$ satisfying $|z| < T$ for all $z \in D$. Since $\beta_i, \beta_i^{'}, \alpha_i, \gamma_i$ $(i = 1, 2)$ are bounded, there are $M, N, M^{'}, N^{'} > 0$ such that

$$\begin{aligned} & \beta(|u_n||v_n| + |x_n||y_n|) < M, \quad \gamma_1|(|u_n|^2 + |x_n|^2) < N, \\ & \beta^{'}(|u_n||v_n| + |x_n||y_n|) < M^{'}, \quad |\gamma_2|(|v_n|^2 + |y_n|^2) < N^{'}. \end{aligned}$$

From (11.1.18),we have

$$\begin{aligned} & |x_{n+1} - u_{n+1}| \\ & < \alpha\left|\frac{1}{1+k}\right| |x_n - u_n| + \left|\frac{1}{1+k}\right| (M+N) \\ & < \alpha\left|\frac{1}{1+k}\right| \left[\alpha\left|\frac{1}{1+k}\right| |x_{n-1} - u_{n-1}| + \left|\frac{1}{1+k}\right| (M+N)\right] + \left|\frac{1}{1+k}\right| (M+N) \\ & = \left(\alpha\left|\frac{1}{1+k}\right|\right)^2 |x_{n-1} - u_{n-1}| + \left|\frac{1}{1+k}\right| (M+N)\left(\alpha\left|\frac{1}{1+k}\right| + 1\right) \\ & < \cdots \\ & < \left(\alpha\left|\frac{1}{1+k}\right|\right)^n |x_1 - u_1| + \left|\frac{1}{1+k}\right| (M+N)\left(\left(\alpha\left|\frac{1}{1+k}\right|\right)^{n-1} \right. \\ & \quad \left. + \cdots + \alpha\left|\frac{1}{1+k}\right| + 1\right) \end{aligned}$$

$$= \left(\alpha \left|\frac{1}{1+k}\right|\right)^n |x_1 - u_1| + \left|\frac{1}{1+k}\right| (M+N) \frac{1-\left(\alpha \left|\frac{1}{1+k}\right|\right)^n}{1-\alpha \left|\frac{1}{1+k}\right|}, \quad (11.1\ 19)$$

when $n \to +\infty$, $|1/(1+k)| \to 0$, that is $k \to \infty$,

$$|x_{n+1} - u_{n+1}| \to 0.$$

Similarly,

$$\begin{aligned} & y_{n+1} - v_{n+1} \\ & = G_2(x_n, y_n) - \left[H_2(u_n, v_n) + \frac{k}{1+k}(G_2(x_n, y_n) - H_2(u_n, v_n))\right] \\ & = \frac{1}{1+k}\left[G_2(x_n, y_n) - H_2(u_n, v_n)\right], \end{aligned}$$

let $\alpha' = |1+\alpha_2|$, $\beta' = \max(|\beta_2|, |\beta_2'|)$, then

$$\begin{aligned} & |y_{n+1} - v_{n+1}| \\ & = \left|\frac{1}{1+k}\right| |G_2(x_n, y_n) - H_2(u_n, v_n)| \\ & \le \left|\frac{1}{1+k}\right| \left[\alpha' |y_n - v_n| + \beta'(|u_n||v_n| + |x_n||y_n|) + |\gamma_2|(|v_n|^2 + |y_n|^2)\right] \\ & < \alpha' \left|\frac{1}{1+k}\right| |y_n - v_n| + \left|\frac{1}{1+k}\right| (M' + N') \\ & < \alpha' \left|\frac{1}{1+k}\right| \left[\alpha' \left|\frac{1}{1+k}\right| |y_{n-1} - v_{n-1}| + \left|\frac{1}{1+k}\right| (M' + N')\right] \\ & \quad + \left|\frac{1}{1+k}\right| (M' + N') \\ & = \left(\alpha' \left|\frac{1}{1+k}\right|\right)^2 |y_{n-1} - v_{n-1}| + \left|\frac{1}{1+k}\right| (M' + N') \left(\alpha' \left|\frac{1}{1+k}\right| + 1\right) \\ & < \cdots \\ & < \left(\alpha' \left|\frac{1}{1+k}\right|\right)^n |y_1 - v_1| + \left|\frac{1}{1+k}\right| (M' + N') \left(\left(\alpha' \left|\frac{1}{1+k}\right|\right)^{n-1} \right. \\ & \quad \left. + \cdots + \alpha' \left|\frac{1}{1+k}\right| + 1\right) \\ & = \left(\alpha' \left|\frac{1}{1+k}\right|\right)^n |y_1 - v_1| + \left|\frac{1}{1+k}\right| (M' + N') \frac{1 - (\alpha' |\frac{1}{1+k}|)^n}{1 - \alpha' |\frac{1}{1+k}|}, \end{aligned} \quad (11.1\ 20)$$

when $n \to +\infty$, $|1/(1+k)| \to 0$, that is $k \to \infty$,

$$|y_{n+1} - v_{n+1}| \to 0.$$

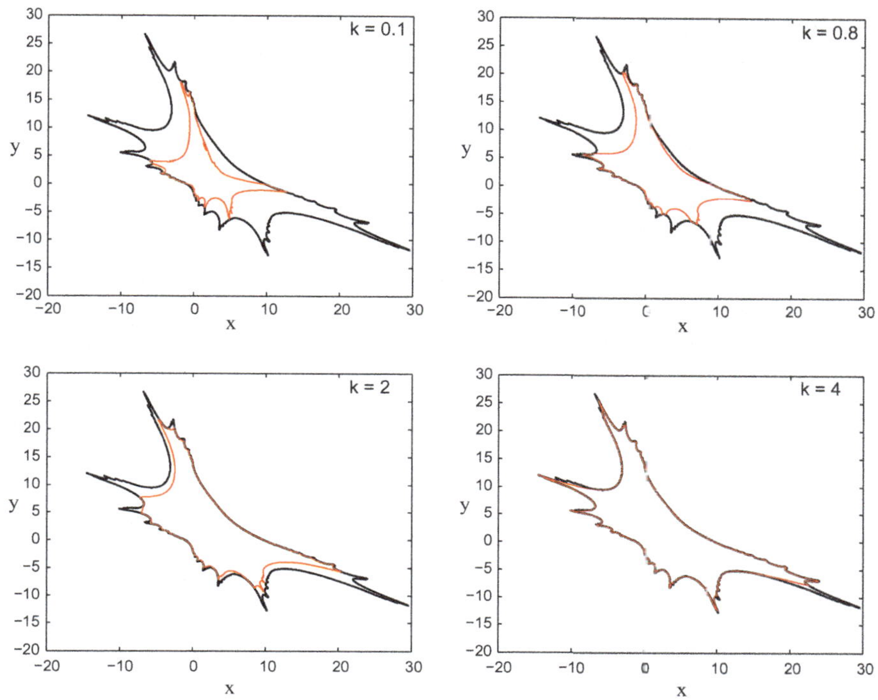


**Fig. 11.6** The changing of the sets of initial densities of H. circularisquama and P. dentatum cells of the controlled system (11.1.17), when $\alpha_1 = 0.2$, $\beta_1' = 0.3$, $\gamma_1 = 0.1$, $\alpha_2 = 0.3$, $\beta_2' = 0.6$, $\gamma_2 = 0.16$

As a result, approximate or same behaviors of the sets of initial cell densities of H. circularisquama and P. dentatum of systems (11.1.2) and (11 1.17) can be achieved, when the control parameter $k$ tends to $\infty$.

Figure 11.6 is the changing of the sets of initial cell densities of H. circularisquama and P. dentatum of the controlled system (11.1.17) with the control parameters. In Fig. 11.6, the red curves are the sets of initial cell densities of H. circularisquama and P. dentatum of the controlled system (11.1.17) with different control parameters and the black curves are the sets of initial cell densities of H. circularisquama and P. dentatum of system (11.1.2) when $\alpha_1 = 0.2$, $\beta_1 = 0.15$, $\gamma_1 = 0.1$, $\alpha_2 = 0.3$, $\beta_2 = 0.25$, $\gamma_2 = 0.16$. It is evident that with the gradual increase of the control parameter $k$, the sets of initial cell densities of H. circularisquama and P.dentatum of the controlled system (11.1.17) more and more approximate the sets of initial cell densities of H. circularisquama and P. dentatum of system (11.1.2).

### 11.1.3 Set of Environmental Factors Disturbing the Growth of H. Circularisquama and P. Dentatum Cells

It must be pointed out that environmental factors in natural waters, such as temperature, salinity, irradiance, pH values, and nutrients are all in the complex dynamic changes, which have a significant effect on the growth and physiological activities to aquatic organisms [490, 491]. In the growth process of H. circularisquama and P. dentatum cells, different environment factors have different influences on their growth. In order to study the influences caused by the environment factors, the set of environmental factors disturbing the growth of H. circularisquama and P. dentatum cells is constructed through introducing two real parameters into system (11.1.2) [492]. Then we get

$$\begin{cases} x_{n+1} = (1+\alpha_1)x_n - \beta_1 x_n y_n - \gamma_1 x_n^2 + c_1, \\ y_{n+1} = (1+\alpha_2)y_n - \beta_2 x_n y_n - \gamma_2 y_n^2 + c_2, \end{cases} \tag{11.1.21}$$

where $c_1, c_2 \in \mathbb{R}$, they are environmental factors disturbing the growth of H. circularisquama and P. dentatum cells, such as temperature, irradiance, oxygen content, PH values, salinity and so on.

**Definition 11.2** The set of environmental factors disturbing the growth of H. circularisquama and P. dentatum cells of system (11.1.21):

$$M = \left\{ (c_1, c_2) \in \mathbb{R}^2 \mid \lim_{k\to\infty} (F^k(x_0, y_0) \nrightarrow \infty \right\},$$

where $F(x, y) = ((1+\alpha_1)x - \beta_1 xy - \gamma_1 x^2 + c_1, (1+\alpha_2)y - \beta_2 xy - \gamma_2 y^2 + c_2)$, and $x_0$, $y_0$ are the initial cell densities of H. circularisquama and P. dentatum of system (11.1.21), respectively.

**Control of Sets of Environmental Factors Disturbing the Growth of H. Circularisquama and P. Dentatum Cells**

In an effort to achieve the goal that make H. circularisquama and P. dentatum cells with persistence have no persistence, control of sets of environmental factors disturbing the growth of H. circularisquama and P. dentatum cells will be discussed in this section. We consider the system as follow:

$$\begin{cases} x_{n+1} = (1+\alpha_1)x_n - \beta_1 x_n y_n - \gamma_1 x_n^2 + tc_1, \\ y_{n+1} = (1+\alpha_2)y_n - \beta_2 x_n y_n - \gamma_2 y_n^2 + tc_2, \end{cases} \tag{11.1.22}$$

where $t \in \mathbb{R}\backslash\{0\}$.

**Theorem 11.2** *The set of environmental factors disturbing the growth of H. circularisquama and P. dentatum cells of system (11.1.22) is*

$$M^* = \left\{\frac{1}{t}(c_1, c_2) : c_i \in M, i = 1, 2\right\}.$$

***Proof*** Let $c_1^{'} = tc_1, c_2^{'} = tc_2$, then the set of environmental factors disturbing the growth of H. circularisquama and P. dentatum cells of system (11.1.22) $M^*$ is same as $M$ about the system parameter $c_1^{'}, c_2^{'}$. Since $c_1 = \frac{1}{t}c_1^{'}, c_2 = \frac{1}{t}c_2^{'}$ and Definition 11.2, we have

$$M^* = \left\{\frac{1}{t}(c_1, c_2) : c_i \in M, i = 1, 2\right\}.$$

From Theorem 11.2, $M^*$ decreases as $t$ increases when $t > 0$ in Fig. 11.7, then the size control of the set of environmental factors disturbing the growth of H. circularisquama and P. dentatum cells of system (11.1.2) is achieved.

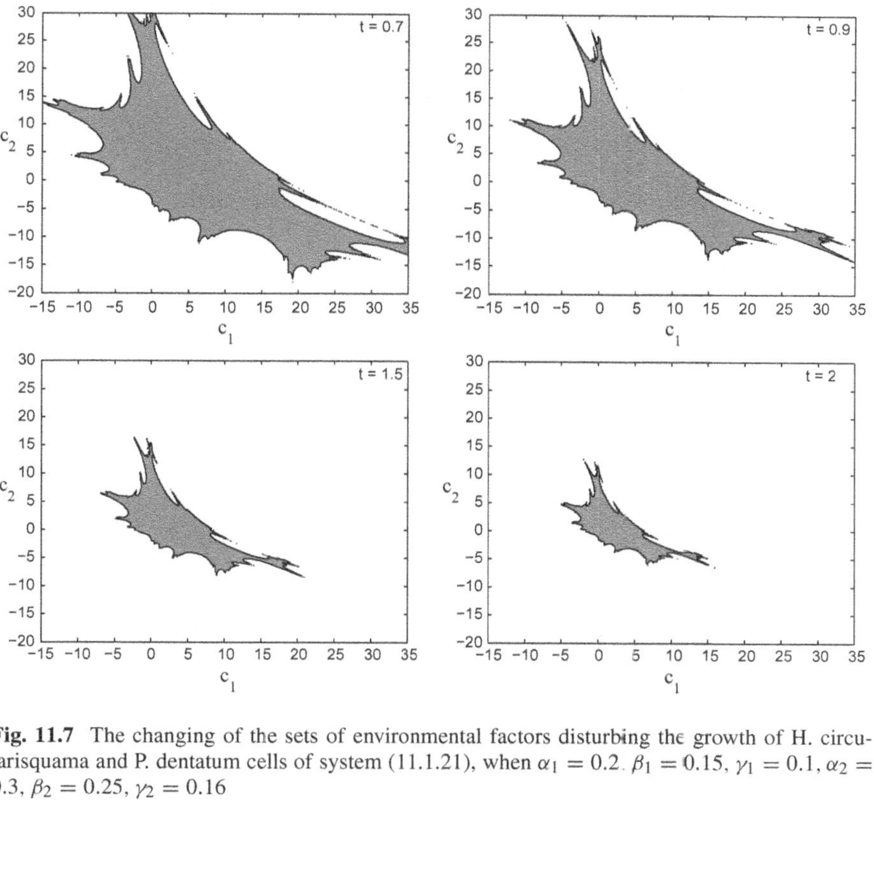


**Fig. 11.7** The changing of the sets of environmental factors disturbing the growth of H. circularisquama and P. dentatum cells of system (11.1.21), when $\alpha_1 = 0.2$, $\beta_1 = 0.15$, $\gamma_1 = 0.1$, $\alpha_2 = 0.3$, $\beta_2 = 0.25$, $\gamma_2 = 0.16$

**Approximate or Same Behaviors of Sets of Environmental Factors Disturbing the Growth of H. Circularisquama and P. Dentatum Cells**
Take a system with the same form to (11.1.2) with different parameters into consideration, whose the set of environmental factors disturbing the growth of H. circularisquama and P. dentatum cells is different from that of system (11.1.2), because the disturbance of environmental factors on H. circularisquama and P. dentatum cells growth in different systems is different. In the experiment, approximate or same behaviors of H. circularisquama and P. dentatum cells growth of different systems with same initial cell densities in different culture environment need to be considered so that we need to take approximate or same behaviors of two different sets of environmental factors disturbing the growth of H. circularisquama and P. dentatum cells into account. For system:

$$\begin{cases} u_{n+1} = (1+\alpha_1)u_n - \theta_1 u_n v_n - \gamma_1 u_n^2, \\ v_{n+1} = (1+\alpha_2)v_n - \theta_2 u_n v_n - \gamma_2 v_n^2, \end{cases} \tag{11.1.23}$$

where $\theta_1 \neq \beta_1$, $\theta_2 \neq \beta_2$. For purpose of getting the set of environmental factors disturbing the growth of H. circularisquama and P. dentatum cells of system (11.1.23), two real parameters are introduced into (11.1.23), then we have

$$\begin{cases} u_{n+1} = (1+\alpha_1)u_n - \theta_1 u_n v_n - \gamma_1 u_n^2 + d_1, \\ v_{n+1} = (1+\alpha_2)v_n - \theta_2 u_n v_n - \gamma_2 v_n^2 + d_2, \end{cases} \tag{11.1.24}$$

where $d_1, d_2 \in \mathbb{R}$. In order to associate the sets of environmental factors disturbing the growth of H. circularisquama and P. dentatum cells of systems (11.1.21) and (11.1.24), coupling items are introduced into system (11.1.24), then we have

$$\begin{cases} u_{n+1} = (1+\alpha_1)u_n - \theta_1 u_n v_n - \gamma_1 u_n^2 + d_1 + g_1, \\ v_{n+1} = (1+\alpha_2)v_n - \theta_2 u_n v_n - \gamma_2 v_n^2 + d_2 + g_2, \end{cases} \tag{11.1.25}$$

here the coupling items are

$$g_1 = r(\theta_1 u_n v_n - \beta_1 x_n y_n + \gamma_1 u_n^2 - \gamma_1 x_n^2) - r(u_n - x_n + d_1 - c_1),$$
$$g_2 = r(\theta_2 u_n v_n - \beta_2 x_n y_n + \gamma_2 v_n^2 - \gamma_2 y_n^2) - r(v_n - y_n + d_2 - c_2).$$

**Theorem 11.3** *Approximate or same behaviors of the sets of environmental factors disturbing the growth of H. circularisquama and P. dentatum cells of systems (11.1.21) and (11.1.25) can be achieved, when the control parameter r tends to 1.*

***Proof*** From (11.1.21) and (11.1.25), we have

$$\begin{aligned} & u_{n+1} - x_{n+1} \\ &= (1+\alpha_1 - r)(u_n - x_n) + (r-1)(\theta_1 u_n v_n - \beta_1 x_n y_n + \gamma_1 u_n^2 - \gamma_1 x_n^2 - c_1 - d_1), \end{aligned}$$

then

$$
\begin{aligned}
&|u_{n+1} - x_{n+1}| \\
&= |(1+\alpha_1 - r)(u_n - x_n) + (r-1)(\theta_1 u_n v_n - \beta_1 x_n y_n + \gamma_1 u_n^2 \\
&\quad - \gamma_1 x_n^2 + c_1 - d_1)|, \\
&\le |1+\alpha_1 - r||u_n - x_n| + |r-1|(\theta_1 |u_n||v_n| + \beta_1 |x_n||y_n| + \gamma_1 |u_n|^2 \\
&\quad + \gamma_1 |x_n|^2 + |c_1| + |d_1|).
\end{aligned}
\tag{11.1.26}
$$

Take 5.2 cells $\text{mL}^{-1}$, 0.5 cells $\text{mL}^{-1}$ as the initial densities of H. circularisquama and P. dentatum cells of system (11.1.21). The sets of environmental factors disturbing the growth of H. circularisquama and P. dentatum cells of system (11.1.21) are bounded based on the Definition 11.2, therefore we can obtain the sets of environmental factors disturbing the growth of H. circularisquama and P. dentatum cells of system (11.1.21) only considering the iteration of points in a bounded region $D$. Since $\alpha_i, \theta_i, \gamma_i$ $(i = 1, 2)$ are bounded, there are $M_0, N_0, M_0^{'}, N_0^{'} > 0$ so that

$$
\begin{aligned}
&\theta_1 |u_n||v_n| + \beta_1 |x_n||y_n| + \gamma_1 |u_n|^2 + \gamma_1 |x_n|^2 < M_0, \\
&\theta_2 |u_n||v_n| + \beta_2 |x_n||y_n| + \gamma_2 |v_n|^2 + \gamma_2 |y_n|^2 < M_0^{'}, \\
&|c_1| + |d_1| < N_0, \quad |c_2| + |d_2| < N_0^{'}.
\end{aligned}
$$

From (11.1.26), we get

$$
\begin{aligned}
&|u_{n+1} - x_{n+1}| \\
&< |1+\alpha_1 - r||u_n - x_n| + |r-1|(M_0 + N_0) \\
&< |1+\alpha_1 - r|\big(|1+\alpha_1 - r||u_{n-1} - x_{n-1}| + |r-1|(M_0 + N_0)\big) \\
&\quad + |r-1|(M_0 + N_0) \\
&= |1+\alpha_1 - r|^2 |u_{n-1} - x_{n-1}| + |r-1|(M_0 + N_0)(1 + |1+\alpha_1 - r|) \\
&< \cdots \\
&< |1+\alpha_1 - r|^n |u_1 - x_1| + |r-1|(M_0 + N_0)(1 + |1+\alpha_1 - r| \\
&\quad + \cdots + |1+\alpha_1 - r|^{n-1}) \\
&= |1+\alpha_1 - r|^n |u_1 - x_1| + |r-1|(M_0 + N_0)\frac{1 - |1+\alpha_1 - r|^n}{1 - |1+\alpha_1 - r|}.
\end{aligned}
\tag{11.1.27}
$$

Since $0 < \alpha_1 < 1$, the limit of the right side of (11.1.27) is $|r-1|(M_0 + N_0)/(1 - |1+\alpha_1 - r|)$ when $n \to +\infty$, $\alpha_1 < r < 1$, thus if $r \to 1$, we have

$$
|u_{n+1} - x_{n+1}| \to 0.
$$

Since

$$
\begin{aligned}
&|v_{n+1} - y_{n+1}| \\
&= |(1+\alpha_2 - r)(v_n - y_n) + (r-1)(\theta_2 u_n v_n - \beta_2 x_n y_n + \gamma_2 v_n^2 - \gamma_2 y_n^2 + c_2 - d_2)| \\
&\le |1+\alpha_2 - r||v_n - y_n| + |r-1|(\theta_2 |u_n||v_n| + \beta_2 |x_n||y_n| + \gamma_2 |v_n|^2 + \gamma_2 |y_n|^2 + |c_2| + |d_2|)
\end{aligned}
$$

$$
\begin{aligned}
&< |1+\alpha_2-r||v_n-y_n|+|r-1|(M_0'+N_0')\\
&< |1+\alpha_2-r|(|1+\alpha_2-r||v_{n-1}-y_{n-1}|+|r-1|(M_0'+N_0'))\\
&\quad +|r-1|(M_0'+N_0')\\
&= |1+\alpha_2-r|^2|v_{n-1}-y_{n-1}|+|r-1|(M_0'+N_0')(1+|1+\alpha_1-r|)\\
&< \cdots\\
&< |1+\alpha_2-r|^n|v_1-y_1|+|r-1|(M_0'+N_0')(1+|1+\alpha_2-r|+\cdots+|1+\alpha_2-r|^{n-1})\\
&= |1+\alpha_2-r|^n|v_1-y_1|+|r-1|(M_0'+N_0')\frac{1-|1+\alpha_2-r|^n}{1-|1+\alpha_2-r|}.
\end{aligned} \tag{11.1.28}
$$

Because of $0 < \alpha_2 < 1$, the limit of the right side of (11.1.28) is $|r-1|(M_0'+N_0')/(1-|1+\alpha_2-r|)$ when $n \to +\infty$, $\alpha_2 < r < 1$, thus if $r \to 1$, we have

$$|v_{n+1}-y_{n+1}| \to 0.$$

As a result, approximate or same behaviors of the sets of environmental factors disturbing the growth of H. circularisquama and P. dentatum cells of systems (11.1.21) and (11.1.25) can be achieved, when the control parameter $r$ tends to 1.

Regardless of the sets of environmental factors disturbing the growth of H. circularisquama and P. dentatum cells are determined when the parameters are given, the following simulation shows that the sets of environmental factors disturbing the growth of H. circularisquama and P. dentatum cells can be changed by taking different coupled parameters $r$. The initial situation of the set of environmental factors disturbing the growth of H. circularisquama and P. dentatum cells of system (11.1.21) as shown in Fig. 11.8, when $\alpha_1 = 0.2$, $\beta_1 = 0.15$, $\gamma_1 = 0.1$, $\alpha_2 = 0.3$, $\beta_2 = 0.25$, $\gamma_2 = 0.15$. Figure 11.9 is the changing of the sets of environmental factors disturbing the growth of H. circularisquama and P. dentatum cells sets of system (11.1.25) with the changing of the control parameters. It is evident that with the gradual increase of the control parameter $r$, $r$ tends to 1, the sets of environmental factors disturbing the growth of H. circularisquama and P. dentatum cells of system (11.1.25) are changing to be the sets of environmental factors disturbing the growth of H. circularisquama and P. dentatum cells of system (11.1.21).

## 11.2 Fractal Analysis and Prediction of Global Warming

### *11.2.1 Background*

Everything in nature is infinitely mysterious, the rolling mountains, the winding coastline, the changing clouds, the rising smoke, the beautiful snowflakes (Fig. 11.10), strange shapes of lightning, thousand miles of rivers, random material cracks, ..., countless. Most of these natural phenomena are unstable, unbalanced and random, and belong to nonlinear state. Behind these complex phenomena, there is a certain regularity. The research object of classical mathematical geometry is regular and smooth geometric configuration, but it is helpless to complex, random and

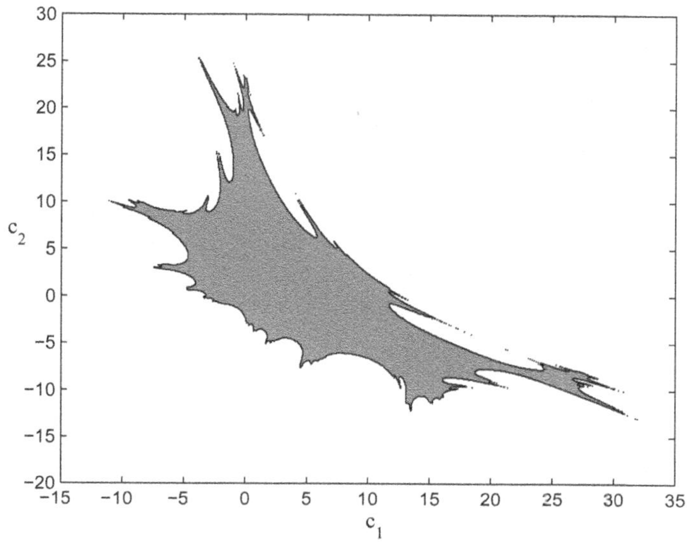


**Fig. 11.8** The original set of environmental factors disturbing the growth of H. circularisquama and P. dentatum cells of the controlled system (11.1.21), when $\alpha_1 = 0.2$, $\beta_1 = 0.15$, $\gamma_1 = 0.1$, $\alpha_2 = 0.3$, $\beta_2 = 0.25$, $\gamma_2 = 0.16$

ruleless nonlinear system in nature [493]. Fractal theory, founded in the mid-1970s, studies disordered and self-similar systems in the natural world, economic society and engineering technology, allowing people to deal with these nonlinear problems with new ideas and methods. Engineering technologies support the national economic construction and development of various industries, its upgrading and innovation have driven the technical progress of the industry. Problems in the field of engineering technologies are often highly complex and random, and fractal attempts to reveal the hidden rules behind complex phenomena, which has become an important theoretical tool to solve these engineering technology problems.

As one of the most classical fractal phenomena mentioned above, the fractal phenomenon of complex dynamical system was proposed by Julia and Mandelbrot [164] and then widely used in physics, biological simulation, image processing, control and other fields. Based on the above analysis, the fractal theory of complex dynamic system is applied to the study of glaciers, ice layers and islands. The concept of temperature parameter is proposed to describe the melting process of glacier and ice layers and the degradation of islands with the change of temperature parameter.

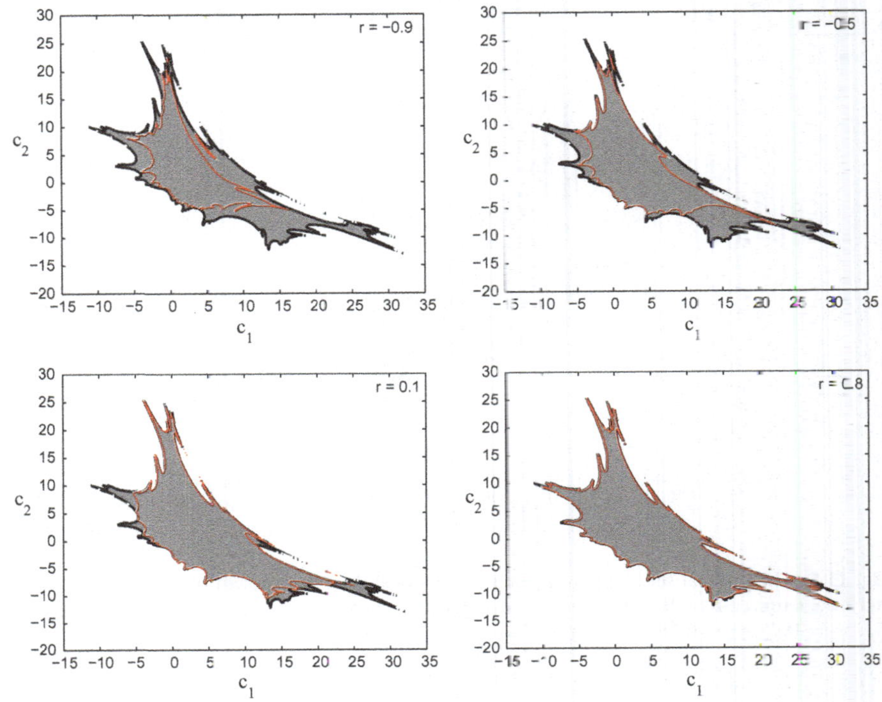


**Fig. 11.9** The changing of the sets of environmental factors disturbing the growth of H. circularisquama and P. dentatum cells of the controlled system (11.1.25), when $\alpha_1 = 0.2, \theta_1 = 0.21, \gamma_1 = 0.1, \alpha_2 = 0.3, \theta_2 = 0.17, \gamma_2 = 0.16$

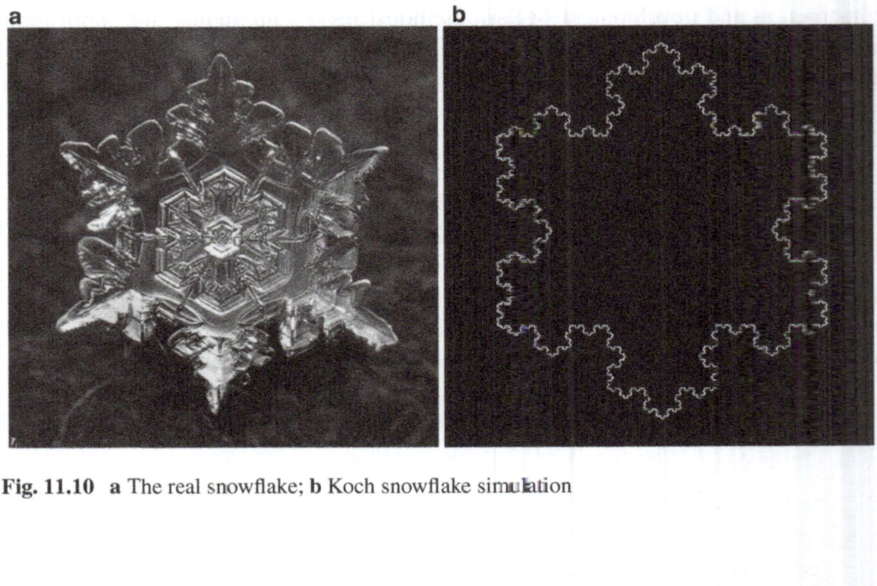


**Fig. 11.10** **a** The real snowflake; **b** Koch snowflake simulation

### 11.2.2 Fractal Characterization and Parameter Adjustment of Islands, Glaciers and Ice Layers

**The Island**

**Definition 11.3** Given a complex system

$$f: \ z_{n+1} = z_n^2 + z_{n-1} + c, \tag{11.2.1}$$

where $z_n, c \in \mathbb{C}$. The filled Julia set of system $f$ is

$$K(f) = \{z_0 \mid f^n(z_0) \nrightarrow \infty\}.$$

Especially, in (11.2.1), a large number of numerical simulations have shown that filled Julia set exactly represents the changes of islands in the ocean, take $z_{-1} = 0.1 + 0.6i$, $c = 0$, the corresponding situation for the New Guinea island is obtained. Other research information about system (11.2.1) can refer to the literatures [187, 208, 230, 428, 469, 494–498] (Fig. 11.11).

In order to describe the evolution process of islands, parameter $\alpha$ is introduced into (11.2.1), $\alpha \in \mathbb{R}^+$, then

$$f: \ z_{n+1} = (\alpha z_n)^2 + z_{n-1} + c, \tag{11.2.2}$$

where $\alpha$ is called the factor about island expansion and contraction.

The Relationship between Island Variation and Fractal Dimension

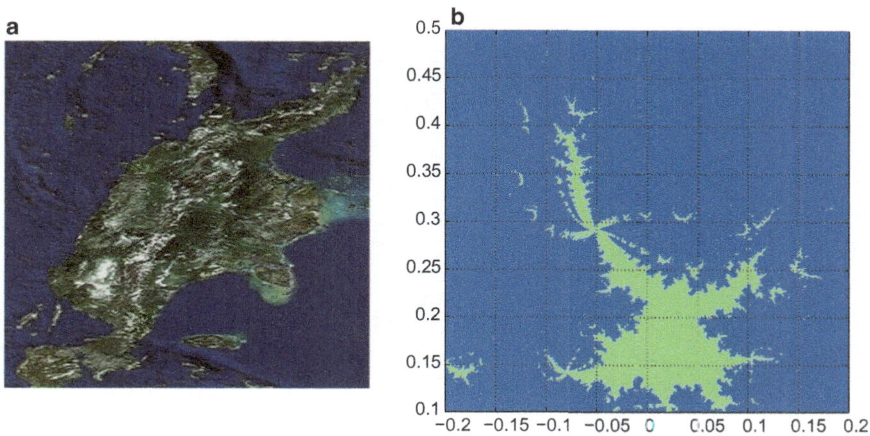


**Fig. 11.11** **a** New Guinea island plate; **b** Numerical simulation of the island plate

**Table 11.3** Data results

| $D_1$ | 1.342 | 1.276 | 1.259 | 1.172 | 1.151 | 1.134 | 1.099 | 0.984 | 1.015 | 0.884 | 0.938 | 0.875 |
|---|---|---|---|---|---|---|---|---|---|---|---|---|
| $N_1$ | 972 | 744 | 688 | 525 | 408 | 363 | 292 | 218 | 171 | 152 | 125 | 87 |

Global climate warming has caused the melting of the arctic sea ice and led to the gradual degradation of islands. From the perspective of the fractal, the process is the degradation of filled Julia set. The fractal dimension characterizes the variation degree of fractal geometry, which means the change of fractal figures will cause the change of fractal dimension, the fractal dimension corresponds to the fractal figure. Based on the discussion of the system (11.2.1), we know the island fractal form, therefore we discuss the relationship between island size and fractal dimension. By adjusting the factor about expansion and contraction, we can get different fractal figures of the island evolution, the box dimension method [486] was used to calculate the corresponding 12 groups of fractal dimensions $D_1$, as shown in Table 11.3:

Where $N_1$ represents the number of small squares covering the island fractal figure when taking the edge length of the small square as 0.02, and the small square belongs to a plane that dividing the island fractal figure, besides, $N_1$ shows the size of the island.

The data in Table 11.3 were fitted to obtain the numerical curve as shown in Fig. 11.12.

$$N_1 = 1.767\exp(4.723D_1). \tag{11.23}$$

As can be seen from Fig. 11.12, when the island fractal dimension decreases gradually, the area of the island also decreases gradually. We can get the changing

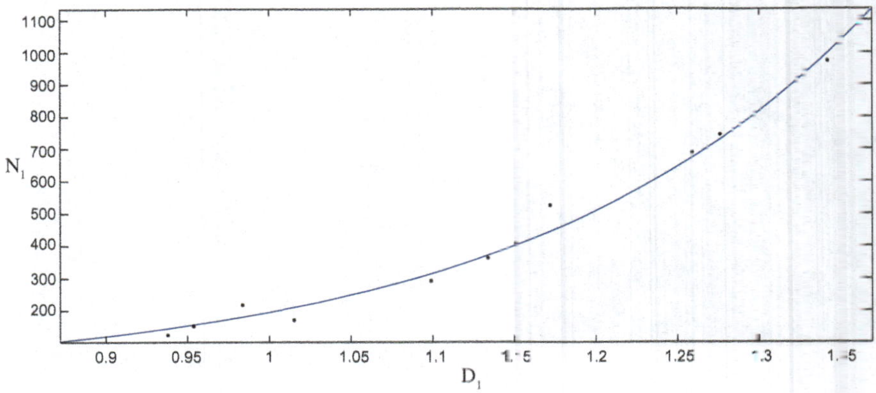


**Fig. 11.12** The size of the island $N_1$ varies with the fractal dimension $D_1$

**Table 11.4** The islands change with the fractal dimensions $D_1$

| $D_1$ | 1.259 | 1.151 | 1.134 | 0.984 |
|---|---|---|---|---|
| | 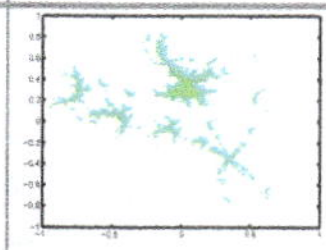 | 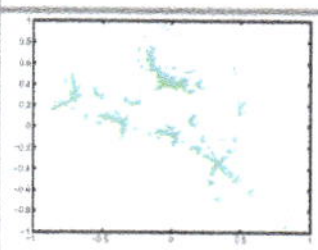 | 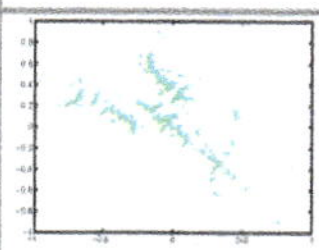 | 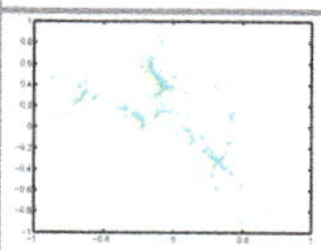 |

situation between the size of the island and the island fractal dimension, as shown in Table 11.4.

Thus, we get the following conclusion: from the perspective of numerical analysis, by using the dimension variation of island fractal geometry, the variation of fractal dimension can be used to predict the degradation of island as temperatures change. In other words, the change of island area can be predicted by the quantitative expression (11.2.3) of fractal dimension.

The above describes the change of the island as parameters change through the fractal point set of the plane complex dynamic system. Next, we expand the complex dynamic system into space to describe the process of the glacier melting.

**The Glacier**

Consider a complex dynamical system with two independent variables in the form of distributed parameters

$$\frac{\partial \omega(z,v)}{\partial z} + a\frac{\partial \omega(z,v)}{\partial \omega} = f((1+a)\omega(z,v)), \tag{11.2.4}$$

where $z = x + yi,\ v = s + ti$, its discrete form is as follows

$$z_{m+1,n} + a z_{m,n+1} = F(c, (1+a)z_{m,n}). \tag{11.2.5}$$

where $F(\cdot)$ is a nonlinear function, $a$ is a real parameter and $a \neq -1$. Take $F(c,(1+a)z_{m,n}) = (1+a)^2 z_{m,n}^2 + c$, the following system is obtained

$$z_{m+1,n} + a z_{m,n+1} = (1+a)^2 z_{m,n}^2 + c. \tag{11.2.6}$$

The concept of the filled Julia set of system (11.2.4) is the same as Definition 11.3. Because $z = x + yi,\ v = s + ti$, the fractal set is described in three-dimensional space, when taking $t = 0$, $a = 0.5$, $c = -0.5 + 0.1i$, the following simulation figure of the glacier structure can be obtained (Fig. 11.13).

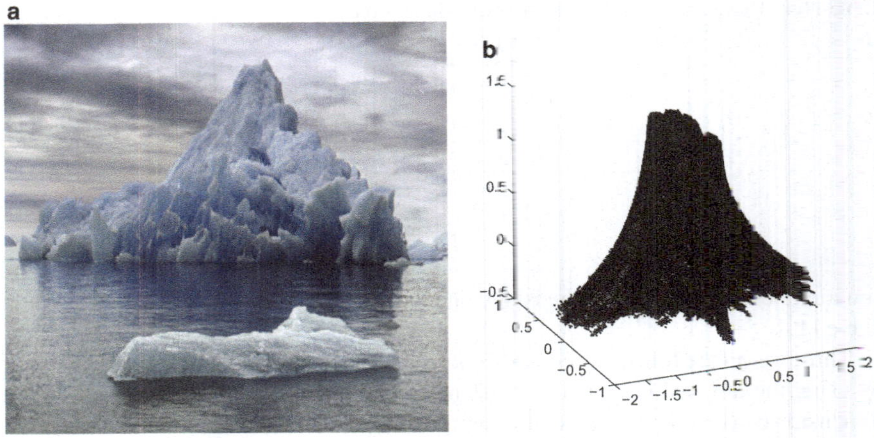


**Fig. 11.13** **a** Actual arctic glacier; **b** Numerical simulation of the glacier

From the result of literature [499], when taking the glacier factor about expansion and contraction as $\beta$, $\beta \in \mathbb{R}^+$, the following controlled complex system can be obtained

$$F: \ z_{m+1,n} + az_{m,n+1} = (1+a)^2(\beta z_{m,n})^2 + c, \tag{11.2.7}$$

therefore the size of filled Julia set $J_\beta(F)$ of controlled system is $\frac{1}{\beta}$ times of original filled Julia set $J(F)$. And other research information related to (11.2.7) can refer to [197, 221, 345, 421, 500–506].

For example, given different glacier factors about expansion and contraction, the following glacier variation with the change of factors can be obtained (Fig. 11.14)

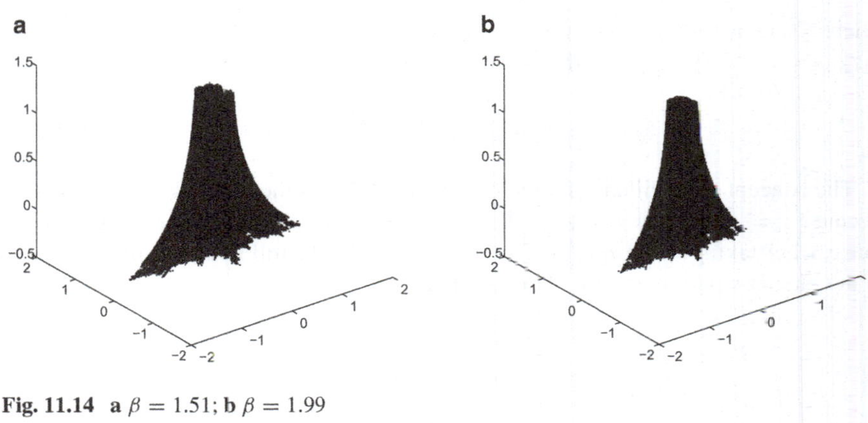


**Fig. 11.14** **a** $\beta = 1.51$; **b** $\beta = 1.99$

**The Ice Layer**

Consider the following real dynamical systems with multiple parameters and two variables

$$\begin{cases} x_{n+1} = (1 + r_1)x_n - s_1 x_n y_n - t_1 x_n^2, \\ y_{n+1} = (1 + r_2)y_n - s_2 x_n y_n - t_2 y_n^2. \end{cases} \tag{11.2.8}$$

The concept of the filled Julia set of system (11.2.8) is the same as definition 1. And research information related to (11.2.8) can refer to literatures [207, 222, 399, 507–513].

As global warming intensifies, the melting rate of arctic ice increases year by year. Figure 11.15a represents the degradation of arctic ice during 1979–2007. As for this degradation process, a melting factor is actually introduced into the system (11.2.8) $\gamma, \gamma \in \mathbb{R}^+$, the controlled complex system is obtained.

$$\begin{cases} x_{n+1} = (1 + r_1)x_n - s_1 \gamma x_n y_n - (t_1 \gamma x_n)^2, \\ y_{n+1} = (1 + r_2)y_n - s_2 \gamma x_n y_n - t_2 y_n^2, \end{cases} \tag{11.2.9}$$

When the melting factor $\gamma$ changes, the changes of ice layers are shown in Fig. 11.15b.

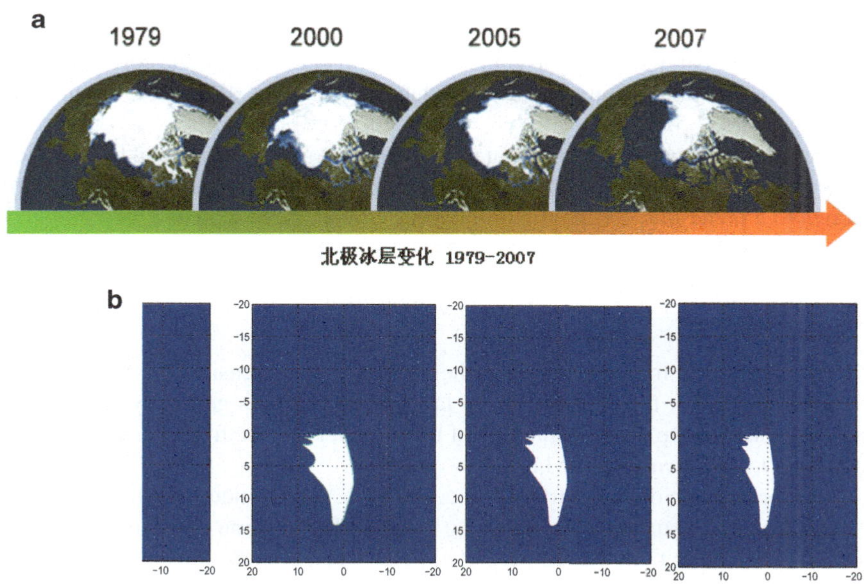


**Fig. 11.15** **a** Ice plates in reality; **b** Numerical simulation of the ice plates when taking $\gamma = 0.71, 1.01, 1.21, 1.41$

**Table 11.5** Data result

| $D_2$ | 1.811 | 1.788 | 1.787 | 1.758 | 1.746 | 1.726 | 1.694 | 1.690 | 1.685 | 1.684 | 1.676 | 1.6[illegible]3 |
|---|---|---|---|---|---|---|---|---|---|---|---|---|
| $N_2$ | 518 | 473 | 432 | 373 | 349 | 333 | 295 | 271 | 261 | 247 | 236 | 228 |

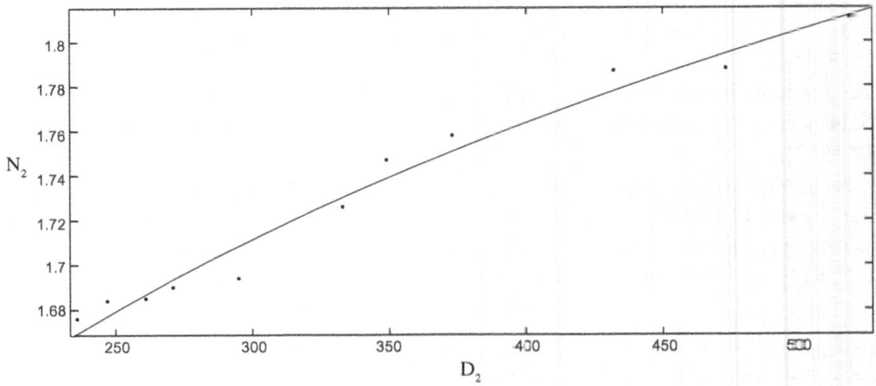


**Fig. 11.16** The variation of ice layer size $N_2$ with fractal dimension $D_2$

The relationship between ice layer variations and fractal dimensions Similar to the previous discussion on the relationship between island changes and fractal dimension, the fractal diagrams of different ice layers variations are obtained by adjusting the parameters in the controlled system (11.2.9), and the corresponding 12 group fractal dimensions $D_2$ are calculated by the box dimension method, as shown in Table 11.5:

Where $N_2$ represents the number of small squares covering the fractal figure of ice layers when taking the edge length of the small square as 0.3, and the small square belongs to a plane that dividing the ice layers fractal figure. The data in Table 11.5 were fitted to obtain the changing curve as shown in Fig. 11.16.

$$N_2 = 0.946 D_2^{0.104} \tag{11.2.10}$$

As can be seen from Fig. 11.16, when the fractal dimension of the ice layer decreases gradually, the area of the ice layer also decreases gradually. The corresponding relationship between the size of the ice layer and the fractal dimension can be obtained, as shown in Table 11.6.

Therefore we can draw the following conclusion: the geometric change of the ice layer on the earth's surface corresponds to the fractal dimension of the ice layer, thus the retreat of the ice plate can be predicted by its fractal dimension.

**Table 11.6** The relationship between the evolution and the fractal dimension of ice layer $D_2$

| $D_2$ | 1.812 | 1.749 | 1.707 | 1.686 |
|---|---|---|---|---|
| | 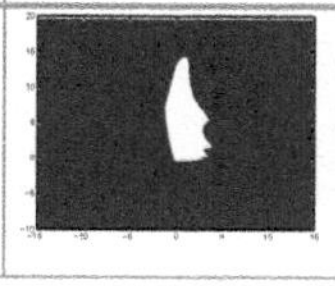 | 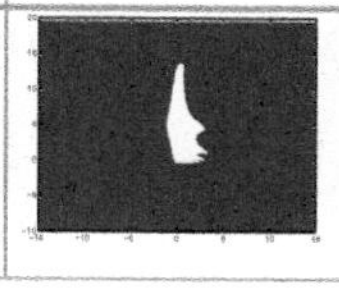 | 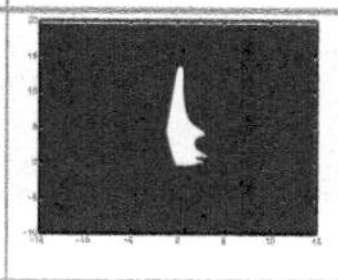 | 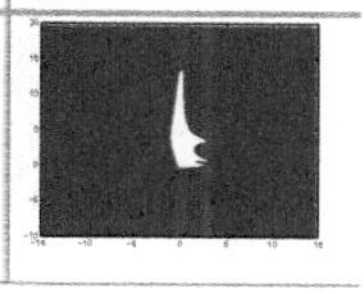 |

### *11.2.3 Fractal Predictions of Land and Ice Layer from Global Warming*

In the previous discussion on island size and fractal dimension, we can also use the island factors about expansion and contraction $\alpha$ to get the relationship between them and dimensions $D_1$, and obtain the data as shown in Table 11.7.

The data in Table 11.7 were fitted to obtain the numerical curve as shown in Fig. 11.17.

Similarly, in the above discussion on the size and fractal dimension of the ice layer, the relationship between the melting factor $\gamma$ and dimension $D_2$ can also be obtained, and the data are shown in Table 11.8.

The data in Table 11.8 were fitted to obtain the numerical curve as shown in Fig. 11.18.

**Table 11.7** Data result

| $\alpha$ | 0.95 | 0.98 | 1.00 | 1.02 | 1.05 | 1.10 | 1.12 | 1.15 | 1.17 | 1.20 | 1.25 | 1.30 |
|---|---|---|---|---|---|---|---|---|---|---|---|---|
| $D_1$ | 1.342 | 1.276 | 1.259 | 1.172 | 1.151 | 1.134 | 1.099 | 0.984 | 1.015 | 0.884 | 0.938 | 0.895 |

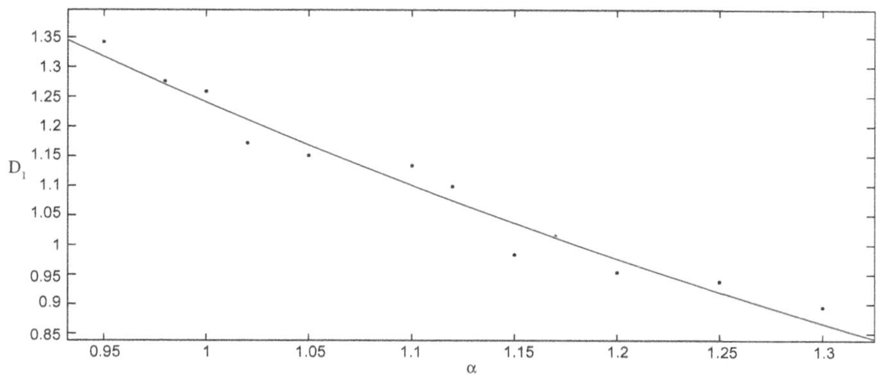


**Fig. 11.17** The variation of island fractal dimension $D_1$ with the factor about expansion and contraction $\alpha$

Table 11.8 Data result

| $\gamma$ | 1.0 | 1.1 | 1.2 | 1.4 | 1.5 | 1.6 | 1.8 | 1.9 | 2.0 | 2.2 | 2.3 | 2.4 |
|---|---|---|---|---|---|---|---|---|---|---|---|---|
| $D_2$ | 1.811 | 1.788 | 1.787 | 1.758 | 1.746 | 1.726 | 1.694 | 1.690 | 1.685 | 1.684 | 1.67[illegible] | 1.5[illegible]3 |

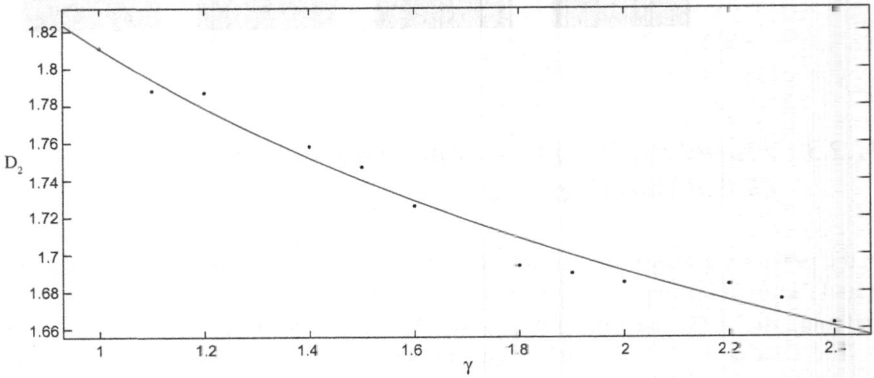


**Fig. 11.18** The variation of ice layer fractal dimension $D_2$ with the melting factor $\gamma$

$$D_2 = 1.811\gamma^{-0.987}. \tag{11.2.11}$$

$$D_1 = 4.102\exp(-1.195\alpha). \tag{11.2.12}$$

From the above analysis, the fractal dimensions of islands and ice layers is a function of the factor expansion and contraction $\alpha$ and the melting factor $\gamma$. With the increase of $\alpha$ and $\gamma$, the corresponding fractal dimension and the island area decrease.

Here we can consider the factor about expansion and contraction $\alpha$ and the melting factor $\gamma$ as temperature $t$, so as to give the following prediction by using the fractal dimension.

Firstly, considering the fractal dimension of islands as

$$D_1 = 4.102\exp(-1.195t). \tag{11.2.13}$$

When the temperature changes, the area of the island also changes accordingly. Here, we take the 2016 year temperature as $t = 1$, and get the corresponding relationship between the island area and temperatures. The prediction effect is shown in Table 11.9.

The whole land plate can be regarded as a big island of the earth. The arctic ice leads to changes in land due to the warming climate, that is, the ice melts, sea level rises, thus the land area shrinks. From the perspective of fractal, it is the degradation

**Table 11.9** The variation of island plates with the temperature parameters $t$

| $t$ | 1.01 | 1.06 | 1.16 | 1.21 |
|---|---|---|---|---|
| | 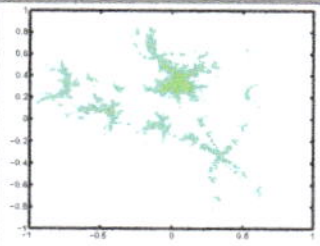 | 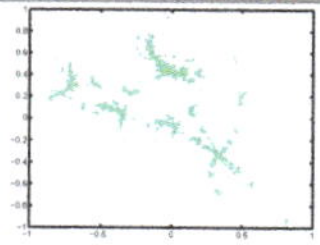 | 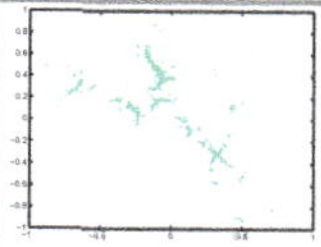 | 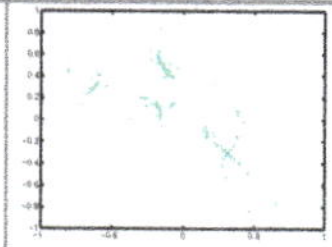 |

**Table 11.10** The variation of ice layers with the temperature parameters $t$

| $t$ | 1.01 | 1.51 | 1.99 | 2.51 |
|---|---|---|---|---|
| | 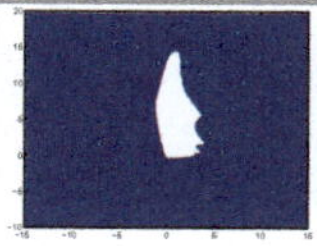 | 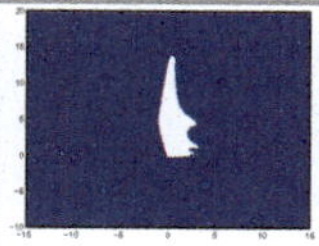 | 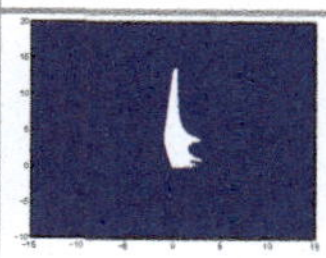 | 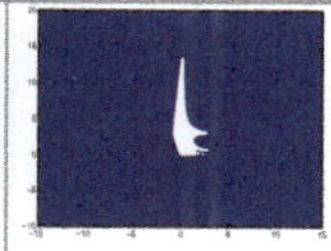 |

process of filled Julia set. This fully proves that the land plate shrinks with the increase of temperature. In addition, it also shows that the sea level rise caused by global climate change has a profound impact on the land.

Secondly, considering the fractal dimension of islands as

$$D_2 = 1.811t^{-0.987}. \tag{11.2.14}$$

When the temperatures change, the size of the ice layer also changes accordingly. We regard the 2016 year temperature as $t = 1$, and get the corresponding relationship between the ice layer area and the temperature. The prediction effect is shown in Table 11.10.

# References


1. Witten TA, Sander LM (1981) Diffusion-limited aggregation, a kinetic critical phenomenon. Phys Rev Lett 47(19):1400–1403
2. Argoul F et al (1988) Self-similarity of diffusion-limited aggregates and electrodeposition clusters. Phys Rev Lett 61(22):2558–2561
3. Wang L, Zheng D (2000) A kind of chaotic neural network optimization algorithm based on annealing strategy. Control Theory Appl 17(1):139–142
4. Yu SM et al (2004) A family of multiple-folded torus chaotic attractors. Acta Phys Sin-Ch Ed 53(7):2084–2088
5. Evans P et al (2006) Precursor-directed control of crystalline type in atmospheric pressure CVD growth of TiO2 on stainless steel. Chem Mater 18(24):5750–5755
6. He QJ, Huang ZL (2007) Controlled growth and kinetics of porous hydroxyapatite spheres by a template-directed method. J Cryst Growth 300(2):460–466
7. Kolesnikov NN et al (2008) Fractal structures of dendrites in GaSe crystals. J Cryst Growth 310(14):3287–3289
8. Yu S, Yu Z (2008) A novel fifth-order hyperchaotic circuit and its research. Acta Phys Sin-Ch Ed 57(11):6859–6867
9. Wang X et al (2011) An improved fast fractal image compression using spatial texture correlation. Chin Phys B 20(10420210)
10. Dias EO (2013) Viscous-fingering minimization in uniform three-dimensional porous media. Phys Rev E 88(0630076)
11. Katyal N et al (2014) Fractal signatures analogs of interplanetary dust particles. J Quant Spectrosc Ra 146(SI):290–294
12. Pecora LM, Carroll TL (1990) Synchronization in chaotic systems. Phys Rev Lett 64(8):821–824
13. Fu J et al (2008) The study on the SIRS epidemic models with vertical infection and continuous vaccination. J Biomath 23(1001-9626(2008)23:2<273:JCZCRH>2.0.TX;2-N2):273–278
14. Yin H, Liang W (2009) Transmission approaches for services of variable Bit-Rate and differentiated QoS using OCDMA. 2009:11
15. King RD et al (2010) Fractal dimension analysis of the cortical ribbon in mild Alzheimer's disease. Neuroimage 53(2):471–479
16. Melesse DY, Gumel AB (2010) Global asymptotic properties of an SEIRS model with multiple infectious stages. J Math Anal Appl 366(1):202–217




S. T. Liu et al., *Fractal Control and Its Applications*,
https://doi.org/10.1007/978-981-15-5459-9

17. Sun Y et al (2010) SIRS epidemic model under real complex circumstances and its control strategy. J Syst Simul 22(1):195–200
18. Misra AK et al (2013) A mathematical model for the control of carrier-dependent infectious diseases with direct transmission and time delay. Chaos Soliton Fract 57:41–53
19. Chen L, Sun J (2014) Global stability and optimal control of an SIRS epidemic model on heterogeneous networks. Phys A 410:196–204
20. Hu Z et al (2014) Stability and bifurcation analysis in a discrete SIR epidemic model. Math Comput Simulat 97:80–93
21. Pan G, Wei J (2015) Design of an adaptive sliding mode controller for synchronization of fractional-order chaotic systems. Acta Phys Sin-Ch Ed 64(0405054)
22. Rao PRS, Kumar MN (2015) A dynamic model for infectious diseases: the role of vaccination and treatment. Chaos Soliton Fract 75:34–49
23. Figueiredo TC et al (2016) Fractal measures of video-recorded trajectories can classify motor subtypes in Parkinson's disease. Phys A 462:12–20
24. Li Z et al (2016) Fractal dimension of the controlled Julia sets of the output duopoly competing evolution model. J Math Comput Sci-JMCS 16(3):461–471
25. Zhang N et al (2016) Flow of top coal and roof rock and loss of top coal in fully mechanized top coal caving mining of extra thick coal seams. Arab J Geosci 9(6):465
26. Gao W et al (2018) Ultimate bound estimation set and chaos synchronization for a financial risk system. Math Comput Simulat 154:19–33
27. Zirkohi MM, Khorashadizadeh S (2018) Chaos synchronization using higher-order adaptive PID controller. Aeu-Int J Electron C 94:157–167
28. Metzler R, Klafter J (2000) The random walk's guide to anomalous diffusion: a fractional dynamics approach. Phys Rep 339(1):1–77
29. Dzielinski A, Sierociuk D (2008) Stability of discrete fractional order state-space systems. J Vib Control 14(9–10):1543–1556
30. Ahmed E, El-Saka HA (2010) On fractional order models for Hepatitis C. Nonlinear Biomed Phys 4:3
31. Guermah S et al (2010) In: Baleanu D et al (eds) A new approach for stability analysis of linear Discrete-Time Fractional-order systems 2010:151
32. Matallana LG et al (2011) Nonlinear dynamic systems design based on the optimization of the domain of attraction. Math Comput Model 53(5–6):731–745
33. Ding Y et al (2012) Optimal control of a Fractional-Order HIV-Immune system with memory. IEEE T Control Syst T 20(3):763–769
34. Yin C et al (2013) Control of a novel class of fractional-order chaotic systems via adaptive sliding mode control approach. Appl Math Model 37(4):2469–2483
35. Muthukumar P, Balasubramaniam P (2013) Feedback synchronization of the fractional order reverse butterfly-shaped chaotic system and its application to digital cryptography. Nonlinear Dynam 74(4):1169–1181
36. Tang Y et al (2013) Fractional order sliding mode controller design for antilock braking systems. Neurocomputing 111:122–130
37. Wu G, Baleanu D (2014) Discrete fractional logistic map and its chaos. Nonlinear Dynam 75(1–2):283–287
38. Xi H et al (2014) Generation and nonlinear dynamical analyses of Fractional-Order Memristor-Based lorenz systems. Entropy-Switz 16(12):6240–6253
39. He J (2014) A tutorial review on fractal spacetime and fractional calculus. Int J Theor Phys 53(11):3698–3718
40. Machado JT (2014) Numerical analysis of the initial conditions in fractional systems. Commun Nonlinear Sci 19(9):2935–2941
41. Xu B et al (2015) Dynamic analysis and modeling of a novel fractional-order hydro-turbine-generator unit. Nonlinear Dynam 81(3):1263–1274
42. Area I et al (2015) On a fractional order Ebola epidemic model. Adv Differ Equ-Ny (278)
43. Franc E, Goufo D (2015) A biomathematical view on the fractional dynamics of cellulose degradation. Fract Calc Appl Anal 18(3):554–564

44. Huo J et al (2015) The effect of vaccines on backward bifurcation in a fractional order HIV model. Nonlinear Anal-Real 26:289–305
45. Atangana A, Koca I (2016) Chaos in a simple nonlinear system with Atangana-Baleanu derivatives with fractional order. Chaos Soliton Fract, 89(SI):447–454
46. Atangana A, Baleanu D (2016) New fractional derivatives with nonlocal and Non-Singular kernel: theory and application to heat transfer model. Therm Sci 20(2):763–769
47. Deng Y et al (2016) Fractal dimension analysis of the julia sets of controlled brusselator model. Discrete Dyn Nat Soc 2016(8234108)
48. Radulescu A, Pignatelli A (2016) Symbolic template iterations of complex quadratic maps. Nonlinear Dynam 84(4):2025–2042
49. Yang X et al (2017) On a fractal LC-electric circuit modeled by local fractional calculus. Commun Nonlinear Sci 47:200–206
50. Abdelaziz MAM et al (2018) Bifurcations and chaos in a discrete SI epidemic model with fractional order. Adv Differ Equ-Ny (44)
51. Lefever R et al (1977) Dissipative structures in a soluble nonlinear reaction-diffusion system. Phys Lett A 60(5):389–391
52. Erneux T, Reiss EL (1983) Brussellator isolas. Siam J Appl Math 43(6):1240–1246
53. Lewis M, Rees DC (1985) Fractal surfaces of proteins. Science 230(4730):1163–1165
54. Vani PK et al (1993) Painleve analysis and particular solutions of a coupled nonlinear reaction diffusion system. J Phys A-Math Gen 26(3):L97–L99
55. Larsen AL (1993) Weiss approach to a pair of coupled nonlinear reaction-diffusion equations. Phys Lett A 179(4–5):284–290
56. Ndayirinde I, Malfliet W (1997) New special solutions of the ‘Brusselator’ reaction model. J Phys A-Math Gen 30(14):5151–5157
57. Rocco A, West BJ (1999) Fractional calculus and the evolution of fractal phenomena. Phys A 265(3–4):535–546
58. Saito Y (2001) Permanence and global stability for general Lotka-Volterra predator-prey systems with distributed delays. Nonlinear Anal-Theor 47(9SI9):6157–6168
59. Xu R, Chen LS (2001) Persistence and global stability for a delayed nonautonomous predator-prey system without dominating instantaneous negative feedback. J Math Anal Appl 262(1):50–61
60. Zamir M (2001) Fractal dimensions and multifractility in vascular branching. J Theor Biol 212(2):183–190
61. Shu J et al (2003) Fractal structure of China stock market. J Southwest Jiaotong Univ 38(0258-2724(2003)38:2<212:ZGGPSC>2.0.TX;2-62):212–215
62. Zhao JD et al (2004) The permanence and global attractivity in a nonautonomous Lotka-Volterra system. Nonlinear Anal-Real 5(2):265–276
63. Bentley PJ (2004) Fractal proteins. Genet Program Evol M 5(1):71–101
64. Lin SQ, Lu ZY (2006) Permanence for two-species Lotka-Volterra systems with delays. Math Biosci Eng 3(1):137–144
65. Jumarie G (2006) Modified Riemann-Liouville derivative and fractional Taylor series of non-differentiable functions further results. Comput Math Appl 51:1367–1376
66. Yonaiguchi N et al (2007) Fractal analysis for VHF electromagnetic noises and the identification of preseismic signature of an earthquake. J Atmos Sol-Terr Phys 69(15):1825–1832
67. Isvoran A et al (2008) Fractal aspects of calcium binding protein structures. Chaos Soliton Fract 35(5):960–966
68. Movahed MS, Hermanis E (2008) Fractal analysis of river flow fluctuations. Phys A 387(4):915–932
69. Lancia MR et al (2008) Homogenization for conductive thin layers of pre-fractal type. J Math Anal Appl 347(1):354–369
70. Jiang H, Gu Z (2009) Stokes’ theorem for domains with fractal boundary. J Math Anal Appl 355(1):164–169
71. Schindler I, Tintarev K (2009) Semilinear equations on fractal blowups. J Math Anal Appl 352(1SI):57–64

72. Chen J, Ngai S (2010) Eigenvalues and eigenfunctions of one-dimensional fractal Laplacians defined by iterated function systems with overlaps. J Math Anal Appl 364(1):222–241
73. Liu X, Hong L (2013) Chaos and adaptive synchronizations in fractional-order systems. Int J Bifurcat Chaos 23(135017511)
74. Zhang Y (2013) Feedback control and synchronization of Mandelbrot sets. Chin Phys B 22(0105021)
75. Pitulice L et al (2014) Monte Carlo simulations of enzymatic reactions in crowded media. Effect of the enzyme-obstacle relative size. Math Biosci 251:72–82
76. Fernandez-Martinez M, Sanchez-Granero MA (2014) Fractal dimension for fractal structures: a Hausdorff approach revisited. J Math Anal Appl 409(1):321–330
77. Pitulice L et al (2016) Fractal dimension of the trajectory of a single particle diffusing in crowded media. Rom J Phys 61(7–8):1276–1286
78. Sun Y et al (2016) Research on characteristics of noise-perturbed M-J sets based on equipotential point algorithm. Commun Nonlinear Sci 30(1–3):284–298
79. Li C et al (2017) An infinite 2-D lattice of strange attractors. Nonlinear Dynam 89(4):2629–2639
80. Wang D et al (2017) A preliminary study on the fractal phenomenon: "Disconnected plus disconnected = connected". Fractals 25(17500041)
81. Jiang C et al (2018) Synchronization and antisynchronization of N-coupled fractional-order complex chaotic systems with ring connection. Mathematical Methods in the Applied Sciences 41:2625–2638
82. Tikhomirov G et al (2017) Fractal assembly of micrometre-scale DNA origami arrays with arbitrary patterns. Nature 552(7683):67–71
83. Jiang Z et al (2017) Multifractal cross wavelet analysis. Fractals 25(17500546)
84. Namazi H et al (2018) Fractal-based analysis of the variations of cutting forces along different axes in end milling operation. Fractals 26(18500895)
85. Namazi H et al (2018) Decoding of steady-state visual evoked potentials by fractal analysis of the electroencephalographic (eeg) signal. Fractals 26(18500926)
86. Tian D et al (2018) Hall-petch effect and inverse hall-petch effect: a fractal unification. Fractals 26(18500836)
87. Wang Q et al (2018) Fractal calculus and its application to explanation of biomechanism of polar bear hairs. Fractals 26(18500866)
88. Wang X et al (2018) Combining pressure-controlled porosimetry and rate-controlled porosimetry to investigate the fractal characteristics of full-range pores in tight oil reservoirs. J Petrol Sci Eng 171:353–361
89. Cai J et al (2018) Shale gas transport model in 3D fractal porous media with variable pore sizes. Mar Petrol Geol 98:437–447
90. Ahmadi-Pajouh MA et al (2018) Fractal-based classification of human brain response to living and non-living visual stimuli. Fractals 26(18500695)
91. Namazi H et al (2018) Decoding of upper limb movement by fractal analysis of electroencephalogram (eeg) signal. Fractals 26(18500815)
92. Alipour ZM et al (2018) Fractal-based analysis of the influence of variations of rhythmic patterns of music on human brain response. Fractals 26(18500805)
93. Namazi H et al (2018) Fractal-based analysis of the influence of cutting depth on complex structure of cutting forces in rough end milling. Fractals 26(18500685)
94. Namazi H et al (2018) Complexity-based analysis of the influence of tool geometry on cutting forces in rough end milling. Fractals 26(18500785)
95. Guariglia E (2018) Harmonic sierpinski gasket and applications. Entropy-Switz 20(7149)
96. Wang F et al (2018) A simple empirical formula of origin intensity factor in singular boundary method for two-dimensional Hausdorff derivative Laplace equations with Dirichlet boundary. Comput Math Appl 76(5):1075–1084
97. He J (2018) Fractal calculus and its geometrical explanation. Results Phys 10:272–276
98. Mozaffarilegha M et al (2018) Complexity-based analysis of the difference in speech-evoked auditory brainstem responses (s-abrs) between binaural and monaural listening conditions. Fractals 26(18500524)

99. Namazi H, Jafari S (2018) Age-based variations of fractal structure of eeg signal in patients with epilepsy. Fractals 26(18500514)
100. Dawes TJW et al (2018) Fractal analysis of right ventricular trabeculae in pulmonary hypertension. Radiology 288(2):386–395
101. Han Y, Deng Y (2018) An evidential fractal analytic hierarchy process target recognition method. Defence Sci J 68(4):367–373
102. Namazi H et al (2018) Fractal-based analysis of the influence of auditory stimuli on eye movements. Fractals 26(18500403)
103. Li X et al (2018) Pattern recognition of mine microseismic and blasting events based on wave fractal features. Fractals 26(18500293)
104. Liu C et al (2018) The absorption Angstrom exponent of black carbon: from numerical aspects. Atmos Chem Phys 18(9):6259–6273
105. Reznikov N et al (2018) Fractal-like hierarchical organization of bone begins at the nanoscale. Science 360(eaao21896388):507
106. Li Y et al (2019) (2018) Classification and fractal characteristics of coal rock fragments under uniaxial cyclic loading conditions. Arab J Geosci 11
107. Wang F et al (2018) Fractal characterization of tight oil reservoir pore structure using nuclear magnetic resonance and mercury intrusion porosimetry. Fractals 26(18400172SI)
108. Zhang S et al (2018) Numerical and experimental analysis of hot dry rock fracturing stimulation with high-pressure abrasive liquid nitrogen jet. J Petrol Sci Eng 163:156–165
109. Zhang K et al (2018) Parameter prediction of hydraulic fracture for tight reservoir based on micro-seismic and history matching. Fractals 26(18400092SI)
110. Kuznetsov NV et al (2018) Finite-time Lyapunov dimension and hidden attractor of the Rabinovich system. Nonlinear Dynam 92(2):267–285
111. Xiao B et al (2018) A fractal model for water flow through unsaturated porous rocks. Fractals 26(18400152SI)
112. Laubach SE et al (2018) Spatial arrangement of faults and opening-mode fractures. J Struct Geol 108:2–15
113. Lai J et al (2018) Investigation of pore structure and petrophysical property in tight sandstones. Mar Petrol Geol 91:179–189
114. Poplin R et al (2018) Prediction of cardiovascular risk factors from retinal fundus photographs via deep learning. Nat Biomed Eng 2(3):158–164
115. Taubert F et al (2018) Global patterns of tropical forest fragmentation. Nature 554(7693):519
116. Corvi F et al (2018) Reproducibility of vessel density, fractal dimension, and foveal avascular zone using 7 different optical coherence tomography angiography devices. Am J Ophthalmol 186:25–31
117. Xia Y et al (2018) A new method for calculating fractal dimensions of porous media based on pore size distribution. Fractals 26(18500061)
118. Lai J et al (2018) Fractal analysis of tight shaly sandstones using nuclear magnetic resonance measurements. Aapg Bull 102(2):175–193
119. Shi X et al (2018) Micrometer-scale fractures in coal related to coal rank based on micro-CT scanning and fractal theory. Fuel 212:162–172
120. Ghai S et al (2018) Effect of rhythmic auditory cueing on parkinsonian gait: a systematic review and meta-analysis. Sci Rep-Uk 8(506)
121. Tao S et al (2018) Material composition, pore structure and adsorption capacity of low-rank coals around the first coalification jump: a case of eastern Junggar Basin, China. Fuel 211:804–815
122. Zhu F et al (2018) Influence of natural regeneration on fractal features of residue microaggregates in bauxite residue disposal areas. Land Degrad Dev 29(1):138–149
123. Tian D et al (2018) Self-assembly of macromolecules in a long and narrow tube. Therm Sci 22(4):1659–1664
124. Wang F et al (2018) A novel fractal Contact-Electromechanical impedance model for quantitative monitoring of bolted joint looseness. IEEE Access 6:40212–40220

125. Zhang P et al (2018) Petrophysical characterization of oil-bearing shales by low-field nuclear magnetic resonance (NMR). Mar Petrol Geol 89(3):775–785
126. Lahmiri S, Bekiros S (2018) Chaos, randomness and multi-fractality in Bitcoin market. Chaos Soliton Fract 106:28–34
127. Wang Z et al (2018) Anisotropic characteristics of low-rank coal fractures in the Fukang mining area, China. Fuel 211:182–193
128. Mozaffarilegha M et al (2019) Complexity-Based analysis of the difference between normal subjects and subjects with stuttering in speech evoked auditory brainstem response. J Med Biol Eng 39(4):490–497
129. Tian D et al (2019) Geometrical potential and nanofiber membrane's highly selective adsorption property. Adsorpt Sci Technol 37(5–6):367–388
130. Namazi H, Jafari S (2019) Decoding of simple hand movements by fractal analysis of electromyography (emg) signal. Fractals 27(19500424)
131. Letu H et al (2019) Ice cloud properties from Himawari-8/AHI Next-Generation geostationary satellite: capability of the AHI to monitor the DC cloud generation process. IEEE T Geosci Remote 57(6):3229–3239
132. Wei B, Deng Y (2019) A cluster-growing dimension of complex networks: from the view of node closeness centrality. Phys A 522:80–87
133. Ma D et al (2019) The role of gangue on the mitigation of mining-induced hazards and environmental pollution: an experimental investigation. Sci Total Environ 664:436–448
134. Alipour H et al (2019) Complexity-based analysis of the relation between moving visual stimuli and human eye movement. Fractals 27(19500243)
135. Namazi H (2019) Fractal-based classification of electromyography (emg) signal in response to basic movements of the fingers. Fractals 27(19500373)
136. Namazi H, Jafari S (2019) Estimating of brain development in newborns by fractal analysis of sleep electroencephalographic (eeg) signal. Fractals 27(19500213)
137. Namazi H (2019) Decoding of hand gestures by fractal analysis of electromyography (emg) signal. Fractals 27(19500223)
138. Xue S et al (2019) Changes in distribution and microstructure of bauxite residue aggregates following amendments addition. J Environ Sci-China 78:276–286
139. Singh J et al (2019) On the local fractional wave equation in fractal strings. Math Method Appl Sci 42(5):1588–1595
140. Guariglia E (2019) Primality, fractality, and image analysis. Entropy-Switz 21(304)
141. Li H et al (2019) Pore structure and multifractal analysis of coal subjected to microwave heating. Powder Technol 346:97–108
142. Seyler KL et al (2019) Signatures of moire-trapped valley excitons in MoSe2/WSe2 heterobilayers. Nature 567(7746):66
143. Alipour H et al (2019) Complexity-based analysis of the influence of visual stimulus color on human eye movement. Fractals 27(19500022)
144. Xiao B et al (2019) A novel fractal model for relative permeability of gas diffusion layer in proton exchange membrane fuel cell with capillary pressure effect. Fractals 27(19500122)
145. Fang D et al (2019) Repeatability, interocular correlation and agreement of quantitative swept-source optical coherence tomography angiography macular metrics in healthy subjects. Brit J Ophthalmol 103(3):415–420
146. Namazi H, Jafari S (2019) Decoding of wrist movements' direction by fractal analysis of magnetoencephalography (meg) signal. Fractals 27(19500012)
147. Thasthakeer AT et al (2019) Analysis of the correlation between fractal structure of cutting force signal and surface roughness of machined workpiece in end milling operation. Fractals 27(19500132)
148. Wang Y, Deng Q (2019) Fractal derivative model for tsunami traveling. Fractals 27(19500172)
149. Ye Q et al (2019) Eigentime identities of fractal flower networks. Fractals 27(19500082)
150. Li X et al (2019) A fractal modification of the surface coverage model for an electrochemical arsenic sensor. Electrochim Acta 296:491–493

151. Tao G et al (2019) A unified fractal model for permeability coefficient of unsaturated soil. Fractals 27(19400121SI)
152. Hirano T et al (2019) Quantifying vascular density and morphology using different swept-source optical coherence tomography angiographic scan patterns in diabetic retinopathy. Brit J Ophthalmol 103(2):216–221
153. Kempkes SN et al (2019) Design and characterization of electrons in a fractal geometry. Nat Phys 15(2):127
154. Dutkay DE et al (2019) Hadamard triples generate self-affine spectral measures. T Am Math Soc 371(2):1439–1481
155. Alibakhshikenari M et al (2019) Mutual coupling suppression between two closely placed microstrip patches using EM-Bandgap metamaterial fractal loading. IEEE Access 7:23606–23614
156. Botero ML et al (2019) Experimental and numerical study of the evolution of soot primary particles in a diffusion flame. P Combust Inst 37(2):2047–2055
157. Harmer GP, Abbott D (1999) Game theory: losing strategies can win by Parrondo's paradox. Nature 402(6764):864
158. Harmer GP, Abbott D (1999) Parrondo's paradox. Stat Sci 14(2):206–213
159. Minor DP (2003) Parrondo's paradox-hope for losers!. College Math J 34(1):15–20
160. Chen L et al (2010) Quantum Parrondo game based on a quantum ratchet effect. Phys A 389(19):4071–4074
161. Almeida J et al (2005) Can two chaotic systems give rise to order? Phys D 200(1–2):124–132
162. Danca M et al (2014) Generalized form of parrondo's paradoxical game with applications to chaos control. Int J Bifurcat Chaos 24(14500081)
163. Danca MF, Tang WKS (2015) Parrondos paradox for chaos control and anticontrol of fractional-order systems. Chin Phys B 25(1):10505
164. Mandelbrot BB (1977) Fractals: form, chance, and dimension. Wiley, San Francisco
165. Mojica NS et al (2009) Cellular "bauplans": evolving unicellular forms by means of Julia sets and Pickover biomorphs. Biosystems 98(1):19–30
166. Dhurandhar SV et al (1993) Analysis of z-plane fractal images from $z \leftarrow z^{\alpha} + c$ for $\alpha<0$. Comput Graph 17(1):89–94
167. Wang XY et al (2000) Analysis of C-plane fractal images from $z \leftarrow z^{\alpha} + c$ for $(\alpha<0)$. Fractals 8(3):307–314
168. Rani M, Kumar V (2004) Superior Julia set. Res Math Educ 8(4):261–277
169. Wang X, Gu L (2008) Research fractal structures of generalized M-J sets using three algorithms. Fractals 16(1):79–88
170. Nazeer W et al (2015) Fixed point results in the generation of Julia and Mandelbrot sets. J Inequal Appl 2015(298)
171. Wang X et al (2008) Noise perturbed generalized Mandelbrot sets. J Math Anal Appl 347(1):179–187
172. Andreadis I, Karakasidis TE (2012) On a closeness of the Julia sets of noise-perturbed complex quadratic maps. Int J Bifurcat Chaos 22(12502219)
173. Wang X, Ge F (2012) Quasi-sine Fibonacci M set with perturbation. Nonlinear Dynam 69(4):1765–1779
174. Beck C (1999) Physical meaning for Mandelbrot and Julia sets. Phys D 125(3–4):171–182
175. Wang XY, Meng QY (2004) Study on the physical meaning for generalized Mandelbrot-Julia sets based on the Langevin problem. Acta Phys Sin-Ch Ed 53(2):388–395
176. Wang X et al (2007) Additive perturbed generalized Mandelbrot-Julia sets. Appl Math Comput 189(1):754–765
177. Wang X et al (2007) Research on Brownian movement based on generalized Mandelbrot-Julia sets from a class complex mapping system. Mod Phys Lett B 21(20):1321–1341
178. Wang X et al (2009) The generalized Mandelbrot set perturbed by composing noise of additive and multiplicative. Appl Math Comput 210(1):107–118
179. Sun W et al (2017) Fractal analysis and control in the predator-prey model. Int J Comput Math 94(4):737–746

180. Rochon D (2000) A generalized Mandelbrot set for biocomplex numbers. Fractals 8(4):355–368
181. Wang X, Sun Y (2007) The general quaternionic M-J sets on the mapping $z \leftarrow z^{\alpha} + c$ (alpha is an element of N). Comput Math Appl 53(11):1718–1732
182. Wang X, Song W (2013) The generalized M-J sets for bicomplex numbers. Nonlinear Dynam 72(1–2):17–26
183. Wang X, Jin T (2013) Hyperdimensional generalized M-J sets in hypercomplex number space. Nonlinear Dynam 73(1–2):843–852
184. Radulescu A, Pignatelli A (2017) Real and complex behavior for networks of coupled logistic maps. 87(2):1295–1313
185. Danca M et al (2009) Alternated Julia sets and connectivity properties. Int J Bifurcat Chaos 19(6):2123–2129
186. Danca M et al (2013) Graphical exploration of the connectivity sets of alternated Julia sets. Nonlinear Dynam 73(1–2):1155–1163
187. Wang D, Liu S (2016) On the boundedness and symmetry properties of the fractal sets generated from alternated complex map. Symmetry 8(2):7
188. Qiu W, Yin Y (2009) Proof of the Branner-Hubbard conjecture on Cantor Julia sets. Sci China S A Math 52(1):45–65
189. Julia G (1918) Memoire sur l'iteration des fonctions rationnelles. J Math Pures Appl 1:47–246
190. Lakhtakia A et al (1987) On the symmetries of the Julia sets for the process $z \leftarrow z^p - c$. J Phys A: Math Gen 20(11):3533–3535
191. Branner B, Hubbard JH (1992) The iteration of cubic polynomials part II: patterns and parapatterns. Acta Math-Djursholm 169(3–4):229–325
192. Wang X, Yu X (2007) Julia sets for the standard Newton's method, Halley's method and Schroder's method. Appl Math Comput 189(2):1186–1195
193. Rani M, Kumar V (2004) Superior Mandelbrot set. J Korea Soc Math Educ Ser D Res Math Educ 8:279–291
194. Negi A, Rani M (2008) Midgets of superior Mandelbrot set. Chaos Soliton Fract 36(2):237–245
195. Sun Y et al (2015) Complex time-delay dynamical systems of quadratic polynomials mapping. Nonlinear Dynam 79(1):369–375
196. Liu S, Zhang Y (2008) Synchronization of Julia sets of complex systems. Acta Phys Sin-Ch Ed 57(2):737–742
197. Liu P, Liu C (2011) Linear generalized synchronization of spatial Julia sets. Int J Bifurcat Chaos 21(5):1281–1291
198. Levin M (2012) Morphogenetic fields in embryogenesis, regeneration, and cancer: non-local control of complex patterning. Biosystems 109(3SI):243–261
199. Sun Y et al (2015) Image compression and encryption scheme using fractal dictionary and Julia set. Iet Image Process 9(3):173–183
200. Romera M et al (2007) Deterministic and random synthesis of discrete chaos. Appl Math Comput 192(1):283–297
201. Wang X, Chang P (2006) Research on fractal structure of generalized M-J sets utilized Lyapunov exponents and periodic scanning techniques. Appl Math Comput 175(2):1007–1025
202. Fisher Y et al (2011) The science of fractal images. Springer Publishing Company, Incorporated, Berlin
203. Bogush AA et al (2001) On symmetry properties of quaternionic analogs of Julia sets. In: Proceedings of 9th annual seminar NPCS-2000. Minsk, Belarus
204. Branner B (1989) The Mandelbrot set. Proc Symp Appl Math 39:75–105
205. Uspensky JV (1948) Theory of equations. McGraw-Hill, New York
206. Falconer K (2004) Fractal geometry: mathematical foundations and applications. Wiley, Chichester
207. Zhang YP, Liu ST (2008) Gradient control and synchronization of Julia sets. Chin Phys B 17(2):543–549

208. Zhang Y et al (2010) Control and synchronization of second Julia sets. Chin Phys B 19(5):146–153
209. Brandt ME, Chen GR (1997) Bifurcation control of two nonlinear models of cardiac activity. IEEE T Circuits-I 44(10):1031–1034
210. Saitoh N et al (1996) An analysis of a family of rational maps containing integrable and non-integrable difference analogue of the logistic equation. J Phys A-Math Gen 29(8):1831–1840
211. Fornaess JE (2006) The Julia set of Henon maps. Math Ann 334(2):457–464
212. Blanchard P et al (2008) A generalized version of the McMullen domain. Int J Bifurcat Chaos 18:2309–2318
213. Blanchard P et al (2009) The rabbit and other Julia sets wrapped in Sierpinski carpets. Complex Dynamics, Families and Friends, pp 277C–296C
214. Qiu W et al (2012) Dynamics of McMullen maps. Adv Math 229(4):2525–2577
215. Morabito M, Devaney RL (2008) Limiting Julia sets for singularly perturbed rational maps. Int J Bifurcat Chaos 18(10):3175–3181
216. Zhang Y (2013) Control and synchronization of Julia sets of the complex perturbed rational maps. Int J Bifurcat Chaos 23(13500835)
217. Helmstedt JM, Getz C (2004) Graphics with mathematica. Elsevier Science, Amsterdam
218. Zeng WQ (2007) Fractal geometry: mathematical foundations and applications. Posts and Telecom Press, Beijing
219. Andreadis I, Karakasidis TE (2010) On a topological closeness of perturbed Mandelbrot sets. Appl Math Comput 215(10):3674–3683
220. Luo XS et al (2003) Hybrid control of period-doubling bifurcation and chaos in discrete nonlinear dynamical systems. Chaos Soliton Fract 18(4):775–783
221. Wang D, Liu S (2015) Synchronization between the spatial Julia sets of complex Lorenz system and complex Henon map. Nonlinear Dynam 81(3):1197–1205
222. Liu P, Liu S (2011) Control and synchronization of Julia sets in coupled map lattice. Commun Nonlinear Sci 16(8):3344–3355
223. Norton A (1982) Generation and display of geometric fractals in 3-D. ACM SIGGRAPH Comput Graph 16(3):61–67
224. Hart JC et al (1989) Ray tracing deterministic 3-D fractals. ACM SIGGRAPH Comput Graph 23(3):289–296
225. Hruska SL (2006) A numerical method for constructing the hyperbolic structure of complex Henon mappings. Found Comput Math 6(4):427–455
226. Biham O, Wenzel W (1990) Unstable periodic orbits and the symbolic dynamics of the complex 1,0,0 Hénon map. Phys Rev A 42(8):4639–4646
227. Shudo A et al (2009) Julia sets and chaotic tunneling: II. J Phys A-Math Theor 42(26510226)
228. Shudo A et al (2009) Julia sets and chaotic tunneling: I. J Phys A-Math Theor 42(26510126)
229. Shudo A et al (2002) Julia set describes quantum tunnelling in the presence of chaos. J Phys A-Math Gen 35(PII S0305-4470(02)34589-X17):L225–L231
230. Liu P, Liu S (2012) Control and coupling synchronization of Julia sets in coupled map lattice. Indian J Phys 86(6):455–462
231. Lorenz EN (1989) Computational chaos-a prelude to computational instability. Phys D 35(3):299–317
232. Frouzakis CE et al (2003) A route to computational chaos revisited: noninvertibility and the breakup of an invariant circle. Phys D 177(PII S0167-2789(02)00751-01-4):101–121
233. Djellit I, Hachemi-Kara A (2011) Weak attractors and invariant sets in Lorenz model. Facta Univ-Ser: Electron Energ 24(2):271–280
234. Elabbasy EM et al (2014) Bifurcation analysis and chaos in a discrete reduced Lorenz system. Appl Math Comput 228:184–194
235. Peitgen HO, Saupe D (1988) The science of fractal images. Springer, New York
236. Chen GR, Liu ST (2003) On generalized synchronization of spatial chaos. Chaos Soliton Fract 15(PII S0960-0779(02)00101-72):311–318
237. Chen HK (2005) Global chaos synchronization of new chaotic systems via nonlinear control. Chaos Soliton Fract 23(4):1245–1251

238. Liu ST, Chen G (2003) On spatial lyapunov exponents and spatial chaos. Int J Bifurcat Chaos 13(5):1163–1181
239. Shougang S (2005) Spatial fractal Julia sets of complex dynamic systems. Jinan
240. Jie S et al (2011) Parameter identification of generalized Julia sets. Acta Phys Sin-Ch Ed 60(7):120–128
241. Sun J, Liu S (2013) A new identification control for generalized Julia sets. Chin Phys B 22(5):50505
242. Mandelbrot BB (1983) The fractal geometry of nature. W. H, Freeman, New York
243. Hooper KJ (1991) A note on some internal structures of the Mandelbrot set. Comput Graph 15(2):295–297
244. Shishikura M (1998) The Hausdorff dimension of the boundary of the Mandelbrot set and Julia sets. Ann Math 147(2):225–267
245. Argyris J et al (2000) On perturbations of the Mandelbrot map. Chaos Soliton Fract 11(7):1131–1136
246. Isaeva OB et al (2001) Mandelbrot set in coupled logistic maps and in an electronic experiment. Phys Rev E 64(5):55201
247. Di Ieva A et al (2014) Fractals in the neurosciences, part i: general principles and basic neurosciences. Neuroscientist 20(4):403–417
248. Gomatam J et al (1995) Generalization of the Mandelbrot set: quaternionic quadratic maps. Chaos Soliton Fract 5(6):971–986
249. Garant-Pelletier V, Rochon D (2009) On a generalized Fatou-Julia theorem in multicomplex spaces. Fractals 17(3):241–255
250. Liu P (2013) Control and synchronization of Mandelbrot sets in coupled map lattice. Nonlinear Dynam 73(1–2):299–310
251. Puente-Baliarda C et al (1998) On the behavior of the Sierpinski multiband fractal antenna. IEEE T Antenn Propag 46(4):517–524
252. Vinoy KJ et al (2001) Hilbert curve fractal antenna: a small resonant antenna for VHF/UHF applications. Microw Opt Techn Let 29(4):215–219
253. Vinoy KJ et al (2003) On the relationship between fractal dimension and the performance of multi-resonant dipole antennas using Koch curves. IEEE T Antenn Propag 51(9):2296–2303
254. Berry MV, Lewis ZV (1980) On the Weierstrass-Mandelbrot fractal function. Proc R Soc Lond Ser A-Math Phys Sci 370(1743):459–484
255. Humphrey JAC et al (1992) On the use of the Weierstrass-Mandelbrot function to describe the fractal component of turbulent velocity. Fluid Dyn Res 9(1–3):81–95
256. Borgnat P, Flandrin P (2003) On the chirp decomposition of Weierstrass-Mandelbrot functions, and their time-frequency interpretation. Appl Comput Harmon A 15(2):134–146
257. Kutner R, Switala F (2003) Stochastic simulations of time series within Weierstrass-Mandelbrot walks. Quant Financ 3(PII S1469-7688(03)33854-03):201–211
258. Yao K et al (2009) On the connection between the order of the fractional derivative and the Hausdorff dimension of a fractal function. Chaos Soliton Fract 41(5):2538–2545
259. Jiang S, Zheng Y (2010) An analytical model of thermal contact resistance based on the Weierstrass-Mandelbrot fractal function. P I Mech Eng C-J Mec 224(C4):959–967
260. Sanchez I, Uzcategui G (2011) Fractals in dentistry. J Dent 39(4):273–292
261. Werner TR et al (2012) Modeling of super-extreme events: an application to the hierarchical Weierstrass-Mandelbrot continuous-time random walk. Eur Phys J-Spec Top 205(1):27–52
262. Liu L et al (2013) On the use of Weierstrass-Mandelbrot function to simulate fractal wind fluctuations. Clim Environ Res 18(1006-9585(2013)18:1<43:JYWMHS>2.0.TX;2-H1):43–50
263. Ibrahim N et al (2013) Diagnostic imaging of trabecular bone microstructure for oral implants: a literature review. Dentomaxillofac Rad 42(20120075 3)
264. Silva DA et al (2013) Fractal dimension and Shannon's entropy analyses of the architectural complexity caused by the inflammatory reactions induced by highly crystalline poly(vinyl alcohol) microspheres implanted in subcutaneous tissues of the Wistar rats. J Biomed Mater Res a 101(2):326–339

265. Sun W, Lv H (2013) Dissipative feedback control and synchronization of Julia sets of the complex standard family. Appl Math Comput 219(15):8026–8032
266. Zhang YP et al (2013) Control and synchronization of Julia sets of complex standard family. Indian J Phys 87(3):271–274
267. Sweldens W (1998) The lifting scheme: a construction of second generation wavelets. Siam J Math Anal 29(2):511–546
268. Lin J, Qu LS (2000) Feature extraction based on Morlet wavelet and its application for mechanical fault diagnosis. J Sound Vib 234(1):135–148
269. Chau T (2001) A review of analytical techniques for gait data. Part 2: neural network and wavelet methods. Gait Posture 13(2):102–120
270. Zhu K et al (2009) Wavelet analysis of sensor signals for tool condition monitoring: a review and some new results. Int J Mach Tool Manu 49(7–8):537–553
271. Schneider K, Vasilyev OV (2010) Annual Review of Fluid Mechanics. Wavelet methods in computational fluid dynamics. 42:473–503
272. Singh CB et al (2010) Wavelet analysis of signals in agriculture and food quality inspection. Food Bioprocess Tech 3(1):2–12
273. Yin J et al (2011) Significance test and precision analysis of Morlet wavelet: application to studying periodic variation of earthquake activity and geomagnetic field. Acta Seismologica Sinica 33(5):663–671
274. Weierstrass K (1895) Uber continuirliche functionen eiens reellen arguments, die fur keinen werth des letzteren einen bestimmten differentia lquotienten besitzen. German, Berlin: Königlich Preussichen Akademie der Wissenschaften, Mathematische Werke von Karl Weierstrass
275. Benedetto JJ, Zimmermann G (1997) Sampling multipliers and the Poisson summation formula. J Fourier Anal Appl 3(5):505–523
276. Nemes G (2010) New asymptotic expansion for the Gamma function. Arch Math 95(2):161–169
277. Mandelbrot BB et al (1984) Fractal character of fracture surfaces of metals. Nature 308(5961):721–722
278. Wang H et al (2012) A multifractal detrended fluctuation analysis (MDFA) of the Chinese growth enterprise market (GEM). Phys A 391(12):3496–3502
279. Cao G et al (2012) Multifractal detrended cross-correlations between the Chinese exchange market and stock market. Phys A 391(20):4855–4866
280. Grech D, Pamula G (2013) Multifractality of nonlinear transformations with application in finances. Acta Phys Pol A 123(3):529–537
281. Hurst HE et al (1965) Long-Term storage: an experimental study. London Constable Company, London
282. Madhavan A (2000) Market microstructure: a survey. J Financ Markets 3(3):205–258
283. Sinha S et al (2010) Econophysics: an introduction. WILEY-VCH Verlag GmbH and Co, KGaA
284. Liu L, Wan J (2012) The relationships between Shanghai stock market and CNY/USD exchange rate: new evidence based on cross-correlation analysis, structural cointegration and nonlinear causality test. Phys A 391(23):6051–6059
285. Petrosky-Nadeau N, Wasmer E (2013) The cyclical volatility of labor markets under frictional financial markets. Amer Econ J-Macroecon 5(1):193–221
286. Onali E, Goddard J (2011) Are European equity markets efficient? New evidence from fractal analysis. Int Rev Financ Anal 20(2):59–67
287. Gronlund A et al (2012) Fractal profit landscape of the stock market. Plos One 7(e339604)
288. Budinski-Petkovic L et al (2014) Fractal properties of financial markets. Phys A 410:43–53
289. He L, Chen S (2010) Are developed and emerging agricultural futures markets multifractal? A comparative perspective. Phys A 389(18):3828–3836
290. Chen S, He L (2013) Bubble formation and heterogeneity of traders: a Multi-Agent perspective. Comput Econ 42(3):267–289

291. Alvarez-Ramirez J et al (2008) Time-varying Hurst exponent for US stock markets. Phys A 387(24):6159–6169
292. Rejichi IZ, Aloui C (2012) Hurst exponent behavior and assessment of the MENA stock markets efficiency. Res Int Bus Financ 26(3):353–370
293. Webel K (2012) Chaos in German stock returns - New evidence from the 0–1 test. Econ Lett 115(3):487–489
294. Wen F et al (2012) Study on the fractal and chaotic features of the shanghai composite index. Fractals 20(2):133–140
295. BenSaida A, Litimi H (2013) High level chaos in the exchange and index markets. Chaos Soliton Fract 54:90–95
296. BenSaida A (2014) Noisy chaos in intraday financial data: evidence from the American index. Appl Math Comput 226:258–265
297. Pichl L et al (2007) Stylized facts in internal rates of return on stock index and its derivative transactions. Phys A 382(1):219–227
298. Delatte A, Lopez C (2013) Commodity and equity markets: some stylized facts from a copula approach. J Bank Financ 37(12SI):5346–5356
299. He L, Wen X (2013) Statistical revisit to the mike-farmer model: can this model capture the stylized facts in real world markets? Fractals 21(13500082)
300. Kuosmanen P, Vataja J (2014) Forecasting GDP growth with financial market data in Finland: revisiting stylized facts in a small open economy during the financial crisis. Rev Financ Econ 23(2):90–97
301. Hurst HE (1951) Long-term storage capacity of reservoirs. Amer Soc Civ Eng 116(1):770–808
302. Chen ZL, Huang MH (2010) Hurst exponent analysis for the intertrade durations of stocks in China's stock market based on the detrended fluctuation analysis. Manag Sci Stat Decis 7(2):23–31
303. He L, Qian W (2012) A Monte Carlo simulation to the performance of the R/S and V/S methods-Statistical revisit and real world application. Phys A 391(14):3770–3782
304. Sanchez-Granero MJ et al (2012) Introducing fractal dimension algorithms to calculate the Hurst exponent of financial time series. Eur Phys J B 85(86)
305. Oh KJ, Kim KJ (2002) Analyzing stock market tick data using piecewise nonlinear model. Expert Syst Appl 22(PII S0957-4174(01)00058-6):249–255
306. Gomes O (2009) A two-dimensional non-equilibrium dynamic model. Struct Change Econ D 20(3):221–238
307. Wu W et al (2010) Complex nonlinear dynamics and controlling chaos in a Cournot duopoly economic model. Nonlinear Anal-Real 11(5):4363–4377
308. McMillan DG (2001) Nonlinear predictability of stock market returns: evidence from non-parametric and threshold models. Int Rev Econ Financ 10(4):353–368
309. McMillan DG (2007) Non-linear forecasting of stock returns: does volume help? Int J Forecasting 23(1):115–126
310. Koulouriotis DE et al (2005) Development of dynamic cognitive networks as complex systems approximators: validation in financial time series. Appl Soft Comput 5(2):157–179
311. Wagner N et al (2008) Forecasting economic time series with the DyFor genetic program model. Appl Financ Econ 18(5):357–378
312. Dai W et al (2012) Combining nonlinear independent component analysis and neural network for the prediction of Asian stock market indexes. Expert Syst Appl 39(4):4444–4452
313. Cheng C et al (2013) OWA-based ANFIS model for TAIEX forecasting. Econ Model 30:442–448
314. Fiedor P (2014) Networks in financial markets based on the mutual information rate. Phys Rev E 89(5):52801
315. Falconer K (1990) Mathematical foundations and applications. Wiley, New York
316. Yu C et al (2014) DFA7, a new method to distinguish between intron-containing and intronless genes. Plos One 9(7):e1013637
317. Zhang L et al (2014) Control effects of Morlet wavelet term on Weierstrass-Mandelbrot function model. Indian J Phys 88(8):867–874

318. He L, Chen S (2011) A new approach to quantify power-law cross-correlation and its application to commodity markets. Phys A 390(21–22):3806–3814
319. Guresen E et al (2011) Using artificial neural network models in stock market index prediction. Expert Syst Appl 38(8):10389–10397
320. Banik S et al (2012) Modeling chaotic behavior of dhaka stock market index values using the neuro-fuzzy model. Recent Patents Comput Sci 5(1):72–77
321. Soliman AS (1996) Fractals in nonlinear economic dynamic systems. Chaos Soliton Fract 7(2):247–256
322. Zhang M, Wei Z (1996) The Application of the fractal theory in describing the regional transportation economy coordination. China Soft Sci 1:112–123
323. Zhu X et al (2006) Fractal analysis applied to spatial structure of chinas vegetation. Chinese Geograph Sci 16(1):48–55
324. Lu F (2011) Fractal theory and its application in economic management. Financ Manag Inst 3:91–93
325. Preiss D (1983) Nondifferentiable functions. (Czech) Pokroky Mat Fyz Astronom 28:148–154
326. Agiza HN et al (2001) The dynamics of Bowley's model with bounded rationality. Chaos Soliton Fract 12(9):1705–1717
327. Du J et al (2009) Analysis of decision-making in economic chaos control. Nonlinear Anal-Real 10(4):2493–2501
328. Wang C et al (2015) Controlling a chaotic resonator by means of dynamic track control. Complexity 21(1):370–378
329. Zhang F et al (2015) Dynamics of two classes of Lorenz-Type chaotic systems. Complexity 21(1):363–369
330. Vieira MD, Lichtenberg AJ (1996) Controlling chaos using nonlinear feedback with delay. Phys Rev E 54(2):1200–1207
331. Liu P (2012) Spatial fractal control and chaotic synchronization in complex dynamical system
332. Leung AYT et al (2004) Resonance control for a forced single-degree-of-freedom nonlinear system. Int J Bifurcat Chaos 14(4):1423–1429
333. Qin Y, Zeng X (1980) Qualitative research of Brusselator oscillator equation in biochemistry. Chi Sci Bull 25:337–339
334. Bashkirtseva IA, Ryashko LB (2000) Sensitivity analysis of the stochastically and periodically forced Brusselator. Phys A 278(1–2):126–139
335. Chen H et al (2013) Synchronization criteria for two boolean networks based on logical control. Int J Bifurcation Chaos 23:135–178
336. Halsey TC et al (1986) Fractal measures and their singularities: the characterization of strange sets. Phys Rev A 33(2):1141–1151
337. Hu B, Lin B (1989) Yang-lee zeros, Julia sets, and their singularity spectra. Phys Rev A 39(9):4789–4796
338. Nauenberg M, Schellnhuber HJ (1989) Analytic evaluation of the multifractal properties of a Newtonian Julia set. Phys Rev Lett 62(16):1807–1810
339. Wang X, Yu X (2009) Julia set of the Newton transformation for solving some complex exponential equation. Fractals 17(2):197–204
340. Wang XY et al (2010) Julia sets of Newton's method for a class of complex-exponential function F(z)=P(z)e (Q(z)). Nonlinear Dynam 62(4):955–966
341. Wang X, He Y (2008) Projective synchronization of fractional order chaotic system based on linear separation. Phys Lett A 372(4):435–441
342. Wang X, Song J (2009) Synchronization of the fractional order hyperchaos Lorenz systems with activation feedback control. Commun Nonlinear Sci 14(8):3351–3357
343. Lin D, Wang X (2010) Observer-based decentralized fuzzy neural sliding mode control for interconnected unknown chaotic systems via network structure adaptation. Fuzzy Set Syst 161(15):2066–2080
344. Wang XY, Ge FD (2010) The quasi-sine Fibonacci hyperbolic dynamic system. Fractals 18(1):45–51

345. Sun W, Zhang Y (2015) Control and synchronization of julia sets in the forced brusselator model. Int J Bifurcat Chaos 25(9):1550113
346. Liu X, Xiao D (2007) Complex dynamic behaviors of a discrete-time predator-prey system. Chaos Soliton Fract 32(1):80–94
347. Pogromsky AY (2015) Introduction to control of oscillations and chaos. Mod Lang J 23(92):1–5
348. Qin Y, Zeng X (1980) Qualitative study of the Brussels oscillator equation in Biochemistry. Chin Sci Bull 25(8):337–339
349. Chirikov BV (1979) Universal instability of many-dimensional oscillator systems. Phys Rep 52(5):263–379
350. Zaslavsky GM (2005) Hamiltonian chaos and fractional dynamics. Oxford University Press, New York
351. Stix TH (1976) Current penetration and plasma disruption. Phys Rev Lett 36(10):521–524
352. Izraelev FM (1980) Nearly linear mappings and their applications. Phys D 1(3):243–266
353. Casati G et al (1986) Search for randomness in the kicked quantum rotator. Phys Rev A 34(2):1413–1419
354. Jensen MH et al (1983) Complete devils staircase, fractal dimension, and universality of mode-locking structure in the circle map. Phys Rev Lett 50(21):1637–1639
355. Jensen MH et al (1984) Transition to chaos by interaction of resonances in dissipative systems. 1. Circle maps. Phys Rev A 30(4):1960–1969
356. Stavans J et al (1985) Fixed winding number and the quasiperiodic route to chaos in a convective fluid. Phys Rev Lett 55(6):596–599
357. Fagella N (1999) Dynamics of the complex standard family. J Math Anal Appl 229(1):1–31
358. Ott E et al (1990) Controlling chaos. Phys Rev Lett 64(11):1196–1199
359. Hunt ER (1991) Stabilizing high-period orbits in a chaotic system-the diode resonator. Phys Rev Lett 67(15):1953–1955
360. Pyragas K (1992) Continuous control of chaos by self-controlling feedback. Phys Lett A 170(6):421–428
361. Ravichandran V et al (2009) Effect of rectified and modulated sine forces on chaos in Duffing oscillator. Indian J Phys 83(11):1593–1603
362. Emami Z, Pakzad HR (2011) Solitons of KdV and modified KdV in dusty plasmas with superthermal ions. Indian J Phys 85(11):1643–1652
363. Ravichandran V et al (2012) Study of nonescape dynamics in Duffing oscillator with four different periodic forces. Indian J Phys 86(10):907–911
364. Liu ZH, Chen SG (1997) Synchronization of a conservative map. Phys Rev E 56(2):1585–1589
365. Lotka AJ (1925) Elements of physical biology. Williams and Wilkins Co, Baltimore
366. Volterra V (1926) Variazioni e fluttuazioni del numero d'individui in spece animali conviventi. Mem R Accad Naz Dei Lincei Ser VI 2
367. Lu Z, Chen L (1995) Analysis of the periodic Lotka-Volterra three-species mixed model. Pure Appl Math 2:81–85
368. Fu S et al (2006) Uniform boundedness and stability of solutions to the Three-Species Lotka-Volterra competition model with self and Cross-Diffusion. Chin Ann Math Ser A 27(1000-8314(2006)27:3<345:DZKSHJ>2.0.TX;2-Q3):345–356
369. Lu H, Ke W (2013) The problem of positive almost periodic solution for two-species nonautonomous diffusion Lotka-Volterra models. J Biomath 18(4):417–422
370. Jackson EA, Kodogeorgiou A (1992) Entrainment and migration controls of 2-dimensional maps. Phys D 54(3):253–265
371. Thompson JM (1996) Global dynamics of driven oscillators: fractal basin and indeterminate bifurcations. Cambridge University Press, Cambridge
372. Gu E (2004) The stable basin of attracting set in control chaos and some global bifurcations of the domains of feasible trajectories in nonlinear ecological systems
373. Yan H et al (2010) Beta Persistence and beta extinction of a Predator-prey Lotka-Volterra model of two species. Chin J Eng Math 27(1005-3085(2010)27:1<139:BSBBSE>2.0.TX;2-31):139–144

374. Chen YM, Zhou Z (2003) Stable periodic solution of a discrete periodic Lotka-Volterra competition system. J Math Anal Appl 277(PII S0022-247X(02)00611-X1):358–366
375. Capone F et al (2013) On the stability of non-autonomous perturbed Lotka-Volterra models. Appl Math Comput 219(12):6868–6881
376. Takenchi Y (1996) Global dynamical properties of Lotka-Volterra systems. World Scientific, Singapore
377. Hofbauer J, Sigmund K (1998) Evolutionary game and population dynamics. Cambridge University Press, Cambridge
378. Liao XY et al (2007) Permanence for a discrete time Lotka-Volterra type foodchain model with delays. Appl Math Appl 186:279–285
379. Ahmed E et al (2007) Equilibrium points, stability and numerical solutions of fractional-order predator-prey and rabies models. J Math Anal Appl 325(1):542–553
380. Das S et al (2009) A fractional Predator-Prey model and its solution. Int J Nonlin Sci Num 10(7):873–876
381. El-Saka HA et al (2009) On stability, persistence, and Hopf bifurcation in fractional order dynamical systems. Nonlinear Dynam 56(1–2):121–126
382. Das S, Gupta PK (2011) A mathematical model on fractional Lotka-Volterra equations. J Theor Biol 277(1):1–6
383. Agrawal SK et al (2012) Synchronization between fractional-order Ravinovich-Fabrikant and Lotka-Volterra systems. Nonlinear Dynam 69(4):2277–2288
384. Elsadany AA, Matouk AE (2015) Dynamical behaviors of fractional-order Lotka-Volterra predator-prey model and its discretization. J Appl Math Comput 49(1–2):269–283
385. Matouk AE, Elsadany AA (2016) Dynamical analysis, stabilization and discretization of a chaotic fractional-order GLV model. Nonlinear Dynam 85(3):1597–1612
386. Nosrati K, Shafiee M (2017) Dynamic analysis of fractional-order singular Holling type-II predator-prey system. Appl Math Comput 313:159–179
387. Podlubny I (1999) Fractional differential equations: an introduction to fractional derivatives, fractional differential equations, to methods of their solution and some of their applications. Academic Press, San Diego
388. Petráš I (2011) Fractional-Order nonlinear systems: modeling, analysis and simulation. Springer, Berlin
389. Wang Y, Sun S (2018) Solvability to infinite-point boundary value problems for singular fractional differential equations on the half-line. J Appl Math Comput 57(1–2):359–373
390. Arfken G, Weber H (2005) Mathematical methods for physicists, 6th edn. Academic Press, San Diego
391. Zhu S et al (2015) A nonlinear and fractional derivative viscoelastic model for rail pads in the dynamic analysis of coupled vehicle-slab track systems. J Sound Vib 335:304–320
392. Spanos PD, Evangelatos GI (2010) Response of a non-linear system with restoring forces governed by fractional derivatives-Time domain simulation and statistical linearization solution. Soil Dyn Earthq Eng 30(9SI):811–821
393. Elaydi S (2005) An introduction to difference equations. Springer, New York
394. Wang D, Liu X (2017) On the noise-perturbed spatial Julia set generated by Lorenz system. Commun Nonlinear Sci 50:229–240
395. Fradkov AL et al (1999) Nonlinear and adaptive control of complex systems. Springer, Dordrecht
396. Andrievskii BR, Fradkov AL (2003) Control of chaos: methods and applications. I. Methods. Automat Rem Contr+ 64(5):673–713
397. Andrievskii BR, Fradkov AL (2004) Control of chaos: methods and applications. II. Applications. Automat Rem Contr+ 65(4):505–533
398. Zhang Y, Guo X (2012) Control of Julia sets of the complex Henon system. Nonlinear Dynam 69(1–2):185–192
399. Zhang M, Zhang Y (2016) Fractal analysis and control of the competition model. Int J Biomath 9(3):1–16

400. Wang Y, Liu S (2019) Fractal analysis and control of the fractional Lotka-Volterra model. Nonlinear Dynam 95:1457–1470
401. Wang Y et al (2019) Fractal dimension analysis and control of Julia set generated by fractional Lotka-Volterra models. Commun Nonlinear Sci Numer Simul 72:417–431
402. Liu S, Wang P (2018) Fractal control theory. Springer, Singapore
403. Argyris J et al (2000) On the Julia set of the perturbed Mandelbrot map. Chaos Solitons Fractal 11:2067–2073
404. Argyris J et al (2002) On the Julia sets of a noise-perturbed Mandelbrot map. Chaos Solitons Fractal 13:245–252
405. Andreadis I, Karakasidis TE (2010) On a topological closeness of perturbed Julia sets. Appl Math Comput 217:2883–2890
406. Ahmed E, Elgazzar AS (2007) On fractional order differential equations model for non-local epidemics. Phys A Stat Mech Appl 379:607–614
407. Rivero M et al (2011) Fractional dynamics of populations. Appl Math Comput 218:1089–1095
408. Hamdan N, Kilicman A (2018) A fractional order SIR epidemic model for dengue transmission. Chaos Solitons Fractal 114:55–62
409. Tarasov V (2018) No nonlocality. No fractional derivative. Commun Nonlinear Sci Numer Simul 62:157–163
410. Barnsley M et al (1988) The science of fractal images. Springer, New York
411. Barbera A, Spagnolo B (2002) Spatio-temporal patterns in population dynamics. Phys A Stat Mech Appl 314:120–124
412. Valenti D et al (2004) Pattern formation and spatial correlation induced by the noise in two competing species. Acta Phys Pol B 1481–1489(35)
413. Fiasconaro A et al (2004) Nonmonotonic behavior of spatiotemporal pattern formation in a noisy Lotka-Volterra system. Acta Phys Pol B 35:1491–1500
414. Chichigina O et al (2005) A simple noise model with memory for biological systems. Fluct Noise Lett 5:243–250
415. Valenti D et al (2008) Environmental metal pollution considered as noise: effects on the spatial distribution of benthic foraminifera in two coastal marine areas of Sicily (Southern Italy). Ecol Model 213:449–462
416. Chichigina O et al (2011) Stability in a system subject to noise with regulated periodicity. Phys Rev E 84:21134
417. Bashkirtseva I et al (2019) Stochastic variability and transitions to chaos in a hierarchical three-species population model. Chaos Solitons Fractal 276–283(119)
418. Yang W (2010) Symmetries of the Julia sets of Newton's method for multiple root. Appl Math Comput 217:2490–2494
419. Mandelbrot B (2004) Fractals and chaos: the mandelbrot set and beyond. Springer, New York
420. Fatou P (1919) Sur les équations fonctionelles. Bulletin de la Société Mathématique de France 47:161–271
421. Wang P, Liu S (2015) The gradient control of spatial-alternated Julia sets. Nonlinear Dynam 80(3):1291–1302
422. Wang Y et al (2019) On the spatial Julia set generated by fractional Lotka-Volterra system with noise. Chaos Solitons Fractal 128:129–138
423. Misiurewicz M (1981) On iterates of $e^z$. Ergod Theor Dyn Syst 1:103–106
424. Devaney R, Krych M (1984) Dynamics of exp(z). Ergod Theor Dyn Syst 4:35–52
425. Devaney R (1991) Exp(z): dynamics and bifurcations. Int J Bifurcat Chaos 1:287–308
426. Bodelon C et al (1999) Hairs for the complex exponential family. Int J Bifurcat Chaos 9:1517–1534
427. Devaney R et al (2005) Indecomposable continua and Misiurewicz points in exponential dynamics. Int J Bifurcat Chaos 15:3281–3293
428. Zhang YP et al (2009) Feedback control in fractal mechanics. Chaos Solitons Fractal 41(1):337–343
429. McMullen C (1987) Area and Hausdorff dimension of Julia sets of entire functions. T Am Math Soc 300:329–342

430. Sun J et al (2015) New identification and control methods of sine-function Julia sets. J Appl Anal Comput 5:220–231
431. Sajid M, Kapoor G (2005) Dynamics of non-critically finite odd transcendental meromorphic function. Indian J Pure Appl Math 36:203–221
432. Gorenflo R et al (2014) Mittag-Leffler functions, related topics and applications. Springer, Berlin
433. Kilbas A et al (2006) Theory and applications of fractional differential equations. Elsevier Science Limited, Amsterdam
434. Khan Y et al (2012) A new fractional analytical approach via a modified Riemann-Liouville derivative. Appl Math Lett 25:1340–1346
435. Mainardi F, Gorenflo R (2000) On Mittag-Leffler-type functions in fractional evolution processes. J Comput Appl Math 118:283–299
436. Jumarie G (2008) Fourier's transformation of fractional order via Mittag-Leffler function and modified Riemann-Liouville derivatives. J Appl Math Inf 26:1101–1121
437. Devi J et al (2013) On fractional trigonometric functions and their generalizations. Dynam Syst Appl 22:337–354
438. Devi J, Nagamani S (2013) On fractional hyperbolic functions and their generalizations. Nonlinear Studies 20:331–348
439. Nagamani S (2014) Generalized fractional hyperbolic functions. Bull Marathwada Math Soc 15:54–61
440. Liang F (2014) The control of a grey rational fractal. Master thesis, Shandong University
441. Hengster-Movric K et al (2013) Synchronization of discrete-time multi-agent systems on graphs using Riccati design. Automatica 49:414–423
442. Liu J, Liu S (2017) Complex modified function projective synchronization of complex chaotic systems with known and unknown complex parameters. Appl Math Model 48:440–450
443. Tikhonov A, Samarskii A (2011) Equations of mathematical physics. Dover Publications, Mineola
444. Herrmann H (1986) Geometrical cluster growth models and kinetic relation. Phys Rep 136:153–224
445. Sander L (2000) Di?usion-limited aggregation: a kinetic critical phenomenon? Contem Phys 41:203–218
446. Chen W (2004) Fractional Laplacian time-space models for linear and nonlinear lossy media exhibiting arbitrary frequency power-law dependency. J Acoust Soc Amer 115:1424–1430
447. Du M et al (2013) Measuring memory with the order of fractional derivative. Sci Rep 3:3431
448. Chen W (2006) Time-space fabric underlying anomalous diffusion. Chaos Solitons Fractal 28:923–929
449. Rossikhin Y, Shitikova M (1997) Applications of fractional calculus to dynamic problems of linear and nonlinear hereditary mechanics of solids. Appl Mech Rev 50:15–67
450. Niemeyer L et al (1984) Fractal dimension of dielectric breakdown. Phys Rev Lett 52:1033–1036
451. Derrida B, Hakim V (1992) Needle models of Laplacian growth. Phys Rev A 45:8759–8765
452. Zhou H, Yang S (2018) Fractional derivative approach to non-Darcian flow in porous media. J Hydrol 566:910–918
453. Sun X, Wu ZQ (2003) Fractal theory and its application. Press of University of Science and Technology of China, Hefei
454. Wang X, Liang Q (2008) Reverse bifurcation and fractal of the compound logistic map. Commun Nonlinear Sci 13(5):913–927
455. Witten TA, Sander LM (1983) Diffusion-limited aggregation. Phys Rev B 27(9):5686–5697
456. Liu Y et al (2010) Fractal model for simulation of frost formation and growth. Sci China Technol Sc 53(3):807–812
457. Shin S et al (2010) Hybrid simulation of miscible mixing with viscous fingering. Comput Graph Forum 29(2):675–683
458. Li RT (1992) Heat transfer theory and method. Tianjin University Press, Tianjin

459. Liu HC (1981) The differential equations of heat transfer for some cases of variable thermal conductivities. Appl Math Mech 2(4):379–385
460. Zhang HJ (1992) Heat conduction. Higher Education Press, Beijing
461. Lu JP, Guan Z (2003) Numerical solution to partial differential equations. Tsinghua University Press, Beijing
462. Qiao W et al (2014) Control of thermal fractal diffusion of space body. Control Theory Appl 31(3):361–365
463. Pietronero L, Wiesmann HJ (1984) Stochastic model for dielectric breakdown. J Stat Phys 36(5–6):909–916
464. Wang L (2003) Shop scheduling with genetic algorithms. Tsinghua University Press, Beijing
465. Jiadeke BK (1989) An introduction to uniform polynomial apporximation of function. Beijing University Press, Beijing
466. Xue ZC et al (2008) Advanced mathematics. Tsinghua University Press, Beijing
467. Qi MY, Wu FT (1999) Generalized functions and equations of mathematical physics. Higher Education Press, Beijing
468. Lv HX, Wang TM (2011) Applied functional analysis. Dalian University of Technology Press, Dalian
469. Qiao W et al (2014) Control of the thermal fractal diffusion of tightly compressed heterogeneous layers of thin plates. Math Probl Eng 2014(672547)
470. Yin H (2009) Dynamics of two epidemic models
471. Rashidinia J et al (2018) On the dynamical complexity of a seasonally forced discrete SIR epidemic model with a constant vaccination strategy. Complexity 2018:7191487
472. Zhang Y et al (2017) Fractal dimension and synchronization of the controlled julia sets of a Reaction-Diffusion system. Asian J Control 19(5):1741–1747
473. Ahmad WM, Harb AM (2003) On nonlinear control design for autonomous chaotic systems of integer and fractional orders. Chaos Soliton Fract 18(4):693–701
474. Maamar B, Rachid M (2014) IMC-PID-fractional-order-filter controllers design for integer order systems. Isa T 53(5SI):1620–1628
475. Atici FM, Eloe PW (2009) Initial value problems in discrete fractional calculus. P Am Math Soc 137(3):981–989
476. Abdeljawad T (2011) On Riemann and Caputo fractional differences. Comput Math Appl 62(3):1602–1611
477. Chen F et al (2011) Existence results for nonlinear fractional difference equation. Adv Differ Equ-Ny 2011:713201
478. Ouyang M, Zhang YP (2019) Julia sets and their control of discrete fractional SIRS models. Complexity 2019(1–10)
479. Girejko E et al (2013) On Solutions to Fractional Discrete Systems with Sequential h-differences. Abstr Appl Anal 2013(1–11)
480. Hu SD, Wang YH (2015) Fractional order discrete logistic mapping synchronization based on parameter adaptive control. Commun Appl Math Comput 29(2):232–239
481. Yamasaki Y et al (2011) Cell contact-dependent lethal effect of the dinoflagellate Heterocapsa circularisquama on phytoplankton-phytoplankton interactions. J Sea Res 65(1):76–83
482. Nagasoe S et al (2006) Effects of temperature, salinity and irradiance on the growth of the red tide dinoflagellate Gyrodinium instriatum Freudenthal et Lee. Harmful Algae 5(1):20–25
483. Wang R (2007) Studies on the allelopathic effects of macroalgae on red tide microalgae
484. Yamaguchi H et al (2010) Effects of temperature, salinity and irradiance on growth of the novel red tide flagellate Chattonella ovata (Raphidophyceae). Harmful Algae 9(4):398–401
485. Kang H, Pesin Y (2005) Dynamics of a discrete brusselator model: escape to infinity and Julia set. Milan J Math 73(1):1–17
486. Zhang L (2008) Construction method in the edge of general M-J sets and its fractal dimension. J Chin Comput Syst 29(7):1338–1342
487. Fei Y (1983) Error theory and data processing. China Machine Press, Beijing
488. Zhang Y (2008) Control and applications of fractals. Jinan

489. Han F et al (2009) Chaotic burst synchronization in heterogeneous small-world neuronal network with noise. Int J Nonlin Mech 44(3):298–303
490. Taylor R et al (2001) Preliminary studies on the growth of selected 'green tide' algae in laboratory culture: effects of irradiance, temperature, salinity and nutrients on growth rate. Bot Mar 44(4):327–336
491. Wang R et al (2006) Comparative studies on the allelopathic effects of Ulva pertusa Kjellml, Corallina pilulifera Postl et Ruprl, and Sargassum thunbergii Mertl O. Kuntze on Skeletonema Costatum (Grev.) Cleve J Integr Plant Biol 48(12):1415–1423
492. Wang X, Luo C (2005) Bifurcation and fractal of the coupled Logistic maps. Acta Mech Sinica-Prc 37(3):346–355
493. Sun X, Wu Z (2003) Fractal principle and its application. University of science and technology of China press, Hefei
494. Liu P, Liu ST (2012) Robust adaptive full state hybrid synchronization of chaotic complex systems with unknown parameters and external disturbances. Nonlinear Dynam 70(1):585–599
495. Liu P, Liu ST (2012) Adaptive modified function projective synchronization of general uncertain chaotic complex systems. Phys Scripta 85(3):35005
496. Jiang CM et al (2015) Generalized combination complex synchronization for fractional-order chaotic complex systems. Entropy-Switz 17(8):5199–5217
497. Wang P, Liu ST (2016) Optimal control and synchronization of alternated Julia sets. Asian J Control 18(5):1698–1705
498. Wang D et al (2016) Control and synchronization of Julia sets generated by a class of complex time-delay rational. J Appl Anal Comput 6(4):1049–1063
499. Sui SG, Liu ST (2005) Control of julia set. Chaos Solitons Fractals 26(4):1135–1147
500. Zhang YP, Sun WH (2011) Synchronization and coupling of Mandelbrot sets. Nonlinear Dynam 64(1):59–63
501. Liu P, Liu ST (2013) Nonlinear generalized synchronization of two different spatial Julia sets. Control Theory Appl 30(9):1159–1164
502. Liu ST et al (2015) Spatial chaos on surface and its associated bifurcation and Feigenbaum problem. Nonlinear Dynam 81(1):283–298
503. Jiang CM (2015) Liu ST (2015) Generalized combination complex synchronization of new hyperchaotic complex Lv-like systems. Adv Differ Equ-Ny 1:214
504. Qiao W et al (2015) Directional region control of the thermal fractal diffusion of a space body. Chin Phys B 24(5):275–280
505. Wang D et al (2016) Control of the spatial Mandelbrot set generated in coupled map lattice. Nonlinear Dynam 84(3):1795–1803
506. Liu J et al (2016) Adaptive complex modified hybrid function projective synchronization of different dimensional complex chaos with uncertain complex parameters. Nonlinear Dynam 83(1):1109–1121
507. Zhang YP et al (2009) Fractals control in article's velocity. Chaos Solitons Fractals 39(4):1811–1816
508. Liu P, Liu ST (2011) Anti-synchronization between different chaotic complex systems. Phys Scripta 83(6):65006
509. Liu ST, Liu P (2011) Adaptive anti-synchronization of chaotic complex nonlinear systems with unknown parameters. Nonlinear Anal Real World Appl 12(6):3046–3055
510. Jiang CM et al (2014) (2014) A new fractional-order chaotic complex system and its antisynchronization. Abstr Appl Anal 1:1–12
511. Liu J et al (2015) Adaptive complex modified projective synchronization of complex chaotic (hyperchaotic) systems with uncertain complex parameters. Nonlinear Dynam 79(2):1035–1047
512. Wang P, Liu ST (2016) Control of Julia sets in generalized alternated system. Asian J Control 18(6):2234–2243
513. Jiang CM et al (2016) Combination complex synchronization among three incommensurate fractional-order chaotic systems. J Math Comput Sci 17:308–323

The manufacturer's authorised representative in the EU is Springer Nature Customer Service Centre GmbH, Europaplatz 3, 69115 Heidelberg, Germany. If you have any concerns regarding our products, please contact ProductSafety@springernature com

Printed and bound by CPI Group (UK) Ltd, Croydon, CR0 4YY
30/07/2026
02181701-0008